25th Summer School and International Symposium on the Physics of Ionized Gases 2010

(SPIG 2010)

Journal of Physics: Conference Series Volume 257

Donji Milanovac, Serbia
30 August-3 September 2010

Editors:

Luka Č Popović **Milorad Kuraica**

ISBN: 978-1-61782-247-6
ISSN: 1742-6588

Printed from e-media with permission by:

Curran Associates, Inc.
57 Morehouse Lane
Red Hook, NY 12571

Some format issues inherent in the e-media version may also appear in this print version.

Printed by Curran Associates, Inc. (2011)

For permission requests, please contact the Institute of Physics
at the address below.

Institute of Physics
Dirac House, Temple Back
Bristol BS1 6BE UK

Phone: 44 1 17 929 7481
Fax: 44 1 17 920 0979

techtracking@iop.org

Additional copies of this publication are available from:

Curran Associates, Inc.
57 Morehouse Lane
Red Hook, NY 12571 USA
Phone: 845-758-0400
Fax: 845-758-2634
Email: curran@proceedings.com
Web: www.proceedings.com

TABLE OF CONTENTS

THE HISTORY OF THE SPIG CONFERENCES

ATOMIC COLLISION PROCESSES

PARTICLE AND LASER BEAM INTERACTION WITH SOLIDS

LOW TEMPERATURE PLASMAS

GENERAL PLASMAS

PREFACE

This volume of the *Journal of Physics: Conference Series* contains the Invited lectures, Topical invited lectures and Progress reports presented at the 25th Summer School and International Symposium on the Physics of Ionized Gases – SPIG 2010. The conference was held in Donji Milanovac, Serbia, from 29 August to 3 September 2010. Since SPIG has a long tradition and this is a jubilee anniversary, the 25th one, we had the opportunity to recall the history of the Conference (see the first paper in this proceedings).

The structure of papers in this Proceedings covers the following sections: Atomic Collision Processes, Particle and Laser Beam Interactions with Solids, Low Temperature Plasmas and General Plasmas. As the four above mentioned topics often overlap and merge in numerous fundamental studies and more importantly applications, SPIG in general serves as a venue for exchanging ideas in the related fields. We hope that this Proceedings will be an important source of information about progress in plasma physics and will be useful, first of all, for students, and also for plasma physics scientists.

The Editors would like to thank the invited speakers for their participation at SPIG 2010 and for their efforts writing contributions for this Proceedings. We also express our gratitude to the members of Scientific and Organizing committees for their efforts in organizing this SPIG. Especially we would like to thank the Ministry of Science and Technological Development of Republic of Serbia for financial support as well as the European Physical Society (EPS) for supporting the participation of three younger scientists.

Luka Č Popović
Milorad Kuraica
October 2010

All papers published in this volume of *Journal of Physics: Conference Series* have been peer reviewed through processes administered by the proceedings Editors. Reviews were conducted by expert referees to the professional and scientific standards expected of a proceedings journal published by IOP Publishing.

SPIG from beginning to today

J M Labat

Faculty of Physics, University of Belgrade
Studentski trg 12, Belgrade, Serbia

jlabat@sezampro.rs

1. How it all begins

It may be of interest to find out how the research in the field of physics of ionized gases (PIG) started and what was the reason that so many scientists in the fifties become interested in the relevant fields. Research in nuclear physics that was in its heights and was heavily supported by the governments in the fifties, and generous funds have been provided. Our country was not an exception since even a relatively small countries as ours, expected various benefits, not only a new source of energy, application of radio isotopes in medicine and agriculture but also, to be honest, some kind of nuclear weapon. The Federal Nuclear Energy Commission was formed and three centers were founded: in Ljubljana (Slovenia), in Zagreb (Croatia) and in Belgrade (Serbia). First two were named after world known physicists (J.Stefan an R.Bošković) while the Belgrade institute was named after active politician and minister in the Yugoslav government Boris Kidrič. A large number of the first class scientists were engaged in the research. It was a dream of physics, physical chemistry and biology students, such as myself, to work after graduation in some of the nuclear institutes. Even the curricula in Science University departments have been altered and adopted to the nuclear physics research requirements. Gradually, research in nuclear physics entered a more advanced phase. At the Belgrade Institute, a very successful group led by prof.Branislava-Brana Perović was developing methods of electromagnetic separation of isotopes, mass spectrometer was introduced by K. Zmbov, interaction of ions with surfaces was studied by B.Navinšek (Ljubljana) and T. Nenadović (Belgrade), gas filled radiation detectors were developed (prof. A.Milojević, M. Petrović, Đ.Bek-Uzarov, V.Urošević, V.Ajdačić and M.Kurepa ,all in Belgrade, M.Srdoč in Zagreb and Đ.Bošan in Niš) All this, and many other fields of research required substantial knowledge on various aspects that take place in ionized gases (to mention Z.Šternberg and V.Vujnović in Zagreb, M.Popović and D.Tošić in Belgrade) .Research and applications of the arc discharges in gases for analytical purposes (V.Vukanović , D.Vukanović, M.Todorović, D.Pešić and M.Marinković in Belgrade) was also successfully progressing. Construction of ion accelerators mastered (A.Strojnik and J.Dekleva in Ljubljana, B.Maršičanin and B.Aničin in Belgrade nad K.Prelec in Zagreb) , but majority of scientists were not trained and not skilled in the field. Some of them already returned after training (Z.Šternberg and V.Vujnović from Zagreb and B.Čobić from Belgrade) while others were still abroad as postgraduate and research students in the field related to the physics of ionized gases (J.Labat and M.Kurepa from Belgrade) in the leading Universities. It is a fact that study of conduction of electricity through gases has been a matter of mainly academic interest during nineteenth and beginning of twentieth century. It undoubtedly leads to several new fields, such as discovery of electrons and X-rays that lay down basics of modern physics. However, methods of experimental and theoretical physics that may be applied to the field of ionized gases, have not been developed. Furthermore, scanty sources of atomic data, particularly on excitation and ionization collisions of electrons with atoms and molecules, has limited the extension of the research .It was therefore decided, on the federal scale, to tackle this problem in various ways.

Journal of Physics: Conference Series **257** (2010) 012001 doi:10.1088/1742-6596/257/1/012001

Firstly: to organize meetings of people working in the field in order to exchange experience, ideas and suggest the other ways to overcome these problems. Secondly: to send certain number of younger students to renowned centers abroad, to provide highly trained personnel capable of quality research in the field. In the meantime, several new ideas appeared, that have partly bearing in the nuclear research. After successful explosion of the hydrogen bomb in 1952, the quest for controlled thermonuclear reaction began, in order to provide practically unlimited supply of energy. Under heading in print "*Sun in the Laboratory* " results have been announced on the toroidal machine named ZETA build in Harvel (UK). Apparently, high electric currents induced in ionized deuterium gas, has heated it to temperature sufficient to fuse the atoms. Similar research has been done also in other countries, particularly in USA and Soviet Union. Based on extension and results of these studies the commercial fusion reactor seemed to be just round the corner. Also, an evidence has been provided that, so called MHD power generation could be easily attainable by use of highly ionized gas as a conducting fluid. Initial results have been very promising so that practically any physical laboratory in the world has started some kind of related research. Everybody expected to find a way to heat and confine the plasma in order to reach the goal - the controlled fusion reaction. A tremendous amount of money has been invested, especially in highly developed countries. The results varied, from highly promising to disappointing. And as the situation today, there are just a few centers in the world left with large scale controlled fusion research: USA, Great Britain, Germany, Japan. It seems that just some international joint project could lead to some kind of commercial fusion reactor. It already exists, as you may know, it is under construction in south France, under the name of ITER. Joint effort of China, EU, India, Japan, Korea, Russia and USA hopefully guarantee that it will follow the original plan: 10 years of construction and 20 years of operation. For certain, it is not the task of small countries, since it definitely requires very large scale devices and highly sophisticated technology. Small countries, as in the case of LHC project, may participate via some of their scientists involved in the project. Scientists from our country contribute also by direct participation the international large projects, or, as majority did, undertake smaller scale experiments or theoretical projects.

A number of scientists in the field gradually increased and good part of projects financed by the state were from the field of physics of ionized gases and plasmas. Therefore, the initial idea to train students and research workers by acquisition and exchanging of knowledge and experience in an organized manner has been justified.

2. How the SPIG was born

2.1. First three meetings

Who, where and when exactly the idea to organize meetings in the field started is not easy to establish. However, one can say that mainly thanks to effort and persistency of Prof. Branislava-Brana Perović from the Nuclear Institute in Belgrade, supported, stimulated and backed enthusiastically by prof. A.Milojević (Belgrade), Z.Šternberg (Zagreb), Đ. Bošan (Niš) and A.Moljk (Ljubljana), the first meeting was organized in Belgrade in 1962, under the title: " *1st Yugoslav Symposium on the Physics of Ionized Gases"(SPIG)*. At this first meeting eight invited lectures were presented by the most experienced local scientists from institutions where experimental and theoretical work has already started. In addition to invited lectures, twenty six original contributions were also presented. It was surprisingly a large number of contributions, since the preference in scientific research was given to nuclear physics. These contributions pointed to the fields of future meetings: Atomic collision physics, Interaction of particles with solids and Physics of ionized gases.

Two years later, in 1964. Zagreb was chosen for the second meeting. In Zagreb, majority of the research in the field of ionized gases was done in "Rudjer Bošković" Institute and just begins in the Department of Physics, Faculty of Sciences and Mathematics. The general pattern of the meeting was similar to the first one. This time only four invited lectures were presented, while the number of original contributions was twenty seven.

The town of Niš was in 1966. place for the next gathering of the third, and the last meeting of this kind. Apart from the Department of Physics at the Faculty of Philosophy and

Journal of Physics: Conference Series **257** (2010) 012001 doi:10.1088/1742-6596/257/1/012001

Faculty of Electronics University of Niš, the main interest in the physics of ionized gases existed in the Niš Electronic Industry , a factory where electronic tubes, X-ray sources and proportional Geiger counters were manufactured. At this meeting five invited lectures were presented, but the number of contributions increased sharply to forty eight! This was indeed an encouraging sign that the effort to organize SPIG meetings was justified. It should be noted that participants were exclusively from Yugoslavia, and that the official language of the conference was Serbo-Croatian.

2.1.1. Intermission: Summer School on Physics of Ionized Gases

Unexpectedly, large number of participants and contributions on the first two SPIG meetings was an encouraging sign that the effort to organize PIG related meetings starts to give positive results. During the first meeting the general conclusion was that learning of different fields related to PIG should be done in a more organized manner. It meant that, in order to do it more efficiently, we should invite some foreign experts in the field and engage them as lecturers and teachers. In 1964. six leading scientists have been invited to deliver lectures to Yugoslav participants, scientists and postgraduate students. On the Adriatic coast, in the town of Herceg Novi the *" Summer School on the Physics of Ionized Gases"* was held. Organizers of this summer school succeeded to bring six world-wide known scientists and University professors to come and teach our, mainly young, research workers. Names of these people are built into the basis of all our further research and should be praised, with dignity and pride. These were:

Prof. J.D.Craggs, University of Liverpool, G. Britain, three lectures,
Prof. A.L.Cullen, University of Sheffield, G. Britain, three lectures,
Prof. Yu.H.Demkov, University of Leningrad, Soviet Union, three lectures,
Prof. A.von Engel, University of Oxford, G. Britain, eight lectures,
Prof. R.Herman, Obs. de Paris France, two lectures,
Prof. J.B.Hasted, University College London, G. Britain, six lectures.

Results of this summer school proved to be excellent. Participants agreed that this venture was highly successful and that they have learned a substantial amount of new knowledge. It was therefore decided that summer school of this kind should be organized on a regular basis, possibly combined with the symposia.

The place chosen for the summer school was ideal for the purpose. The Yugoslav Atomic Energy Commission had at the time in Herceg Novi , a small town on the Adriatic coast, an object assigned for scientific meetings. The building consists of cozy lecture theater, restaurant and accommodations. Most valuable was the beautiful garden with Mediterranean vegetation, where attendees could rest between lectures in a shadow of the trees. And the beach was less than 100 meters from the place! During the summer school participants were all the time together (except for those at the beach), discussing with lecturers and among themselves, exactly as organizers wished. Maybe it was also due to a highly stimulating atmosphere of the surrounding that this summer school was so successful!

3. Finally: Summer School and International Symposium on Physics of Ionized Gases

After the success of the two symposia and the summer school, the idea was born that these two should be combined and held biannually on a regular basis. As a result of this proposal, the first meeting in a long series, was held in 1968 under the title: "Yugoslav Symposium and International Summer School on the Physics of Ionized Gases", now known world wide (without any modesty) as SPIG. It was rather ambitious title, because at that time no one was sure how much interest such a meeting could attract. It turns out that the meeting and its location was a hit! Mainly, young scientists and postgraduate students from number of European countries started to take part, attracted mainly by first class lecturers, an to be honest, also by the site. Their number was increasing, so that later they represented approximately 1/3 of all participants, no counting the invited lecturers. Except for 16th and 17th SPIG when turmoil in our country was at it highest, this ratio remains the same.

The 1968 meeting held also in Herceg-Novi, was distinct by introduction of the three sections: Atomic collision processes, Particle interaction with solids and Phenomena in plasmas. Number of the invited lectures was very large, and were given, mainly by foreign participants, there were as much as 23 lectures given by 13 lecturers. It is of particular importance that official language becomes English. The organizers were aware that this may cause difficulties to younger participants when orally presenting their contributions what was a regular way at the time. But it turns out to be a very good stimulation to them to improve heir knowledge of English that was gradually becoming a kind of official language in science.

The following, 5th SPIG meeting was held in 1970 again in Herceg Novi. The lecture theatre was still large enough to accommodate all the participants. However, the tourist industry started sudden rise at the Adriatic coast. It was followed by erection of new large hotels, so that the conference site was surrounded by many new objects and crowd of tourists. The quietness and exclusiveness of the conference site garden has been lost. As a consequence, it was decided that the next meeting should be held somewhere else on the Adriatic coast. However, the Herceg-Novi meetings will be remembered as special place where numerous friendships have been founded and collaborations started. The lectures as well as lecturers greatly influenced the further development of research in the field of physics of ionized gases in Yugoslavia. We quote their names always with great respect. Their names with affiliations for first 20 meetings are given in Appendix A.1. in ref.1.

The next, 6th SPIG meeting was held in 1972. in Split, a town on the middle of Adriatic, actually in a brand new hotel few miles out of the city center. This meeting will be remembered by an appreciably large number of participants, and because of an oral presentation of contributions, three parallel sessions were held. One of the conference rooms was in fact the hotel night club! Due to the large distance to the city center the participants were bound to the meeting site , which seems to improve discussions, conversations and friendships. As during the previous meetings, excursions were organized to some places of natural and historical interest. These excursions by bus or by boat were welcomed by the participants, particularly to Split that was erected around AD 300 by Diocletian as his seat in this part of the Roman empire. His palace is a place of interest to be visited and seen even now. And of course, there was a beach close to the hotel, what was very much appreciated by the participants as in the case of Herceg –Novi.

The 7th SPIG was held in the town of Rovinj, the lovely place at the northern Adriatic. The meeting cite was a small island called St.Catherina. The see was not 100 meters from the meeting place but 100 meters all around it! The favorite excursion destinations were two islands at the entrance of the Rovinj bay.

The meeting cite of the next 8th SPIG in 1976 has been moved to the most beautiful place of Adriatic coast, the ancient Dubrovnik. The town has completely taken hearths of participants, so that the also following three meetings (9th SPIG 1978, 10thSPIG 1980, and 11th SPIG 1982) were also held there. The quality of the invited lectures was on a very high level although the number of lecturers and lectures was reduced to about thirty. The meeting halls and accommodation for the participants was provided by one of the nicest hotels called Libertas. The traditional conference excursions were improved in number and quality and the hotel beach was again very close!

Strangely enough, in 1990. the 15th SPIG meeting was also held in Dubrovnik. At that time it was almost certain that Yugoslavia is approaching an extremely dangerous period. Some think that this was the last SPIG meeting, and it was felt that the occurrences expected in near future announces that this was the last occasion for meeting our friends and colleagues from all parts of Yugoslavia.

Successive Dubrovnik SPIG meetings in 1978, 1980. and 1982. are remembered by further improvement of the quality of organization. It was felt that this meeting has got its own face, being always different from other similar meetings. And, as pointed out by most of the participants, the Dubrovnik has completely taken their hearts. Dubrovnik has long and exciting history and was an independent state during many centuries. It is surrounded by walls and has some special unforgettable atmosphere. The participants could attend evening concerts at "Dubrovnik Summer Festival", or just drink coffee in one of many typical local restaurants. The meetings were an exceptional mixture of high level physics, relaxation on the beach and highest cultural performances. Conference excursions were organized mainly by boats, to visit near-by islands of Lokrum and Lopud and nicest and biggest nearby island of Mljet.

The 1984.(XII) and 1986.(XIII) SPIG meetings were held in the town of Šibenik, a lovely middle Adriatic town, with plenty remains from middle ages. Among them the most representative is Gothic-Renaissance cathedral. Meetings were held again in a nearby hotel on the seaside. The number of invited lectures was rather large, between 40 and 50, more than usual at the previous meetings. Both of the meetings were noted by the large number of participants as well as large number of original contributions. It was therefore decided that the contributions will be presented on the panels, that was already a praxis in the other international conferences , so that this way of presentation was easily accepted by participans on these and all the other successive SPIG meetings.

Organizers of the XIV SPIG meeting from the University of Sarajevo, feel that the meeting should be moved from the seaside to mountains. They have chosen mountain Igman, that become famous during the Sarajevo winter Olympic games in 1984. The number of participants and also lecturers was again reasonably high, proving that not much the site but the quality of meeting was attractive factor for attendees.

As already mentioned, the XV SPIG was again held in Dubrovnik, shortly before the former Yugoslavia disintegrates. It was the end of gathering of people from all Yugoslav centers and a fruitful mutual collaboration.

3.1. *Intermission 2. Related international conferences*

The oldest international conference on physics of ionized gases was started in 1953. in Oxford by professor A. von Engel. Initially, until 1963 conference in Paris it was called "International Conference on Ionization Phenomena in Gases". However, several new fields have been encompassed gradually, so that after recommendation of the International Scientific Committee, already the next conference held in 1965 in Belgrade becomes "International Conference on Physics of Ionized Gases" (ICPIG). It exists until today, the last, 29th was held in Cancun, Mexico 2009. In recognition for our activity in the field, this conference has been organized twice in Belgrade (Yugoslavia), 7th ICPIG in 1965 organized by B.Perović and 19th ICPIG, organized in 1989 by J.Labat and J.Purić. Also, three of us were members of the twelve-member Conference International Scientific Committee, each with 6 years election period. These included J.Labat, J.Purić and B.Milić (all from Belgrade) of which J. Puric chaired this scientific committee form 1989. to 1991. (20th ICPIG).

Purely Europen Coference of slightly narower scope started in 1973. It is the " European Conference on the Atomic and Molecular Physics of Ionised Gases " (ESCAMPIG). Our scientists have organized this conference also twice: 4th in Dubrovnik in 1980 organized by R.Janev and 20th in Novi Sad organized in 2010. by Z.Petrović, both from Belgrade.

Finally, "International Conference on Photonic, Electronic and Atomic Collisions" (ICPEAC) that started in 1953 in New York, its last, 26th was held 2009. in Kalamazoo, Michigan (USA).

It was a great honor given to some of us receiving invitations by the International Scientific Committees of these conferences to present general invited lectures from the particular field of interest on the results that were partially published and recognized by the scientific community.

I have deliberately excluded from this review conferences devoted to various aspects of fusion, although they stem from the research in the field of physics of ionized gases, and may be regarded as highly specialized. The same applies also to the "International Conference on Spectral Line Shapes" and also to "Serbian Conference on Spectral Line Shapes".

3.2. Meetings after disintegration of Yugoslavia

Beginning of nineties Yugoslavia fell apart. Serbia and Montenegro formed new Federal Republic of Yugoslavia. Distressed, and depressed physicists though that the SPIG meetings should be discontinued. Fortunately the strong tradition of the meetings prevailed and the renewed 16th meeting was organized in Belgrade in 1993. Many of scientists were a little surprised by this decision owing to the international insulation of the country, including air traffic, closed borders and high inflation. The number of contributed papers was below the average, however, several invited speakers have sufficient courage to come to Belgrade and were cordially welcomed.

In order to hold the further meetings during the even years, the next 17th SPIG meeting was held also in Belgrade already in 1994. Due to surrounding war and largely unsettled situation in the country the number of invited speakers dropped to 15, although the number of local participants was reasonably large.

The general situation in the country gradually improved, so that the next 18th SPIG meeting was held in a town of Kotor in 1996. Medieval town on the south Adriatic provided very convenient cite, and the number of participants was again very large, and so was with the invited lecturers (35). The town itself lies in the well preserved fortress in the bay of Kotor, famous for brave marine captains and sailors and as a tourist resource. The participants enjoy very much this site and it was felt that the places on the Adriatic coast will prevail for further meetings.

This experience seemed not to be sufficient argument for the next 19th SPIG meeting organizers, since they have decided to move to the mountains. The mountain of Zlatibor was chosen for the reason of convenience, in terms of its position, transport, possibilities of accommodation and beautiful surroundings. However, this time, probably due to the lack of financial support, the number of invited lecturers dropped to 27. Regardless of this, the meeting was highly successful and has given a strong support to the idea where to hold the SPIG meetings in the future.

As much as three future meetings were held on the mountains and one in a spa resort. The mountain Zlatibor was the place of jubilar 20th SPIG meeting in 2000. Following success of previous meeting it was rich with the invited lectures (27) and number of participants. The next 21st meting was held in Soko Banja in 2002, the spa resort, with 25 invited lectures and 14 progress report presented. The 22nd SPIG was held at mountain Tara, Bajina Bašta in 2004, another mountain tourist resort in Serbia. The Montenegro has become in 2006 an independent country and so did Serbia. This greatly limits the choice of places, since the Serbia has no exit to the sea. Therefore the tourist resort- mountain Kopaonik, was chosen for the site of 23rd SPIG in 2006. It was rich with the invited lectures and 25 progress reports, but possibly due to transport problems, the number of participants was smaller than usual. However, the next 23rd SPIG was held in town of Novi Sad in 2008. The meeting was noted for high number of participants (160 local and 71 from abroad), 26 invited lectures and 14 progress reports. The reason was most probably more convenient transport, and possibility to reach the place easily by the car. Undoubtedly this meeting was very successful, the SPIG has again reached its full form. The people from Belgrade institutes have proven to be not only first class scientists, but also excellent organizers. This, 25th meeting will surely be remembered not only as 25th in a row, but also by the quality of lectures and number of participants. Hopefully, the SPIG meetings will continue to be successfully organized by highly skilled people for the benefit of Serbian science.

Chronology of SPIG meetings with the names of places and organizers is given in *Appendix.*

4. Scientific interests and publications

4.1 Sections

First few meetings have only three sections. It was felt that is sufficient knowing the extend of the scientific activity in the early sixties, that was just gaining its initial momentum. Following the evidence in the fields of interests it was decided that only three sections withouth subsections should be sufficient. As an example, for the 4th SPIG in 1968 the sections were:

1. Atomic collision processes,
2. Particle interactions with solids,
3. Processes in plasmas.

Very soon, activities in the field have greatly evolved, number of contributions sharply increased, and it was of necessity to elaborate their classification.

At the 10th SPIG in 1980 three sections were of slightly different titles and each of them has few subsections:

1. Atomic collision processes
   1.1. Scattering of electrons
   1.2. Heavy particle collisions

2. Particle interactions with solids
   2.1. Scattering and emission processes
   2.2. Penetration and sputtering
   2.3. Surface changes
   2.4. Experimental techniques

3. Phenomena in ionized gases
   3.1. Interaction of laser radiation with plasma
   3.2. Non-ideal plasmas
   3.3. Plasma spectroscopy
   3.4. Waves and instabilities in plasmas
   3.5. Electric gas discharges

The subsections indicate much broader interest in the fields related to the physics of ionized gases and also the general trends in the physics. Possibly that also influenced by the sections and subsections at the international ICPIG conferences, already 14th SPIG meeting in 1988 has four sections with different subsections indicating even newest directions of research:

1. Atomic collision processes
   1.1. Electron and photon interactions with atomic particles
   1.2. Heavy particles and photon interactions
   1.3. Swarm and transport processes

2 Particle and laser beam interaction with solids
   2.1 Atomic collision processes in solids
   2.2. Sputtering and deposition
   2.3. Laser and plasma interaction with surfaces

3. Low temperature plasmas
   3.1. Plasma spectroscopy and other diagnostic methods
   3.2. Gas discharges
   3.3. Plasma applications and devices

4. General plasma theory

Since then the situation seems to stabilize, so that only the fourth section has been further elaborated. Also, the non-ideal plasmas do not seemed to be of further interest. Already, from 19th SPIG in 1998 until today, the sections and subsections are:

1. Atomic collision processes
   1.1 Electron and atom interactions with atomic particles
   1.2 Heavy particle collisions
   1.3 Swarm and transport phenomena

2. Particle and laser beam interactions with solids
   2.1. Atomic collisions with solids
   2.2. Sputtering and deposition
   2.3. Laser and plasma interaction with surfaces

3. Low temperature plasmas
   3.1. Plasma spectroscopy and other diagnostic methods
   3.2. Gas discharges
   3.3. Plasma applications and devices

4. General plasmas
   4.1. Fusion plasmas
   4.2. Astrophysical plasmas
   4.3. Collective phenomena

4.2. SPIG related publications

Every SPIG meeting including the Summer school was followed by one or two publications. Except for summer school, the SPIG activities comprise of two separate books: invited lectures and original contribution. Initially, the books were printed as internal reports of the Belgrade Nuclear Institute in Serbo-Croatian language.

The summer school lectures were recorded and printed as transcripts after approval by the lecturers. Up to now, each meeting has its book of invited lectures printed by the local organizers, and, with certain difficulties distributed around the world. Some of them were published by international publishing houses, so they reached the scientific centers much easier.

All the SPIG invited lectures were printed on over twelve thousand pages. For research and postgraduate students in Yugoslavia and we believe also in other countries, these book were valuable textbook as well as sources of information. As a support of this statement is the fact that one can find these books in large number in laboratories around the world. Wherever an experimental and theoretical work in areas included in topics of SPIG meetings, book of invited lectures, as well as books of contributed papers can be found on the bookshelves. Number of pages of the books of contributed papers summed up to 10000.

The complete list of publication related to SPIG activities for first eleven meetings is given in Appendix A.2. in ref [1].

5. Influence of SPIG on international collaboration

As a result of SPIG meetings, in particular the initial ones, several kinds of cooperation have been started. Direct contacts with invited lecturers and later also with foreign participants were extremely fruitful and strongly influenced further development of related fields in science in our country. The atmosphere at the meetings was always relaxed, invited lecturers were extremely friendly and helpful in many ways. In particular, I would like to mention professor John D. Craggs from Department of Electrical Engineering and Electronics University of Liverpool, who was willing to participate a couple of times as a lecturer. He was willing to provide financial support to number of Yugoslavs, mainly younger scientists, to work in his Department as research students. Shorter period of time in some of the Departmental laboratories worked Z.Šternberg from Zagreb, B.Čobić and M.Kurepa from Belgrade. PhD degree have obtained V.Vujnović from Zagreb, J.Labat, N.Konjević, M.Platiša and S.Božin from Belgrade and M.Pavlov from Novi Sad. His warm reception to all of us will never be forgotten. Friendly contacts were established also with other professors and scientists from the Department. Although Prof. Craggs's relations with us was unique and deserves highest appreciation, there were also other numerous scientists all over the world, that contributed in building up highly trained scientific personnel for our institutes.

Finally I would like to mention just a few scientists from Belgrade that grow up together with SPIG and become very successful leading scientists in various fields of PIG. Eventually they led large groups of younger people introducing them to new problems and organizing research mostly by themselves and sometimes with the help of some leading scientists abroad. SPIG meetings in this way give very fruitful results.

Let it be allowed to me to mention some names that, according to my opinion, contributed most to the prestige of our science in Serbia. Their international collaboration with the dignified institutions gave numerous results recognized by the world scientific community. These are: Milan Kurepa, Marko Popović, Nikola Konjević, Jagoš Purić, Zoran Petrović and many others.

This is not the end of the SPIG story. It is a continuing story of the efforts and skill of many people, my colleagues and friends that have organized, lectured or at least participated the meetings. I am very happy to know that there are so many youngsters able to continue their efforts and I wish them a lot of energy in successful organization of the next SPIG-s.

Acknowledgement
I am grateful to Mrs Patricia Broedehl who spent some of her valuable time to correct my manuscript in order to make it acceptable for publishing.

References

[1] M Kurepa and J Labat: " *History of SPIG Meetings*" in : The Physics of Ionized Gases: Invited Lectures and Progress Reports: Dedicated to the memory of Prof. Milan Kurepa / 20th SPIG, September 4.- September 8. 2000, Zlatibor , Yugoslavia, page 483
Also:
SFIN Year XIV No. A1 (2001)
Editors: N.Konjević, Z.Lj.Petrović and G.Malović

Appendix: Chronology of SPIG meetings

Symposium on Physics of Ionized Gases

I 1962	Beograd	B.Perović (B)
II 1964	Zagreb	Z.Šternberg (Z)

Summer School on Physics of Ionized Gases

1964	Herceg Novi	B.Perović(B)	
III 1966	Niš	Dj Bošan(N)	Dj.Bošan(N)

Summer School and International Symposium on Physics of Ionized Gases

IV 1968	Herceg Novi	B.PerovićB)	B.PerovićB)
V 1970	Herceg Novi	B.Navinšek(Lj)	B.Čobić(B)
VI 1972	Split	M.Kurepa(B)	--
VII 1974	Rovinj	V.Vujnović(Z)	--
VIII 1976	Dubrovnik	B.Navinšek(Lj)	--
IX 1978	Dubrovnik	R.Janev (B)	B.PerovićB)
X 1980	Dubrovnik	B.Čobić(B)	H.Drawin(France)
XI 1982	Dubrovnik	G.Pichler (Z)	B.Aničin(B)
XII 1984	Šibenik	M.M.Popović (B	M.M.Popović(B)
XIII 1986	Šibenik	J.Purić(B)	M.Kurepa (B)
XIV 1988	Sarajevo	L.Tanović(S)	N.Konjević(B)
XV 1990	Dubrovnik	D.Veža(Z)	I.Terzić(B)
XVI 1993	Beograd	M.Milosavljević(B)	Z.Petrović(B)

XVII 1994	Beograd	B.Marinković(B)	Z.Petrović(B)
XVIII 1996	Kotor	B.Vujičić , S.Djurović(NS)	J.Purić(B)
XIX 1998	Zlatibor	N.Konjević(B)	S.Djurović(NS)
XX 2000	Zlatibor	Z.Petrović, N.Bibić,(B)	M.Kuraica(B)
XXI 2002	Soko Banja	M.Radović(N)	N.Bibić(B)
XXII 2004	Bajina Bašta	Lj.Hadžjevski(B)	T.Grozdanov(B)
XXIII 2006	Kopaonik	N.Simonović, P.Marinković,	T.Grozdanov(B)
XXIV 2008	Novi Sad	L.Popović, M.Dimitrijević,	G.Malović(B)
XXV 2010	Donji Milanovac	M.Kuraica (B)	L.Popović(B)

(B)-Belgrade, (Z)-Zagreb, (N)-Niš, (Lj)-Ljubljana, (S)-Sarajevo, (NS)-Novi Sad. Second column: place of the conference, third column: chairman of the Organizing committee, fourth column: chairman of the Scientific committee.

On the calculation of resonances by means of analytic continuation in coupling constant

J Horáček[1], **I Paidarová**[2]

[1] Charles University, Faculty of Mathematics and Physics,
V Holešovičkách 2, CZ-180 00 Praha 8, Czech Republic
[2] J. Heyrovský Institute of Physical Chemistry of the ASCR, v.v.i.,
Dolejškova 3, CZ-182 23 Praha 8, Czech Republic

E-mail: horacek@mbox.troja.mff.cuni.cz

Abstract. The method of analytic continuation in the coupling constant in combination with the use of statistical Padé approximation designed to determine resonance parameters is introduced. It is shown that standard quantum chemistry codes provide accurate data which can be used for the process of analytic continuation in coupling constant. Resonance parameters, both the energy and the width, can be inferred for real molecules with accuracy comparable to other more elaborated methods.

1. Introduction

In plasma physics many scattering processes are responsible for the determination of the state of the plasma. In addition to elastic scattering the processes of dissociative electron attachment (DA)

$$\mathrm{e} + \mathrm{H}_2(\mathrm{v_i}, \mathrm{J_i}) \rightarrow \mathrm{H} + \mathrm{H}^-, \tag{1}$$

associative detachment (AD)

$$\mathrm{H} + \mathrm{H}^- \rightarrow \mathrm{e} + \mathrm{H}_2(\mathrm{v_f}, \mathrm{J_f}), \tag{2}$$

vibrational excitation (VE)

$$\mathrm{e} + \mathrm{H}_2(\mathrm{v_i}, \mathrm{J_i}) \rightarrow \mathrm{e} + \mathrm{H}_2(\mathrm{v_f}, \mathrm{J_i}) \tag{3}$$

and dissociative recombination (DR)

$$\mathrm{e} + \mathrm{H}_2^+ \rightarrow \mathrm{H} + \mathrm{H} \tag{4}$$

are important. These processes play a decisive role in many other fields. In astrophysics, for example, it is assumed [1] that the process of AD is responsible for the creation of the hydrogen molecule at early stages of the development of the Universe. The existence of the H_2 molecule represents a very efficient cooling mechanism of the hot materials which eventually leads to the creation of stars, galaxies, etc. It was suggested [2, 3] that DA to vibrationally excited molecular hydrogen plays an important role in molecular activated recombination in fusion divertor plasmas. This process is also believed to be the primary source of negative-ion production in low-density hydrogen plasmas, see e.g. [4, 5, 6]. All the processes mentioned above are of resonance nature: they are efficient in a restricted range of energy. This is demonstrated

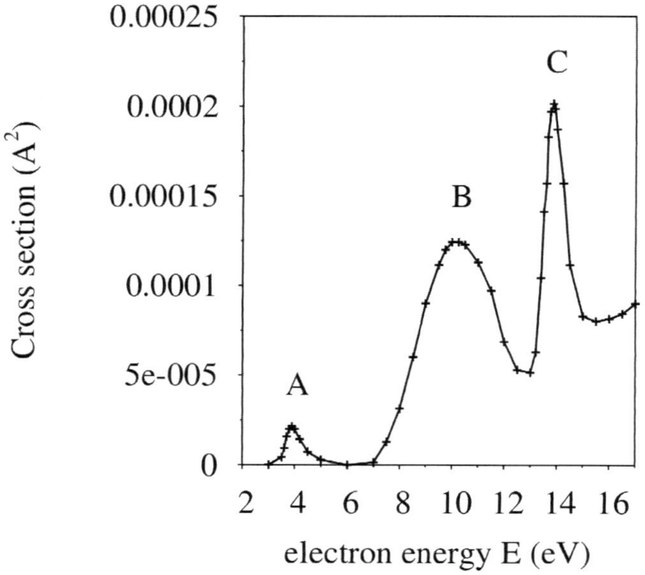

Figure 1. H_2 dissociative electron attachment cross section. Feature A: $^2\Sigma_u^+$ shape resonance, feature B: $^2\Sigma_g^+$ shape resonance, feature C: formation of excited atomic state $e + H_2 \rightarrow H^- + H(2s)$.

in Figure 1 where the dissociative attachment cross section is shown for the e - H_2 system. Each peak in the figure corresponds to a different resonance. For determination of the cross sections it is important to know the resonance energy and its width. It is the purpose of this note to describe a simple method for determination of those parameters.

2. Determination of resonance parameters

In practical applications several more or less accurate definitions of a resonance were used. For example one can fit measured scattering cross section by the Breit-Wigner or Fano formula and derive the resonance parameters. This works well for narrow resonances with small background. Mathematically correct definition of a resonance is related to the S-matrix. Let us assume that the interaction between colliding particles is described by a potential $U(r)$ which is negligible at large distances $U(r) = 0, r > R$. Then the solution of the Schrödinger equation, $\psi_l(r)$ has the asymptotic form (for zero angular momentum this expression is exact for any $r \geq$ R).

$$\psi_l(r) \approx N(e^{-ikr} + S_l(k)e^{ikr}), \qquad r \rightarrow \infty. \tag{5}$$

The function $S_l(k)$ - the S-matrix - is known to have poles in the complex plane of k [7]. If the pole occurs on the positive imaginary axis, $k = i\kappa, \kappa > 0$, the wave function $\psi_l(r)$ behaves at large distances as

$$\psi_l(r) \approx e^{-\kappa r} \tag{6}$$

and describes bound states. For poles at negative imaginary axis, $k = -i\kappa, \kappa > 0$ we have

$$\psi_l(r) \approx e^{\kappa r}. \tag{7}$$

This pole describes the so called virtual states. Finally poles at $k = k_1 - ik_2, k_1, k_2 > 0$ describe resonances which we are looking for. The resonance wave function behaves at large distances as

$$\psi_l(r) \approx e^{ik_1 r} e^{k_2 r}, \tag{8}$$

i.e. the resonance wave function oscillates and its amplitude exponentially increases at increasing r. This feature indicates that to calculate the resonance wave function is a difficult task whereas the calculation of bound states is easy because the wave function is so simple.

3. Analytical continuation

In what follows we introduce a method which for determination of resonance parameters uses only bound state energy data. The method - the so called method of analytical continuation in coupling constant (ACCC) is based on the following observation: It is well known that if we make the potential $U(r)$ describing our system more attractive by adding a perturbation $\lambda V(r)$,

$$U(r) \to U(r) + \lambda V(r), \tag{9}$$

the bound states get more bound and resonances move closer to the origin. At some value of $\lambda = \lambda_0$ both resonance poles merge and transform into a pair of bound and virtual states. If we calculate bound state energy $E(\lambda)$ for several values of λ we can construct a frequently used polynomial approximation to $E(\lambda)$, e.g.,

$$E(\lambda) = E_0 + E_1\lambda + E_2\lambda^2 + \dots \tag{10}$$

and extrapolate this function to $\lambda \to 0$. But since $E(\lambda)$ attains only real values - the bound state energies are of course real - the result of the extrapolation is a real quantity which in some sense represents the resonance energy but gives no information on the imaginary part of the resonance energy. It is a common situation in physics that some quantity which can be measured or calculated in a restricted range of some parameter is required outside this range. Let us consider for example the function $\frac{1}{1-x}$ represented as a power series

$$f(x) = \frac{1}{1-x} = 1 + x + x^2 + x^3 + \dots. \tag{11}$$

Since the radius of convergence of this series is equal to 1, it can be used in practice only for $x < 1$. The radius of convergence is determined by the nearest singularity, here the simple pole at $x = 1$. There exist several methods how to use the expansion (11) even at $x > 1$. Here we mention just one of the approaches - the Padé approximation. As Padé approximation [8] we mean the representation of a function $f(x)$ in the form

$$f(x) \approx f^{[N,M]}(x) = \frac{P_N(x)}{Q_M(x)}, \tag{12}$$

where $P_N(x)$ and $Q_M(X)$ are polynomials

$$P_N(x) = \sum_{i=0}^{N} p_i x^i, \qquad Q_M(x) = 1 + \sum_{i=1}^{M} q_i x^i. \tag{13}$$

Three ways how to determine the coefficients of the polynomials are in common use. As suggested by Padé the function $f^{[N,M]}(x)$ can be expanded into a Taylor series and the obtained coefficients compared with the expansion coefficients of the function $f(x)$. If this is done say for $N = 1, M = 1$ we obtain

$$f^{[1,1]}(x) = \frac{1}{1-x}, \tag{14}$$

i. e. from the knowledge of several expansion coefficients we obtain the exact functional form of $f(x)$. The obtained expression can now be used even for $x > 1$. This is not fortuitous, it is known that the Padé approximation carries out the analytical continuation automatically [8]. The Padé approximation discussed above, the so-called Padé I approximation, is not useful for practical application. The reason is that our data are always of limited accuracy. Instead we propose to use the so-called statistical Padé approximation, Padé III approximation, defined as follows: again

$$f(x) \approx f^{[N,M]}(x) = \frac{P_N(x)}{Q_M(x)}, \tag{15}$$

but the coefficients of the polynomials are determined as the least square approximation to the data obtained by minimizing the χ^2functional

$$\chi^2 = \frac{1}{L}\sum_{l=1}^{L}\frac{1}{\epsilon_l^2}\left|\frac{P_N(x_l)}{Q_M(x_l)} - f_l\right|^2. \tag{16}$$

In the above equation $\{\epsilon_l\}$ indicates the inaccuracy of the data used. The calculation of the Padé III approximation is a nonlinear problem; an efficient solution is described in [8] and [9]. As discussed in [8] the Padé approximation has two very important features (among others): 1. as already mentioned above it carries out the analytical continuation automatically; and 2. it is flexible enough to approximate functions with singularities (poles and cuts).

3.1. Analytic continuation in the coupling constant

As shown in [8] and references therein the function $k(\lambda)$, $E(\lambda) = k^2(\lambda)$, has for nonzero angular momentum l a singularity at a point λ_0 where $k(\lambda_0) = 0$. The singularity is of the square root type

$$k(\lambda) \approx a\sqrt{\lambda - \lambda_0} + ... \tag{17}$$

This indicates that the expansion (10) cannot represent the resonance energy for λ below λ_0. In accordance with (17) we represent $k(\lambda)$ in the form of Padé approximation in a new variable y

$$k(y) \approx \frac{P_N(y)}{Q_M(y)}, \qquad y = \sqrt{\lambda - \lambda_0}. \tag{18}$$

This representation takes into account the singularity at $\lambda = \lambda_0$ exactly and the Padé approximation carries out the analytical continuation. At values $\lambda < \lambda_0$, $k(\lambda)$ gets complex and the resonance energy acquires its complex part. To calculate the bound state energies and their square integrable wave functions is now a routine task and many commercial programs are available. This makes this method very attractive. This ACCC method was proposed in the field of nuclear physics by Krasnopolsky and Kukulin [10] and [11] in seventieths and is in detail described in the book [8]. This approach has found applications mostly in nuclear physics, see e.g. [12], [13], [14]. In those applications the so called Padé II approximation was used which is based on interpolation of the data. Because interpolation is sensitive to small changes in the data a high accuracy of the input data is required [15]. This limits the use of ACCC mostly to model analytic problems.

In this work we will follow the work of Nestmann and Peyerimhoff [16] and use the Coulomb potential for the analytical continuation. This is not the best choice from the point of the scattering theory but it is very convenient for the use of commercial quantum chemistry codes.

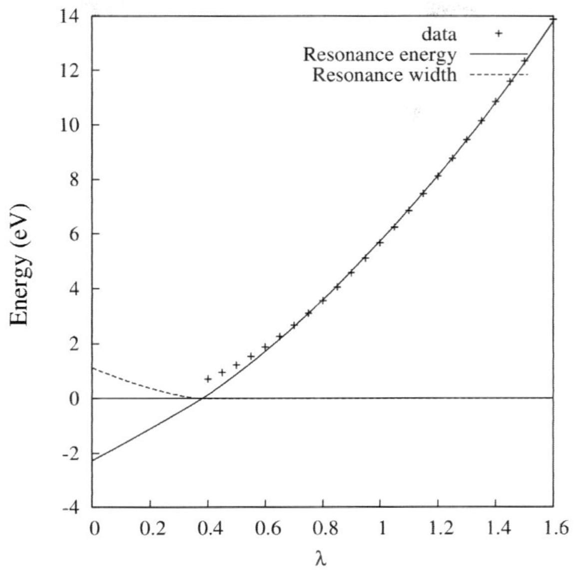

Figure 2. Analytic continuation (Padé 1/1) for the $^2\Sigma_u^+$ state of H_2^-: Solid line - the resonance energy, the dashed line - the resonance width, crosses - the calculated bound state energies.

3.2. Determination of the parameter λ_0.
Correct determination of the branch point λ_0 is of decisive importance for the accuracy of the ACCC approach. A possible way how to find it is proposed in [8]. For nonzero values of the angular momentum (and this is the most important case) the bifurcation point λ_c coincides with λ_0, $\lambda_c = \lambda_0$. Since molecular s-wave resonances are very rare we restrict here to the case of nonzero angular momentum. A rough approximation of λ_0 can be obtained by simple extrapolation of $\lambda(E)$ to zero, $E \rightarrow 0$. Here, however, when using the statistical Padé approximation, we have a new tool at hand: provided the data are accurate enough the function χ^2 as a function of the parameter λ_0 possesses a sharp minimum at the true value of λ_0, see. [17]. In addition the knowledge of $\chi^2(\lambda_0)$ can serve as a tool to select an optimal value of the order N/M of the Padé approximation. If N and M are too high the function $\chi^2(\lambda_0)$ becomes erratic because the approximation reproduces the noise.

4. Application

The ACCC approach was recently successfully applied to the treatment of $^2\Pi_g$ state of N_2^- [18]. It was shown there that the resonance width was determined with a very high accuracy $\approx$ 50 meV and the resonance energy $\approx$100 meV. Preliminary results were also obtained for CO^- and H_2^- resonance states. The Σ_u of H_2^- resonance is an example of a very broad resonance which is hardly to be discerned in scattering calculation [19]. The bound state data available for this state did not allow us to use higher order Padé approximation; only the lowest [1/1] approximation provided stable results. For more accurate determination of the resonance parameters of this state more extensive data are needed. In Figure 2 we plot the results of the analytical continuation by means of Padé 1/1 for this state. The bound state energy is positive in this Figure, the calculated resonance energy negative. In Figure 3 we compare the present results with other calculations, [20] and [21].

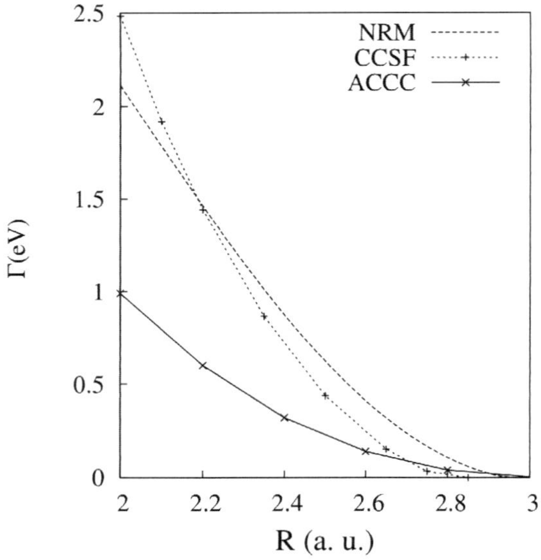

Figure 3. Resonance width of the $^2\Sigma_u^+$ state of H_2^- calculated by three different approaches: The dashed line - empirically fitted resonance width [21], dotted line - complex self consisted field calculation [20], solid line - present calculation.

5. Conclusions

We conclude that the ACCC method can be applied to the calculation of low-energy electron-molecule resonances even while using the long-range Coulomb potential for the transformation of resonances into bound states. The obtained data agree well with the results of other, often much more elaborate, methods [18]. The great advantage of the present method is that, without changes, it could be used also to polyatomic molecules relevant to plasma physics. This will be discussed in our future publications. It is however necessary to say that the method has its limitations. The most important limitation is that in the present form it is applicable to low-energy isolated shape resonances only. If the resonance is located very far from the origin (e.g. very broad resonance) the method looses its accuracy because the continuation must be done to very large distances.

Acknowledgments

This work was supported by Záměr MSM0021620860, by the Center of Theoretical Astrophysics LC06014 of the Ministry of Education, Youth and Sports of the Czech republic and the COST Action CM0601 ECCL Electron Controlled Chemical Lithography No. OC09079.

References

[1] Lepp S, Stancil P C and Dalgarno A 2002 *J. Phys. B: At. Mol. Opt. Phys.* **35** R57
[2] Post D E 1995 *J. Nucl. Mater.* **220** 143-57
Krasheninnikov S I, Pigarov A Yu and Sigmar D J 1996 *Phys. Lett.* A **214** 285
[3] Ohno N, Ezumi N, Takamura S, Krasheninnikov S I and Pigarov A Yu 1998 *Phys. Rev. Lett.* **81** 818
[4] Hiskes J R 1987 *Comments At. Mol. Phys.* **19** 75
[5] Hiskes J R and Karo A M 1989 *Appl. Phys. Lett.* **54** 508
[6] Skinner D A, Bruneteau A M, Berlemont P, Courteille C, Leroy R and Bacal M 1993 *Phys. Rev.* E **48** 2122
[7] Newton R G 2002 *Scattering Theory of Waves and Particles Second edition* (Dover publications)

[8] Kukulin V I, Krasnopolsky V M and Horáček J 1998 *Theory of Resonances: Principles and Applications* (Dordrecht/Boston/London: Kluwer Academic Publishers)
[9] Horáček J and Zejda L 1993 *Comp.Phys.Comm.* **74** 187-92
[10] Kukulin V I and Krasnopolsky V M 1977 *J.Phys.*A *: Math. Gen.* **10** L33-7
[11] Krasnopolsky V M and Kukulin V I 1978 *Phys. Lett.* A **69** 251-4
[12] Tanaka N, Suzuki Y and Varga K 1997 *Phys. Rev.* C **56** 562-5
[13] Aoyama S 2002 *Phys. Rev. Lett.* **89** 052501
[14] Funaki Y, Horiuchi H and Tohsaki A 2006 *Progress of Theor. Phys.* **115** 115-27
[15] Chao J S-Y, Falcetta M F and Jordan K D 1990 *J. Chem. Phys.* **93** 1125-35
[16] Nestmann B and Peyerimhoff S D 1985 *J. Phys.*B*: At. Mol. Phys.* **18** 615-26
[17] Horáček J 2009 *Numerical Analysis and Applied Mathematics (Sep 18-22 Rethymno)* Book Series: AIP Conf. Proc. **1168** ed Simos T E, Psiohyois G et al 139-42
[18] Horáček J, Mach P and Urban J 2010 accepted for publication in *Phys. Rev.* A
[19] Stibbe D and Tennyson J 1998 *J. Phys.* B **31** 815-44
[20] McCurdy C W and Mowrey R C 1982 *Phys.Rev.* A **25** 2529
[21] Bardsley J N , Herzenberg A and Mandl 1996 *Proc. Phys.Soc.* **89** 305

Journal of Physics: Conference Series **257** (2010) 012003 doi:10.1088/1742-6596/257/1/012003

Isotopic Effects in Ion-Molecule Reactions

Ronald McCarroll

Laboratoire de Chimie Physique-Mat'ere et Rayonnement, Unite Mixte de Recherche n°7614 , Universite Pierre et Marie Curie and CNRS,
11 Rue Pierre et Marie Curie, 75231 Paris Cedex 05, France.
E-mail: ronald.mac_carroll@upmc.fr

Abstract. A review is presented of theoretical methods for the treatment of isotopic exchange in exothermic low energy ion-molecule reactions. These processes play a fundamental role in our understanding of the chemical evolution of interstellar space. While many experimental measurements of the rate coefficients have been carried out over the past 20 years for a wide variety of molecular species, most of these were for temperatures exceeding 200K. At lower temperatures there are very few precise measurements for most of the reactions of interest. And, while the existence of a substantial isotopic effect has been recognized there are even fewer quantitative estimations of the dependence of the rate coefficient on temperature. In this work a statistical model is developed which takes account fully of energy, angular momentum and symmetry constraints to calculate both collision cross sections and rate coefficients.

1. Introduction

This talk is concerned with low energy reactive collisions between ions and molecules, in particular those which involve isotope exchange. In many astrophysical and planetary atmospheres the abundance of molecules containing one or more of the rarer isotopic constituents yields very useful information on the formation mechanisms of the different molecular species [1]. The simplest and most studied example of the problem concerns the formation of the HD isotope [2, 3, 4]. The observational astrophysical evidence indicates that the rate of formation of HD is much more rapid than that of H_2. The formation of H_2 takes place via the association of two H atoms on the surface of an interstellar grain. A similar mechanism will of course be effective for HD formation, but the corresponding rate is insufficient to explain the observations. A more rapid way to form HD is via an indirect chain reaction involving a weak non thermal source of cosmic rays which is responsible for a small but significant fraction (of the order of 10^{-4}) of H^+ ions in the interstellar medium. Subsequently, the D atoms are rapidly ionized by the resonant charge exchange reaction with H^+ ions

$$H^+ + D \rightarrow H + D^+ \tag{1}$$

followed by the ion-molecule reaction

$$D^+ + H_2 \rightarrow HD + H^+ \tag{2}$$

We observe that since the internal energy of HD is 0.0358 eV lower than that of H_2 the exothermic reaction (2) can occur with a rate coefficient of the order of 10^{-9} cm^3 s^{-1} even at the lowest temperature. Similar overabundance of many more complex isotopic deuterated species have been observed. The effect can also be found with heavier species such as ^{13}CO. For this reason the isotopic abundances in space, derived from observations of CO may not yield a correct estimation of the ^{12}C/^{13}C abundance ratio.

The problem of isotopic substitution in ion molecule reaction has attracted little detailed interest in the past 10 years. This stems in part form the difficulty of laboratory experiments at the very low temperatures (10-300K) required for astrophysical applications. Ideally, we would require collision cross section for molecules in a specific rotation state (at the temperatures of interest, most molecules of interest are in their ground vibration state). But even at temperatures as low as 100K, many rotation states can be populated. Most laboratory experiments are carried out at pressures which ensure a thermal distribution of rotation states and only rate coefficients can be determined. It is clear that the extraction of a state selective collision rate from experimental measurements is at best a highly risky process.

However, in the very low energy regime, theory offers many attractive features, which allows us to investigate state selective cross sections with some degree of confidence. But first of all it is of interest to recall some aspects of statistical mixing models, originally introduced by Langevin [5] proposed for the study of ion mobility in a gas, and first applied to ion-molecule reactions in the 1970's. See review by Watson [6].

2. Statistical ion-molecule models

First of all, let us recall the basic hypothesis of the simplest Langevin type model, when applied to exothermic reactions involving molecules with no internal rotation. Such a theoretical model is expected to be valid only for temperatures $T < B/k$ where k is the Boltzmann constant and B the rotational constant of the molecule. For relative kinetic energies of the reactants up to about 100 meV, the reactive cross sections collisions are governed by the long range interaction between the reactants. (1 meV corresponds roughly to a temperature of 10K). In the classical statistical model [6] the initial state of the system is characterized by the relative kinetic energy of the reactants E_t and the impact parameter b. When, in addition, it is assumed that the interaction potential depends only on the ion-molecule distance R, the effective radial potential may be written as

$$V_{\text{eff}}(b, E_t, R) = V(R) + \frac{E_t b^2}{R^2} \tag{3}$$

For R greater than about 5 Bohrs, $V(R)$ is attractive of the form $V(R) = -C_4/\mathrm{R}^4$. For low collision energies (of the order of around 100 meV or less), such a potential will lead to orbiting of the reactants at some critical distance $R = R_{\text{mx}}$.

The usual procedure is to calculate, for some given E_t the value of $R = R_{\text{mx}}$ for which $V_{\text{eff}}(E_k, b, R)$ is maximum. Then considering R_{mx} as a function of E and b, *we* determine the value of b=b_c for which $V_{\text{eff}}(R_{mx}) = E_t$. We observe now that b_c is a function of E_k. The critical assumption of statistical models, is that the reaction proceeds with unit probability if $b < b_c$ but is forbidden if $b > b_c$. In other words, the centrifugal barrier can only be crossed if $b < b_c$. Then, the reactive cross section is given by

$$\sigma(E_t) = 2\pi \int_0^\infty bP(E_t, b)db = 2\pi \int_0^{b_c} bdb = \pi b_c^2 \tag{4}$$

This calculation is elementary for interaction potentials of the form R^{-4}, when expressions both R_{mx} and b_c can readily be obtained in analytic form.

However, if, the aim is to determine rate coefficients, it is more convenient to characterize the initial state of the collision system by E_t and its angular momentum by $L = b\sqrt{2\mu E_t}$ where μ is the reduced mass.

$$V_{eff}(L, R) = V(R) + \frac{L^2}{2\mu R^2} \tag{5}$$

The advantage of this procedure is that it is readily adaptable to a semi classical quantification of the relative angular momentum and, as will become apparent later, makes it much simpler to enforce angular momentum conservation.

As before, we determine, for a given value of L the radial distance R_{mx} for which the effective potential $V_{\text{eff}}(L,R)$ is maximum. In conformity with the capture model, the reaction for any given L is assumed to proceed with unit probability if $E_t > V_{\text{eff}}(L,R_{mx})$ but is forbidden if $E_t < V_{\text{eff}}(L,R_{mx})$. To obtain the cross section, then if L is treated as a continuous variable, we integrate over all possible L to obtain

$$\sigma(E_t) = \frac{\pi}{\mu E_t}\int_0^\infty L P_L(E_t)\,dL \tag{6}$$

where

$$\begin{aligned} P_L &= 1 \ \text{ if } \ E_t \geq V_{eff}(L, r_{mx}) \\ &= 0 \ \text{ if } \ E_t < V_{eff}(L, r_{mx}) \end{aligned} \tag{7}$$

For molecules with no internal rotation, a classical description of L is satisfactory since the values of L which contribute most to the cross section are large (of the order of 50- 150 a.u.). But when we treat molecules with internal rotation, the rotational quantum number j is quite small (in the range from 0 to15 a.u.) for which a semi-classical quantification is preferable. So, treating L as a semiclassical quantized variable $L^{sc} = L + 1/2$, the effective radial potential may be defined as

$$V_{\text{eff}}^{\text{sc}}(L_{\text{sc}}, R) = V(R) + \frac{L_{\text{sc}}^{\ 2}}{2\mu R^2} = V(R) + \frac{(L+1/2)^2}{2\mu R^2} \approx V(R) + \frac{L(L+1)}{2\mu R^2} \tag{8}$$

Then the total cross-section can be expressed as a sum over all possible values of L.

$$\sigma(E_t) = \frac{\pi}{2\mu E_t}\sum_{L=0}^{\infty}(2L+1)P_L(E_t) \tag{9}$$

The expressions (4) and (9), which are not significantly different, correspond to the Langevin capture model. For reactions of ions with non-polar molecules this model predicts qualitatively rate coefficients at room temperature (300-400K). But at lower temperatures (10-200 K), it fails to reproduce correctly the measured temperature dependence. Also, it is unsatisfactory to describe reactions between neutral molecules.

One obvious weakness of the simple Langevin model is the neglect of the anisotropic contribution to the interaction potential. A more precise representation of the interaction potential is given by [7, 8]

$$V(R,\theta) = -\frac{\mu_D}{R^2}\cos\theta + \frac{Q(3\cos^2\theta - 1)}{2R^3} - \frac{1}{6R^4}\left[3\alpha + (\alpha_{\parallel} - \alpha_{\perp})((3\cos^2\theta - 1)\right] \tag{10}$$

where μ_D is the dipole moment of the molecule, Q is the quadrupole moment, α is the isotropic polarizability and $\alpha_{\parallel}, \alpha_{\perp}$ are the anisotropic components of the polarizabulity parallel and perpendicular to the molecular axis. The anisotropic terms of (10) can be taken into account using the statistical adiabatic channel model of Troe [7, 8] or variants therof [9-14]. The main corrective term in (10) is only present in the case of polar molecules. In this work, we shall only be concerned with non-polar molecules such as H_2, the main corrective term arises from the quadrupole term. This term is not expected to be dominant, but its inclusion will give a measure of the accuracy of the theoretical model.

3. Calculation of the rate coefficient at very low temperatures

The rate coefficient can always be defined, but it is only when the reactants are in thermal equilibrium that a practical calculation can be made. In this section, we determine the rate constant when only the ground rotational state ($j = 0$) of the molecule is significantly populated. In this case the kinetic energy of the reactants, $E_t = E$.

We first define the transition rate $K_0(E)$ at a specific energy E,

$$K_0(E) = \sqrt{(2E/\mu)}\sigma_0(E) \tag{11}$$

The rate coefficient at a temperature T is then given by integrating (11) over the Maxwell distribution of E. Then, we obtain, using (8)

$$K_0(T) = \frac{1}{2\mu}\frac{1}{kT}\left(\frac{8}{\pi\mu kT}\right)^{1/2}\int_0^\infty \pi\sum_{L=0}^{\infty}(2L+1)P_L(E)\exp\left(-\frac{E}{kT}\right)dE \tag{12}$$

and permuting the integration over E with the sum over L, we have

$$K_0(T) = \left(\frac{2\pi}{\mu^3 kT}\right)^{1/2}\sum_{L=0}^{\infty}(2L+1)\exp(-E_L^*/kT) \tag{13}$$

This expression (13) was first used to calculate reaction rates in [15]. We may remark that it is identical to the result of the statistical capture model [7, 8] based on the assumption that the reaction proceeds via an energetically accessible transition state. For every value of L, there exists a quasi-stationary transition state, and if accessible, the reaction proceeds with unit probability. However, the formula (13) is only valid at very low temperatures T for which $T < B/k$. So even for a molecule such as H_2, which has a large rotation constant of 60.4 cm^{-1} (13) is not expected to be valid for temperatures exceeding 50K when the population of excited rotational states becomes significant.

4. Calculation of the rate coefficients including rotation

Other considerations become important when the reactant molecule has internal rotational energy. In section 3, the reactant molecule has no internal rotation energy and the relative angular momentum of the system **L** is conserved. But when the molecule possesses an internal angular momentum **j**, the conservation of the total angular momentum imposes additional constraints. For a general discussion of this problem see [16-18] and for an application to reactive collisions involving neutral molecules see [19].

Fortunately, in the asymptotic region, where the interaction potential is purely radial, the internal rotation is not coupled to the relative angular momentum and so, in this region the initial relative angular momentum L is still conserved. Consequently, as in the preceding section, the transition states can be labelled by a value of L from 0 up to L_{mx} and their energies are determined uniquely by the asymptotic potential and L. All transition states which contribute effectively to the reaction are defined by the asymptotic potential. Energy conservation is ensured by the condition that the collision energy must exceed the centrifugal barrier. So, when the energy E and the angular momentum L are such that the centrifugal barrier can be crossed, and the reactive region, where the interaction is no longer isotropic, becomes accessible. But, whereas the j, L states are uncoupled in the asymptotic region, they become strongly coupled in the reactive region. Then, neither j nor L is conserved in the reaction process.

So let us now consider the cross section when the reacting molecule is in a specific initial rotation state j. If no restriction is introduced on the nature of the coupling between states j and L, we might expect statistical mixing of the $(2j+1)(2L+1)$ degenerate states of the entry channel. But this is incorrect since such a statistical mixing of all the $(2j+1)(2L+1)$ states of the (j, L) entry channel would violate angular momentum conservation. . The (j, L) entry channel does not correspond to just one

Journal of Physics: Conference Series **257** (2010) 012003 doi:10.1088/1742-6596/257/1/012003

specific total angular momentum (as in section 3). Rather, it corresponds to a superposition of (2*j*+1) total angular momentum states *J* where

$$|L-j| \leq J \leq L+j \tag{14}$$

The net results of this constraint that since there is only a probability of $1/(2j+1)$ of forming a complex of total angular momentum *J* and since there is only a statistical mixing of states with a given J, M_J we find that

$$\begin{aligned} P_L^j &= \frac{1}{2j+1} \text{ if } E_t \geq V_{eff}(L, r_{mx}) \\ &= 0 \text{ if } E_t < V_{eff}(L, r_{mx}) \end{aligned} \tag{15}$$

where E_t is the relative kinetic energy of the system, initially in rotational state *j*. It then follows that

$$\bar{\sigma}_j(E_t) = \frac{\pi}{2\mu E_t} \frac{1}{2j+1} \sum_{L=0}^{\infty} (2L+1) P_L^j(E_t) \tag{16}$$

We may observe that this result is due to the fact that there is only a fractional probability of forming a *J* state from any given couple (*j*, *l*). No probability is lost since other *J* states can be formed from other (*j*, *l*) couples. And, when we sum over *L*, it is easy to verify that all possible values of *J* have been taken into account. The result (16) is, of course, only valid for randomly oriented molecules. It would require modification in the case of an aligned reactant.

The rate constant for a reactant in state *j* is then given by

$$K_j(T) = \frac{1}{2\pi Q_T} \frac{1}{2j+1} \int_0^{\infty} \sum_{L=0}^{\infty} (2L+1) P_L^j(E_t) \exp\left(-\frac{E_t}{kT}\right) dE \tag{17}$$

where Q_T is the partition function of the translation motion given by:

$$Q_T = \left(\frac{2\pi}{\mu kT}\right)^{-3/2} \tag{18}$$

We note that when there is no internal rotation, that is for j =0, the result is identical to expression (10). As before, permuting the integration over E_t with the sum over *L*, we obtain

$$K_j(T) = \frac{1}{2\pi Q_T} \frac{1}{2j+1} \sum_{L=0}^{\infty} (2L+1) \exp\left(-E_L^*/kT\right) \qquad E_L^* = V_{eff}(L, R_{mx}) \tag{19}$$

The total reaction rate, summed over all *j*) is then given by:

$$K(T) = \frac{1}{Q_r(T)} \sum_{j=0}^{\infty} K_j(T)(2j+1) \exp(-\varepsilon_j/kT) = \frac{S_r(T)}{2\pi Q_T Q_r(T)} \sum_{L=0}^{\infty} (2L+1) \exp(-E_L^*/kT) \tag{20}$$

where

$$S_r(T)=\sum_{j=0}^{\infty}\exp-\left(\varepsilon_j/kT\right) \tag{21}$$

and $Q_r(T)$ is the partition function of the reactant molecule. We observe that the rate coefficient given by formula (20) differs from the result of Liao and Herbst [20]. To our knowledge, the formulae (20) and (21) do not seem to have been used before to calculate rate coefficients.

In the case of a diatomic molecule whose electronic ground state is of Σ symmetry, $Q_r(T)$ is defined as

$$Q_r(T)=\frac{1}{\sigma_r}\sum_{j=0}^{\infty}(2j+1)\exp-\left(E_j/kT\right) \tag{22}$$

where E_j is the energy of the j th rotational level and σ_r =2 for symmetric diatomic molecules and is unity for non symmetric species.

It may be observed that an analogous expression to (20) has been successfully used by a number of applications [21, 22] to take account of the statistical mixing of fine structure levels in atom -molecule reactions at low temperatures. It may also be remarked that the expression (20) ensures detailed balance.

6. Calculation in the statistical adiabatic channel method (SACM)

The modifications introduced in the SACM approach can be implemented in a relatively simple way. The main initial requirement is the determination of the adiabatic channel potentials which are given by the eigenvalues of the ion molecule system at fixed *R*. These eigenvalues are labelled as $V_{j,|m|}(R)$ where *j,m* are the rotational quantum numbers of the molecule. So it then becomes necessary to compute the orbiting radius, not only for each L, but all possible values of *L, j,m*.

The calculation of $V_{j,|m|}(r)$ for all *j,m* over the range of *r* is much simpler than it might at first appear, since we only require $V_{j,|m|}(r)$ in the asymptotic region, where perturbation theory can be applied. The first order term vanishes for *j*=0, but for $j\geq 1$ it gives rise to a term varying as R^{-3}. The main contribution comes from the second order term, which is proportional to Q^2/B and varies as R^{-6}. So in the SACM approach we may write $V_{j,|m|}(r)$ in the form

$$V_{j,|m|}(r)=-\frac{\alpha}{2R^4}+\frac{C_3(j,|m|)}{R^3}+\frac{C_6(j,|m|)}{R^6} \tag{23}$$

The results can be summarized in Table 1, where we give the C_3 and C_6 coefficients for the H_2 molecules with the following values of the quadrupole moment and rotation constant: Q=0.9462 a.u., B=60.4 cm^{-1}.

A cursory examination of Table 1 indicates that the inclusion of the anisotropic potential has only a fairly weak influence on the rate coefficient at very low temperatures. Not only are the C_3 coefficients small, but they also tend to be positive for $m\leq j/2$ but negative for large $m>j/2$. As for the C_6 coefficients they are only negative for $m=j$ or $j-1$. So the net effect of the anisotropic potential is only of any real importance for $j<2$ Of course the influence on the state selective reactive cross sections for $j<2$ does indeed, increase as higher collision energies. But, then, as the temperature increases, more highly excited rotation states contribute to the rate coefficient. So the net influence on the averaged rate coefficient tends to remain weak at thermal energies.

Table 1. Coefficients $C_3(j,|m|)$ and $C_6(j,|m|)$ in atomic units

$j,\|m\|$	$C_3(j,\|m\|)$	$C_6(j,\|m\|)$
0,0	0	-107.5
1,0	0.378	-49.7
1,1	-0.189	-33.2
2,0	0.270	-73.6
2,1	0.135	-28.2
2,2	-0.270	-14.1
3,0	0.252	23.9
3,1	0.189	9.90
3,2	0.000	-16.3
3,3	-0.315	-7.24
4, 0	0.245	12.8
4, 1	0.209	8.60
4, 2	0.098	-1.55
4, 3	-0.086	-10.0
4,4	-0.344	-4.20

5. Applications

We give here a brief discussion of one of the most studied reactions involving isotope exchange, namely the reaction $D^+ + H_2 \to HD + H^+$. On the experimental side the reaction rates were first obtained by ion flow tube techniques in the 200-300K range by Henchman et al. [2] and in the 400-2000K by Villinger et al. [3]. Both these experiments find the rate coefficient to increase as the temperature decreases. More recently ion-trap experiments by Gerlich [4] using specially designed to produce very low energy ions do not seem to exhibit any very strong temperature dependence, of the rate coefficient. But in such ion-trap experiments, the rotation temperature of the molecule may be quite different from the ion temperature. So the measured rate coefficient does not correspond to a truly thermalized rate. Of course, it is easily possible to deduce the corresponding rate if the rotational temperature of the molecule is known. But for this reason, it is difficult to make a direct comparison of theory and experiment.

A selection of theoretical results is summarized in the figure. Results are given separately for para-H_2 , ortho-H_2 , and for a thermal average of para and ortho H_2. Under most laboratory conditions, the ortho and para states are populated statistically. But it should be borne in mind that this condition may not be satisfied in all applications. For example, under interstellar conditions, the conversion of ortho to para H_2 proceeds via the reaction $\mathrm{H}^+ + \text{ortho-H}_2 \to \mathrm{H}^+ + \text{para-H}_2$, rather analogous to the reaction studied here.

There are several interesting features of these results. First of all, we observe that the rate coefficient for p-H_2 exhibits a fairly rapid decline by a factor of 2 between 50 and 200K, whereas the o-H_2 rate coefficient remains fairly constant. This is mainly a consequence of the fact that the j=2 level of H_2 is significantly populated for T greater than 50K, whereas the first excited ortho state is not significantly populated until much higher temperatures. This explains why the average over para and ortho H_2, the temperature variation is less pronounced than for p-H_2.

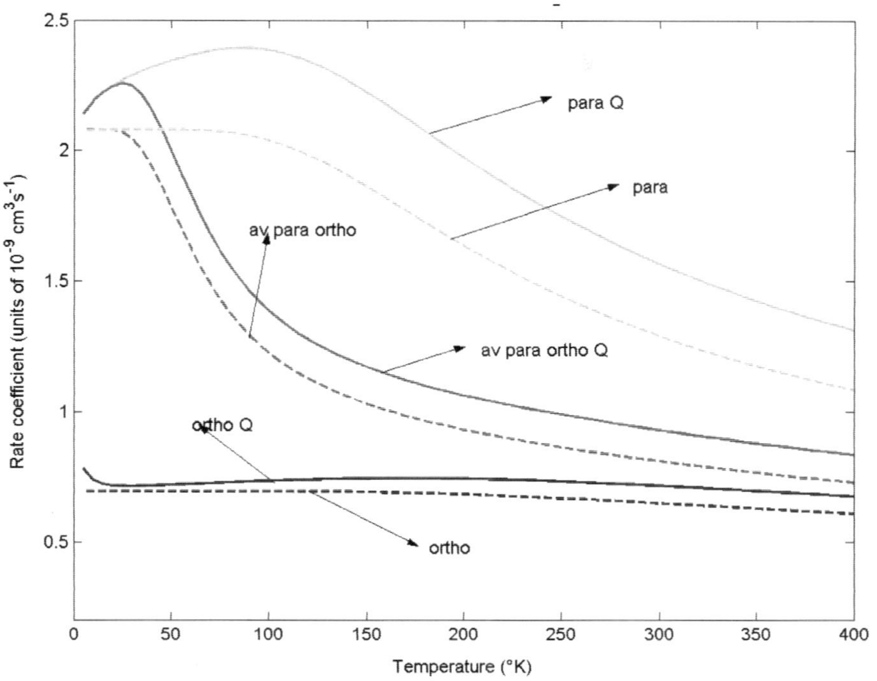

Figure 1. Isotope exchange reaction $D^+ + H_2 \rightarrow H^+ + HD$

On the other hand the inclusion of the anisotropic terms on the potential leads to a significant increase of the rate coefficient by less than 20% but the temperature dependence is similar. So the effect of the anisotropic contribution is minor compared with the case of a reaction of a polar molecule with a permanent dipole moment.

Finally a few comments should be made on the interaction potential used in this work. The potential used in this work is asymptotically correct and should be accurate for distances greater than about 5-6 Bohr. This is not the case for many ab-initio calculated potentials, such as the one used in recent quantum calculations [23] on the H^++D_2 reaction presents a non-physical potential barrier around 10 Bohr. We have therefore consistently verified that at all temperatures given here, the orbiting resonance corresponding to the largest value of L which contributes to the reaction rate occurs at values of R, where the asymptotic representation (23) is accurate. Of course the inner part of the potential would be required to verify the validity of the strong statistical mixing approximation of the allowed reaction channels. However, the excellent agreement between the theoretical and experimental rate coefficients provides strong evidence that statistical mixing approximation is satisfactory for collision energies investigated here.

We may also observe that the procedure outlined here to obtain cross sections or rate coefficients is exceedingly rapid. Of course the calculations involve a sum over a large number of j,L angular momentum states (generally up to about j=6 and L up to about 80 au.) but the convergence with respect to j,L and the dependence of the computed rate coefficients on the interaction potential can be easily tested.

Acknowledgements
The author wishes to his numerous discussions with Tasko Grozdanov on different aspects of the statistical mixing approximation.

Journal of Physics: Conference Series **257** (2010) 012003 doi:10.1088/1742-6596/257/1/012003

References

[1] D. Gerlich and S. Schlemmer, Planetary and Space Science **50**, 1287 (2002)
[2] M.J. Henchman, N.G. Adams and D. Smith, J. Chem. Phys **75,** 1201 (1981)
[3] H. Villinger, M.J. Henchman and W. Lindinger, J. Chem. Phys. **76**, 1590 (1982)
[4] D. Gerlich, J. Chem. Soc. Faraday Trans. **89**, 2199 (1993)
[5] P. Langevin, Ann. Chim. Physique, **5**, 245 (1905)
[6] W.D. Watson, Rev. Mod. Phys.**48**, 513 (1976)
[7] J. Troe, Chem. Phys. Lett. **122**, 425 (1987)
[8 J. Troe, J. Chem Phys. **87**, 2773 (1987)
[9] K. Sakimoto, Chem. Phys. **63**, 419 (1981)
[10] D.R. Bates, Proc. R. Soc. Lond. Ser. A **384**, 289 (1982)
[11] D.C. Clary, Mol. Phys. **54**, 605 (1985)
[12] D.C. Clary, in *Rate Coefficients in Astrochemistry*, edited by T.J. Millar and D.A. Williams (Kluwer, Dordrecht, 1988)
[13] .M.L. Dubernet and R. McCarroll, Z. Phys.D **13**, 255 (1989)
[14] M.L. Dubernet, M. Gargaud and R. McCarroll, Astron. Astrophys. **259**, 373 (1992)
[15] M. Ramillon and R. McCarroll, J. Chem. Phys. **101**, 8697 (1994)
[16] J.C. Light, Disc. Faraday Soc. **44**, 14 (1967)
[17] M. Quack, Mol. Phys. **34**, 477 (1977)
[18] R.D. Levine, *Molecular Reaction Dynamics*, (Cambridge University Press 2009)
[19] R. McCarroll, J. Phys. Chem.A, **113**, 14845 (2009)
[20] Quing Liao and E. Herbst, Astrophys. J. **444**, 694 (1995)
[21] M.M. Graff and A.F. Wagner, J. Chem. Phys **92**, 2423,(1990)
[22] F. Dayou and A. Spielfiefel, J. Chem. Phys. **119**, 4237 (2003)
[23] P.G. Jambrina, F.J. Aoiz, N. Bulut, S.C. Smith, G.G. Balint-Kurti and M. Hankel, Phys. Chem. Chem. Phys. **12**, 1102 (2010)

Journal of Physics: Conference Series **257** (2010) 012004 doi:10.1088/1742-6596/257/1/012004

Antimatter transport processes

D P van der Werf[1], G B Andresen[2], M D Ashkezari[3], M Baquero-Ruiz[4], W Bertsche[1], P D Bowe[2], C C Bray[4], E Butler[1], C L Cesar[5], S Chapman[4], M Charlton[1], J Fajans[4], T Friesen[6], M C Fujiwara[6,7], D R Gill[6], J S Hangst[2], W N Hardy[8], R S Hayano[9], M E Hayden[3], A J Humphries[1], R Hydomako[6], S Jonsell[1,10], L V Jørgensen[1], L Kurchaninov[7], R Lambo[5], N Madsen[1], S Menary[11], P Nolan[12], K Olchanski[7], A Olin[7], A Povilus[4], P Pusa[11], F Robicheaux[13], E Sarid[14], D M Silveira[9,15], C So[4], J W Storey[7], R I Thompson[6], J S Wurtele[4] and Y Yamazaki[15]
(ALPHA Collaboration)

[1] Department of Physics, Swansea University, Swansea SA2 8PP, United Kingdom
[2] Department of Physics and Astronomy, Aarhus University, DK-8000 Aarhus C, Denmark
[3] Department of Physics, Simon Fraser University, Burnaby BC, V5A 1S6, Canada
[4] Department of Physics, University of California, Berkeley, CA 94720-7300, USA
[5] Instituto de Física, Universidade Federal do Rio de Janeiro, Rio de Janeiro 21941-972, Brazil
[6] Department of Physics and Astronomy, University of Calgary, Calgary AB, T2N 1N4, Canada
[7] TRIUMF, 4004 Wesbrook Mall, Vancouver BC, V6T 2A3, Canada
[8] Department of Physics and Astronomy, University of British Columbia, Vancouver BC, V6T 1Z4, Canada
[9] Department of Physics, University of Tokyo, Tokyo 113-0033, Japan
[10] Fysikum, Stockholm University, SE-10691, Stockholm, Sweden
[11] Department of Physics and Astronomy, York University, Toronto, ON, M3J 1P3, Canada
[12] Department of Physics, University of Liverpool, Liverpool L69 7ZE, United Kingdom
[13] Department of Physics, Auburn University, Auburn, AL 36849-5311, USA
[14] Department of Physics, NRCN-Nuclear Research Center Negev, Beer Sheva, IL-84190, Israel
[15] Atomic Physics Laboratory, RIKEN, Saitama 351-0198, Japan

E-mail: `D.P.van.der.Werf@Swansea.ac.uk`

Abstract. A comparison of the 1S-2S transitions of hydrogen and antihydrogen will yield a stringent test of CPT conservation. Necessarily, the antihydrogen atoms need to be trapped to perform high precision spectroscopy measurements. Therefore, an approximately 0.75 T deep neutral atom trap, equivalent to about 0.5 K for ground state (anti)hydrogen atoms, has been superimposed on a Penning-Malmberg trap in which the anti-atoms are formed. The antihydrogen atoms are produced following a number of steps. A bunch of antiprotons from the CERN Antiproton Decelerator is caught in a Penning-Malmberg trap and subsequently sympathetically cooled and then compressed using rotating wall electric fields. A positron plasma, formed in a separate accumulator, is transported to the main system and also compressed. Antihydrogen atoms are then formed by mixing the antiprotons and positrons. The velocity of the anti-atoms, and their binding energies, will strongly depend on the initial conditions of the constituent particles, for example their temperatures and densities, and on the details of the mixing process. In this paper the complete lifecycle of antihydrogen atoms will be presented, starting with the production of the constituent antiparticles and the description of the manipulations necessary to prepare them appropriately for antihydrogen formation. The latter will also be described, as will the possible fates of the anti-atoms.

1. Introduction

It has become routine in recent years to produce antihydrogen at low energies via the controlled mixing of antiproton and positron clouds. Currently, much of the effort is devoted to attempts to trap antihydrogen in a magnetic minimum neutral atom trap in preparation for spectroscopic interrogation of its states and comparisons with hydrogen; (see e.g. [1, 2]). Complex apparatus [3, 4, 5], involving traps for the charged particles with superimposed atom traps, have been assembled for this purpose and are presently in development and operation.

In order to promote antihydrogen capture in the state-of-the-art, sub-K deep, neutral traps available to us, the positrons and antiprotons have to be collected and carefully manipulated before they are mixed together to promote combination. Many of the techniques that have been developed to achieve this involve transport of the antiparticles. Typically this will include collisions with buffer gas and/or interactions with clouds of electrons, often in the presence of time-varying electric fields designed to manipulate the radial profile of the charged plasmas. Transport is also a feature of the antihydrogen formation process itself, since the repeated formation and break-up of loosely bound anti-atoms can result in the antiprotons being transported to the outskirts of the positron cloud, or beyond [6, 7].

This article will contain a brief introduction to the techniques involved in antiparticle manipulation for antihydrogen formation with particular emphasis on aspects involving charged particle transport. A recent summary of positron transport phenomena in gases can be found in [8].

2. Traps and Plasmas

Most of the processes and techniques described in this paper take place in Penning traps, which are a common workhorse for charged particle confinement. This is achieved by arranging a series of (typically) cylindrical electrodes along the axis of a solenoid. The combination of the magnetic field of magnitude B (which provides the radial confinement for the particles) and the electric fields due to the voltages applied to the electrodes, results in three-dimensional confinement. The motion of a single particle, or a small collection of them, in such a trap is well understood (see e.g. [9]). Simplifying somewhat, a particle of mass, m, and charge, q, will undergo an axial (conventionally the z-direction) bounce motion with an angular frequency $\omega_z = \sqrt{2qV_0/(md^2)}$, where V_0 and d are characteristic trap voltages and dimensions, respectively. The radial component is composed of two superimposed motions: the familiar cyclotron motion with an angular frequency $\omega_c = qB/m$ and a slower magnetron drift of frequency $\omega_m = \omega_z^2/(2\omega_c) = V_0/(Bd^2)$.

It is commonplace, however, that the antiparticle clouds are sufficiently dense that they form so-called single-component plasmas. Here the Debye screening length, $\lambda_D = \sqrt{k_B T \epsilon_0/(n_e q^2)}$ (where n_e and T are the density and temperature of the cloud, respectively) is smaller than all the physical dimensions of the cloud. Note that λ_D is the length over which external electric fields are shielded by particles at the edge of the plasma. Collective plasma oscillations can be important (see e.g [10]) and can even be used as plasma diagnostics [11, 12].

Of particular note is the self electric field of the plasma, which is radial in nature with, for spheroidal plasmas, a magnitude $E_r = n_e qr/(2\epsilon_0)$. This field effectively replaces the shielded trap electric field and modifies the magnetron, or $\mathbf{E} \times \mathbf{B}$, drift which now has a characteristic angular frequency [10] $\omega_D = E_r/B = n_e q/(2\epsilon_0 B)$. Note that this implies that an antiproton at a radial distance r will have a transverse speed associated with this motion of $v_r = \omega_D r$, which will contribute to the kinetic energy of the atom if the antiproton forms antihydrogen. It is easy to see, for the parameters common in antihydrogen experiments (see e.g. [7]), that this kinetic energy can exceed that due to the thermal energy of the plasmas. This "drift" kinetic energy is proportional to r^2 and can be reduced by working with clouds of small radii. This is one of the main motivating factors behind the manipulation and transport schemes developed for antihydrogen experimentation, which we now outline.

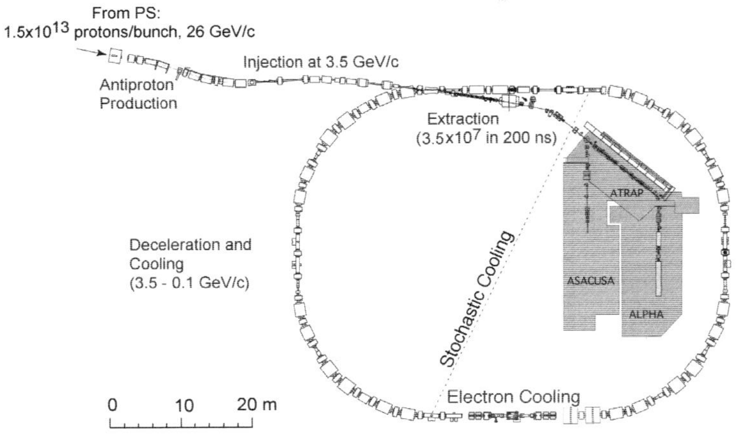

Figure 1. Schematic illustration of the AD ring.

3. Antiproton Capture, Cooling and Compression

The Antiproton Decelerator (AD) facility at CERN [13] is presently the only place in the world producing relatively low energy antiprotons. Figure 1 shows a schematic of the AD ring and the positions of the experiments. Every 100 seconds, protons with a momentum of 26 GeV/c hit a fixed target, producing antiprotons via $p + p \rightarrow p + p + p + \overline{p}$. A bunch of about 10^7 $\overline{p}$'s approximately 200 ns long, emanating from the target at around 3.5 GeV/c, is captured in the AD ring. To prevent the bunch from blowing up during deceleration, the antiparticles are cooled in four stages using stochastic [13] and electron [14] cooling techniques.

The antiprotons entering the ALPHA apparatus are still far too energetic for antihydrogen formation. Thus, a thin foil is mounted at the entrance of the ALPHA trap (degrader in figure 2a) and a small fraction of the antiprotons are slowed down to trappable energies and enter the electrode stack, which has a 4 kV potential applied at one end (figure 2b). The antiprotons bounce back and before they are able to leave the trap a 4 kV potential is raised on an electrode near the degrader, thereby confining the antiparticles (figure 2c).

Although now trapped, there is no mechanism for the antiprotons to cool down. Therefore, to enable cooling, a cloud of electrons is injected in the trap before the antiprotons arrive. The electrons cool down by synchrotron cooling in about 0.4 s when immersed in a magnetic field of 3 T and in turn they cool the antiprotons via the Coulomb interactions in about 20 s. Note that the synchrotron cooling rate is proportional to m^{-3} so that it would take of the order of a century for the antiprotons to cool down via this process.

When trapped and cooled, the antiproton cloud does not yet have the optimum shape (radius, length) and/or density for antihydrogen experimentation. However, it is possible to compress plasmas by applying a rotating electric dipole or quadrupole field [15]; the so-called rotating wall technique. This can be implemented by azimuthally segmenting one of the confining electrodes and applying a sinusoidal voltage to the segments, with each having a different phase with respect to the others.

The antiprotons were compressed by applying the rotating wall technique on the electron plasma containing the antiprotons [16] for a certain time. The radial profile of both the electrons and antiprotons were subsequently determined by ejecting the particles onto a MCP/Phosphor screen assembly residing outside the main ALPHA system (see figure 4a) [17]. Figure 3 shows

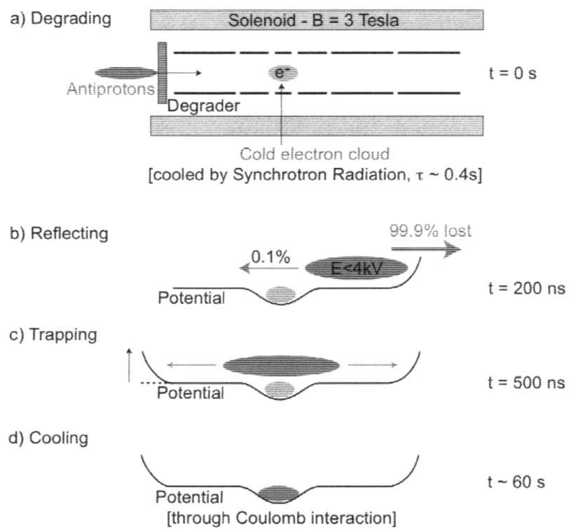

Figure 2. (a) Antiprotons arrive from the AD (to the left) with 5.3 MeV kinetic energy and are decelerated in the degrader, which is a metal foil of about 0.1 mm thickness. Electrons have been loaded previously into a centrally located trap. (b) An electrostatic wall, formed by an electrode held at high voltage is erected before the antiprotons arrive. Antiprotons with low enough energy are reflected back towards the degrader. (c) About 500 ns after the antiprotons arrive the entrance is closed by erecting a similar wall, and the antiprotons are trapped. (d) About 60 s later the antiprotons have cooled down through collisions with the cold (and self-cooling) electrons.

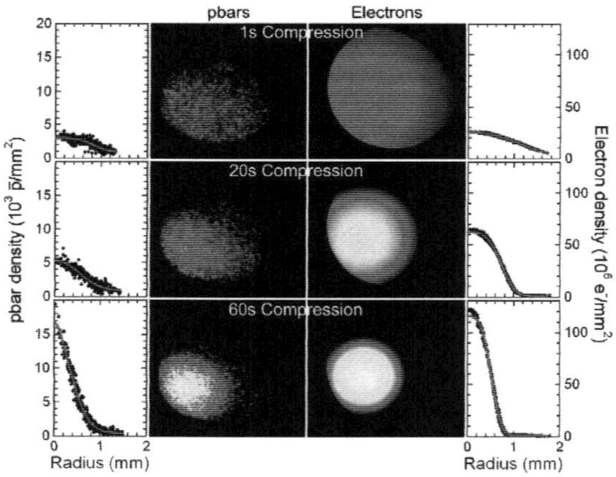

Figure 3. Antiproton and positron images showing the effects of compression, and the resulting radial profiles. The solid (red) lines are Gaussian-like (i.e., $\exp(-[r/r_0]^k)$, where $k \approx 2$) fits to the radial profiles.

examples of the phosphor screen image after applying the rotating wall for 1, 20 and 60 seconds. The antiproton images are shown on the left side, with the outer left column displaying the radial profile. The right side shows the corresponding electron data. It can be observed that the radial extent of both species reduces as the rotating wall is applied for longer durations.

Both the formation rate and the trapping probability of antihydrogen are dependent on the

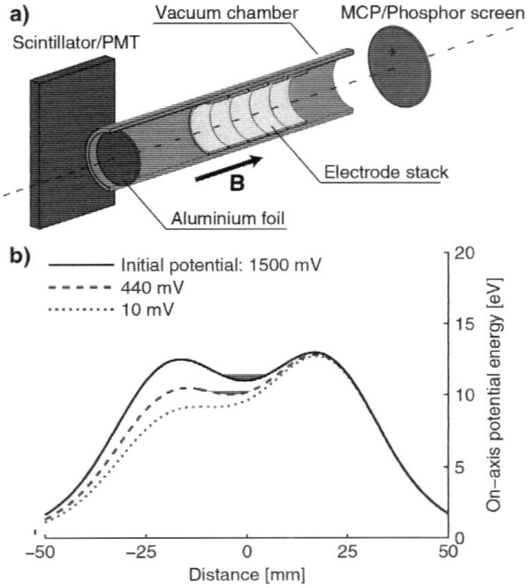

Figure 4. (a) Simplified schematic of the Penning-Malmberg trap used to confine the antiprotons and of the two diagnostic devices used. The direction of the magnetic field is indicated by the arrow. (b) Potential wells used to confine the antiprotons during the evaporative cooling ramp. The antiprotons are indicated at the bottom of the potential well (red), and the different wells are labelled by their on-axis depth.

temperature of the constituent particles. Therefore, ALPHA applied the evaporative cooling technique, well known for its use on atoms, for the first time to a cryogenic, non-neutral plasma, in this case consisting of antiprotons [18]. The temperature of the cloud was determined by lowering the potential on one side, thus allowing antiprotons to leave and hit an aluminium foil (see figure 4a) where they annihilated. The annihilation products were detected using the external scintillator/PMT arrangement. If the particles are in thermal equilibrium, the energy distribution of the first particles to escape, found from mapping the arrival time at the foil to the well depth, will follow the Boltzmann distribution [19].

For each experiment 45,000 antiprotons were prepared in a plasma with a radius of 0.6 mm and density of 7.5×10^6 cm^{-3}. Subsequently, the potentials were linearly lowered to a well with a certain depth (a few examples are shown in 4b) and the temperature was measured using the method above. In figure 5, six measurements are shown and the lowest temperature achieved was (9 ± 4) K using a 10 mV deep well, when (6 ± 1) % of the antiprotons remained trapped.

4. Positron Accumulation, Compression, Transfer and Cooling

Positrons are accumulated using a 3-stage Surko-type buffer gas device (see figure 6) [20]. Positrons emanating from a 2 GBq ($\approx$ 75 mCi) ^{22}Na source are slowed using a solid neon moderator [21, 22]. The neon is condensed directly onto the source which is mounted on a cold finger connected to a closed cycle helium refrigerator capable of reaching temperatures down to 4 K. The moderated positrons are then accelerated to 80 eV and transported into the first stage of the electrode stack. Nitrogen gas is admitted into this stage at a pressure of about 10^{-3} mbar, and, due to the increased diameters of following stages, pressures of 10^{-4} and 10^{-6} mbar are established in, respectively, the second and third stages. The relatively high pressure in the first stage causes the positrons to lose enough kinetic energy in an inelastic collision with the

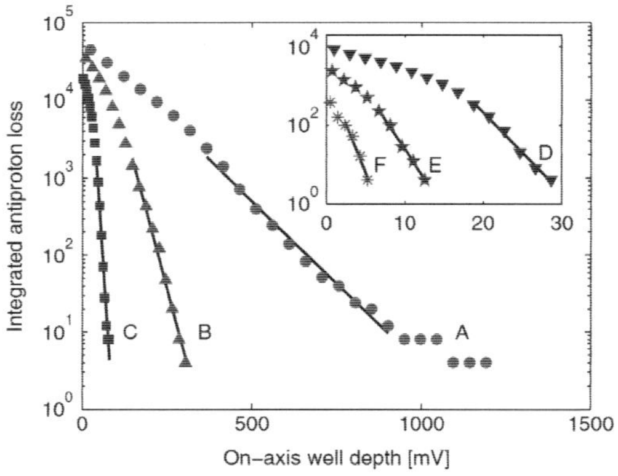

Figure 5. The number of antiprotons lost from the well as its depth is reduced is integrated over time and plotted against the well depth. The well depth was ramped from high to low; thus, time flows from right to left in the figure. The measured number was corrected for the 25 % detection efficiency. The curves are labelled in decreasing order of the temperatures extracted from an exponential fit, shown as the solid lines. The temperatures extracted are: A: 1040, B: 325, C: 57, D: 23, E: 19, and F: 9 K. As the antiprotons get colder, fewer can be used to determine their temperature, an effect described in Ref. [19].

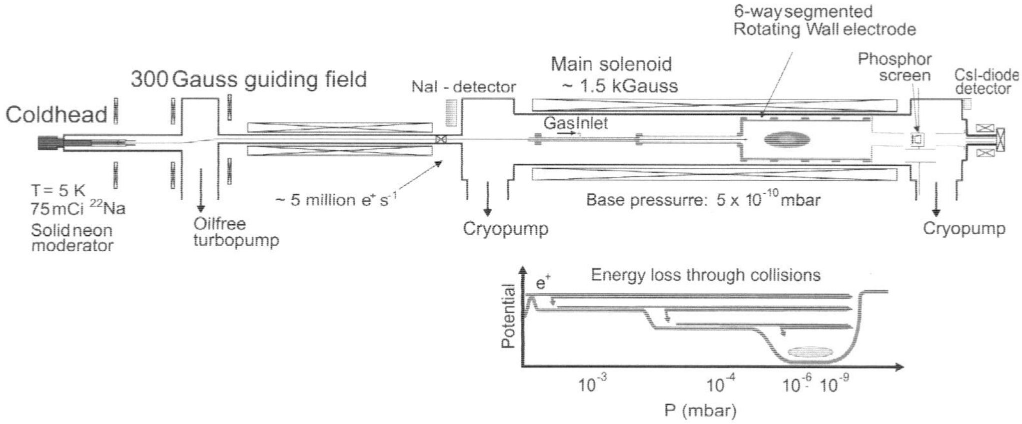

Figure 6. Schematic illustration of the positron beamline and buffer gas accumulator. The lower panel is a representation of the axial electrical potential of the trap and shows how collisions, progressively in each stage, result in accumulation in the third stage. When the nitrogen line is closed, the gas is pumped out promptly [20] in readiness for transfer of the positrons to the main ALPHA system.

gas before they can escape the trap. As shown on the lower panel of figure 6, further collisions in stages two and three results in positron accumulation in the latter. The low pressure in stage three reduces the annihilation probability and the rate of radial transport of the trapped positrons such that lifetimes of around 100 s can be achieved routinely.

It is advantageous to reduce as far as possible the radial extent of the accumulated positrons since, on transfer, they must pass through a small pumping restriction which separates the main

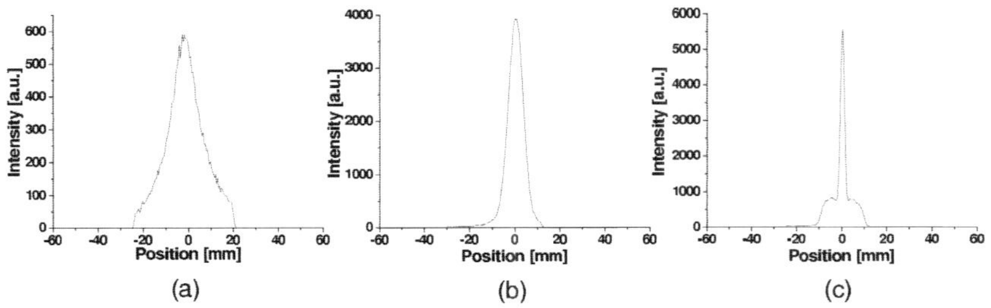

Figure 7. One dimensional projection of the phosphor screen image of the ejected positron cloud following dipolar rotating wall compression. (Note that the overall distribution is ellipsoidal.) (a) no rotating wall; (b) rotating wall with N_2 gas only and (c) rotating wall with N_2 and added CO_2 cooling gas. The integrated yields are around 29, 38 and 30 million positrons for (a), (b) and (c) respectively.

ALPHA system from the accumulator. Thus, a rotating wall is employed in the accumulator to compress the positrons and figure 7 shows radial profiles measured using the phosphor screen (figure 6) with and without the rotating wall. Using the rotating wall during accumulation at a fixed frequency of 700 kHz narrows the radial profile (figure 7b), and plasmas sizes of 7 mm diameter have been obtained. Interestingly, inserting a cooling gas such as CO_2 [23], the radial profile can be changed dramatically as shown in figure 7c. More research is needed to understand these data, which seem to indicate that there are two radial distributions within the accumulator.

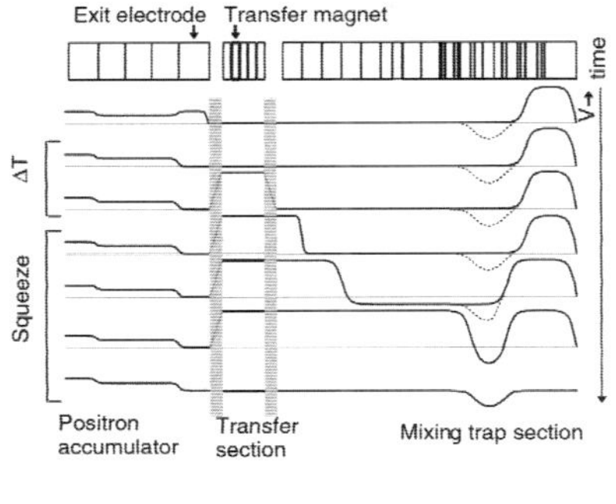

Figure 8. Schematic of the electrode system used to transfer and re-trap the positrons. The time sequence of the potentials is shown. A pulsed magnetic field of around 1 T for 1 s was applied using the so-called transfer magnet to squeeze the positrons through a narrow pumping restriction.

After accumulation and subjecting the positrons to the desired rotating wall manipulation, the nitrogen buffer gas is pumped out and the positrons are transported into the main ALPHA system [24]. Figure 8 illustrates how the potential changes in time during the transport procedure. The upper three lines show the potential of the accumulator exit electrode dropping rapidly to allow the positrons to escape and then the first electrode in the main system closing after a time ΔT to re-trap them. The well size was then slowly squeezed until the positrons cooled (in the 1 T magnetic field) into a small well in the mixing section. Once held there, the process could be repeated and stacking of positrons achieved [24]. In the cryogenic environment of the

ALPHA system, the positron lifetime is extremely long (and is more-or-less immeasureable), such that long timescale operations extending over many hours can be undertaken if required. Dedicated experiments by the ATHENA collaboration [24, 25], working under similar conditions to those of ALPHA, stacked 40 positron plasmas containing, in total, around 1.2×10^9 particles and with overall densities up to $2.6 \times 10^{10}\mathrm{cm}^{-3}$.

5. Antihydrogen Formation

Once the antiprotons and positrons are held together in the ALPHA mixing region (see figure 9) they are merged to form antihydrogen. In ATHENA, this was achieved by injecting the antiprotons into the positrons with several eV of kinetic energy [26]. It was found, however, that this produced antihydrogen before the antiprotons had fully slowed in the positrons, [27]; i.e. with kinetic energies above thermal, and which would preclude trapping. Accordingly, ALPHA has developed new mixing methodologies, which will be described elsewhere [28], to allow the antiprotons to enter the positron plasma with minimal addition to their kinetic energy.

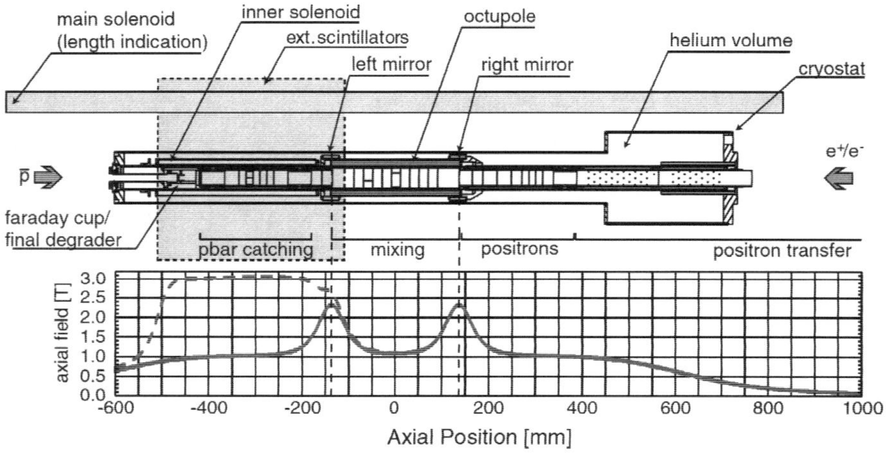

Figure 9. Schematic diagram of the ALPHA apparatus. The graph shows the on-axis longitudinal magnetic field due to the solenoids and mirror coils. The blue (red) curve is the field with (without) the inner solenoid. The positron accumulator (not pictured) is located to the right of the apparatus. The positrons and antiprotons are held and manipulated in the marked regions before transfer to the mixing section, where further manipulations, including evaporative cooling, are performed.

At the positron temperatures and densities used in most antihydrogen experiments to date it is expected that the three-body formation reaction (TBR), $\mathrm{e}^+ + \mathrm{e}^+ + \overline{\mathrm{p}} \rightarrow \overline{\mathrm{H}} + \mathrm{e}^+$, will dominate over the spontaneous radiation recombination process, $\mathrm{e}^+ + \overline{\mathrm{p}} \rightarrow \overline{\mathrm{H}} + \mathrm{h}\nu$. This is supported by some direct evidence, such as the rate of antihydrogen production, which is many times that expected from the radiative reaction [18, 29] and the abundant production of weakly bound states [18, 30], which is a characteristic of the TBR.

The TBR has textbook dependencies on T and n_e and as $n_e^2 T^{-9/2}$, and is expected to occur with a rate of $\approx 4 \times 10^{-9} n_e^2 T^{-9/2}$ s^{-1} (with n_e in units of cm^{-3} and T in K), which applies in thermodynamic equilibrium, with zero external fields and for plasmas of essentially infinite extent, such that none of the species can reach a boundary. However, the manner in which antiprotons and positrons are mixed in antihydrogen experiments is far from this ideal.

Following simulation work by Robicheaux and co-workers [31, 32] it has become apparent that the conditions which apply in the antihydrogen experiments, and in particular the finite size of the positron plasmas, have a profound influence on the distributions of binding energy and speed of the antihydrogen, as well as on the fate of the antiprotons. This has been highlighted by recent detailed simulations of the ATHENA experiment [6].

A particular finding, also mentioned by Robicheaux [31], is that it is misleading to consider the straightforward TBR as a "one-way street" to antihydrogen formation. Indeed, weakly bound antihydrogen is formed readily in the positron plasma, but in most cases is ionized in further collisions with the positrons. This can proceed very rapidly since the cross sections for collisions of such Rydberg-like species can be very high ($> 10^{-13}$ cm^2). Thus, it may be more appropriate to write a three-body scenario as follows:

$$\begin{aligned} e^+ + e^+ + \overline{p} \rightarrow \overline{H}^{**} + e^+ &\rightarrow \overline{H}^{*} + e^+ \\ &\quad \text{or} \\ &\rightarrow e^+ + e^+ + \overline{p}, \end{aligned} \qquad (1)$$

where the asterisks denote degrees of internal excitation. The branching ratio of the two outcomes is overwhelmingly in favour of re-ionization such that the antiproton is recycled. Here we denote $\overline{H}^*$ as an excited antihydrogen atom which is produced from a higher state in collision. Notwithstanding, in order for the antihydrogen to be detected via its annihilation on the electrode walls of the Penning trap, it must survive the trap fields and the electric field of the plasma. Recall from section 2 that the latter has a radial dependence as $E_r \propto r$ such that the field will be highest at the plasma outer edge. (The detailed mechanism of field ionization by the radial field in the plasma has been elucidated in [6].) The effects of antihydrogen atoms field ionized at large radial distances (outside the positron plasma) were also recently observed by the ALPHA experiment [18]. In this case the remnant antiprotons were swept to the electrode walls due to the presence of an octupolar magnetic field used to promote anti-atom trapping.

The repeated formation and ionization of antihydrogen via the reaction schema (1) has a marked effect on the radial distribution of the antiprotons. When they are neutralised as antihydrogen the antiprotons are no longer tightly pinned by the magnetic field and the nett effect is that they are transported across the field lines and can quickly reach the edge of the plasma. Thus, antiprotons can reside in a shell on the outskirts of the plasma where they might eventually form antihydrogen but which will gain extra kinetic energy from the radial drift which, for many combinations of n_e and r, will make trapping impossible. Therefore, it is important to understand and control this mechanism of antihydrogen formation-mediated antiproton transport in order that the fraction of antiprotons that form trappable antihydrogen can be maximized. The effect is illustrated by the results of the simulations [6] as shown in figure 10. Here the yield of antihydrogen, both detected via annihilation and field ionized, is plotted versus the difference, Δr, between the formation position of antihydrogen and the position at which it leaves the positron plasma, for two different values of n_e. Note the logarithmic scale, the marked differences between the distributions at the two densities and the very large fraction of ionized antihydrogen at small Δr. It is these antihydrogen atoms which result in the antiprotons residing on the edge of the positron plasma due to the aforementioned transport. Much experimental effort has been expended to control the effects of this by manipulating the radial profiles and the densities of the plasmas prior to mixing; added insights from transport theories and simulations would be welcome.

6. Conclusion

We have given a brief overview of aspects of antihydrogen formation, most particularly for the ALPHA experiment which aims to trap antihydrogen in a magnetic minimum neutral atom trap

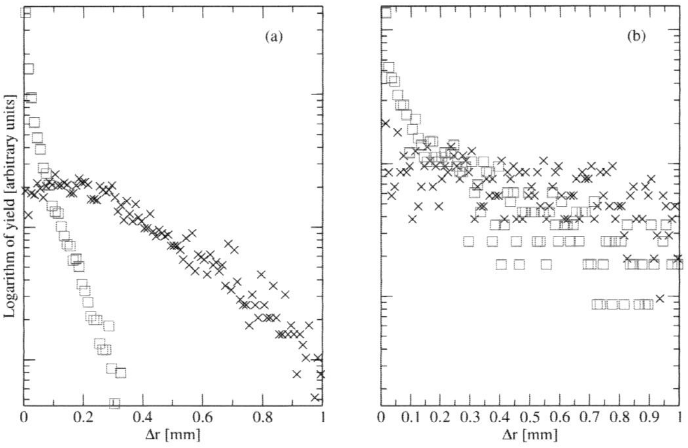

Figure 10. The distribution of the difference in radius between the point of formation of antihydrogen atoms and the point at which they left the positron plasma, for (×) detected and (□) ionized antihydrogen, (a) $n_e = 10^9$ cm^{-3}, (b) $n_e = 5 \times 10^7$ cm^{-3}.

to promote spectroscopic comparisons of the anti-atom with hydrogen. Many of the techniques that have been developed to achieve this involve transport phenomena, and we have highlighted a few examples. Further effort is required to develop antihydrogen into a tool to probe new physics, and much is based upon our understanding and manipulation of the underlying reaction dynamics; theoretical input has always been an important element of this emerging field [33], and there would be great added value from an enhanced understanding of antimatter transport physics.

7. Acknowledgements

Our work has been supported by a number of national funding agencies. We are grateful to: EPSRC and the Leverhulme Trust (UK); CNPq and FINEP (Brazil); ISF (Israel); FNU (Denmark); VR (Sweden); NSERC, NRC/TRIUMF and AIF (Canada); DOE and NSF (USA). We are very appreciative of the support we receive at CERN.

8. References

[1] Charlton M, Jonsell S, Jørgensen L V, Madsen N and van der Werf D P 2008 *Cont. Phys.* **49** 29
[2] Madsen N 2010 *Proc. Roy. Soc.* **368** 3671
[3] Bertsche W *et al* (ALPHA collaboration) 2006 *Nuc. Inst. Meth. in Phys. Res. A* **566** 746
[4] Andresen G B *et al* (ALPHA collaboration) 2007 *Phys. Rev. Lett.* **98** 023402
[5] Gabrielse G *et al* (ATRAP collaboration) 2008 *Phys. Rev. Lett.* **100** 113001
[6] Jonsell S, van der Werf D P, Charlton M and Robicheaux F 2009 *J. Phys. B: At. Mol. Opt. Phys.* **42** 215002
[7] Andresen G B *et al* (ALPHA collaboration) 2010 *Phys. Lett. B* **685** 141
[8] Charlton M 2009 *J. Phys. Conf. Series* **162** 012003
[9] Major F G, Gheorghe V N and Werth G 2005 *Charged Particle Traps: Physics and Techniques of Charged Particle Field Confinement* (Springer, Berlin)
[10] Davidson R C 1990 *An Introduction to the Physics of Nonneutral Plasmas* (Addison-Wesley)
[11] Amoretti M *et al* (ATHENA collaboration) 2003 *Phys. Rev. Lett.* **91** 055001

[12] Amoretti M *et al* (ATHENA collaboration) 2003 *Phys. Plasmas* **10** 3056
[13] Maury S 1997 *Hyperfine Interact.* **109** 43
[14] Budker G I and Skrinskii A N 1978 *Sov. Phys. Usp.* **21** 277
[15] Anderegg F, Hollman E M and Driscoll C F 1998 *Phys. Rev. Lett.* **81** 4875
[16] Andresen G B *et al* (ALPHA collaboration) 2008 *Phys. Rev. Lett.* **100** 203401
[17] Andresen G B *et al* (ALPHA collaboration) 2009 *Rev. Sci. Inst.* **80** 123701
[18] Andresen G B *et al* (ALPHA collaboration) 2010 *Phys. Rev. Lett.* **105** 013003
[19] Eggleston D L, Driscoll C F, Beck B R and Hyat A W 1992 *Phys. Fluids B* **4** 3432
[20] Surko C M, Greaves R G and Charlton M 1997 *Hyperfine Interact.* **109** 181
[21] Mills Jr A P, and Gullikson E M 1986 *Appl. Phys.* **49** 1121
[22] Khatri R, Charlton M, Sferlazzo P, Lynn K G, Mills Jr A P and Roellig L O 1990 *Appl. Phys. Lett.* **57** 2374
[23] Al Qaradawi I, Charlton M, Borozan I and Whitehead R 2000 *J. Phys. B: At. Mol. Opt. Phys.* **33** 2725
[24] Jørgensen L V *et al* (ATHENA collaboration) 2005 *Phys. Rev. Lett.* **95** 025002
[25] Funakoshi R *et al* (ATHENA collaboration) 2007 *Phys. Rev. A* **76** 012713
[26] Amoretti M *et al* (ATHENA collaboration) 2002 *Nature* **419** 456
[27] Madsen N *et al* (ATHENA collaboration) 2005 *Phys. Rev. Lett.* **94** 033403
[28] Andresen G B *et al* (ALPHA collaboration) 2010 in preparation
[29] Amoretti M *et al* (ATHENA collaboration) 2004 *Phys. Lett. B* **578** 23
[30] Gabrielse G *et al* (ATRAP collaboration) 2002 *Phys. Rev. Lett.* **89** 233401
[31] Robicheaux F 2004 *Phys. Rev. A* **70** 022510
[32] Robicheaux F and Hanson J D 2004 *Phys. Rev. A* **69** 010701
[33] Robicheaux F 2008 *J. Phys. B: At. Mol. Opt. Phys.* **41** 192001

Recent advances in studies of positron swarms in electric and magnetic fields in gases

A. Banković

Institute of Physics, University of Belgrade, Pregrevica 118, POB 68, Zemun, Serbia

ana.bankovic@gmail.com

Abstract. A Monte Carlo simulation technique has been used for the calculation of positron transport properties in neutral gases under the influence of electric and magnetic fields. The high magnitude of the cross section for positronium (Ps) formation process comparing to other collisional processes and its strong energy dependence are two major factors for the development of many interesting kinetic phenomena observed in the positron transport. One of the most striking phenomena is the existence of negative differential conductivity (NDC) in the bulk drift speed component and the absence of any sign of this phenomenon in the profile of the flux component. Along similar lines we have observed that the differences between the flux and bulk components of various transport coefficients, originating from the non-conservative nature of Ps formation, are much higher than those observed for electrons. The influence of magnetic field in a crossed field configuration (**E**×**B**) is also investigated. Due to the presence of a magnetic field the number of transport coefficients is increased. Since the longitudinal and transverse components of the drift velocity exhibit different sensitivities with respect to the magnetic field strength, it is found that the NDC effect in a crossed field configuration can be controlled by the magnetic field. The magnetic cooling effect has also been observed – the mean energy of the swarm is a decreasing function of the magnetic field for all electric field strengths considered in this work. In that sense, positrons behave in exactly the same way as electrons. The results of the Monte Carlo simulations are compared with those obtained by the multi-term theory for solving the Boltzmann equation. The excellent agreement between these two entirely independent techniques supports the numerical integrity of our Monte Carlo code.

1. Introduction

Positrons have applications in many research areas ranging from gamma-ray astronomy to biomedicine. In the last few decades there has been growing interest for studying positrons both theoretically and experimentally [1]. One of the major motivational factors for this was the revolutionary improvement of buffer-gas, Penning Malmberg traps for positrons made by Surko and co-workers [2, 3]. The improved positron traps have provided high-flux, low-energy positron beams and thus the new era of positron physics was born. The first response came from the atomic physicists with the provision of accurate cross sections for positron interactions with different gases [4-7], which has opened the possibility to model the transport processes of positrons in neutral gases and soft-condensed matter [8-10, 11], with the principal idea of improving the positron based technologies such as those employed in positron emission tomography. In this paper we investigate the transport properties of positron swarms in neutral molecular gases in a crossed electric and magnetic field configuration. We investigate the positron transport in crossed electric and magnetic fields as possible applications may use combined fields to achieve a better control of properties of the ensemble. It is well known from the recent comprehensive electron studies that the transport properties of electrons can be significantly affected by the magnetic field, particularly under conditions when the cyclotron frequency dominated the collision frequency [12]. Using these arguments as a background, we have been motivated to check the basic features of positron transport in electric and magnetic fields.

In calculations of transport coefficients it is necessary to have a complete set of cross sections [13]. In Figure 1 the compilation of the best theoretical [14 – 16] and experimental [17 – 19] cross sections for positrons in hydrogen available from the literature is given. There are three main differences between electron and positron interactions with matter. First, the absence of the resonances for positrons leaves a very small non-resonant vibrational excitation. Second, the absence of the exchange interactions leads to a smaller number of electronic states that can be excited by positron impact. Finally, the Ps formation channel, a non-conservative process unique to positrons, often has a significant cross section.

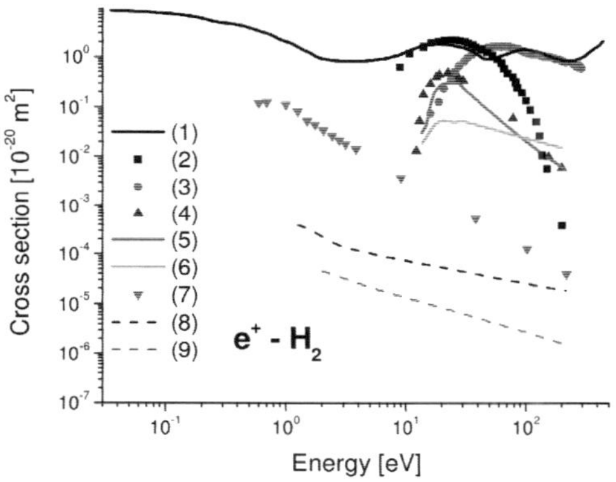

Figure 1: Complete cross section set for positrons in hydrogen: (1) elastic = total [14] – all the others, (2) Ps formation [17], (3) ionization [17], electronic excitations: (4) B1Σ [18], (5) X-C [15], (6) X-E [15]; vibrational excitations: (7) v1 [19], (8) 02 [16] and (9) 01 [16].

The transport properties of positron swarms moving in neutral gases have been studied using a multi-term theory for solving the Boltzmann equation [20] and a Monte Carlo simulation technique [21]. Both techniques have been developed for studying the electron swarms in varying configurations of electric and magnetic fields under hydrodynamic and non-hydrodynamic conditions. Numerical codes have passed all well-established benchmarks with high accuracy levels. In the framework of recent positrons studies, however, special attention has been paid to the correct treatment of positronium (Ps) formation. Ps formation is a non-conservative process unique to positrons which often has a cross section larger than the cross sections for elastic and various inelastic collisions and as a consequence it selectively removes the higher energy positrons. The kinetic effects induced by the non-conservative nature of collisions manifest themselves as a difference between the flux and bulk transport properties [22]. This duality is an inherent property of all transport coefficients, although it has been systematically ignored for many years in plasma modeling community and related electron studies. The flux transport properties, which are input parameters in fluid plasma models, are defined through the flux-gradient relation and they can be calculated from the spatially uniform average quantities. These quantities are not measurable. On the other hand, bulk transport properties, quantities measurable in swarm-type experiments, are defined through diffusion equation and they can be calculated as the rate of changes of the appropriate averages of the positions of the swarm particles, in the configuration space. The difference between flux and bulk transport coefficients exists only if the number of particles in the swarm changes with time. The differences between the flux and bulk transport properties observed in positron transport are often more pronounced than those for electrons, particularly for those gases with a large Ps formation cross-section. One of the most important kinetic phenomena observed both in electron and positron transport is negative differential conductivity (NDC). This phenomenon can be defined as a decrease in drift velocity of the swarm with increasing driving electric field. Both the nature and the manifestation of this phenomenon are different for electrons and positrons. In case of electrons [23-25], NDC is present in both the flux and bulk velocity components while for positrons it has been observed only in the bulk drift velocity. The differences between flux and bulk drift velocities can be more than two orders of magnitude [8-10], while for electrons they are generally 15-20% [12, 20]. Also, NDC is much more pronounced for positrons than for electrons and it originates purely from the non-conservative nature of Ps formation. The interesting situation occurs when the inelastic channels start to open in the vicinity of the threshold of Ps formation. Sometimes, for example in nitrogen [9], the competition between these channels leads to the disappearance of NDC, even though the cross section for Ps formation is as high as it is for the gases with very pronounced NDC. In what follows these issues are illustrated and discussed for positrons in molecular hydrogen and water vapour.

2. Negative differential conductivity (NDC)

The most interesting kinetic effect in positron transport is the existence of the negative differential conductivity (NDC). It represents a decrease in the drift velocity with increasing external electric field. In positron transport this effect appears only in the bulk drift velocity profiles. It has been shown that the origin of the phenomenon lies in the non-conservative nature of positron collisions associated with Ps formation [8] and is quite different in origin to the better known NDC effect in electron transport [23 - 25]. In Fig. 2 we show the bulk and flux drift velocities for both positrons and electrons in molecular hydrogen. In this energy range for electrons in molecular hydrogen the flux and bulk drift velocities are the same, which means that the effects of non-conservative processes are insignificant. On the other hand, the flux and bulk drift velocities for positrons are markedly different, in both quantitative and qualitative way. The maximum difference is of the order of magnitude, similar as for positrons in argon [8], and it is interesting to note that such a huge difference between two types of transport properties has never been observed in electron transport. The NDC effect in bulk drift velocity is well pronounced for positrons, with no sign of it in the profile of the flux velocity.

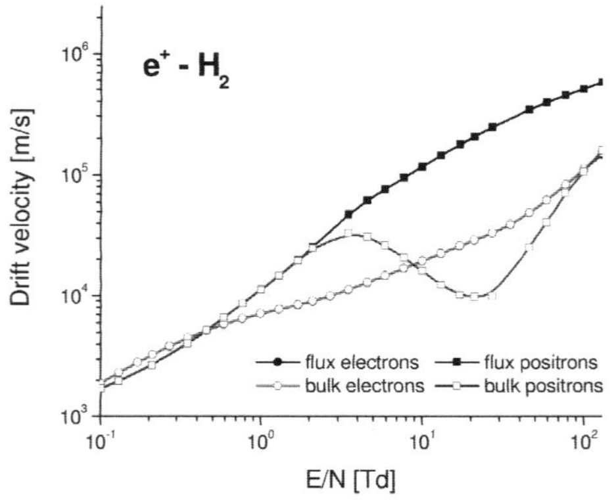

Figure 2: Drift velocities of electrons and positrons in pure hydrogen. The data for electrons were obtained from one of the standard sets of cross sections [26].

There is nothing in our experience with electrons that could prepare us for this situation: strong bulk NDC with no sign of flux NDC. We used a standard formula defining the difference between the flux and bulk properties to see if it works well for positrons [25]. The formula is

$$W = w - \frac{2\varepsilon}{3e}\frac{d\nu_{PF}}{dE}, \qquad (1)$$

where ν_{PF} is Ps formation rate, ε is the mean energy, e elementary charge and E is the electric field. We start from the MC flux drift velocity and add to it the second term from this equation where we apply the MC determined rates of Ps formation. The comparison is presented in Figure 3. As can be seen we obtained a qualitatively good agreement.

The calculations of the bulk drift velocity W were made both directly from MC and by modifying the results of Eq. (1) with calculated Ps formation rates. We also show as vertical lines the ranges of NDC predicted on the basis of the condition given by the Eq. (2).

Another criteria against which we tested our results is NDC criteria developed by Vrhovac and Petrović [25], which was the first to consider NDC due to non-conservative collisions. The formula is

$$\frac{dW}{dE}=\frac{dw}{dE}-\frac{1}{e}\frac{2}{3}\left[\frac{d\varepsilon}{dE}\frac{d\nu_{PF}}{dE}+\varepsilon\frac{d^2\nu_{PF}}{dE^2}\right]<0. \quad (2)$$

It predicts the region where the drift velocity falls. In Figure 3 this region is marked by vertical lines and we can say that the agreement is better for the onset than for the end of the range but overall the qualitatively prediction is good. In case of electrons it is well known fact that if NDC exist it must be seen in velocity space [25]. For positrons it was observed that this criterion is never satisfied.

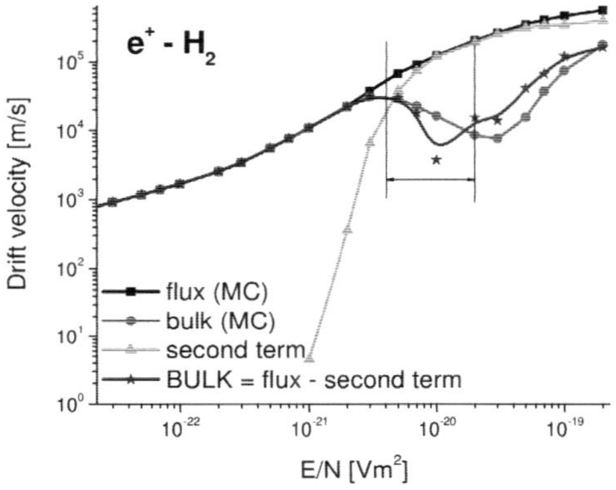

Figure 3: Comparison between bulk drift velocities calculated using our MC code and predicted by Eq. (1) for positron transport in pure hydrogen [27].

3. Transport in a crossed field configuration

The number of transport properties in crossed electric and magnetic field configuration is increased comparing to the case of pure electric field. The drift velocity has two components: the longitudinal w_E and transverse $w_{E\times B}$ which behaves in a different manner. If a swarm of positrons is drifting under the influence of electric field only, then the diffusion tensor has three non-zero diagonal elements: one longitudinal component and two transverse components which are by symmetry equal. In a crossed electric and magnetic field configuration there are five non-zero elements of the diffusion tensor. First, there are three diagonal diffusion elements along the **E** (D_E), the **E×B** ($D_{E\times B}$), and **B** direction (D_B), respectively. Second, there are also two individual off-diagonal elements which are not the same, but experimentally can not be individually measured. They form the so-called Hall diffusion coefficient and are responsible for the existence of the fluxes known as the Hall current which is experimentally detectable.

3.1. Is there NDC in E×B fields?

As emphasized previously, when both the electric and magnetic fields are present, the drift velocity has two components: the longitudinal drift velocity component describes the drift along the electric field direction while the transverse component describes the drift along the **E×B** direction. Therefore, the drift can now be represented in terms of a drift speed and its direction is at some angle to the electric field (magnetic deflection angle). An interesting question arises: how does the magnetic field affect the drift speed under conditions when the cyclotron frequency is much less, of the same order, and much greater than the collision frequency? Figure 4 displays the flux and bulk components of the drift speed as a function of E/N for different values of B/N in these different regimes for positrons in molecular hydrogen. NDC effect has disappeared from the profiles for non-zero magnetic fields. This behavior has been explained in [28]. The longitudinal and transverse components of the drift speed behave in completely different manner so they have to be examined separately. The easiest way to explain the absence of NDC effect in the profiles of the drift speed is by looking at the profiles of the magnetic deflection angle (see Fig 3 of [28]). For B/N of 100 Hx, the flux magnetic deflection angle stays around 45° for the entire range of E/N considered, while the bulk value goes to 90° in the energy region where the Ps formation channel is open. In the energy region where NDC exists for magnetic field free case, for the non-zero magnetic field profiles, the behavior of the drift speed is dominated by the component along the **E**×**B** direction. This is the unique situation where due to the nature of collisions the flux and bulk drift velocities differ not only in magnitude, but also in direction.

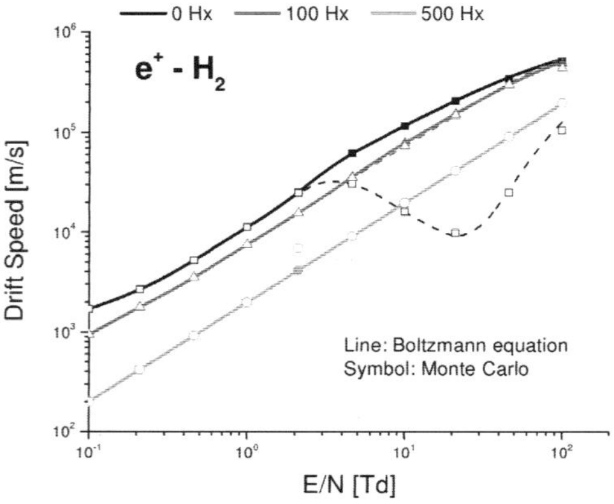

Figure 4: Drift speed of a positron swarm in molecular hydrogen as a function of E/N for various B/N. Flux values are represented by full lines and symbols, and bulk by dashed lines and open symbols.

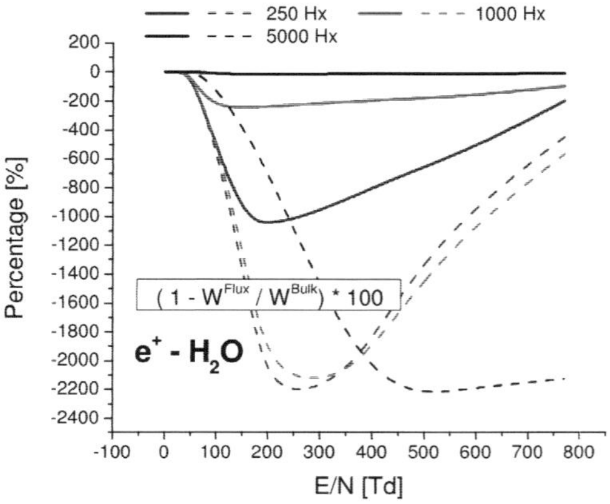

Figure 5: The differences between flux and bulk components of the drift speed for **E**×**B** configuration and the effective field approximation case (Tonks' theorem – dashed lines) for positrons in water vapour.

At this point we asked ourselves a question: can these effects be predicted by using the concept of effective field? The basics of this concept have been given in [29]. We used the Tonks' theorem to calculate effective electric field:

$$E_{eff}(\varepsilon) = E\sqrt{\frac{1+(\Omega/\nu_m)^2\cos^2\Psi}{1+(\Omega/\nu_m)^2}}\,. \tag{3}$$

Using this effective field we have iteratively calculated the mean energy of the swarm, ε, and the drift velocity W:

$$\varepsilon(E,B,\Psi) = \varepsilon(E_{eff},0,0),$$
$$W(E,B,\Psi) = W(E_{eff},0,0), \tag{4}$$

since equations (3) and (4) represents a system of non-linear equations.

In Fig. 5 the differences between flux and bulk components of the drift speed are compared for **E**×**B** configuration, and the effective field approximation case (Tonks' theorem) for positrons in water vapour. We may observe that effective field concept predicts much larger differences between two components of the drift speed. This is more evident for higher values of magnetic field. The lesson from this is that the concept of effective field is not generally appropriate for studying positron transport.

3.2. On diffusion of positrons in gases

In this section we investigate the influence of the non-conservative nature of the Ps formation process on the diagonal elements of the diffusion tensor and also the effects induced by the presence of the magnetic field in a crossed field configuration. Generally speaking, it is hard to fully understand the behavior of diffusion coefficients in electric and magnetic fields since many parallel factors affect them significantly. Among these factors are those associated with the anisotropic nature of the temperature tensor and those associated with the magnetic and electric field anisotropies that are coupled producing different fluxes of positrons along different directions [20]. In addition, the collisions and complex energy dependence of collision frequency further complicate this issue.

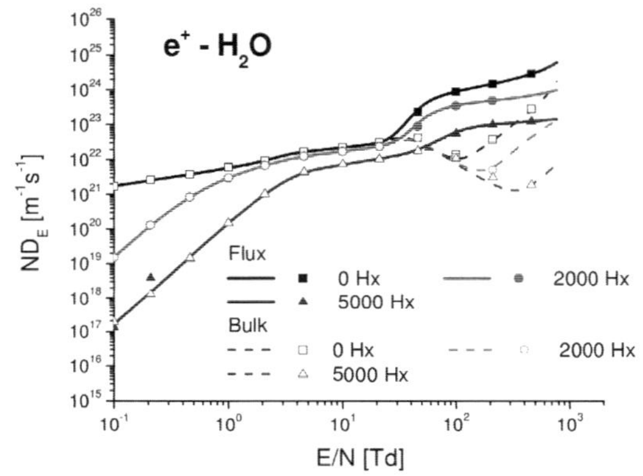

Figure 6: Longitudinal diffusion coefficient as a function of E/N for various B/N for positron swarm in water vapour.

Different diagonal elements of the diffusion tensor show different sensitivities with respect to magnetic field, non-conservative collisions and generally to the energy dependence of the cross sections. The diffusion coefficients along the **E** and **E**×**B** directions can vary up to five orders of magnitude in the limit of low values of E/N where the magnetic field controls the behaviour of the swarm. The longitudinal diffusion coefficient shows the highest sensitivity with respect to the presence of Ps formation, i.e. differences between flux and bulk values are much higher for this diffusion coefficient than those observed for $D_{E\times B}$ and D_B components. D_B component shows the weakest sensitivity to the changes of magnetic field. Essentially, it follows the variation of the mean energy with both E/N and B/N, and hence the thermal effects play the most important role in the behaviour of this transport property. In Fig. 6 the longitudinal diffusion coefficient is shown for positrons in water vapour. In conclusion, better understanding of the synergetic effects of the magnetic field and non-conservative collisions on the diffusion in **E** and **B** fields requires knowledge of the spatially resolved data, particularly those associated with the second order variations of the average energy along the swarm [20].

4. Towards application of positrons in medicine

One of the most important applications of positrons is their use in medicine in positron emission tomography (PET). PET is a medical imaging technique, widely and mostly used for tumour diagnostics, but also for studying brain metabolism and brain functions [30]. In spite of the fact that this technique is well developed, the atomic and molecular processes which take place in the human body due to positron interactions are still not well understood and quantified. Currently the PET application is based on models which assume that positrons behave similarly to electrons while interacting with human tissue. This is clearly untrue and in order to have a complete and accurate understanding of PET and other positron based technologies, one must also have a complete knowledge of other secondary processes involved in positron interactions with matter. Investigation of the low energy positron interactions with biologically relevant molecules is of special importance. Certainly the first and the most important among them is the water molecule. Only recently the accurate measurements of the positronium (Ps) formation and total cross section for positron interactions with water molecule at low energies have become possible (see Fig. 7 [7]). As a consequence, for the first time a reasonably complete set of cross sections for positrons in water became available.

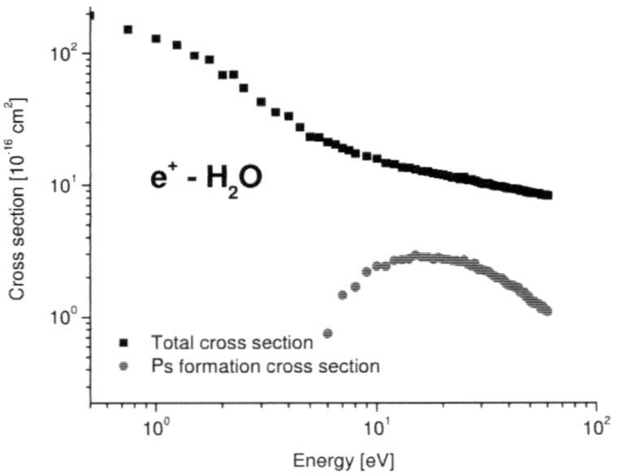

Figure 7: Total cross section and Ps formation cross section for positrons in water vapour [7].

Modelling of tracks for radiation therapy is one of the interfaces between medical applications and atomic and molecular collision physics. The tracks may be used to envisage the ranges and how therapy proceeds but even for the exactly the same set of cross sections very different tracks may occur depending on the sequence of random numbers. Testing of Monte Carlo codes in such cases proceeds by calculation of standardized benchmark results for averaged properties. The tracks of individual particles are analogous in physics of electron, ion and positron collisions in gaseous background to swarms of particles ensembles of non-interacting particles that do not perturb the background gas. The swarms are best described by transport properties and in the presence of non-conservative collisions the definition of transport properties is ambiguous and one should be aware which property is being calculated. Effects of non-conservative processes on tracks would be profound but difficult to describe in a quantitative manner so one needs averaged properties. The best choices for radiation therapy would be diffusion coefficient/length, range of particles and radial spread. In Figure 6 the longitudinal component of the diffusion tensor for positrons in water vapour is given. In applications to modelling of realistic applications one should be aware that all particles start with very large energies, produce a lot of secondary particles and then reach low energies as well. In other words hydrodynamic definition of transport coefficients may not be valid, although comparisons with such benchmarks are a valid test of the code. In order to test the performance for non-local transport one would need also to use non-hydrodynamic benchmarks such as models of Frank Hertz experiment, Steady state Townsend experiment, Penning-Malmberg-Surko trap and similar experiments.

5. Conclusion

In this work we have calculated various transport properties of positrons in neutral gases in crossed electric and magnetic fields using a multi-term theory for solving the Boltzmann equation and Monte Carlo simulation technique. The results obtained with these two different techniques are in a very good agreement. We hope that present results and the multitude of interesting kinetic phenomena, induced by the synergism of magnetic field and Ps-induced non-conservative collisions, can be a good motivation for a further development of positron swarm experiments.

Acknowledgements
The author is grateful to Z. Lj. Petrović, S. Dujko, S. J. Buckman, R. D. White, M. Šuvakov, J. P. Marler, R. E. Robson, C. Makochekanwa, J. P. Sullivan and G. Malović for collaboration on this project and also to C. Surko and M. Charlton for useful comments. This work was supported by the MNTRS project 141025 and partly by the Australian Research Council and Centre for Antimatter-Matter Studies.

References

[1] Charlton M and Humberston J 2000 Positron Physics (New York: Cambridge University Press)
[2] Surko CM, Passner A, Leventhal M, Wysoki FJ, 1988 *Phys. Rev. Lett.* **61** 1831
[3] Murphy TJ, Surko CM, 1992, *Phys. Rev. A* **46** 5696
[4] Sullivan JP, Gilbert SJ, Marler JP, Greaves RG, Buckman SJ, Surko CM, 2002 *Phys. Rev. A* **66** 042708
[5] Marler JP, Sullivan JP, Surko CM, 2005 *Phys. Rev. A* **71** 022701
[6] Sullivan JP, Makochekanwa C, Jones A, Caradonna P, Buckman S J, 2008 *J. Phys. B: At. Mol. Opt. Phys.* **41** 081001
[7] Makochekanwa C, Banković A, Tattersall W, Jones A, Caradonna P, Slaughter DS, Nixon K, Brunger MJ, Petrovic ZLj, Sullivan JP, Buckman SJ, 2009 *New J. Phys.* **11** 103036
[8] Šuvakov M, Petrović ZLj, Marler JP, Buckman SJ, Robson RE, Malović G, 2008 *New J. Phys.* **10** 053034
[9] Banković A, Marler JP, Šuvakov M, Malović G, Petrović ZLj, 2008 *NIMB* **266** 462
[10] Marler JP, Petrović ZLj, Banković A, Dujko S, Šuvakov M, Malović G and Buckman SJ, 2009 *Physics of Plasmas* **16** 057101
[11] White RD, Robson RE 2009 *Phys. Rev. Lett.* **102** 230602
[12] Dujko S, Raspopović ZM, Petrović ZLj 2005 *J. Phys. D: Appl. Phys.* **38** 2952
[13] Makabe T, Petrović ZLj, Plasma Electronics: Applications in Microelectronic Device Fabrication Taylor and Francis, CRC Press,New York, 2006.
[14] Hoffman KR, Dababneh MS, Hsieh YF, Kauppila WE, Pol V, Smart JH, Stein TS 1982 *Phys. Rev. A* **25** 1393
[15] Arretche F, Lima MAP 2006 *Phys. Rev. A* **74** 042713
[16] Gianturco FA, Mukherjee T 2001 *Phys. Rev. A* **64** 024703
[17] Fromme D, Kruse G, Raith W, Sinapius G 1988 *J. Phys. 8: At. Mol. Opt. Phys.* **21** L261
[18] Sullivan JP, Marler JP, Gilbert SJ, Buckman SJ, Surko CM 2001 *Phys. Rev. Lett.* **87**, 073201
[19] Sullivan JP, Gilbert SJ, Surko CM 2001 *Phys. Rev. Lett.* **86** 1494
[20] S. Dujko, Ph. D. Thesis, James Cook University, Townsville, Australia, 2009
[21] Šuvakov M, Ristivojević Z, Petrović ZLj, Dujko S, Raspopović ZM, Dyatko NA, Napartovich AP 2005 *IEEE Trans. Plasma Sci.* **33** 532
[22] Robson RE 1991 *Aust. J. Phys.* **44** 685
[23] Petrović ZLj, Crompton RW, Haddad GN 1984 *Aust. J. Phys.* **37** 23
[24] Robson RE 1986 *J. Chem. Phys.* **85** 4486
[25] Vrhovac SB, Petrović ZLj 1996 *Phys. Rev. E* **53** 4012
[26] Yoon J-S, Song M-J, Han J-M, Hwang SH, Chang W-S, Lee BJ, Itikawa Y 2008 *J. Phys.Chem. Ref. Data* **37** 913
[27] Banković A, Petrović ZLj, Robson RE, Marler JP, Dujko S, Malović G 2009 *NIMB* **267** 350
[28] Banković A, Dujko S, White RD, Malović G, Buckman SJ and Petrović ZLj 2010 *J. Phys. Conf. Series* (accepted for publication – SLOPOS12)
[29] Li B, White RD, Robson RE, Ness KF 2001 *Ann. Phys.* **292** 179
[30] Muehhlehner G, Karp J S 2006 *Phys. Med. Biol.* **51** 117

Gas-phase spectroscopy of a protein

A. R. Milosavljević[1,2,*], A. Giuliani[1,3], C. Nicolas[1], J-F Gil[1], J. Lemaire[4], M. Réfrégiers[1] and L. Nahon[1]

[1]*Synchrotron SOLEIL, L'Orme des Merisiers, Saint Aubin, B.P. 48, 91192 Gif-sur-Yvette, France*

[2]*Laboratory for Atomic Collision Processes, Institute of Physics, University of Belgrade, Pregrevica 118,11080 Belgrade, Serbia*

[3]*Cepia, Institut National de la Recherche Agronomique (INRA), B.P. 71627, 44316 Nantes Cedex 3, France*

[4]*Laboratoire de Chimie Physique, UMR 8000 du CNRS, Université Paris-Sud, Bâtiment 350, 91405 Orsay cedex, France*

*e-mail: vraz@ipb.ac.rs

Abstract. We present preliminary results on gas phase photoionization of electrospray-produced multiply protonated cytochrome *c* protein (104 amino acids;≈12.4 kDa), which has been achieved with an experimental system for spectroscopy of electrosprayed ions in the ion trap using synchrotron radiation. The present results are the first reported on photoionization of kDa species in the gas phase and are valuable regarding fundamental interest of accessing physical properties of large biological species isolated *in vacuo.*

1. Introduction

The investigation of proteins in the gas phase, where they are free of the influence of counterions and solvent molecules, offers a possibility to understand their intrinsic molecular properties. However, due to limited both ion density and available photons flux, the use of synchrotron radiation for the trapped ions spectroscopy is a rather challenging task. The feasibility of coupling a Fourier transform ion cyclotron resonance ion trap with soft x-ray synchrotron beamline and the first successful use of synchrotron radiation for spectroscopy of electrosprayed negative ions stored in a three-dimensional quadrupole ion trap have been demonstrated only recently [1,2].

The possible efficiency for the ionization of large molecules has attracted considerable discussion some years ago. Shlag *et al* (1992) argued that ionization efficiency strongly diminished with increasing the mass of the molecule suggesting that "a very large molecule can act at its own solvent". They have suggested that in the large molecules, the ionization occurs through the intermediate formation of charge pairs which can ionize in a unimolecular-like fashion; nevertheless, a more probable route was considered for the charge pair to recombine. Shlag *et al* [3] also presented the experimental observation that the ionization efficiency rapidly decreases with increasing the weight of the molecule, whether one uses electron impact, single-photon or multi-photon ionization. The least efficient process was found for the single-photon ionization. It should be noted, however, that the experiment used only 10.5 eV photons (118 nm generated by frequency tripling the third harmonic of a Nd:YAG laser), while the cross sections were not discussed. Later theoretical work by Thoss and Domcke [4] on a model study of nearthreshold PI of large molecules indicated that, indeed, vibrational relaxation may be an important mechanism suppressing the ionization process. A more profound discussion and a review on the issue of the ionization efficiency of large molecules can be found in a later paper by Berkowitz [5]. The author concludes that the ionization of large molecules should be possible but indeed with increased difficulty. Note, finally, that removing an electron from a positive ion, which is the present case, is expected to be even more difficult due to the attractive Coulomb field.

According to our knowledge, the first experimental evidence that irradiation of large polyatomic ions by electrons at energies above the ionization threshold causes the ionization of the latter species has been presented by Zubarev group at the beginning of 2000s [6,7]. The authors have measured the ionization energies, obtained by tandem ionization in Fourier transform mass spectrometers, for different polyprotonated peptides (up to about 3.5 kDa) and different charge states of the precursor ions. They reported a dependence of the ionization energies of protonated polypeptides on the number of charges, where a clear trend of the increase in the ionization energy with the charge was found, the likely reason for which is the Coulombic effect. When put on a plot, the IE data strongly suggested a linear dependence. Most recently, the same group reported electron ionization of multiply charged ubiquitin protein (≈8.6 kDa) cation, in the frame of promoting a new MS2 technique – electron ionization dissociation (EID) [8]. The later paper by Fung *et al* [8] neither discusses the ionization efficiency curves nor the threshold for the electron ionization of ubiquitin. Finally, it should be noted that Kalcic *et al* [9] has recently reported femtosecond laser-induced ionization/dissociation of protonated peptides (angiotensin II and GAILpTGAILK – up to about 1 kDa), obtained by coupling an Ti:Al2O3 laser with a modified Thermo Scientific LCQ DECA XP Plus ion trap mass spectrometer. This photoionization process involves the use of ultrashort (<35 fsec) laser pulses of 28 nm bandwidth centered around 800 nm, resulting in a nonergodic dissociation.

In the present paper, we report on the experimental evidence of a single photon ionization of a large polyprotonated biopolymer in the gas phase. The measurements were performed by means of coupling a commercial linear ion trap with a VUV synchrotron beam line. A special care has been taken to design a system which would allow a perfect alignment and large overlap between the trapped ion volume and the photon beam, which resulted in an increased sensitivity of the experiment and large signal to noise ratio.

2. Experimental setup

The experimental setup for spectroscopy of electrosprayed ions was coupled with the DESIRS beamline[§] at the SOLEIL synchrotron radiation facility (Fig. 1). The DESIRS undulator emission allows measurement in the 4-40 eV photon energy range, although for the present study we only used the 4-20 eV range. The wavelength is selected by a normal incidence monochromator, using a grating of 200 gr mm^{-1} which provides a high photon flux (10^{12} – 10^{13} photons/s) with a photon resolution of typically 12 meV at 10 eV photon energy with a 200 μm exit slit. The photon beam can be filtered out for high harmonics using a gas filter.

The experimental system was based upon a commercial linear quadrupole ion trap ("Thermo scientific LTQ XL"), equipped with the ESI probe. The electrosprayed ions are introduced from the front side of the spectrometer and guided through the system of multipole ion lenses into a two-dimensional quadrupole ion trap [10]. The trapping region is formed as a cylindrical volume, which allows improved overlap with the synchrotron beam with respect to Paul ion traps. The synchrotron beam is introduced into the trap through the back lens of the spectrometer. A special frame has been constructed to allow fine-tuning of the position of the spectrometer, i.e. of the trapping region, with respect to the light beam. The mechanical frame is made of five different plates, so the mass spectrometer that is mounted on the top of the system can be positioned with respect to four independent movements: direction along the beam, 2 directions normal to the beam and rotation of the cylindrical trapping region in the horizontal plane.

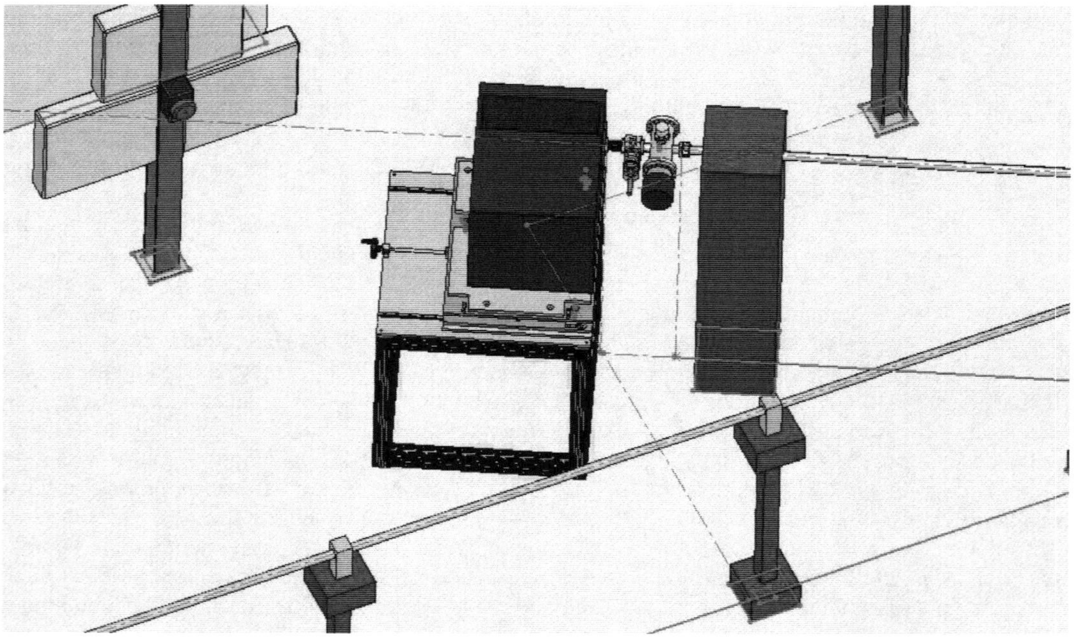

Figure 1: Coupling of the DESIRS beamline at SOLEIL synchrotron facility with the LTQ XL mass spectrometer (Thermo Scientific).

The vacuum manifold with a turbo pumping stage has been designed to accommodate pressure difference between the beamline (10^{-8} mbar) and LTQ (10^{-5} mbar). A turbo pump of 300 L/s was used for the present purpose. The manifold also includes a suprasil window for additional filtering of high order radiation at lower photon energies. A photodiode for the photon flux measurements and the photon beam shutter are placed inside a small cube, which is connected to the back plate of the LTQ mass spectrometer through a flexible bellow allowing the alignment. A new design of the shutter, with an electro-motor (from the brand "KUHNKE") inside the vacuum attached to a copper heat-sink, allowed achieving short (about 1 ms) and reproducible chopping time under high-vacuum conditions, with a good reliability during all the experimental time.

An acquisition system was made to synchronize the beamline energy scanning with the trapped ions activation, thus allowing an efficient collection of large amount of data (a number of tandem mass spectra averaged over desired time interval for a desired photon energy). A home-made PC application synchronizes the ESI injection, mass spectra recording and the beam line photon energy change. Multiply protonated cytochrome *c* molecules are generated in the ESI source from water/ACN (75:25) solution (with 0.2% of pure ammonia) at 10 μM bovine cytochrome *c* (Sigma Aldrich).

The geometry of the linear quadrupole ion trap and the experimental procedure is shown in Fig. 2. The sequence of events is as follows. Electrosprayed ions are injected, selected and stored in the trap; when the desired ion capacity is reached, the beam shutter opens, thus starting the irradiation; after desired time of irradiation, the shutter intercept the beam; the mass spectrum is recorded; the monochromator is set to the next wavelength and the procedure is repeated. The recorded mass spectra can be processed during the measurements or afterwards.

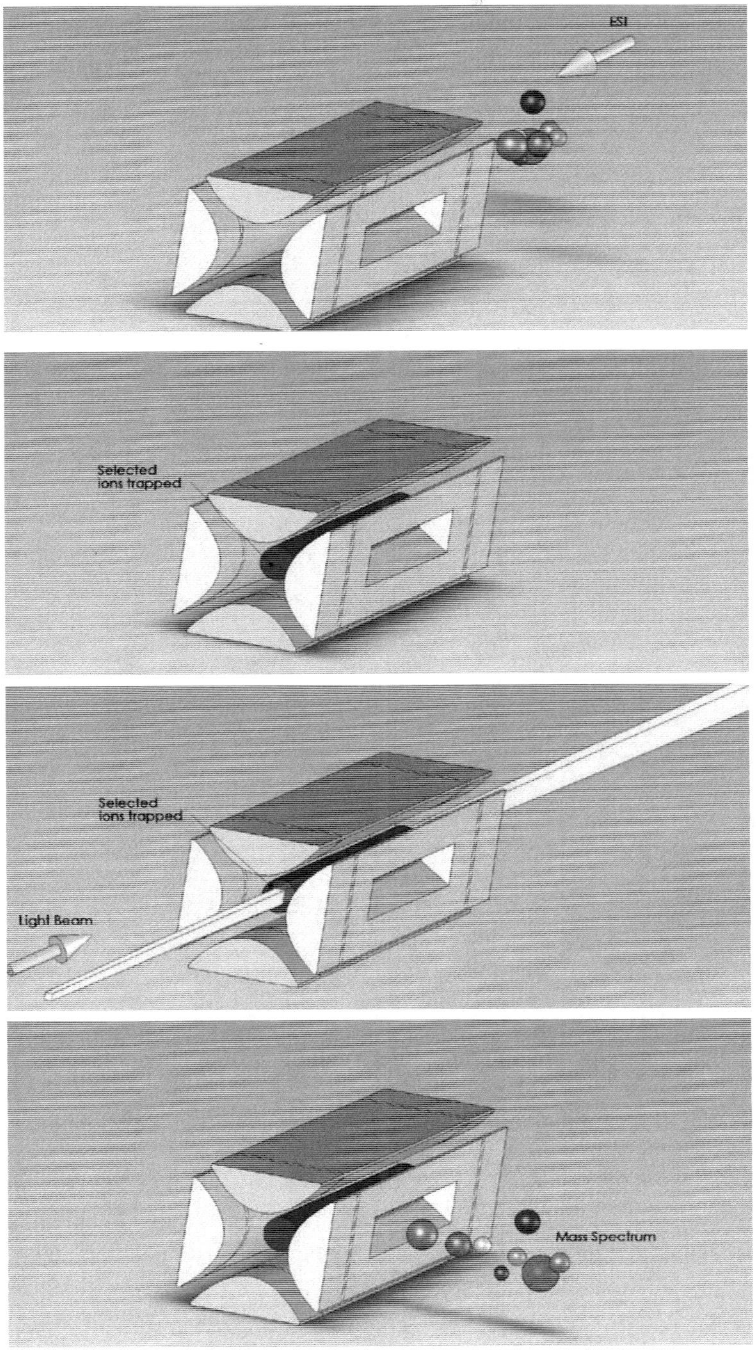

Figure 2: Schematic drawing of the experimental procedure.

Journal of Physics: Conference Series **257** (2010) 012006 doi:10.1088/1742-6596/257/1/012006

3. Results

The ESI/photon-activation tandem mass spectrum recorded after about 80 ms irradiation of the $[M+4H]^{4+}$ isolated protonated protein ions, at the photon energy of 15.5 eV, is shown in Fig. 3. The dominant peak at about m/z=3080 corresponds to the precursor ion, selected from the ESI spectrum. Nevertheless, an additional feature peaking at about m/z=2460 that corresponds to an odd electron radical product cation $[M+4H]^{\bullet 5+}$, formed via photoionization of the $[M+4H]^{4+}$ precursor, is clearly seen.

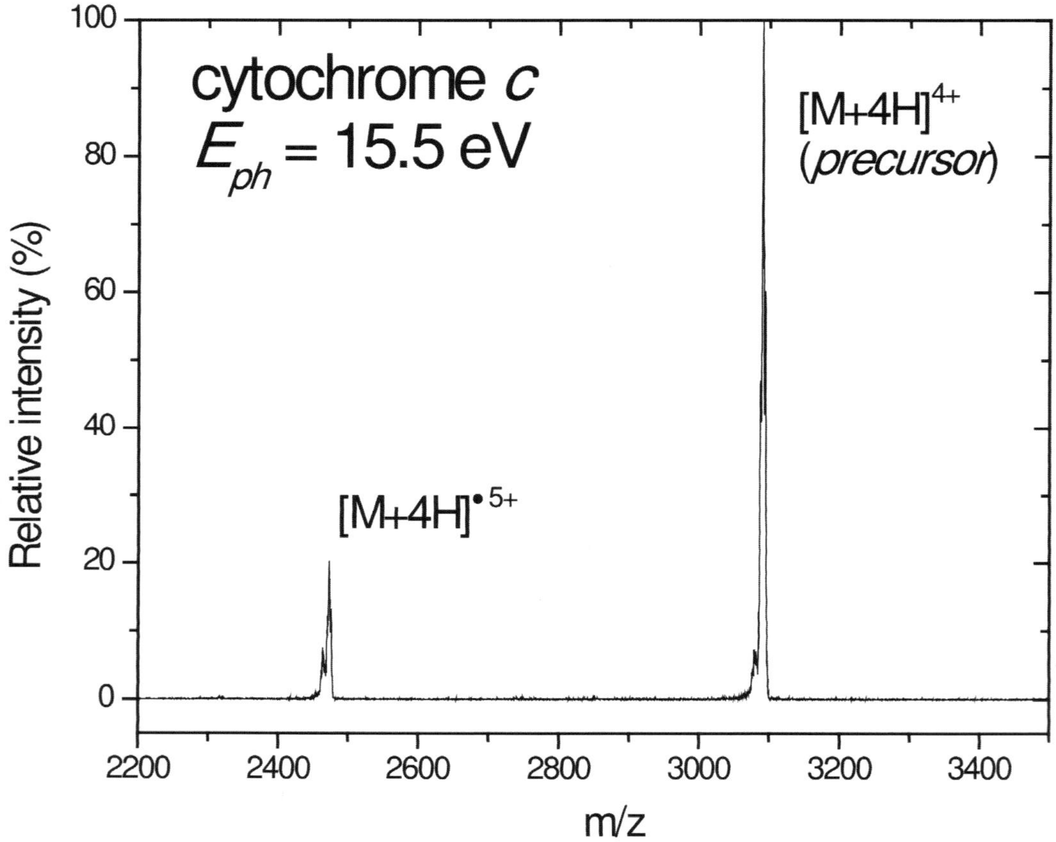

Figure 3: ESI/photon-activation tandem mass spectrum of cytochrome *c* protein obtained with 80 ms of irradiation time, at the photon energy of 15.5 eV.

The presented result shows a very efficient photoionization of the polyprotonated cytochrome *c* protein upon VUV radiation. The measurements of the ionization efficiency as a function of the irradiation time (in the range from about 50 ms up to 400 ms) showed a linear dependence, thus confirming a single photon process. Further work will be focused on the photoionization efficiency measured as a function of the photon energy, which will allow to determine ionization thresholds for different protein charge states. The results will be valuable to ascertain fundamental physical properties of the polyprotonated biopolymers isolated *in vacuo*.

Acknowledgements

This work is supported by the Agence Nationale de la Recherche Scientifique, France, under the project ANR-08-BLAN-0065. We gratefully acknowledge the help of Jocelyn Dupuis from Thermo Scientific, France and August Specht from Thermo Scientific, USA. A.R.M. acknowledges support by the Ministry of Science and Technological Development of Republic of Serbia under Project No. 141011. We acknowledge the SOLEIL Synchrotron Radiation Facility for providing beamtime and we are very grateful to Dr Gustavo Garcia from the DESIRS beamline for his help in setting up the experiment and useful discussion.

References

[1] Thissen R, Bizau J M, Blancard C, Coreno M, Dehon C, Franceschi P, Giuliani A, Lemaire J and Nicolas C, Phys. Rev. Lett. **100**, 223001 (2008)
[2] Giulliani A, Scuderi D, Lemaire J, Dehon C, Thissen R, Nicolas C, Duflot D, Maitre P and Nahon L, Proc. 57th ASMS Conf., Philadelphia (2009)
[3] Schlag E, Grotemeyer J and Levine R D, Chem. Phys. Lett. **190**, 521 (1992)
[4] Thoss M and Domcke W, J. Chem. Phys. **106**, 3174 (1997)
[5] Berkowitz J, Physics Essays **13**, 248 (2000)
[6] Budnik B A and Zubarev R A, Chem. Phys. Lett. **316**, 19 (2000)
[7] Budnik B A, Tsubin Y O, Håkansson P and Zubarev R A, J. Mass. Spectrom. **37**, 1141 (2002)
[8] Fung Y M E, Adams C M and Zubarev R A, J. Am. Chem. Soc. **131**, 9977 (2009)
[9] Kalcic C L, Gunaratne T C, Daniel Jones A, Dantus M and Reid G E, J. Am. Chem. Soc. **131**, 940 (2009)
[10] Shwartz J C, Senko M W, Syka J E P, J. Am. Soc. Mass. Spectrom. **13**, 659 (2002)

§ http://www.synchrotron-soleil.fr/Recherche/LignesLumiere/DESIRS

Journal of Physics: Conference Series **257** (2010) 012007 doi:10.1088/1742-6596/257/1/012007

Plasma-surface Interactions in Material Processing

Satoshi Hamaguchi
Center for Atomic and Molecular Technologies, Osaka University,
2-1 Yamadaoka, Suita, Osaka 565-0871, Japan.

E-mail: hamaguch@ppl.eng.osaka-u.ac.jp

Abstract. In plasma processes such as reactive ion etching and thin film deposition for microelectronics device fabrication, atomic-level control of surface morphologies and compositions of processed materials has become increasingly important as the device sizes diminish to the nano-meter range. While various species such as ions, neutral radicals, electrons and photons simultaneously hit the material surface in a plasma, the plasma-surface interactions can be best understood if individual elementary processes such as interaction of specific species with the surface at specific incident energy are studied separately under well controlled conditions. In this article, a molecular dynamics (MD) simulation technique is reviewed as a means to analyse plasma-surface interactions in such a manner and some sample simulation results for polymer etching and diamond-like carbon (DLC) deposition are presented.

1. Introduction

As the dimensions of semiconductor devices diminish, surface irregularities in less-than-a-nanometer scale have increasingly higher chances to affect the performance of such devices. In plasma processing used for semiconductor fabrication, therefore, etching and deposition processes that are less damaging to the surface are now in high demand. One of the most significant causes of surface damages in plasma etching processes is the kinetic energy of ions bombarding the surface. For example, in a plasma process that contains hydrogen in the gas phase, hydrogen ions, even if their kinetic energies are relatively low, can penetrate deeply into the material and damage or modify the material structures (e.g., the crystalline structure of silicon at the gate channel) below the surface. In the semiconductor industry, such damages are now carefully studied mostly with the use of plasma processing tools that are similar to, or essentially of the same type as, those used in the manufacturing lines. In such a plasma processing tool, however, it is generally extremely hard, if not impossible, to control ion and/or neutral fluxes generated from the plasma so that one could obtain detailed information on plasma surface interactions. If one wishes to understand better the atomic processes on or near the surface during processing, more carefully controlled studies on surface processing are needed.

As a means for studies on plasma-surface interactions, molecular dynamics (MD) simulation techniques are discussed in this article. Although some approximations and assumptions are used for interatomic potential functions, classical MD simulations are generally considered to be a useful tool to analyze atomic-level phenomena associated with plasma surface interactions. The results of such MD simulations can be directly compared with results obtained from corresponding beam experiments and are often found in good agreement. In the rest of this article, after briefly reviewing MD simulation techniques and describing some of its historical development up to today, MD simulation

Journal of Physics: Conference Series **257** (2010) 012007 doi:10.1088/1742-6596/257/1/012007

studies on reactive ion etching (RIE) of organic polymer and formation of sp^3 structures during diamond-like carbon (DLC) deposition processes will be presented as sample simulations.

2. Molecular dynamics simulations

The first MD simulation study that the author is aware of on the effect of energetic impact of incoming atoms on a solid material is that by Gibson, Goland, Milgram, and Vineyard in 1960 [1], where the "radiation damage" of metallic copper, i.e., how Cu atoms in crystalline sites are displaced by an incoming ion, was examined. In 1968 the first study on direct MD simulation of sputtering (i.e., particle emission from a surface by injected particles) was reported by Harrison, Levy, Johnson, and Effron [2], where Cu sputtering yields by Ar impact were evaluated from the simulations. Subsequently a series of MD simulations studies on Cu sputtering by Ar impact based on the same or similar potential functions were published by Harrison and his coworkers, as in [3] and [4].

The studies mentioned above are exclusively on physical sputtering, in the sense that neither incoming atoms/ions nor substrate atoms form new chemical bonds among them that did not exist prior to the beam injections. In 1980, Winograd, Garrison, and Harrison examined interaction of incoming carbon monoxide (CO) with a Ni substrate [5] that involves some surface chemical reactions.

Until late 80s, essentially all atomic interaction models used in MD simulations for sputtering studies were of two-body short range. However, since mid 80s, various multi-body classical interatomic potential functions have been developed for more accurate MD simulation studies on material properties. For example, in 1984, Daw and Baskes developed a method to represent multi-body interaction among metallic atoms called Embedded Atom Method (EAM), which is based on density functional theory for atomic orbitals [6]. The EAM potential functions were later generalized to represent interactions among non-metallic atoms [7]. Based on cluster expansion, Stillinger and Weber developed functional forms for multi-body interactions for covalent bonds in 1985 [8-11]. On the other hand, in 1988, Tersoff proposed different multi-body interatomic function models to represent covalent bonds [12-14], based on a general expression for bond energies derived by Abell [15]. Multi-body interatomic potential functions models used today in classical MD simulations for beam-surface interactions mostly have functional forms similar to any one or combination of those mentioned above.

To the best of the author's knowledge, the first published MD simulation study for sputtering based on multi-body interatomic potential functions was that by Reimann, Walzl, El-Maazawi, Deaven, Garrison, and Winograd in 1988 [16], in which energetic impact of Ar on Rh was examined using EAM interatomic potential functions.

On the other hand, the first MD simulation study on semiconductor sputtering using multi-body potential functions was reported by Stansfield, Broomfield, and Clary in 1989 [17], where Si sputtering by energetic Ar impact was examined based on Stillinger-Weber-type potential functions. A similar MD simulation study was also reported by Smith, Harrison, and Garrison [18] in the same year, in which Ar impact on Si was studied based on Tersoff-type potential functions.

As to surface chemical reactions associated with semiconductor etching, Schoolcraft and Garrison studied reactions of fluorine atoms impinging on a Si surface at very low energies (in the range of 0.5 – 3eV) in early 90s [19,20], using MD simulations based on the potential functions proposed by Stillinger and Weber [10]. Using similar MD simulations with a much higher injection energy (200eV), Feil, Dieleman, and Garrison [21] reported in 1993 MD simulation of reactive ion etching for Cl passivated Si surface by energetic Ar beams, using Stillinger-Weber type potential functions developed by the same authors for the system consisting of Si, Cl, and Ar atoms.

From mid 80s to mid 90s many experimental data were accumulated on surface reactions associated with reactive ion etching of Si surfaces by Cl_2 gas and energetic Ar ion bombardment, obtained from well controlled beam experiments [22-27]. These experimental results prompted the interest in the mechanisms of reactive ion etching in general and the simulation studies mentioned above demonstrated that classical MD simulation is an ideal tool to elucidate such mechanisms at the

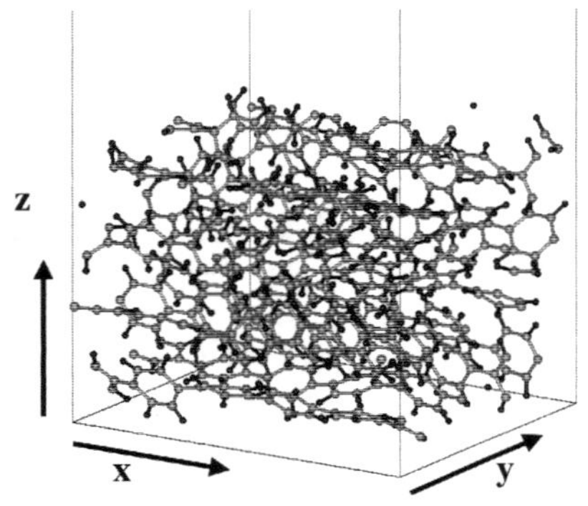

Figure 1. A model substrate of polyparaphenylene (PPP) in thermal equilibrium at 300K used in MD simulations presented in Sections. 3 and 4. White and black spheres represent carbon and hydrogen. [46, 49,50]

atomic level. In addition, the proliferation of inexpensive but powerful computers drove the cost of high-performance computing significantly down in 90s. Due to all such combined factors, the time became ripe in mid 90s for extensive classical MD simulation studies for nonequilibrium surface reactions caused by impinging beams.

Various MD simulation studies on Si etching by Cl or F based plasmas began to be performed from early 90's by Weakliem, Wu, and Carter [28,29], and by Barone and Graves [30] for Si-F systems, Feil [22,31] for Si-Cl-Ar systems, by Athavale and Economou [32] for layer-by-layer etching of Cl passivated Si surface by Ar ion injections. Other studies include those by Hanson, Voter, and Kress [33-35], Helmer and Graves [36,37], Kubota, Economou, and Plimpton [37]. On the other hand, SiO_2 etching were studied by Ohta and Hamaguchi [39,40] for Cl or F based plasma processes and by Smirnov *et al.* [41-43] and by Ohta and Hamaguchi [44] for fluorocarbon (C_xF_y) plasma processes.

Similarly etching and deposition processes for organic polymers have been studied with MD simulations since early 90s. Due to the limited space, we here only refer to Ref. [45] for references on work performed by mid 90s. In the following sections, some examples of MD simulations by the author and co-workers for polymer etching and deposition will be presented.

3. Outline of MD simulations

Figure 1 shows an example of substrate material used for MD simulation [46]. In this example, polyparaphenylene (PPP) is placed in a simulation box, whose perpendicular edges are aligned to the *x*,*y*, and *z* coordinates. In MD simulations, motion of each atom in the simulation is obtained from the integration of its equation of motion. In this particular example, the interatomic potential functions based on Brenner [47] are employed for covalent bonds and Lennard-Jones type two-body potential functions for weak Van der Waals interactions [46]. For the simulation box shown in Fig. 1, periodic boundary conditions are imposed in the horizontal directions and atoms near the bottom were assumed immobile so that the substrate will not drift when it is subject to ion bombardment. The substrate system is initially in thermal equilibrium at 300K.

Interatomic potential functions used in the MD simulation studies presented in this article are those for charge-neutral atoms and therefore all particles are assumed to be charge neutral for the sake of simplicity. In other words, energetic ions that are typically seen in plasmas are represented in this MD model by fast neutrals. This assumption may be justified under some conditions, where impinging ions are neutralized right before the impact by the transfer of electrons from the surface (Auger neutralization) [48]. However, the validity of use of charge neutral interatomic potential functions for ions, which has been commonly practiced in MD simulation studies for beam-surface interactions, has not been closely examined. In what follows, however, we do not address this question directly, either, and instead examine the validity of MD simulations by comparing simulation results with

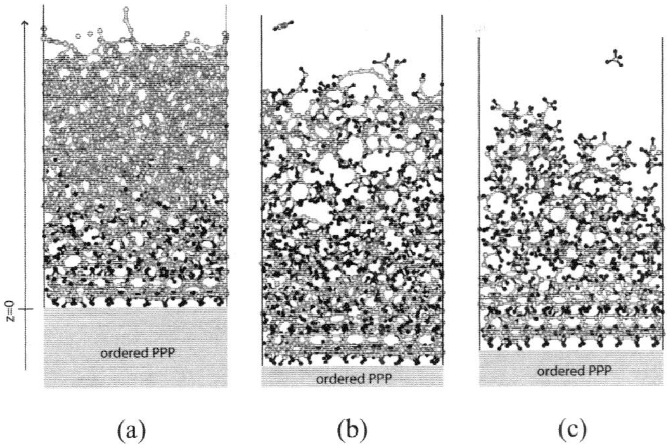

Figure 2. Atomic structures of PPP near their surfaces obtained from MD simulations. The injected species are (a) carbon atoms C, (b) hydrogen molecules H_2, and (c) methane molecules CH_4 . The injection dose and energy of each case are 2.4×10^{16} cm^{-2} and 50 eV and the angle of incidence is normal to the surface. The z coordinate represents the height of the substrate and z = 0 represents the bottom position of the initial substrate. The portion indicated as "ordered PPP" in the figure represents PPP that are assumed to be unaffected by beam injections. [46]

experimental observations obtained from corresponding systems if such experimental data are available.

In MD simulations, atoms or clusters of atoms are injected into the model substrate. After each injection, the system is left to evolve under the microcanonical conditions (i.e., conditions where the total energy of the system is set to be constant) for about 1 pico second (in the simulation time scale) and then rapidly cooled to a room temperature. Once the substrate system reaches the initial temperature, the next injection process will start. Such cycles of injections are repeated for hundreds or thousands of times and then physical properties such as sputtered species, surface roughness, atomic compositions near the top surface, etc. obtained from the simulations are examined. Especially, by counting the number of atoms removed from the surface, one can obtain sputtering yields of the system.

4. Sample simulations: polymer etching

Figure 2 shows atomic morphologies of the surface after injecting (a) C atoms, (b) H_2 molecules, and (c) CH_4 molecules into the PPP model substrate shown in Fig. 1 at the kinetic energy of 50eV [49,50]. The angle of incidence was 90° and the injection dose in each was 2.4×10^{16} cm^{-2}. It is seen in the case of (a), injected C atoms are simply accumulated on the surface whereas, in the case of (b), hydrogen molecules are separated into hydrogen atoms at impact and break π bonds of phenyl rings and roughen the surface. Since the mass of hydrogen is so low that the momentum it carries is also low, resulting little sputtering of the substrate at this energy. In the case of (c), on the other hand, the combined effect of strong impact by heavier carbon atoms and scission of bonds by hydrogen atoms, which are brought to the surface by injected CH_4 molecules, causes sputtering of the surface. Sputtering caused by such combination of physical and chemical effects is called reactive ion etching (RIE).

The "sputtering yield" (for an atomic species A) means the number of A atoms removed from the surface per injection. For example, if a CH_4 molecule is injected into a surface and a single C atom is emitted from a surface, then the sputtering yield for C is 1 although there is no net decrease of the number of C atoms from the surface in this case. Following Refs. [46,49,51], we define the term "net

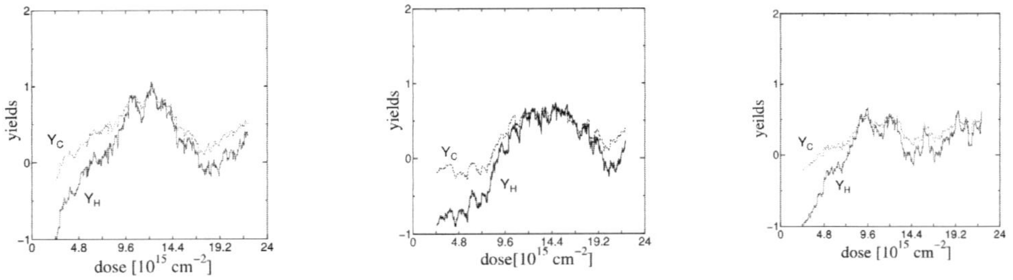

(a) (b) (c)

Figure 3. The net erosion yields for carbon and hydrogen Y_C and Y_H as functions of the injection dose for 50eV CH_4 molecules. The injection conditions for (a) (b) and (c) are the same as those given of Fig. 2(c). The only difference among the three cases (a)-(c) here is the random number sequence employed in each case that determines the lateral locations of injection. [50]

erosion yield" for an atomic species A as the number obtained by subtracting the number of injected A atoms from the corresponding sputtering yield.

Figure 3 shows the net erosion yields for carbon and hydrogen as functions of injection dose in the case where a PPP substrate is subject to CH_4 bombardment at incident energy of 50eV [50]. In (a), (b), and (c), all macroscopic physical parameters are set to be the same and the only difference among them is the sequence of random numbers employed in the simulations that determines the position of impact on the surface for the injected species CH_4. Since the way the surface roughens depends on the specific sequence of injection locations, the yield of the narrow region of the surface is also expected to depend on the sequence of impacts, as shown in Fig. 3. The macroscopic yield is, of course, the average over the entire surface. Figure 4 shows the average of 9 separate simulations based on different sequences of impact with the same macroscopic parameters. (Three of such examples are shown in Fig. 3). The yield dependence on the dose shown in Fig. 4 is expected to be close to the experimentally observable macroscopic yield dependence.

Similar polymer etching simulations by nitrogen-based NxHy radicals are found in Ref. [51].

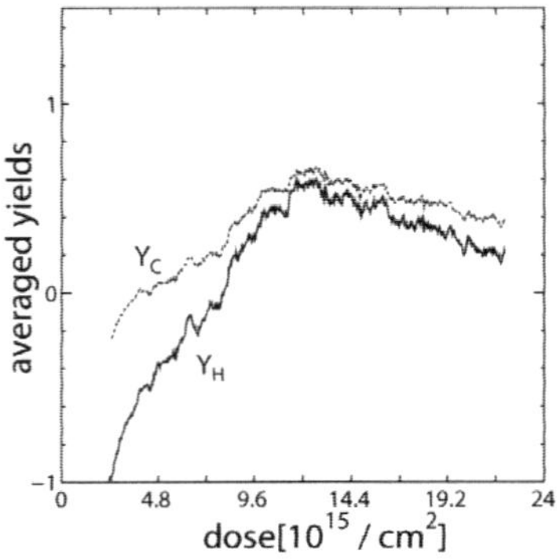

Figure 4. The net erosion yields for carbon Y_C and hydrogen Y_H as functions of injection dose averaged over 9 cases with different random number sequences (as give in Fig. 3). The other conditions are the same as those of Fig. 3. [50]

5. Sample simulations: diamond like carbon deposition

Amorphous carbon with a high content of sp^3 structures is often called diamond-like carbon (DLC) and known to have extraordinary material properties such as high mechanical hardness, high wear resistance, low friction coefficients, high chemical inertness, good infrared (IR) transparency, high electrical resistivity, good field emission properties, and low dielectric coefficients [52-58]. Heterogeneous growth of DLC films on various materials at low deposition temperatures can be achieved by plasma enhanced chemical vapor deposition (PECVD). In most PECVD processes, the growth of DLC, as opposed to amorphous carbon, can be achieved by applying a certain amount of bias voltage to the substrate, which makes ionized deposition precursors reach the surface with finite kinetic energies. Typically, if the impact energy is too low, the deposited carbon is amorphous whereas, if the energy is too high, sputtering of the surface becomes significant. DLC differs from amorphous carbon as it contains a large number of sp^3 hybridized bonds. The formation mechanism of sp^3 hybridized bonds in carbon deposition process is of major interest in this section. Especially we focus on the effects of impact energy and presence of hydrogen.

As in the previous section, MD simulations can reveal some aspects of the formation of sp^3 hybridized bonds at the atomic level when carbon deposition processes are simulated. In actual PECVD based on hydrocarbon gases, various ions and free radicals can form deposition precursors. Here we simplify the process and simulate only single-species incidence of CH_3 or CH on an a-C film that is initially hydrogen-free [59,60]. To further simplify the study, we also focus just on the initial phase of deposition. A better understanding of individual reaction processes such as those examined here will serve as the basis for a more complete understanding of real PECVD processes.

The interatomic potential functions used in the simulations shown here are those given in Ref. [61], which are similar to Stillinger-Weber functions with bond-order corrections. In this set of potential functions, the bond order of a carbon-carbon (CC) bond is determined from the number of other carbon (C) and hydrogen (H) atoms that are present in the neighborhoods of the carbon atoms that forms the bond in consideration. For example, if there are three H atoms in each neighborhood of the C atom, the CC bond is a single bond whereas if there are two H atoms in each neighborhood, the CC bond is a double bond. A function that counts the neighboring atoms is built in each interatomic potential function.

The model substrate used in the simulation is a hydrogen-free amorphous carbon (a-C) cube with an edge length of about 2nm. The system is in thermal equilibrium at 300K. The initial substrate consists mainly of sp^2 hybridized bonds; the ratio of the number of sp^3 C atoms to that of sp^2 atoms in the model substrate is about 0.4%. As in the model in the previous section, the periodic boundary conditions are imposed in the lateral directions of the system. In the bottom layer of the substrate cube, the atoms are also set immobile.

(a) (b)

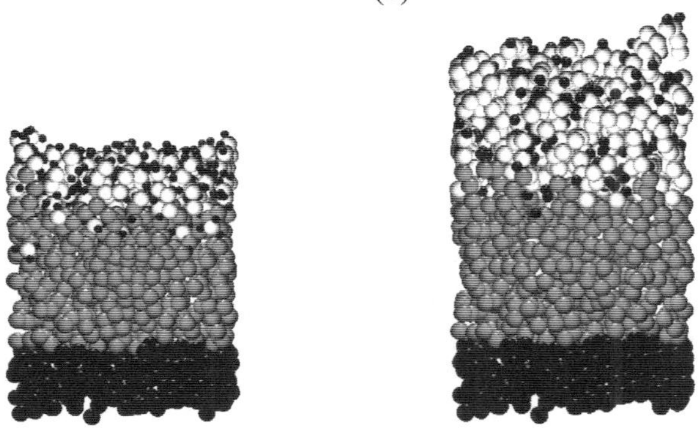

Figure 5. Atomic structures of the substrate and deposited atoms obtained from MD simulations. In each case of (a) CH_3 and (b) CH incidences, the beam dose and kinetic energy are $2.0\times10^{16}cm^{-2}$ and 20eV. The large black and grey spheres represent immobile C atoms in the bottom layer and mobile C atoms of the initial substrate. The large white and small black spheres represent injected C and H atoms. [59]

Journal of Physics: Conference Series **257** (2010) 012007 doi:10.1088/1742-6596/257/1/012007

Figure 5 shows atomic structures of the system at an injection dose of 2.0×10^{16}cm^{-2} of (a) CH_3 and (b) CH at 20 eV impact energy [59]. Here large black spheres represent C atoms in the bottom layer which are assumed to be immobile. The grey spheres represent mobile C atoms that initially constitute the substrate. The large white spheres represent injected C atoms. Injected H atoms are represented by small black spheres. Clearly the deposited film is thicker for CH incidence as CH has a higher sticking probability than CH_3.

Figure 6 shows the depth profiles of carbon and hydrogen atomic species at a dose of 2.0×10^{16}cm^{-2} for CH_3 incidence with incident energies of (a) 20eV and (b) 50 eV [59]. The profiles of substrate C atoms, deposited C atoms, and deposited H atoms are given by the thick solid, broken, and thin solid curves, respectively. The horizontal axis represents the film thickness measured from the bottom of the model substrate. In this figure, the atomic density at each depth is obtained from the total number of the atoms that are present in a 1Å thick layer at the indicated depth. The figure clearly shows that deposited C and H atoms penetrate the substrate more deeply at higher energy. Furthermore, H atoms penetrate the substrate more deeply than C atoms because a H atom has a smaller interaction length (i.e., atomic radius) than a C atom.

In this study, the "carbon-to-carbon (C-to-C)" coordination number is defined as the number of C atoms that exist in the neighborhood of a particular C atom that we consider. For example, in crystalline diamond, the C-to-C coordination number of every C atom is 4. If the C-to-C coordination number of a C atom is 4, then the C atom forms sp^3 hybridized bonds with the surrounding C atoms. Similarly, if the C-to-C coordination number for a C atom is 3, it forms sp^2 hybridized bonds with the surrounding C atoms. Figure 7 shows the ratio of the number of C atoms with each C-to-C coordination number deposited on the substrate to the number of all deposited C atoms as functions of the incident energy at a dose of 2.5×10^{16}cm^{-2} for (a) CH_3 and (b) CH beam incidences [59]. Especially, the ratio for the C-to-C coordination number being 4 represents the fraction of fully carbon terminated sp^3 C atoms (each of which is bonded with 4 other C atoms only, not more and not less) for all deposited C atoms. This fraction represents the "diamond-likeness" of the deposited film. Note that C atoms that constitute the initial substrate (which are mostly of sp^2, as mentioned earlier) are not counted in Fig. 7. However, in the C-to-C coordination numbers listed here, bonds connecting deposited C atoms and substrate C atoms are counted. The "others" in Fig. 7 include deposited C atoms bonded with either H atoms or those bonded with five or more other C atoms (although the latter are scarce).

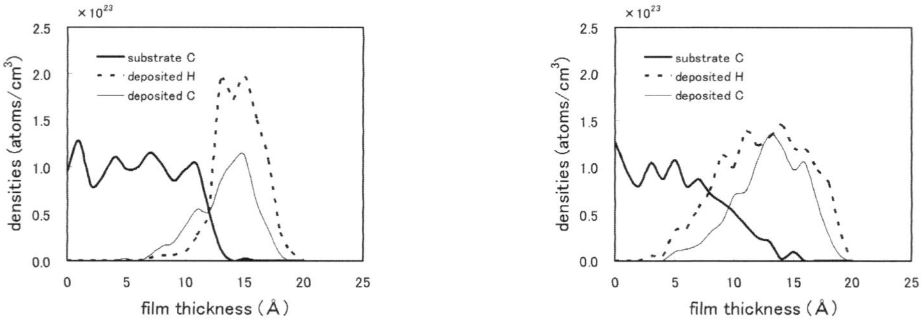

Figure 6. Depth profiles of atomic species at a dose of 2.0×10^{16}cm^{-2} of CH_3 incidence with incident energies of (a) 20eV and (b) 50 eV, obtained from MD simulations. The profiles of substrate C atoms, deposited C atoms, and deposited H atoms are given by the thick solid, broken, and thin solid curves, respectively. The horizontal axis represents the film thickness measured from the bottom of the model substrate. [59]

It is clearly seen in Fig. 7 that, for CH_3 incidence, deposited C atoms are more likely to form sp^3 hybridized bonds with other C atoms at higher incident energy under the conditions studied here. A similar tendency is also seen for CH incidence up to 20eV. However, at 50eV for CH incidence, the sp^3 fraction is significantly lower, which is likely caused by relative scarcity of hydrogen in the substrate. The sticking coefficient of H atoms for CH incidence is low at high energy.

Figure 8 shows the density profile of fully carbon terminated sp^3 C atoms at a dose of $2.0\times10^{16}cm^{-2}$ for 50 eV CH_3 incidence (depicted by the grey curve denoted by "sp3 C") [59]. For reference, the density profiles of C and H atoms are also plotted here, which are the same as those given in Fig. 6(b). Note that "the fully carbon terminated sp^3 C atom" means a C atom that is bonded with four other C atoms only, as defined in Fig. 7. It is clearly seen that the sp^3 C atoms are mostly located in a level below the top surface (i.e., subsurface level), typically around 1 nm below in this case.

6. Conclusions

A technique to apply MD simulation for the analysis of plasma-surface interactions has been reviewed with some sample simulation results. For example, in semiconductor manufacturing, sizes of electronic devices on a chip, such as transistors and interconnect wires, continue to diminish and damages on a processed surface affect device performance more severely as the miniaturization continues. Atomic-scale analyses for plasma-surface interactions may provide useful information for the development of technologies that alleviate such problems.

The simulation results presented in this article have not been compared with experimental results as there has been no experiment data with which the simulation data given here can be directly compared. However, in many cases, MD simulations have been found to provide macroscopic data such as sputtering yields and atomic compositions on the surface that are in reasonably good agreement with experimental observations (e.g., [39,40]). Some may find this counterintuitive as chemical reactions on the surface often play important roles in interaction between a chemically reactive plasma and a surface whereas a classical model of atomic interactions is known to be nearly incapable of representing actual chemical reactions in thermal equilibrium. The reason that classical MD simulations can predict some characteristics of chemical sputtering and/or deposition processes is that, in such processes, the injection energy (which is in the order of a 100eV or so) is far higher than bond energies and therefore the outcome, including some chemical reactions, is less sensitive to the detailed structures of interatomic potential functions. It is mostly likely that only a small number of parameters

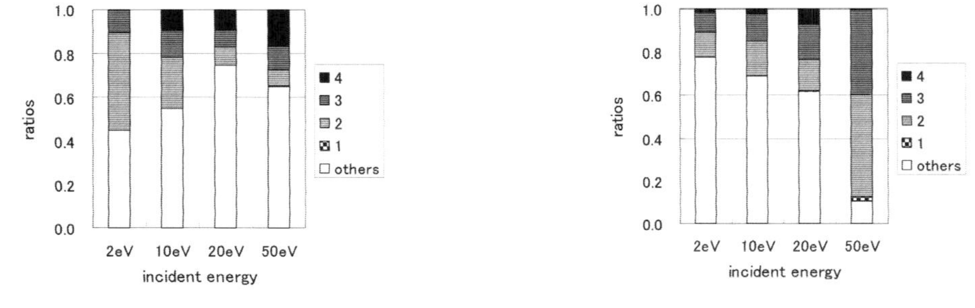

Figure 7. The ratio of the number of C atoms with each C-to-C coordination number deposited on the substrate to the number of all deposited C atoms as functions of the incident energy at a dose of $2.5\times10^{16}cm^{-2}$ for (a) CH_3 and (b) CH beam incidences, obtained from MD simulations. The "others" include deposited C atoms bonded with one or more H atoms or those bonded with five or more other C atoms (although the latter are scarce). [59]

such as the bond energy and bond length are the essential parameters that determine macroscopic outcomes of sputtering or deposition phenomena.

In this article, only the cases of ion injections (i.e., injections of energetic species) into the surface have been discussed and no neutral radical flux effect has been presented. However, effects of neutral radical fluxes can be incorporated in MD simulations in a relatively straightforward manner by the addition of low-energy fluxes of reactive species, as shown in, e.g., Refs. [62-64].

Experiments that can be directly compared with MD simulation are beam experiments. The reader is referred to, e.g., Refs. [65-69], in which mass-selected ion beam experiments and some puttering yield data are presented in detail.

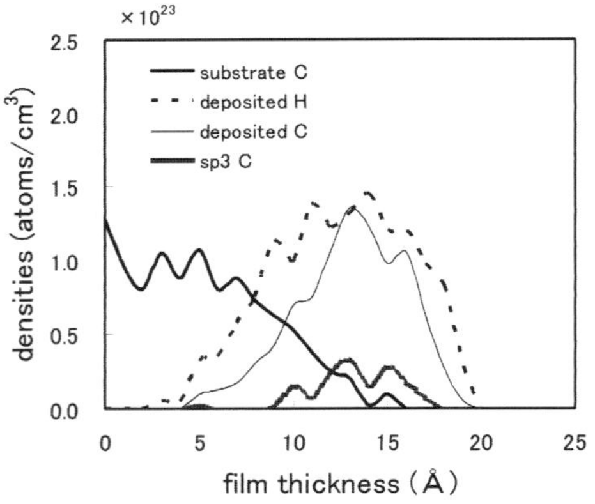

Figure 8. The depth profile of fully carbon terminated sp^3 C atoms at a dose of $2.0\times10^{16}cm^{-2}$ for 50 eV CH_3 incidence (depicted by the grey curve denoted by "sp3 C"), obtained from MD simulations. For reference, the profiles of C and H atoms are also plotted here, which are the same as those in Fig. 6(b). [59]

Acknowledgements
The author thanks Dr. Masashi Yamashiro of Nihon University and Mr. Yasuo Murakami of Canon Inc. for useful discussions. The work is partially supported by Grants-in-Aid for Scientific Research from Ministry of Education, Culture, Sports, Science and Technology (MEXT) Japan.

References

[1] J. B. Gibson, A. N. Goland, M. Milgram, and G. H. Vineyard, Phys. Rev. **120**, 1229 (1960).
[2] D. E. Harrison, Jr., N. S. Levy, J. P. Johnson, III, and H. M. Effron, J. Appl. Phys. **39**, 3742 (1968).
[3] D. E. Harrison, Jr., and C. B. Delaplain, J. Appl. Phys. **47**, 2252 (1976).
[4] D. E. Harrison, Jr., P. W. Kelly, B. J. Garrison, and N. Winograd, Surf. Sci. **76**, 311 (1978).
[5] N. Winograd, B. J. Garrison, and D. E. Harrison, Jr., J. Chem. Phys. **73**, 3473 (1980).
[6] M. S. Daw and M. I. Baskes, Phys. Rev. B**29**, 6443 (1984).
[7] M.I. Baskes, Phys. Rev. B**46**, 2727 (1992).
[8] F. H. Stillinger and T. A. Weber, Phys. Rev. B**31** 5262 (1985).
[9] F. H. Stillinger and T. A. Weber, J. Chem. Phys. **88**, 5123 (1988).
[10] F. H. Stillinger and T. A. Weber, Phys, Rev. Lett. **62**, 2144 (1989).
[11] A. Weber and F. H. Stillinger, J. Chem. Phys. **92**, 6239 (1990)
[12] J. Tersoff, Phys. Rev. B**37**, 6991 (1988).
[13] J. Tersoff, Phys. Rev. B**38**, 9902 (1988).
[14] J. Tersoff, Phys. Rev. B**39**, 5566 (1989).
[15] G. C. Abell, Phys. Rev. B**31**, 6184 (1985)
[16] C. T. Reimann, K. Walzl, M. El-Maazawi, D. M. Deaven, B. J. Garrison, and N. Winograd, J. Chem. Phys. **89**, 2539 (1988).

[17] R. A. Stansfield, K. Broomfield, and D. Clary, Phys. Rev. B **39**, 7680 (1989).
[18] R. Smith, D. E. Harrison, and B.J. Garrison, Phys. Rev. B **40**, 93 (1989).
[19] T. A. Schoolcraft and B. J. Garrison, J. Vac. Sci. Tech. A **8**, 3496 (1990).
[20] T. A. Schoolcraft and B. J. Garrison, J. Am. Chem. Soc., **113**, 8221 (1991).
[21] H. Feil, J. Dieleman, and B. J. Garrison, J. Appl. Phys. **74**, 1303 (1993).
[22] J. Dieleman, F. H. M. Sanders, A. W. Kolfshoten, P. C. Zalm. A. E. de Vries, and A. Haring, J. Vac. Sci. Technol. B 3, 1384 (1985).
[23] R. A. Rossen and H. H. Sawin, J. Vac. Sci. Technol. A **3**, 881 (1985).
[24] J. van Zwol, J. van Laar, A. W. Kolfschoten, and J. Dieleman, J. Vac. Sci. Technol. B **5**, 410 (1987).
[25] R. A. Rossen and H. H. Sawin, J. Vac. Sci. Technol. A **5**, 1595 (1987).
[26] D. J. Oostra, A. Haring, R. P. van Ingen, and A. E. de Vries, J. Appl. Phys. **64**, 315 (1988).
[27] H. F. Winters and J. W. Coburn, Surf. Sci. Rep. **14**, 161 (1992).
[28] P. C. Weakliem, C. J. Wu, and E. A. Carter, Phys. Rev. Lett. **69**, 200 (1992).
[29] P. C. Weakliem and E. A. Carter, J. Chem. Phys. **98**, 737 (1993).
[30] M. E. Barone and D. B. Graves, J. Appl. Phys. **77**, 1263 (1995).
[31] H. Feil, Phys. Rev. Lett. 74, 1879 (1995).
[32] S. D. Athavale and D. J. Economou, J. Vac. Sci. Technol. A **13**, 966 (1995).
[33] D. E. Hanson, A. F. Voter, and J. D. Kress, J. Appl. Phys. **82**, 3552(1997).
[34] D. E. Hanson, J. D. Kress and A. F. Voter, J. Chem. Phys. **110**, 5983 (1999).
[35] D. E. Hanson, J. D. Kress and A. F. Voter, J. Vac. Sci. Technol. A **17**, 1510 (1999).
[36] B. A. Helmer and D. B. Graves, J. Vac. Sci. Technol. A **15**, 2252 (1997).
[37] B. A. Helmer and D. B. Graves, J. Vac. Sci. Technol. A **16**, 3502 (1998).
[38] N. A. Kubota, D. J. Economou, and S. J. Pimpton, J, Appl, Phys. **83**,4055(1998).
[39] H. Ohta and S. Hamaguchi, J. Chem. Phys. **115**, 6679 (2001).
[40] H. Ohta and S. Hamaguchi, J. Vac. Sci. Technol. A**19**, 2373 (2001).
[41] V. V. Smirnov, A. V. Stengach, K. G. Gaynullin, V. A. Pavlovsky, S. Rauf, P. J. Stout, and P. L. G. Ventzek, J. Appl. Phys. **97**, 093302 (2005).
[42] V. V. Smirnov, A. V. Stengach, K. G. Gaynullin, and V. A. Pavlovsky, S. Rauf, P. Stout, and P. L. G. Ventzek, J. Appl. Phys. **97**, 093303 (2005).
[43] V. V. Smirnov, A. V. Stengach, K. G. Gaynullin, V. A. Pavlovsky, S. Rauf and P. L. G. Ventzek, , J. Appl. Phys. **101**, 053307(2007).
[44] H. Ohta and S. Hamaguchi, J. Plasma Fusion Res. SERIES **6**, 399 (2004).
[45] B. J. Garrison, P. B. S. Kodali, and D. Srivastava, Chem. Rev. **96**, 1327 (1996)
[46] H. Yamada and S. Hamaguchi, J. Appl. Phys.**96**, 6147 (2004).
[47] D. W. Brenner, Phys. Rev. B **42**, 9458 (1990).
[48] H. D. Hagstrum, P. Petrie, and E. E. Chaban, Phys. Rev. B **38**, 10 264 (1988).
[49] M. Yamashiro, H. Yamada, and S. Hamaguchi, J. Appl. Phys.**101**, 046108 (2007).
[50] M. Yamashiro, H. Yamada, and S. Hamaguchi, Jpn. J. Appl. Phys.**46**, 1692 (2007).
[51] M. Yamashiro, H. Yamada, and S. Hamaguchi, Thin Solid Films **516**, 3449 (2008).
[52] J. Robertson, Surf. Coating Technol. **50**, 185 (1992).
[53] W. Jacob, and W. Möller, Appl. Phys. Lett. **63**, 1771 (1993).
[54] Y. Lifshitz, Diamond Relat. Mater. **5**, 388 (1996).
[55] J. Robertson, Phil. Mag. B **76**, 335 (1997).
[56] A. Grill, Diamond Relat. Mater. **8**, 428 (1999).
[57] J. Robertson, Mat. Sci. Eng. **R37**, 129 (2002).
[58] C. Casiraghi, A. C. Ferrari, and J. Robertson, Phys. Rev. B **72**, 085401 (2005).
[59] Y. Murakami, S. Horiguchi, and S. Hamaguchi, Phys. Rev. E. **81**, 041602 (2010).
[60] Y. Murakami, K. Hosaka, and S. Hamaguchi, J. Phys.:Conf. Series **232**, 01200 (2010) 7.
[61] H. Yamada and S. Hamaguchi, Plasma Phys. Control. Fusion **47**, A11 (2005).
[62] M. Taguchi and S. Hamaguchi, J. Appl. Phys. **100**, 123305 (2006).

[63] M. Taguchi and S. Hamaguchi, Jpn. J. Appl. Phys. **45**, 8163 (2006).
[64] M. Taguchi and S. Hamaguchi, Thin Solid Films **515**, 4879 (2007).
[65] K. Hine, S. Yoshimura, K. Ikuse, M. Kiuchi, J. Hashimoto, M. Terauchi, M. Nishitani, and S. Hamaguchi, Thin Solid Films **517**, 835 (2008).
[66] K. Hine, S. Yoshimura, K. Ikuse, M. Kiuchi, J. Hashimoto, M. Terauchi, M. Nishitani, and S. Hamaguchi, Jpn. J. Appl. Phys. **46**, L1132 (2007).
[67] K. Yanai, K. Karahashi, K. Ishikawa, and M. Nakamura, J. Appl. Phys. **97**, 053302 (2005).
[68] K. Karahashi, K. Yanai, K. Ishikawa, H. Tsuboi, K. Kurihara, and M. Nakamura, J. Vac. Sci. Technol. A **22**, 1116 (2004).
[69] K. Ishikawa, K. Karahashi, H. Tsuboi, K. Yanai, and M. Nakamura, J. Vac. Sci. Technol. A **21** L1 (2003).

Adsorption dynamics of water on the surface of TiO2 (110)

N Bundaleski*[1], A G Silva[1], U Schröder[2], A M C Moutinho[1] and O M N D Teodoro[1]

[1]CeFITec, Department of Physics, Faculdade de Ciências e Tecnologia, Universidade Nova de Lisboa, Campus de Caparica, P-2829-516 Caparica, Portugal
[2]Department of Physics, Technical University of Darmstadt, Karolinenplatz 5, D-64289 Darmstadt, Germany

E-mail: n.bundaleski@fct.unl.pt

Abstract. Rutile titanium dioxide TiO_2 is used in a number of technological areas. Therefore, in surface science, it has become the most studied oxide surface. Water adsorption on rutile TiO_2 (110) has been investigated using the X-ray photoelectron spectroscopy (XPS) and the work function study (WF): water adsorption induces formation of a dipole layer, which locally changes the work function. This can be experimentally observed as the onset shift of the secondary electron energy spectrum. While XPS seems to be insufficiently sensitive to monitor water adsorption on TiO_2, there is a clear work function change undoubtedly attributed to the water adsorption. The measurements were done for different water vapour pressures, exposure times, sample temperatures and general surface conditions. Time evolutions of the work function change and the H_2O partial pressure, enable us to successfully model the adsorption dynamics and help us understand the observed results. The analysis clearly shows existence of at least three different adsorption sites. Their interplay governs the work function time evolution, while the relative contributions depend on the surface temperature and, presumably, its topography. These results will be discussed in the light of several recent experimental and theoretical studies of this system done by other authors.

1. Introduction

TiO_2 is one of the most studied metallic oxides due to the possible applications in photocatalysis, heterogeneous catalysis, as a gas sensor, solar cell, biocompatible material etc. Besides, it is investigated as a model metallic oxide surface. Rutile TiO_2 (110) has attracted particular attention as the most stable TiO_2 surface. The chemical activity of this surface is mainly governed by the bridging oxygen vacancies (BOV) which act as the adsorption sites for different molecules. Water adsorption on TiO_2 (110) is thoroughly investigated using various surface sensitive experimental techniques among which STM particularly shed light to this very complex system [1, 2].

The vast majority of H_2O/TiO_2 (110) studies have been performed on the unreconstructed (1x1) surface, shown in Figure 1(a). It practically always contains small relative concentration of BOV N_{BOV} (up to 10-12 %) [3]. Further increase of BOV concentration leads to the (2x1) surface reconstruction, the latter being able to sustain wide range of $N_{BOV} \leq 55$ % [4]. Although the question concerning the corresponding atomic arrangement is still open, it is generally accepted that (2x1) reconstruction is

* To whom any correspondence should be addressed.

formed from the nonreconstructed surface by adding Ti_2O_3 rows on the top of every second Ti row of the (1x1) surface (see Figure 1b).

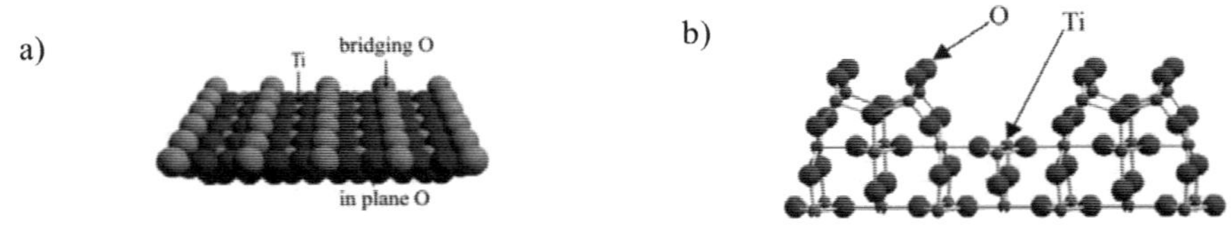

Figure 1 The most common structures of rutile TiO_2 (110) surfaces: a) bulk terminated TiO_2 (110)-(1x1) surface; b) reconstructed TiO_2 (110)-(2x1) surface, model proposed by Onishi et al. [2].

Concerning the water adsorption on rutile TiO_2 (110)-(1x1) surface, two adsorption sites have been mainly identified: BOV on which the dissociative adsorption takes place yielding in formation of two hydroxyls with hydrogen atoms facing the vacuum, and the non-dissociative adsorption on titanium rows with two hydrogen atoms facing the vacuum (it is accepted that the in plane oxygen atoms are not active adsorption sites). The latter is desorbed at 265 K while the recombination of the hydroxyl groups that lead to desorption from oxygen vacancies takes place at about 500 K [5]. It should be stressed that even at vacuum conditions of about $3 \cdot 10^{-10}$ mbar of residual gas surfaces with N_{BOV} of about 5 % can be fully hydroxylated within a few minutes (the exposures of about 0.1 L!) [3] This peculiar result can be explained only by the high mobility of water molecules adsorbed at the surface which allows them to inspect more than one adsorption site: when adsorbed at Ti rows, water molecules travel along the rows until they find BOV on which the dissociative adsorption takes place [5]. There are only few works related to the water adsorption on (2x1) reconstructed rutile TiO_2 (110) surface [6, 7]. The main conclusion is that the adsorption takes place at BOV.

Adsorptions of water or hydroxyl molecules, which are both polar, make a dipole layer at the rutile surface reducing the work function. According to the well-known Topping model [8] the work function change is directly proportional to the dipole concentration as long as the concentration is not too high. The work function change i.e. the dipole concentration can be simply followed by monitoring change of the onset position of the secondary electron distribution. The corresponding experimental technique is known as Work Function Study (WF). To the best of our knowledge, WF has been employed to monitor the water adsorption dynamics of on rutile TiO_2 (110) [6, 7, 9] but not for the unreconstructed surface.

2. Experimental

The experiments were performed in the multipurpose surface analysis apparatus described elsewhere [10]. The sample color is dark blue indicating O-deficient bulk reduced sample. It was cleaned using several sputtering/annealing cycles (3 keV Ar^+ at about 20° with respect to the surface plane, 773 K for 10 min) until only Ti and O peaks were present in X-ray Photoelectron Spectroscopy (XPS) survey spectra. Secondary electron emission was stimulated by non-monochromated X-rays emitted from the Mg anode of an X-ray source, routinely used for XPS. The sample was biased to -30 V to allow the detection of electrons with low kinetic energies. If the energy of secondary electrons with respect to the sample vacuum level is E_k^S the measured energy will be $E_k^A = 30\text{ V} + E_k^S + WF^S - WF^A$ where WF^S and WF^A are the work functions of the sample and the analyzer, respectively. Lowering WF^S will shift the measured secondary electron energy spectrum towards the lower energies. Water used in the adsorption experiments has been purified using several freeze-pump-thaw cycles. Its partial pressure during the adsorption experiments has been followed using the pressure rise detected by the ionization gauge. The purity of the introduced gas has been checked by the residual gas analyzer.

3. Adsorption dynamics modeling

Before we present a simple model for the water adsorption dynamics on unreconstructed rutile TiO_2(110), let us clear out several points about the results one can get from WF measurements [11,

12]. Energy distribution of the secondary electrons is generally a superposition of different contributions associated to the surface regions described by their local work functions. Let us recall that WF actually provides the information only about the parts of the surface having the smallest work function i.e. the highest dipole concentration in our case. As already stated, the dipoles can be formed on BOV (as pairs of hydroxyls) and on the Ti rows (molecular adsorption). The hydroxyl mobility is much smaller than that of the water molecules, which can readily diffuse along the Ti rows [5]. Hence, the highest dipole concentration should be in the regions close to the hydroxyl dipoles i.e. adsorption sites for the molecular adsorption of interest are the Ti atoms in the vicinity of the hydroxyl dipoles. Having that in mind and according to the Topping model for the low dipole concentration [8], the work function shift should be $\Delta WF = -(e/\varepsilon_0)\cdot[n_{vac}\cdot 2\mu_{OH}\theta_{vac}+n_{Tieff}\cdot\mu_{H2O}\theta_{Tieff}]$, where μ_{OH} and μ_{H2O} are the dipole moments of adsorbed OH and H_2O molecules respectively, while n_i and θ_i ($0< \theta_i <1$) are the local surface concentrations and coverages of the adsorption site 'i', respectively. The factor two in the first term is due to the fact that the adsorption on BOV contributes to the formation of two dipoles. The fact that this adsorption is dissociative does not influence its dynamics while BOV concentration is much smaller than the concentration of the bridging oxygen atoms: we assume that the vacancies are always surrounded by the bridging oxygen atoms so the hydrogen atom has always an available bridging oxygen where it can hope.

The overall concentration of Ti atoms and the bridging oxygen atoms of this surface is the same – $n_{Ti} = 5.2\cdot 10^{14}$ cm^{-2}. Therefore, we can rewrite the last equation as

$$\Delta WF(eV) = -1.96\left(N_{vac}\cdot 2\mu_{OH}\theta_{vac} + N_{Tieff}\cdot \mu_{H2O}\theta_{Ti}\right), \tag{1}$$

where N_{vac} and N_{Tieff} are the surface concentrations relative to n_{Ti} and the dipole moments are given in Debye. The adsorption dynamics is governed by the time evolution of θ_i. The adsorption rate should be equal to the product of the particle flux Φ and the probability that the adsorption will take place (sometimes called the sticking coefficient): $dn/dt=\Phi\cdot P_{adsorption} - \beta\cdot n$, where n is the concentration of adsorbed species and β is the desorption rate. Let us know assume that a molecule has a chance to be adsorbed only if it 'falls' on the adsorption site i.e. on a specific surface atom. This case corresponds to the water molecule adsorption on the Ti rows. The adsorption probability is then equal to the probability that the atom hits an empty adsorption site ($n_a(1-\theta_a)/n_s$, n_s is the surface concentration of all atoms and n_a (θ_a) is the surface concentration (coverage) of potential adsorption sites) multiplied by the probability that the adsorption is actually going to happen. The latter can be related to the orientation of the adsorbing molecule or the existence of adsorption energy, but it should be only temperature dependent. We shall describe these effects by a single coefficient α(T): $n_a\cdot d\theta_a/dt = \Phi\cdot\alpha\cdot n_a(1-\theta_a)/n_s - \beta\cdot n_a\cdot\theta_a$ i.e.

$$\frac{d\theta_a}{dt} = \Phi\cdot\alpha\cdot\frac{(1-\theta_a)}{n_s} - \beta\cdot\theta_a\cdot \tag{2}$$

In the case of the hydroxyl adsorption at bridging oxygen vacancies, the diffusion along the Ti rows plays an important role. The direct consequence is that a water molecule 'probes' M different adsorption sites in average. If the probability of finding an empty adsorption site in a single trial is $P_1 = n_a(1-\theta_a)/n_s$, the probability that the molecule will find the adsorption sites by maximum M trials is $1-(1-P_1)^M$. If $P_1<< 1$, which is certainly the case, it can be shown that

$$\frac{d\theta_a}{dt} = \frac{\Phi}{n_a}\cdot\left[1-\exp\left(-M\alpha\frac{n_a(1-\theta_a)}{n_s}\right)\right] - \beta\cdot\theta_a\cdot \tag{3}$$

The equations (2) and (3) are used to model the adsorption dynamics of water on Ti rows and bridging oxygen vacancies, respectively.

3. Results

WF results clearly show two different behaviors in terms of water adsorption dynamics which reflect the surface structure. The difference is present in the XPS results as well. We denote the two surfaces non-reconstructed and partially reconstructed due to the reasons given below. In both cases the overall work function change never exceeds 0.2 eV, which is the reason why the results are quite noisy. Still, by applying strict and careful experimental procedure they are reproducible and can be generally used (in correlation with XPS measurements and modeling) to yield considerable information about the water adsorption on rutile TiO_2 (110) surface.

3.1. Non-reconstructed surface

3.1.1. XPS measurements

The XPS measurements has been performed prior and after each water adsorption experiment. The main XPS lines of oxygen (O 1s) and titanium (Ti 2p) were recorded in order to monitor the surface composition and the peak shapes.

The measured O:Ti composition ratio in the case of the unreconstructed surface is 72:28, with the uncertainty of ±1. Water adsorption does not show any influence to the composition. The discrepancy from the expected 66:33 can be due to different reasons such as inadequate relative sensitivity factors used for the composition analysis (since they are not provided by the manufacturer the data from [13] are used). Additionally, small matrix effects [14] and material depth non-homogeneity could be also partly responsible for the discrepancy.

The peak shapes are also influenced by the surface structure. The Ti 2p peak (Figure 2a) clearly consists of two contributions modeled by pseudo-Voigt profiles. Assuming that the main peak corresponds to Ti from TiO_2 (so called Ti^{4+}) the other one is related to the bridging oxygen vacancies (the corresponding peak shifts are 1.7 eV for Ti $2p_{3/2}$ and 2.4 eV for Ti $2p_{1/2}$) [1]: it is generally assumed that by the creation of BOV two Ti atoms are left with unsaturated bonds which is seen in the XPS as the peak corresponding to Ti_2O_3 bond (Ti^{3+} state). Since the surface concentration of Ti is the same as that of the bridging oxygen, relative magnitude of the Ti^{3+} contribution is two times the relative concentration of oxygen vacancies. The strange feature concerning the Ti^{3+} peak is that it does not disappear after the water adsorption (which should saturate the dangling bonds): the difference between the XPS spectra taken before and after the adsorption experiment is negligible. The same effect was also observed in [7] and [9]. The Ti^{3+} contribution could be related to the concentration of the bridging oxygen vacancies on the freshly prepared surface i.e. before any adsorption, including that from the residual gas, takes place (cf. Section 4). The relative contributions of the Ti^{3+} peak and the estimated BOV concentration are given in Table 1. However, one should have in mind that titanium atoms at interstitial sites can generally be in Ti^{3+} and Ti^{4+} states [1].

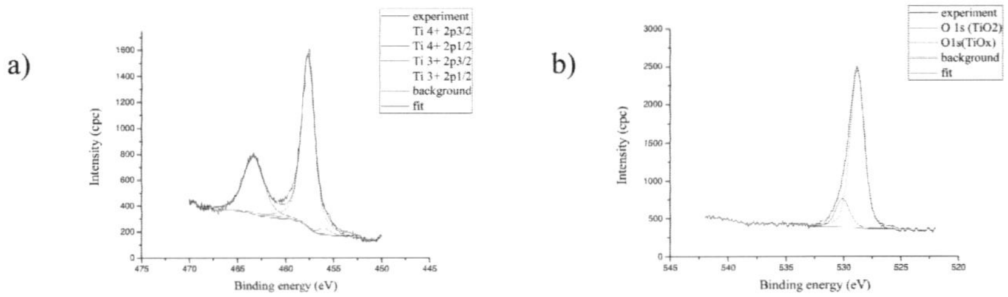

Figure 2 XPS spectrum of TiO_2 (110) sample taken before the water adsorption: a) Ti 2p line; b) O 1s line. The sample is at room temperature.

Figure 2b shows O1s peak, also modeled by 2 pseudo-Voigt profiles, the main one corresponding to TiO_2 and the other one, identified as TiO_x, (the peak shift is 1.6 eV) usually associated to the native oxides of Ti [15]. The peak shift for the Ti_2O_3 is only 0.1 eV so it could not be reliably resolved, while the oxygen from water (the peak shift should be about 3±1 eV [16]) was not observed. The relative contribution of the TiO_x peak is about 11 % with the fluctuations less than 1 % (see Table 1).

3.1.2. Work function change measurements

Two energy spectra of secondary electrons, one taken before and one during the water desorption at room temperature, are presented in Figure 3. While the onset shift is clearly visible, the rest of the energy spectra fully overlap clearly indicating the local character of the work function change. Concerning the time evolution of the work function change (Figure 4a), initial steep drop of the work function followed by the much slower decrease can be clearly observed. This behavior cannot be explained by the existence of a single adsorption site in the frame of the proposed model: if the fast adsorption takes place, the coverage increases quickly until it saturates into the equilibrium between the adsorption and desorption rates (cf. figure 4b). Therefore, the additional slow increase of the coverage can only be explained by the second contribution from another kind of adsorption sites having slower adsorption rate. The water partial pressure decrease contributes only to the work function recovering corresponding to the 2nd contribution. We denote the contributions and appropriate adsorption sites as primary (corresponding to the steep drop of the work function) and secondary (corresponding to the slower drop of the work function followed by the work function recovering). At room temperature, desorption readily takes place from the secondary adsorption sites, while the primary sites are not affected by it. Hence, the equilibrium coverage for the primary adsorption sites should be 1. Besides, the primary sites are occupied very quickly (for the exposure less then 1 L). According to the experimental and theoretical investigations done by different groups on unreconstructed rutile TiO_2 (110) (cf. Introduction) we identify the primary adsorption sites as the BOV and the secondary sites as the Ti atoms. Therefore we apply the approach presented in Section 3 to model the water adsorption dynamics.

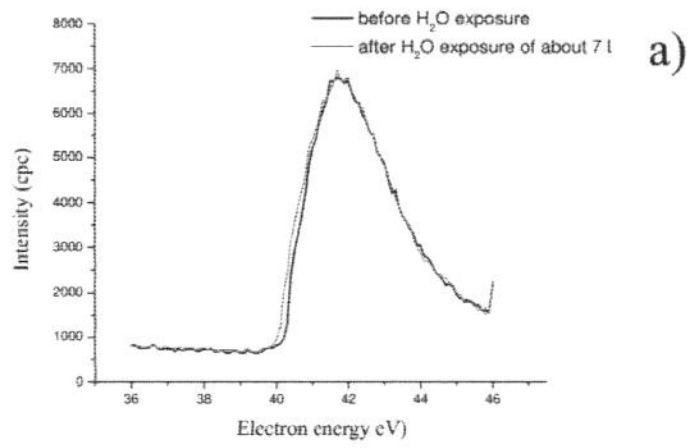

Figure 3 Energy distribution of secondary electrons emitted from TiO_2 (110) before and after the water exposure of about 7 L

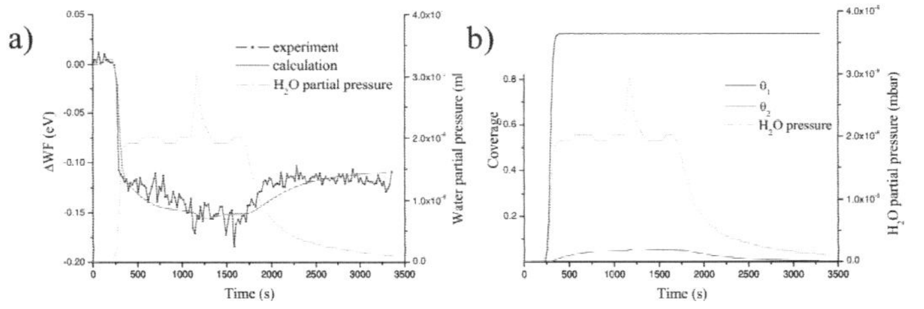

Figure 4 a) Time evolution of the work function change due to the water exposure at room temperature to the unreconstructed surface and the modelling result; b) calculated time evolution of the coverages and the H_2O partial pressure time evolution

From the comparisons to different experimental results it appears that the argument of the exponential term in equation (3) is much smaller than 1. Therefore, since $e^{-x} \approx 1 - x$, (3) can be approximated by $d\theta_a/dt = \Phi \cdot M \cdot \alpha(1 - \theta_a)/n_s - \beta \cdot \theta_a$. Consequently, we assume that the concentration of bridging oxygen vacancies (i.e. the primary adsorption sites) does not influence the adsorption rate but only the overall change of the work function. By 'condensing' the unknown parameters (i.e. by their multiplication) we come to the following set of equations used to model the experimental results:

$$\begin{aligned} \Delta WF(eV) &= -1.96(I_1 \cdot 2\theta_1 + I_2\theta_2) \\ \frac{d\theta_1}{dt} &= 1.7865 \cdot 10^5 \cdot p(mbar) \cdot A_1(1-\theta_1) \\ \frac{d\theta_2}{dt} &= 1.7865 \cdot 10^5 \cdot p(mbar) \cdot A_2(1-\theta_2) - \beta_2 \cdot \theta_2 \end{aligned} \qquad (4)$$

where p is the partial pressure of water, $I_1 = N_{vac} \cdot \mu_{OH}$, $I_2 = N_{Tieff} \cdot \mu_{H2O}$, $A_1 = M_{vac} \cdot \alpha_{vac}$, $A_2 = \alpha_{Tieff}$ and $\beta_1 = 0$. Since the water partial pressure is time dependent, the differential equations can be solved only numerically. Obviously, parameters (A_i and β_i) govern the water adsorption dynamics, while I_i is the magnitude of the i-th contribution. The parameters A_i and β_i can be adjusted and kept constant to provide the best fit to all measurements while I_i change depending on the experimental conditions. Their magnitudes are given in Table 1. Water adsorption has been performed for the samples at room temperature, as well as at 336 K and 373 K. Very good match with the experiments is obtained in all cases, as it is illustrated in Figure 4a, keeping A_i and β_i unchanged.

3.2. Partially reconstructed surface

3.2.1. XPS analysis

Air exposure of the sample followed by the long term annealing at 600 K and residual gas pressure in the 10^{-7} mbar range (due to the baking of the vacuum system) contributes to the considerable changes in the XPS spectra even after the standard cleaning procedure described in Section 2. First of all, there is a considerable relative change of the surface composition – O:Ti composition ratio changed from 72:28 for nonreconsructed surface to 87:13 (the uncertainty is again ±1). The relative contribution of Ti^{3+} peak increased also to 20-30 % depending on the surface conditions. It should correspond to the BOV concentration of about 10-15 %, which sounds confusing knowing that the surface concentration of oxygen increased as well. The relative contribution of the TiO_x bond of the O1s peak generally increased and it varied in the range 10-17 %.

3.2.2. Work function change measurements

As in the case of the XPS results, sample treatment after the air exposure had substantional influence to the WF results as well i.e. to the water adsorption dynamics. A typical result is presented in Figure 5. Besides the two contributions present for the nonreconstructed surface, we observe the third one characterized by slow dynamics and the lack of desorption at room temperature. The relative magnitude of the third contribution depends on the surface preparation: it is more pronounced after the first cleaning right after the air exposure; afterwards, its relative magnitude generally decreases but still varies considerably depending on the sputtering/annealing time. The third contribution can be very efficiently described using the adopted model by introducing the third adsorption sites through the corresponding equation $d\theta_3/dt = 1.7865\cdot 10^5\cdot p(\text{mbar})\cdot A_3(1-\theta_3)$ and modifying the expression (1) which now becomes

$$\Delta WF(eV) = -1.96(I_1 \cdot 2\theta_1 + I_2\theta_2 + I_3 \cdot 2\theta_3) \tag{5}$$

Since we observe increase of the BOV concentration, and knowing that the desorption from these sites is negligible at room temperature, we address the 3rd contribution to BOV showing different dynamical behavior in terms of water desorption as compared to the 1st contribution. More arguments that further justify this assumption will be given in Section 4. The parameter A_3 was tuned to properly describe the experimentally observed water adsorption dynamics keeping the parameters A_2 and β_2 the same as for the unreconstructed surface (while A_1 had to be changed as well, cf. Section 4). Finally, the magnitudes of the three contributions have been adjusted for every experimental result. The match (Figure 5a) with the experiment is excellent. The corresponding time evolution of θ_i (i = 1-3) as well as of the water partial pressure evolution is given in figure 5b.

At the end of the water adsorption experiment i.e. after the water molecules are desorbed from the Ti rows, only hydroxyl dipoles are present at the surface. For this surface, we measured the temperature dependence of the work function shift in the range from the room temperature to about 473 K and back (Figure 6). When increasing the temperature, the work function change decreases relatively slowly until it reaches about 373 K. Then the recovering becomes steeper until the sample temperature reaches about 393 K when a kind of saturation takes place i.e. the work function does not change with further temperature increase. The work function is now even higher than at the beginning of the experiment. If the hydroxyls were desorbed at elevated temperature one should expect that after cooling some of the BOV should be again occupied by the water from the residual gas so that the work function returns to about the same value as in the beginning of the experiment. However, when the same sample is cooled down to the room temperature, there is only slight decrease of the work function. This is expected since the hydroxyls from rutile TiO_2 (110) begin to desorb at temperatures of about 500 K [2]. The result can be explained by the surface diffusion of hydroxyls, which should decrease the work function change by decreasing the hydroxyl concentration. This is in accordance to the earlier conclusion that the observed WF change is due to the formation of areas with low work function at the sample surface.

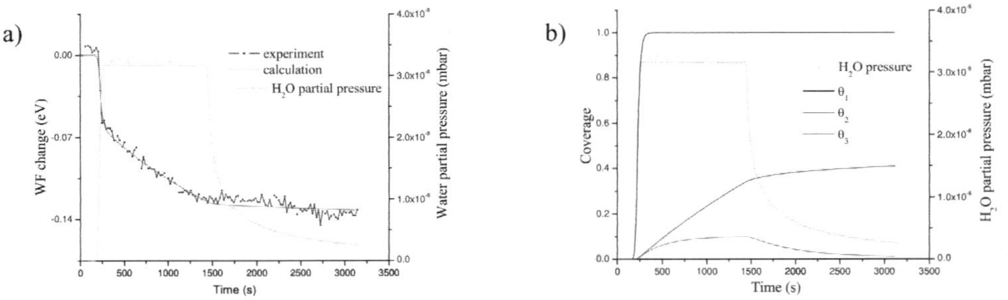

Figure 5 a) Time evolution of the work function change due to the water exposure at room temperature to the partially reconstructed surface and the modelling result; b) calculated time evolution of the coverages and of the H_2O partial pressure

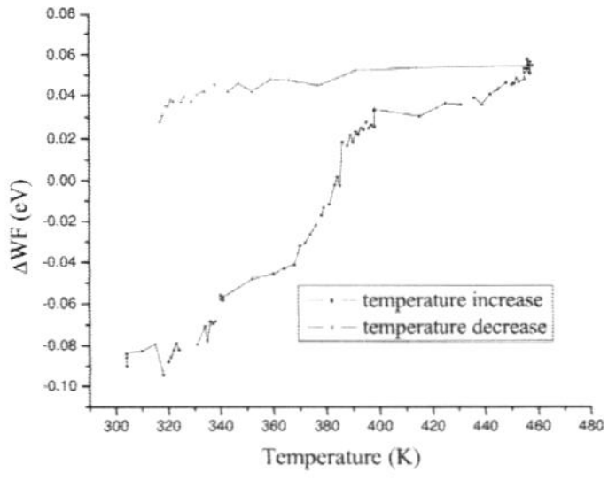

Figure 6 Temperature evolution of the work function of change of TiO_2 (110) surface previously exposed to H_2O

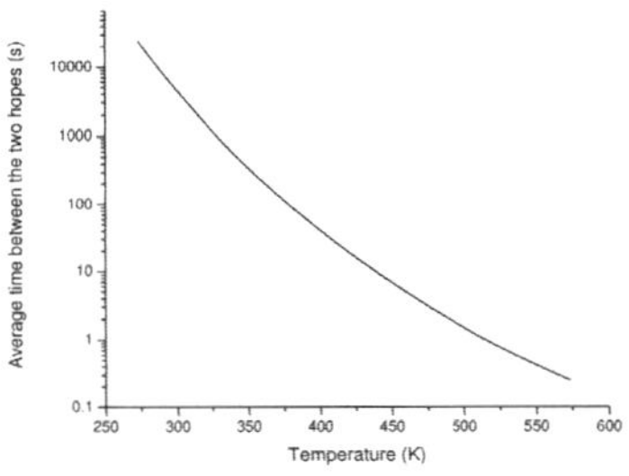

Figure 7 Average time between two hopes of hydrogen atoms along bridging oxygen rows of TiO_2 (110) surface [19]

Table 1 The coefficients obtained from the modelling of WF change in time and the XPS results

Surface condition	Unreconstructed rutile TiO_2 (110) surface			Partially reconstructed rutile TiO_2(110) surface at room temperature
	$T = 300$ K	$T = 326$ K	$T = 373$ K	
I_1	0.037	0.033	0.022	0.02 ± 0.01
A_1	13	13	13	3-12
I_2	0.25	0.25	0.1	0.06 ± 0.03
A_2	0.06	0.06	0.06	0.06
β_2	0.003	0.003	0.003	0.003
I_3	0	0	0	0.07 ± 0.04
A_3	0	0	0	0.05
x_{max}	0	0	0	0.8 ± 0.1
I_2/I_1	6.7	7.5	4.5	4 ± 2
N_{Ti3+} (%)	8-11	9.1	7.4	25 ± 5
N_{BOV} (%)	4-5.5	4.55	3.7	12.5 ± 2.5

4. Discussion

Rutile with the (110) orientation is the most stable TiO_2 surface: it is usually not a problem to obtain it unreconstructed (i.e. with the reconstruction (1x1)). However, ion sputtering being the standard cleaning procedure introduces BOV. After the sample annealing, the surface again becomes with (1x1) reconstruction but creation of BOV in the bulk is the price one has to pay. Each cleaning cycle increases BOV bulk concentration. This crystal is sometimes denoted as 'reduced'. The direct consequence is that TiO_2 changes the electrical properties from a good insulator into N-type semiconductor (which is convenient for the surface characterization techniques that usually use charged particles as a probe). The optical properties change as well: initially transparent, with the increase of BOV TiO_2 becomes pale blue, then darker, and finally black. The color of the crystal is frequently used to describe the state of the sample [1].

Although Ti interstitials can be in Ti^{3+} state, it is accepted that the concentration of BOV equals half of the Ti^{3+} surface concentration. The relative change of oxygen concentration due to the water adsorption cannot be observed in XPS composition measurements since the technique is sensitive to first ten atomic layers. However, the fact that XPS profile of the Ti 2p peak does not change after the H_2O exposure is puzzling. Recent paper of the Aarhus group [17] reveals this paradox: the creation of BOV does not leave two Ti atoms with unsaturated bonds because these atoms do not stay on their original positions! They migrate into the bulk on interstitial sites instead. Latter, when BOV is filled-in with a hydroxyl, they do not return back right away due to the limited mobility. Having this in mind it seems reasonable to conclude that the contribution that we identify as Ti^{3+} corresponds to Ti interstitials, still representing a measure of the BOV concentration. This also explains why the peak does not disappear after the adsorption takes place. Besides, one should be careful when estimating BOV concentration from Ti^{3+} signal since Ti interstitials (i.e. BOV) are present in the bulk as well in spite of their limited mobility.

When the crystal becomes heavily reduced, it is not possible anymore to obtain (1x1) surface reconstruction. The surface becomes usually (2x1) reconstructed instead having strong influence to the water adsorption dynamics BOV (being the dominant water adsorption sites in this case [6, 7]) are farther from the Ti rows as compared to the nonreconstructed surface (cf. Figure 1). This could suppress H_2O molecules adsorbed on Ti rows to 'reach' BOV. Indeed, WF measurements done on TiO_2 (110)-(2x1) surface show very slow response to the water adsorption [7], qualitatively very much alike the third contribution identified in this work. Moreover, our earlier WF measurements on heavily reduced TiO_2 (110) [11] also show very similar behavior in terms of adsorption dynamics. The total WF change in these measurements was 0.4 eV or even more, strongly suggesting high BOV concentration and probably (2x1) reconstruction. At first site, the observed relative increase of Ti^{3+} signal can be understood as the increase of BOV concentration (about 15 %) leading to the (2x1) reconstruction. If so, three questions are still open: 1. How did we abruptly introduce such a high quantity of BOV and provoked (2x1) reconstruction; 2. From WF measurements it seems that besides (2x1), (1x1) reconstructions are still present in spite of increased BOV concentration; 3. In the case of (2x1) reconstruction Ti^{3+} signal partially corresponds to Ti atom in Ti_2O_3 (which is also in Ti^{3+} state) so it cannot be used as a measure of BOV concentration.

In order to understand the obtained results we refer to the investigation of the TiO_2 (110)-(1x1) exposure to oxygen (p_{O2} in 10^{-6}-10^{-5} mbar range) at elevated temperatures (600-800 K) (cf. [1] and references therein). The main conclusion is that at elevated temperatures Ti interstitials come to the surface and react with oxygen from the atmosphere building TiO_2 (110)-(2x1) reconstructed islands on the top of the unreconstructed TiO_2 (110) surface. We assume that similar process took place during the baking process: the sample was at about 600 K for few days exposed to the residual gas in the 10^{-7} mbar range consisting mainly of water. This would explain the three open questions above mentioned: 1. The surface concentration of BOV was actually not increased; 2. The sample surface is nonreconstructed on which (2x1) reconstructed islands are grown the latter being responsible for the 3rd contribution in the model; besides, the lowest work function regions are larger than the grown islands; 3. Ti^{3+} signal growth is mainly due to the Ti_2O_3 rows in (2x1) reconstructed islands.

If the (2x1) is made by adding Ti_2O_3 rows above the nonreconstructed surface, one should expect increase of the Ti surface composition being in contrast to our measurements. Besides, if the atomic arrangement of the (2x1) reconstructed surface is valid we should expect stronger suppression of the 1st contribution than of the 2nd one (in the frame of the (2x1) islands there should still be some Ti rows), which is again in contrast to our observations (see below). Impact collision ion scattering spectroscopy measurements of rutile TiO_2 (110)-(2x1) did not detect Ti rows in the first layer [18]. The authors suggest that the rest of the Ti rows (those not covered by the Ti_2O_3 rows) are actually covered by some kind of oxygen structure. Here discussed results including the increase of the TiO_x contribution of the O 1s peak for the partially reconstructed surface strongly support their conclusions.

Let us now focus to the results of modeling. In the case of the unreconstructed surface, it is easy to decouple the contributions due to the lack of desorption for the 1st and intense desorption for the 2nd one. Moreover, this enables to get reliable determination of the coefficients I_1 and A_1 (the first defining the magnitude and the second the dynamics of the primary contribution) in contrast to the those describing the secondary contribution: the existence of the desorption makes possible to get a good fit for different sets of parameters. Additionally, the determined magnitudes of the coefficients A_i and β_i are not reliable due to the questionable precision of the water partial pressure measurement (the appropriate system was not calibrated for H_2O). Nevertheless, the obtained values confirm the basic assumptions on which the model is based: $A_1 > 1$ i.e. $M_1 > 1$ (since α_i, being a probability, has to be in the interval [0, 1]), confirming the idea that due to the surface diffusion several surface atoms are inspected in the search for the adsorption site. In contrast, A_2 and A_3 are both less than 1 implying the lack of the surface diffusion. Parameters that define the water dynamics of the 2nd and 3rd contribution should not depend on the surface structure since the interaction between the H_2O molecule and the adsorption site is mainly 'local'. This is the main reason why we kept them constant. On the other hand, change from (1x1) reconstruction to the partially reconstructed surface reduces the length of Ti rows and thus decreases A_1, which is exactly what we observe (cf. Table 1).

In the case of the partially reconstructed surface determining coefficients I_3 and A_3 also looks very reliable. In conclusion, we can try to yield some physical parameters starting from the coefficients I_1 and I_3. But before that, the coefficients I_i have to be redefined in order to incorporate the assumptions of the existence of (1x1) and (2x1) regions. We introduce parameters x and y in the range [0, 1] as abundance of (2x1) islands and the abundance of Ti rows, respectively. Then, $I_1 = N_1.\mu_{OH}.(1-x)$, $I_2 = N_2.\mu_{H2O}.y$ and $I_3 = N_3.\mu_{OH}.x$. Depending on the structure of the (2x1) islands, y can be equal to $1-x$ (no Ti rows on (2x1) islands), or to $(1-x)+x/2$ (with Ti rows on (2x1) islands). Since the parameters A_2 and β_2 are fixed, the ratio I_1/I_2 for different structures should help us to estimate if the (2x1) islands contain Ti rows. From the data given in Table 1 it seems that I_1/I_2 does not depend too much on the surface structure (i.e. $y \approx 1-x$) which further supports our assumption about the structure of (2x1) islands. Assuming that relative concentration of BOV at (2x1) islands is the same as for the nonreconstructed surface ($N_1 = N_3$) we can estimate parameter x from the ratio $I_3/I_1 = (x/(1-x))$. This equality does not hold: the adsorption of water from the residual gas on BOV is going to take place mainly on the nonreconstructed regions because $A_1 >> A_3$. Therefore, $x = I_3/(I_3 + I_1 \cdot N_3/N_1)$ and since $N_1 < N_3$ the approximate expression will give overestimated value for x. From the data in Table 1 we estimate that x should be smaller than 0.8.

To the best of our knowledge, there is no experimental or theoretical data of the dipole moment of hydroxyl adsorbed on rutile TiO_2 (110). Assuming that the Ti^{3+} signal corresponds mainly to BOV in the case of the nonreconstructed surface its estimation might look straightforward: the parameter I_1 (which we address to the hydroxyl adsorption at BOV) for unreconstructed surface should be divided by the estimated relative BOV concentration obtained from the XPS measurements. By doing that we get the value of 0.95 D at room temperature which seems to be too high: H_2O adsorption on Ti rows of this system reduces the dipole moment of the molecule from 1.84 D to 0.47 D and the dipole moment of the OH bond is 1.51 D. One of the problems arises from the fact that not all BOV contribute to the measured work function change, since some of them are occupied by the water from the residual gas before the measurement actually began. Knowing that the base pressure is in the low 10^{-10} range, this effect can be significant [3]. However, its direct consequence would be underestimating the dipole moment. The main problem is actually related to the fact that the measured BOV concentration is averaged to the whole surface while the WF signal is related to the low work function areas where the concentration is certainly higher. An elegant way to resolve this problem is to heat the sample and increase the mobility of hydrogen atoms. After some time, depending on the temperature, the dipole distribution should become homogeneous. Indeed, this is what we observe: the calculated dipole moment drops to 0.8 D for 336 K and 0.6 D for 373 K (we did not observe any evidence of hydroxyl desorption for these temperatures). From the work function dependence on the sample temperature it seems that the convenient temperature would be about 423 K. Additional advantages of the sample heating would be strong suppression of the surface contamination as well as of the 2nd contribution.

In order to better understand the temperature dependence of the work function (Figure 6), we refer to the work of the Austin group concerning the hydrogen migration along the bridging oxygen rows of rutile TiO_2 (110)-(1x1) surface [19]. Based on their measurements and modeling the temperature dependence of the average hopping time can be calculated, which is presented in Figure 7. At the temperature of about 400 K the average hopping time drops to few tens of seconds, which probably provides enough mobility to obtain homogeneous surface distribution of hydroxyls in several minutes. Having in mind that this is the typical time scale in our experiments, the steep work function increase at 390 K (cf. Figure 6) quantitatively agrees very well with the results of the Austin group.

5. Summary

Water adsorption dynamics on rutile TiO_2 (110) surface has been monitored using the time evolution of the work function change for different sample temperatures. A model is presented which successfully describes the obtained data. XPS sample analysis provided estimation of the bridging oxygen vacancy concentration. Using the proposed model, three adsorption sites are identified: bridging oxygen vacancies in (1x1) and (2x1) reconstructed regions and the Ti rows. The upper limit of the abundance of (1x1) islands was estimated as well. The temperature dependence of the work function change can be very well explained by the investigations of the hydrogen mobility along bridging oxygen rows, done by the Austin group. Finally, on the basis of this investigation we propose a novel method to estimate the dipole moment of hydroxyls formed on the bridging oxygen rows.

References

[1] Diebold U 2003 *Surf. Sci. Report* **48** 53
[2] Pang C L, Lindsay R, Thornton G 2008 *Chem. Soc. Review* **37** 2328
[3] Wendt S et al. 2005 *Surf. Sci.***598** 226
[4] Dulub O et al 2007 *Science* **317** 1052
[5] Wendt S et al. 2006 *Phys. Rev. Lett.* **96** 066107
[6] Maksymovych P, Mezhenny S, Yates Jr J T 2003 *Chem. Phys. Lett.***382** 270
[7] Valentin C D et al. 2005, *J. Am. Chem. Soc.*, **127** 9895
[8] Topping J 1927, *Proc. R. Soc. London*, **A114** 67
[9] Marques H P, Canario A R, Moutinho A M C, Teodoro O M N D 2009 *Appl. Surf. Sci.* **255** 7389
[10] Teodoro O, Silva O, Moutinho A 1995, *Vacuum* **46** 1205
[11] Bachmann G, Oechsner H, Scholtes J, Fresenius 1987 *Z. Anal. Chem.* **329** 195
[12] Teodoro O M N D, Moutinho A M C 2002, *Key Engineering Materials* **230-232** 165
[13] Wagner C D et al. 1981 *Surf. Interface Anal.* **3** 211
[14] Seah M P 1999 *J. Elec. Spect. Rel. Phen.* **100** 55
[15] Crist B V 1999 *Handbooks of Monochromatic XPS Spectra* vol. 1, XPS International Inc.
[16] http://srdata.nist.gov/xps/default.aspxWendt et al. 2008 *Science* **320** 1755
[17] Wendt S et al. 2008 *Science* **320** 1755
[18] Asari E, Souda R, 1999 *Phys. Rev. B* **60** 10719
[19] Li S C et al, 2008 *J. Am. Chem. Soc.* **130** 9080

25th Summer School and International Symposium on the Physics of Ionized Gases—SPIG 2010 IOP Publishing
Journal of Physics: Conference Series **257** (2010) 012009 doi:10.1088/1742-6596/257/1/012009

Generation and probing of warm dense matter (Al) created by laser-accelerated proton beams

Ana Mančić

Faculty of Sciences and Mathematics, University of Niš, P. O. B. 224, 18000 Niš

anam@pmf.ni.ac.rs

Abstract. A study of isochoric heating of Al foil by laser-accelerated proton beam is presented. The proton source that induces the heating has been characterized in details. The heated sample conditions at different times during heating were inferred by coupling a time- and space- resolved interferometry diagnostic with a 1-dimensional hydrodynamic code (that includes proton stopping) that uses the previously characterized proton source as input. In this way, we have minimized the number of free parameters in the modeling of the heating. We have shown that heating was isochoric and almost uniform along the target thickness with a maximum temperature of ~ 18 eV. Finally, a X-ray Near Edge Absorption Spectroscopy (XANES) diagnostic has been used to study the structural modification of the warm dense aluminum sample. We have observed a progressive loss, within ~ 10 ps, of lattice ordering within solid density Al samples as the temperature rises from 300 K to > 10^4 K.

1. Introduction

The possibility of producing, in a controlled way, matter at solid density (1-10 g/cm^3) while maintaining it at high temperature (1-100 eV) has been for a long time a desired goal. At the boundary between condensed matter and plasma physics, the study of matter in the warm dense matter (WDM) [1] regime involves, nowadays, many laboratories worldwide. Indeed, the matter in this state is found a large array of physical environments: in astrophysical objects, in the transition from solid density matter to the plasma state found in numerous plasma generators and during the implosion of an inertial confinement fusion capsule. Accurate modeling of inertial plasmas requires physical data, equation of states and transport coefficients in this WDM regime where classical theories such as plasma kinetic theory and condensed matter theory fail as WDM is defined by ion-ion correlation parameter and Fermi degeneracy parameter both of the order of the unity. The lack of data in the WDM regime motivates the need for the design of experiments able to produce with high fidelity and very good characterization the warm dense matter sample in copious quantity.

Different methods and approaches have been used to produce matter in this regime: laser-driven shock heating [2], x-ray heating [3, 4], direct heating with subpicosecond laser pulses [5]. These methods, though relatively fast (10^{-9}-10^{-11} s) either produce heating that is longer than the typical hydrodynamic expansion time scale (10^{-12} – 10^{-11} s) or result in highly nonuniform heating through the sample.

Recent progress in laser matter interaction makes it now possible to produce high proton currents with protons energy spanning spectrum from few hundreds of electronvolt up to several tens of MeV [6]. These proton beams present a promising source for volumetric heating. What makes them good candidate for this is their (protons') longer penetration depth in matter in comparison to x-rays or laser light together with the fact that they deposit most of their energy at the end of the range in the material (so called *Bragg peak*). Also, being produced in a very short time (few ps), protons can heat the sample before it expands. The isochoric heating by laser-accelerated proton beam been demonstrated by P. Patel [7].

In this paper we present the results of a study of isochoric heating of Al foil by laser-accelerated proton beams. By coupling the experimental measurements of heated target expansion and self-consistent simulations of the same experiment, we have inferred the heated target conditions at different times during heating. In order to achieve this, we have characterized experimentally the proton source that induces the heating with the emphasis on the low energy part of the proton spectrum (< 10 % of the proton maximum energy, i. e. < 2 MeV in our case) [8] which plays a dominant part in the heating process (for target thickness of the order of μm) and which was not studied in detail in previous experiments conducted in similar conditions [9]. The experimental results have been compared with the results of 1-D hydrodynamic simulations of aluminium foil expansion that use the measured proton beam characteristics as an input for the energy deposition. In this way, we have eliminated free parameters in the modeling of the heating.

Once the induced heating was characterized, giving us an insight into the target conditions, we have studied the structural modification of the heated sample. We exploit the technique of ultrafast X-ray Absorption Near-Edge Spectroscopy (XANES) [10] which is related to the interpretation of the modulations observed above a photoabsorption edge. The XANES absorption spectrum is very sensitive to the local atomic arrangement and can provide direct insight into the structural properties of matter. X-ray photoabsorption was previously used in slow (ns) shock-compression experiments [11, 12] and calculations [13] but the main focus was on the position and the width of the K-edge as the density and temperature increased. However the full structure of the spectrum was not exploited to get structural information about the phase transition.

Experimentally, we observe the smoothing of the XANES structure of solid sample (aluminum) as it is isochorically heated from room temperature to several eVs. The observations are correlated to Quantum Molecular Dynamic (QMD) calculations as well as a simpler liquid description [14, 15, 16, 17], and good agreement, qualitative as well as quantitative, is found. Such comparison allows conclusively to relate the observed XANES spectral modifications to structural disordering of the heated material. We can thus, for the first time put an upper limit of ~10 ps on the loss of short range ordering during this solid to plasma transition.

2. Experimental set-up

The experiment was performed using the LULI 100 TW laser facility working in the chirped pulse amplification mode at a wavelength λ_0=1.057 μm. The experimental set-up is shown in figure 1(a). The main laser pulse had a duration of 320 fs and irradiated 10 μm thick Au flat foil ('proton source' target) at normal incidence with a peak intensity of ~3 x 10^{19} W/cm^2. This created a pulsed (few ps), highly directional, laminar proton beam containing about 10^{13} protons and having multi-MeV energies up to 20 MeV. This beam was incident on flat Al foils ('heated' target) of different thickness (ranging from 0.5 to 15 μm) which were placed 200 μm away from the proton source target. Protons coming from the 'proton source' target deposited energy within the 'heated' target. The energy deposition varies with incident proton energy and with the heated target thickness producing isochoric heating.

In order to characterize the proton source we used a magnetic spectrometer equipped with an image plate [18] (IP), which was absolutely calibrated as a proton detector [19], in its dispersion plane in order to obtain the continuous energy spectrum of the proton beam and stacks of radiochromic films [20] (RCF) to obtain the transverse characteristics of the beam [see figure 1(b)]. The IP was placed in the magnetic spectrometer in a way that enabled the measurement of the low energy part of the proton spectrum (down to E_{min} ~ 0.13 MeV), which plays a dominant role in the heating, as well of as the high energy one. Examples of measured proton spectra are shown in figure 2(a). Note that there is a shot to shot variation of the proton spectrum which can influence the heating. From the RCFs the data regarding the proton beam's divergence angle were obtained [figure 2(b)]. This represented the first characterization of the low-energy part of a high-energy laser-accelerated proton source [8].

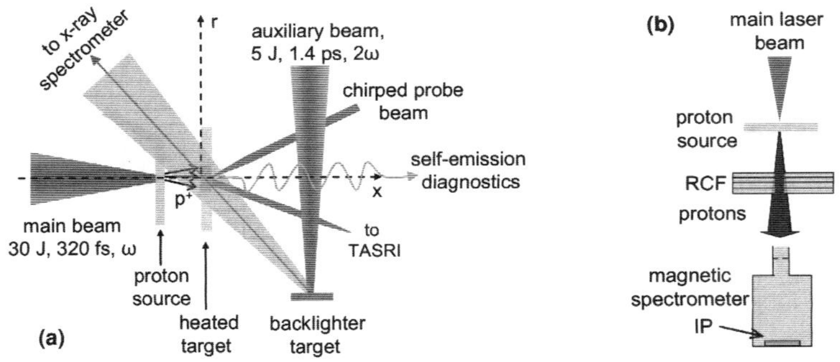

Figure 1. (a) Schematic of the experimental set-up, (b) Set-up for the proton source characterization.

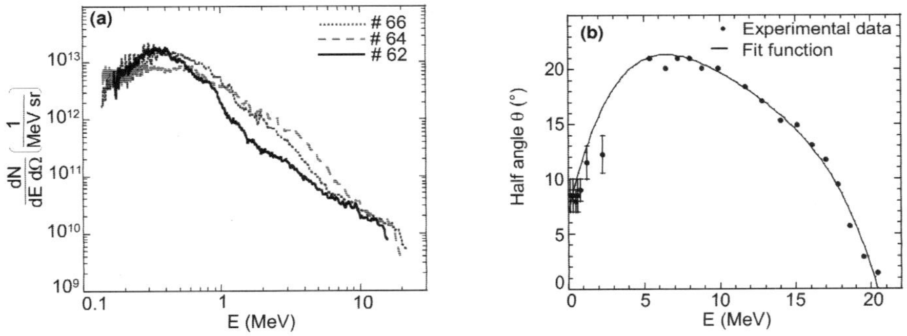

Figure 2. (a) Measured proton spectra from three different shots obtained by irradiating a 10 μm thick Au flat foil with a main laser beam; (b) Proton beam divergence half-angle versus proton energy. Dots represent the measured data that were fitted with a function of a type (energy = x): Halfangle(x)=$p_0+p_1\cdot x+p_2\cdot x^2+p^3\cdot x^3+p_4\cdot x^4$.

The main diagnostic used to characterize the induced heating by the proton beam was time- and space- resolved interferometry [21] (TASRI) of a chirped probe beam (1.057 μm) of 50 ps duration, reflecting from the rear (nonirradiated) surface of the 'heated' target [8]. It allows us to measure the expansion velocity of the heated target critical density interface. For the targets that are heated, since the probe is reflected from the expanding rear surface, there is an induced phase shift compared to the case when the beam is reflected from an unperturbed surface. This shift is related to the induced heating in the target. Two other diagnostics were implemented in the experiment: time- and space-resolved self-emission measurement of the heated target in the visible range (450-600 nm) [22], and X-ray absorption spectroscopy.

3. Accurate description of warm dense Al

To assess the warm dense matter condition of the heated aluminum sample, we have adopted the methodology that is illustrated in figure 3: from the TASRI diagnostic, we retrieve information about the phase-shift of the probe beam reflecting from the rear surface of the heated target, while from the self-emission diagnostic, we retrieve the information about the emissivity that has been collected.

Then, we perform simulations using as input the proton spectra measured during the same experiment. At the output, we obtain heated target conditions as well as rear surface expansion velocity that can be compared to the measured ones. Also, using the calculated temperature, density and ionization degree we can calculate the emissivity that should be produced by this target. This computed emissivity is then compared with the measured emissivity.

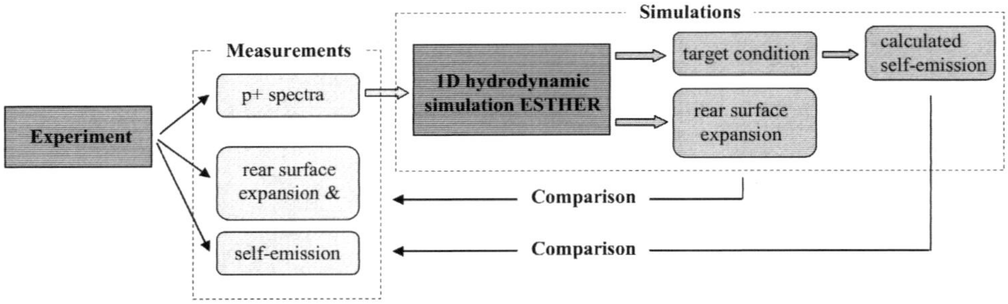

Figure 3. Schematic of the methodology adopted to retrieve the heated target condition using both a simulation code and diagnostics.

We have performed simulations of the heating and expansion of a thin aluminium foil irradiated by a laminar high-energy proton source using a one-dimensional hydrodynamics code ESTHER [23]. This code solves, according to a Lagrangian scheme, the fluid equations for the conservation of mass, momentum and energy. The target material is described either by the Bushman-Lomonosov-Fortov (BLF) multiphase equation of state (EOS) [24] or the SESAME EOS [25]. We sample the experimentally measured proton spectra [see figure 2(a)] and use it as an input parameter in the simulation. For the proton stopping power, the values given by the SRIM database are used [26]. The simulation assumes electron-ion equilibrium, i. e. $T_e=T_i=T$. A laser probe diagnostic is implemented in the code in order to model the TASRI diagnostics. Thus, we obtain, as in the experiment, the phase after reflection off the heated target surface. For this post-processing of the hydrodynamic data the interaction between the laser field and the target is given by the solution of the Helmholtz wave equation. Details on the simulations parameters can be found in [8].

Figure 4 shows the result of the simulation of heating of a 10 μm thick Al target. Density normalized to the solid density and temperature profiles are presented for different simulation times (time '0' corresponds to the moment when first, most energetic protons leave the proton source target). "Warm dense matter" conditions, i.e., degeneracy parameter equal to one half and ion-ion correlation parameter of the order of two, are obtained after 25 ps [8]. It can be seen that the bulk of the target stays at solid density while the density decreases at the edges due to expansion induced by heating. Also, in the early stage, one can notice a small shock in the density profile facing the incident proton beam that is due to energy deposition in a thin layer of the target by electrons co-moving with proton beam [27]. Note that these simulations are performed self-consistently using, where possible, measured parameters as inputs.

In order to assess the validity of these simulations, we have compared the measured probe beam phase shift with the simulated one and found a good agreement between the two [8]. This is illustrated in figure 5(a) for 10 μm thick Al target where several experimental probe beam phases obtained from different shots are plotted. The fluctuation of these phases is likely resulting from shot-to-shot fluctuations of the incident proton beam irradiating the target [see figure 2(a)]. Indeed, simulations that have been performed using as input a range of measured proton spectra, which were measured on separate shots, give the shaded areas in figure 5(a) that indicate the resulting range of probe beam phases derived from the simulations. We observe that the simulated phase spread is of the same order as the experimental phase spread. There are two shaded areas corresponding to the two equations of state that can be used in the code (the BLF EOS -area with vertical lines and the SESAME table - area with diagonal lines).

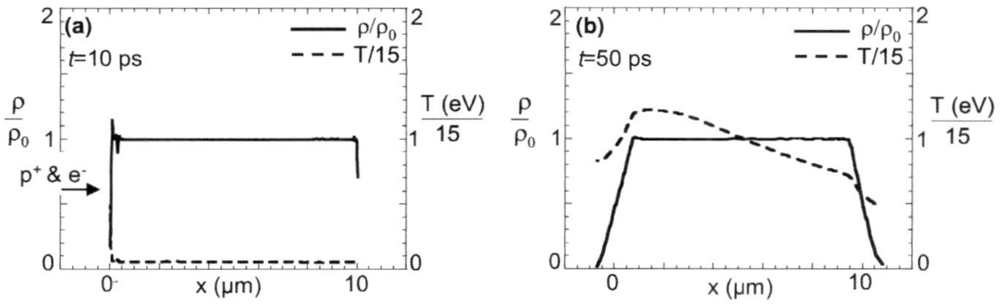

Figure 4. Density (solid line) normalized to the solid density and temperature (dotted line) profiles (normalized to the maximum value 15 eV in the target center) as a function of target thickness plotted for different simulation times: (a) 10 ps; (b) 50 ps. The time '0' corresponds to the moment when first protons leave the 'proton source' target.

Similarly, the measured values of emissivity in time can be compared to the emissivity calculated by using the simulated target conditions at different times by the code ESTHER and the equation of radiative transfer in an expanding plasma [8]. This is illustrated in figure 5(b) for a 10 μm thick Al target. A difference between the measured and calculated emissivity can be attributed to the fact that the target was modeled as a black body radiating at a certain temperature where in reality the target is partially reflective and not perfectly absorbing.

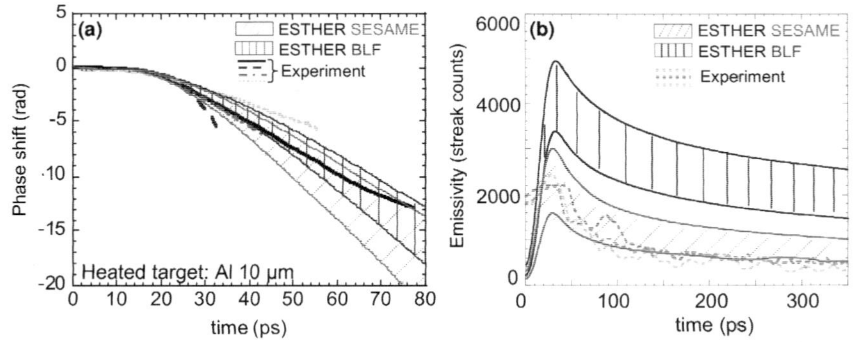

Figure 5. Comparison of the measured and computed (with two different EOS describing the target material) probe beam phase shift (a) and emissivity (b) from the rear surface of the heated 10 μm thick Al target (b). Shaded area with vertical lines corresponds to BLF EOS, area with diagonal lines corresponds to SESAME EOS.

This work is an advance on previous studies where the proton energy deposition was not experimentally measured but, rather, fit to the heating diagnostics measurements. While in [28] the proton spectrum was scaled to reproduce the measured temperature and then used as an input to hydrodynamic simulation, here the measured proton spectrum was directly used as an input for the simulation and then the output compared with the measurements. Similarly, in [7], it was shown that in the proton beam there is enough energy to induce isochoric heating of a material, while the temperature was deduced by comparing the measured rear surface emission with the simulated one assuming the material to be heated to some given initial temperature.

Journal of Physics: Conference Series **257** (2010) 012009 doi:10.1088/1742-6596/257/1/012009

4. Time-resolved study of structural properties of cold to warm dense matter transition

Once the proton source and the induced heating were characterized, giving us an insight into the target conditions, we studied the local atomic structure of warm dense aluminium sample using X-ray Absorption Near-Edge Spectroscopy (XANES).

The ultrashort broadband x-ray probe was created with a synchronized (with a main laser pulse), 5 J energy, 1.4 ps duration, 0.53 μm wavelength beam focused on an erbium (Er) target. To collect the transmitted spectra through the heated sample and the direct emission of the x-ray source (and deduce absorption curves), a specific x-ray absorption spectrometer (presented in figure 6) composed of two distinct conical KAP crystals was used. This configuration releases the results from shot-to-shot fluctuations of the x-ray source. The thickness of the Al sample (d=0.5 or 1.6 μm) was chosen to optimize the signal-to noise ratio in the absorption spectra. Experimental constraints required averaging the x-ray data radially over 170 μm in the heated target, starting from the axis of the proton beam irradiation (with an error of +/- 30 μm). As it was shown, a maximum temperature of 15 eV at solid density could be reached, however the mean temperature integrated over the radial dimension (normal to the axis of irradiation) of the heated area is in the eV range [29].

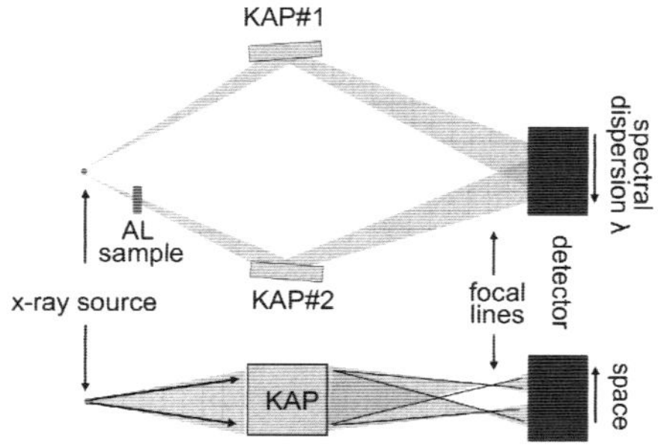

Figure 6. Schematic of the double x-ray spectrometer.

Controlling the delay between the two laser pulses, we probed the sample in different states. Due to shot-to-shot fluctuations, the use of two different Al sample thicknesses and proton beam variations, we collected the XANES spectra for different temperatures. Some results are presented in figure 7(a). The averaged temperatures of the sample are estimated using a simple model based on Fermi-Dirac statistics where a one-to-one correspondence between the *K* edge slope and the temperature is found [29].

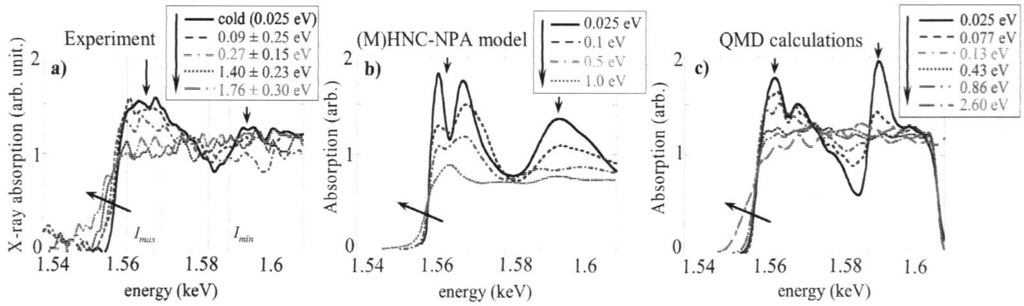

Figure 7. (a) Experimentally measured XANES spectra near the K-edge. The spectra are normalized to the mean value without the XANES structure. The 'cold shot' corresponds to a shot without heating proton beam; (b-c) Calculated K-edge XANES spectra (normalized to the mean value without XANES structures) of aluminum as a function of the temperature (solid density): (b) (M)HNC-NPA model, (c) QMD data.

The measured spectra were compared to two theoretical approaches of modelling. The first one is a liquid-metal approach based on the Neutral Pseudoatom (NPA) concept of dense matter coupled with a specific finite difference modeling of XANES structures [14]. The second one uses QMD simulations in the framework of density functional theory [16, 17]. The absorption spectra calculated for different temperatures using these two approaches are given in figure 7(b) and (c). Both the experimental results and the calculations exhibit a K edge broadening with increasing temperature and the progressive smoothing of the absorption spectra just after the K edge. According to the theory of XANES [30], the modulation observed in the spectrum is directly related to short-range order of the probed material. Consequently, the smoothing of the XANES structures is due to a progressive loss of ion-ion correlation.

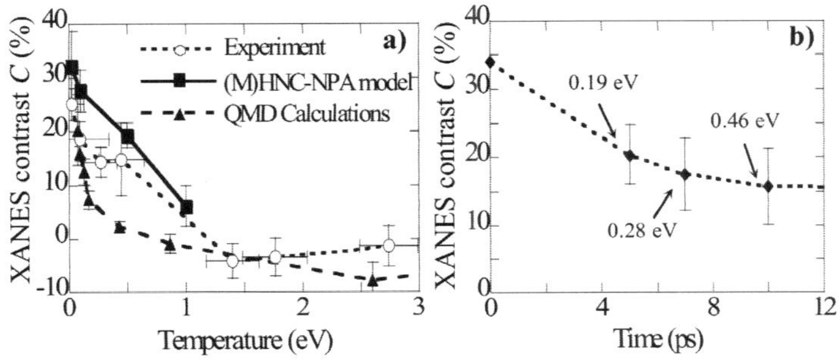

Figure 8. Contrast C(%) of the XANES modulations (a) as a function of the temperature for experimental and theoretical data and (b) as a function of time after the first protons have left the proton source (the arrival of the first protons in the heated Al target starts at t ~ 3 ps) with indicated average temperatures.

To quantitatively compare experimental and calculated XANES profiles, we have calculated a 'contrast' C for each profile as a function of temperature: $C(\%)=(I_{max} - I_{min})/(I_{max} + I_{min})$, where I_{max} and I_{min} are average values of the absorption signal around the first XANES maximum and minimum, respectively. Results are reported in figure 8(a). The evolution of both experimental and calculated contrasts is similar. The information about the temporal dynamics of the observed ordering loss we obtain by following in time the simulated target conditions (using the ESTHER code) and use them to compute averaged XANES spectra similarly as in the experiment. For this, we exploit the QMD spectra calculated at solid-density for various temperatures and calculate the contrast of the simulated average spectrum at each time (details are given in [29]), as shown in figure 8(b). The contrast falls rapidly, in 10 ps, while the average temperature increases. At this point, according to the theoretical models, the medium becomes structure-less and we can put an upper bound of 10 ps on the loss of short range ordering [29].

5. Conclusion

In summary, we have presented the results of the study on characterization of the warm dense aluminum plasma created by laser-accelerated proton beam. We have demonstrated that we can produce large volumes (~ 10^6 μm^3) of well defined warm dense matter as well as that it can be used to study the ionic structure of this state by x-ray absorption measurements.

Using a previously characterized proton source as input, the target heating and expansion were simulated self-consistently, allowing a direct comparison with experimental measurements. In this way, we have minimized the number of free parameters in the modeling of the heating. We have shown that heating was isochoric and almost uniform along the target thickness.

Using ultrafast x-ray probing, we have experimentally observed a progressive loss of ordering within solid density sample as the temperature rises from 300 K to > 104 K. The loss of short-range ordering was detected through the progressive smoothing of the time-resolved X-ray Absorption Near Edge Spectroscopy (XANES) structure. The observation were correlated to Quantum Molecular Dynamic calculation as well as a simpler liquid description and good agreement, qualitative as well as quantitative was found. Such comparison allowed to conclusively relate the observed XANES spectral modifications to structural disordering of the heated material.

The next step would be to study the stopping power of ions in WDM which is still not well known and it is very important for the inertial fusion research. As the electron structure is known to play an important role in SP, we expect to observe variations in the stopping power.

References

[1] Lee R W, Moon S J, Chung H-K *et al.* 2003 *J. Opt. Soc. Am.* B **20**, 1
[2] Huser G *et al* 2005 *Phys. Plasmas* **12**, 060701
[3] Zhang J *et al.* 2009 *Phys. Rev.* E **79**, 016401
[4] Rochau G A, Baley J E and MacFarlane J J 2005 *Phys. Rev.* E **72**, 066405
[5] Audebert P *et al.* 2002 *Phys. Rev. Lett.* **89**, 265001
[6] Borghesi M, Fuchs J, Bulanov S V, Mackinnon A J, Patel P and Roth M 2006 *Fusion Sci. Technol.* **49**, 412
[7] Patel P K, Mackninnon A J, Key M H *et al.* 2003 *Phys. Rev. Lett.* **91**, 125004
[8] Mančić A *et al.* 2010 *High Energy Density Phys.* **6**, 21
[9] Cuneo M, private communication
[10] Bressler C and Chergui M 2004 *Chem. Rev.* **104**, 1781
[11] Godwal B K *et al.* 1989 *Phys. Rev.* A **40**, 4521; Bradley D K *et al.* 1987 *Phys. Rev. Lett.* **59**, 2995
[12] Hall T A *et al.*1998 *Europhys. Lett.* **41**, 495
[13] Perrot F and Dharma-wardana M W C 1993 *Phys. Rev. Lett.* **71**, 797
[14] Peyrusse O 2008 *J. Phys.: Condens. Matter* **20**, 195211
[15] Peyrusse O *et al.* 2009 *Atomic Processes in Plasmas*, edited by K. Fournier, AIP Conf. Proc. No. 1161 (AIP, New York, 2009), p. 200..
[16] Mazevet S and Zérah G 2008 *Phys. Rev. Lett.* **101**, 155001
[17] Recoules V and Mazevet S 2009 *Phys. Rev.* B **80**, 064110
[18] http://www.fujifilm.com/products/life_science/si_imgplate/whatis02.html;
[19] Mančić A, Fuchs J, Antici P, Gaillard S A and Audebert P 2008 *Rev. Sci. Instrum.* **79**, 073301
[20] Klassen N V, Zwan L and Cygler J 1997 *Med. Phys.* **24**, 1924; http://www.ispcorp.com/products/dosimetry/content/gafchromic/index.html.
[21] Geindre J P *et al.* 2001 *Optics Lett.* **26**, 1612
[22] Kodama R, Okada K and Kato Y 1999 *Rev. Sci. Instrum.* **70** 625
[23] Colombier J P, Combis P, Bonneau F, Harzic R Le and Audouard E 2006 *Phys. Rev.* B **71**, 165406
[24] Lomonosov I N, Bushman A V, Fortov V E 1993 *High Pressure Science and Technology*, edited by Schmidt S C, Shaner J W, Samara G A and Ross M, AIP Conf. Proc. No. 309 (AIP, New York 1994) 117.
[25] Holian K S 1986 *J. Appl. Phys.* **59**, 149; Lyon S P, Johnson J D 1992 "SESAME: The Los Alamos National Laboratory Equation of State Database", LA-UR-92-3407
[26] Ziegler J F, Biersack J P and Littmark U 1996 *The Stopping and Range of Ions in Solids* (Pergamon, New York)
[27] Robiche J and Rax J M 2004 *Phys. Rev.* E **77**, 046405 (2004); Brambrink E, Schlegel T, Malka G *et al.* 2007 *Phys. Rev.* E **75**, 065401(R)
[28] Dyer G M *et al.*2008 *Phys. Rev. Lett.* **101**, 015002
[29] Mančić *et al.* 2010 *Phys. Rev. Lett.* **104**, 035002
[30] Rehr J J and Albers R C 2000 *Rev. Mod. Phys.***72**, 621

Journal of Physics: Conference Series **257** (2010) 012010 doi:10.1088/1742-6596/257/1/012010

Formation and decay of the Rydberg states of multiply charged ions interacting with solid surfaces

M. A. Mirković[1], **N. N. Nedeljković**[1], **and D. K. Božanić**[2]

[1] *University of Belgrade, Faculty of Physics, P.O. Box 368, Belgrade, Serbia*

[2] *Vinča Institute of Nuclear Sciences, P. O. Box 522, Belgrade, Serbia*

E-mail: gmirkomarko@sezampro.rs

Abstract. Processes of formation and decay of the Rydberg states of multiply charged ions escaping solid surfaces with intermediate velocities ($v \approx 1$ a.u.) represent complex quantum events that require a detailed quantum description. We have developed a two-state vector model for the population process, with the functions Ψ_1 and Ψ_2 for definition of the state of a single active electron. The electron exchange between the solid and the moving ion is described by a mixed flux through a plane positioned between them. For the low values of the angular momentum quantum numbers l the radial electronic coordinate ρ can be neglected, whereas for the large-l values a wide space region around the projectile trajectory was taken into account. The reionization of the previously populated states is considered as a decay of the wave function Ψ_2. The corresponding decay rates are obtained by an appropriate etalon equation method: in the large-l case the radial electronic coordinate ρ is treated as a variational parameter. The theoretical predictions based on that population-reionization mechanism are compared with the available beam-foil experimental data, as well as the experimental data obtained in the interaction of multiply charged ions with micro-capillary foil. Generally, the model reproduces the experimentally observed non-linear trend of the l distributions from $l = 0$ to $l_{\max} = n - 1$.

1. Introduction

It is well known that the Rydberg states of the multiply charged ions can be populated during the interaction with conducting solid surface. The beam-foil type of experiments [1-3] performed at intermediate velocities of the ionic projectiles ($v \approx 1$ a.u.), provide non-linear electron capture probability distributions for both principal and angular momentum quantum numbers, n and l, of the final bound states (n, l). Recently, a new type of experiment [4] for the multiply charged ions interacting with micro-capillary foils provided the evidence of the l distributions in the large-l region. A series of experimental data suggests that the formation of Rydberg states of multiply charged ionic projectiles escaping from solid surfaces represents a complex quantum event that can be considered as a specific interplay of the neutralization and reionization processes.

Most of the theoretical studies performed so far have not been focused on the cited experiments. The neutralization (electron capture) and ionization (recapture) are treated separately using rather different theoretical tools such as the classical over-barrier model [5, 6] and its extended dynamic version [7, 8], the standard perturbative quantum descriptions [9, 10], the non-perturbative coupled-angular-mode method [11], the complex scaling method [12], the stabilization method [13] and the time dependent close-coupling technique [14]. Recently, the wave packet propagation simulation [15, 16] has also been proposed. A common feature of these models is the use of a single wave function for the description of the electron behavior at intermediate stages of the ion-surface interaction. Also, the models are devoted mainly to the low ionic velocities ($v \ll 1$ a.u.). On the other hand, the theoretical analysis [17] (classical transport theory) of the l distributions concerns the Rydberg-state population of multiply charged ions traversing foil bulk at high velocities ($v \gg 1$ a.u.).

In our theoretical treatment of the process, two specific models are used for the population of the low-l and the large-l Rydberg states, and both the neutralization and reionization are taken into account. The probability distributions for the low-l Rydberg states ($l = 0, 1$ and 2) have been obtained within the framework of two-state-vector model (TVM) [18, 19]. The model has also been extended to the case of large-l Rydberg states ($l \geq 3$) [20]. Using the TVM, our prime intention has been to learn information about the behavior of the active electron which left the solid surface at the time t_{in}, but which will be detected in the final bound state (n, l) at the time t_{fin}. To this end, we describe the single active electron simultaneously using the wave function $\Psi_1(\vec{r}, t)$, evolving from a fixed initial state, and an additional "teleological" wave function $\Psi_2(\vec{r}, t)$, evolving causally towards a fixed final state (n, l) of the ionic projectile [18-23]. In particular, such a description is relevant for the intermediate velocity region ($v \approx 1$) where the adiabatic approximation breaks down [18, 19]. Also, the concept of the mixed flux through the moving "Firsov plane" S_F, used in the calculation of the transition probabilities, offers the possibility to treat the functions $\Psi_1(\vec{r}, t)$ and $\Psi_2(\vec{r}, t)$ in the "interaction-free" region which is far from both the ion and the surface, for large ion-surface distances R. As a result we obtain the intermediate neutralization probabilities $P^N_{\nu_A}(t)$ and neutralization rates $\Gamma^N_{\nu_A}(t)$, where ν_A is the set of the intermediate spherical quantum numbers.

In the Refs. [18], [19] and [20] we considered the electron exchange exclusively as an electron capture from the foil conduction band into the moving Rydberg state (n, l) of the ionic projectile, i.e., the reionization was neglected. In few cases the experimental lower-n values of the electron capture probability $P_{nl} = \lim_{t \to t_{fin}} P^N_{\nu_A}(t)$ near $l = l_{max} = n - 1$ have been overestimated by the proposed model. Moreover, in the case of larger-n states the theory predicted an oscillatory l-behavior of the P_{nl} curves with large amplitudes in the large-l region. Just in this region, the absence of the experimental data in the beam-foil geometry [1, 2, 3] has been previously considered as the existence of the thresholds at $l = l_{thr} < l_{max}$. However, the new experiments with microcapilary foil [4] give the nonzero population probabilities in this region, but with lower values then those obtained in the cited TVM.

The limitations of the TVM are removed by taking into account the fact that the electron capture process is in competition with the reionization at intermediate stages of the ion-surface interaction. Effects of reionization in the low-l case have been included in the population process in Ref. [22]. The reionization of the large-l Rydberg states can be described by the same method as in the low-l case: the wave function $\Psi_2(\vec{r}, t)$ has to be "renormalized", i.e., multiplied by the corresponding decay factor $\mathcal{E}_{\mu_A}$, where μ_A denotes the set of intermediate parabolic quantum numbers (which give the main contribution to the electron capture process) [20],[22]. The decay factor $\mathcal{E}_{\mu_A}$ is expressed via the ionization rate $\Gamma^I_{\mu_A}$, which is calculated using the complex eigenenergy model combined with the etalon equation method (EEM). The parabolic symmetry in the low-l case appears as a consequence of the fact that the radial electronic coordinate ρ can be neglected in comparison to the ion-surface distance R. In contrast to the low-l case, in the case of large-l Rydberg states, the electron transitions are mainly localized far from the ionic trajectory. Within the framework of the large-l EEM, the quantity ρ is considered as a variational parameter; the parabolic symmetry is preserved in this procedure. Using the large-l reionization rate $\Gamma^I_{\mu_A}$, we calculate the relevant decay factor $\mathcal{E}_{\mu_A}$, from which the experimentally verifiable "renormalized" population probabilities $\overline{P}_{nl} = \mathcal{E}^2_{\mu_A} P_{nl}$ can be find.

2. The renormalized population probabilities $\overline{P}_{nl}$

2.1. The population-reionization mechanism

We consider the population-reionization (capture-recapture) process for Rydberg states of multiply charged ions escaping solid surfaces with velocity v, Fig 1(a).

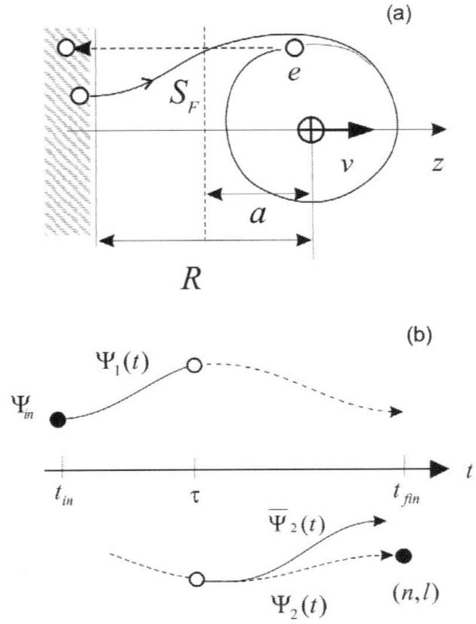

Figure 1. (a) Schematic description of the population-reionization mechanism for the Rydberg states of multiply charged ions escaping the solid surface and (b) the corresponding TVM-scheme. The spherical shape of the formed Rydberg state corresponds to the large-l case. In the low-l case the electronic state is characterized by the large eccentricity. In the large-l case the neutralization process (electron capture from the solid into the ionic field) can be suppressed by the reionization (electron recapture by the solid). Accordingly, if the reionization is remarkable, only a fraction of the formed Rydberg states survives and can be detected as a final state (n, l) at the time $t = t_{fin}$. The reionization and neutralization processes begins practically at the same ion-surface distances, $R = R_\tau \approx R_g$.

The process under consideration has the following TVM interpretation [20, 22, 23], Fig. 1(b). The state $\Psi_1(\vec{r}, t)$ evolves from an initial state Ψ_{in} labeled by the set of parabolic quantum numbers $\mu_{M,in}$ at the time $t_{in} = 0$. The population process represents a "transition" $\Psi_1(\vec{r}, t) \to \Psi_2(\vec{r}, t)$ at the intermediate time $t \in (t_{in}, t_{fin})$. If the reionization could be neglected, the state $\Psi_2(\vec{r}, t)$ would evolve into a final bound state determined by the spherical quantum numbers $\nu_{A,fin} = (n, l, m)$ at the time $t_{fin} \to \infty$. The reionization is included in the model by introducing the "renormalized" wave function $\overline{\Psi}_2(\vec{r}, t) = \mathcal{E}_{\mu_A}(t)\Psi_2(\vec{r}, t)\Theta(z) + \chi(\vec{r}, t)\Theta(-z)$ for $t > \tau$, where $\tau = R_\tau/v$ [22]. For $t < \tau$ we take $\overline{\Psi}_2(\vec{r}, t) = \Psi_2(\vec{r}, t)$. The regions inside and outside the solid ($z < 0$ and $z > 0$) are defined by the Heaviside function $\Theta(z)$. The function $\chi(\vec{r}, t)$ stands for the outgoing wave inside the solid.

The state $\overline{\Psi}_2$ represents a decaying state, with a decay factor

$$\mathcal{E}_{\mu_A}(t) = \exp\left[-\frac{1}{2}\int_\tau^t \Gamma^I_{\mu_A}(t)dt\right], \tag{1}$$

where $\Gamma^I_{\mu_A}(t)$ is the reionization rate for the relevant intermediate parabolic quantum numbers $\mu_A = \{n_1 = 0, n_2 = n - 1, m = 0\}$. The experimentally verifiable renormalized population probability $\overline{P}_{nl}$ is given by [22]

$$\overline{P}_{nl} = \mathcal{E}^2_{\mu_A} P_{nl}, \tag{2}$$

where $\mathcal{E}^2_{\mu_A} = \lim_{t\to\infty} \mathcal{E}^2_{\mu_A}(t)$, whereas P_{nl} represents the electron capture probability (without reionization) into the final Rydbeg state (n, l).

2.2. Calculation of P_{nl}

In order to calculate the population probability P_{nl} within the framework of TVM, we define the two-state probability amplitude $A_{\mu_M,\nu_A}(t) = \langle\Psi_2(t)|\hat{P}_A(t)|\Psi_1(t)\rangle$, where $\hat{P}_A(t) = \int_{V_A} |\vec{r}_A\rangle\langle\vec{r}_A|dV$ is the projection operator onto the ionic region $V_A = \{z; z \geq R - a\}$, see Fig. 1(a), and calculate the intermediate neutralization probability $P^N_{\nu_A}(t)$; the final probability P_{nl} follows for $t \to \infty$ [20, 23].

The intermediate neutralization probability $P^N_{\nu_A}(t)$ is defined as an integral over the solid conduction band energy parameter γ_M and a sum over the first parabolic quantum number n_{1M} and magnetic quantum number m_M of the quantity $|A_{\mu_M,\nu_A}(t)|^2$:

$$P^N_{\nu_A}(t) = \int \sum_{n_{1M},m_M} |A_{\mu_M,\nu_A}(t)|^2 d\gamma_M. \tag{3}$$

The two-state probability amplitude can be expressed via mixed flux $I_{\mu_M,\nu_A}(t)$ through the Firsov plane S_F (with surface element $d\vec{S}$) separating the region V_A from the complementary region:

$$A_{\mu_M,\nu_A}(t) = \int_{t_g=R_g/v}^{t} I_{\mu_M,\nu_A}(t)dt, \tag{4}$$

$$I = \frac{i}{2}\int_{S_F}\left[\frac{\vec{\nabla}\Psi_1}{\Psi_1} - \frac{\vec{\nabla}\Psi_2^*}{\Psi_2^*} - 2i\vec{v}\left(1-\frac{da}{dR}\right)\right]\Psi_2^*\Psi_1 \cdot d\vec{S}. \tag{5}$$

The kinematics of the Firsov plane is determined by the variational requirement $\delta P_{\nu_A}(t)/\delta a(t) = 0$ where $a = a(t)$ is the position of the S_F plane with respect to the ion. The former condition reflects the fact that the probability $P^N_{\nu_A}(t)$ should be independent of the small variations of the S_F plane position.

The wave functions Ψ_1 and Ψ_2 can be expressed as space-time modifications of the eigenfunctions Φ_{MA,μ_M} and Φ_{AM,ν_A} of the Hamiltonians $\hat{H}_i$, $i = 1, 2$, of the first and second scenarios, respectively [18, 19, 23]: $\Psi_1(\vec{r}, t) = \Phi_{MA,\mu_M} \exp\left(f_M + i\gamma_M^2 t/2\right)$ and $\Psi_2(\vec{r}, t) = \Phi_{AM,\nu_A} \exp\left(ivz - iv^2t/2\right) \exp\left(f_A + i\gamma_A^2 t/2\right)$. By $f_M = f_M(\vec{r}, t)$ and $f_A = f_A(\vec{r}, t)$ we denoted the corresponding correction factors, whereas $\exp(ivz - iv^2t/2)$ represents the Galilei factor. The explicit expressions of the eigenfunctions Φ_{MA,μ_M} and Φ_{AM,ν_A}, and the correction factors f_M and f_A are obtained using the different asymptotic methods for the low-l ($l = 0, 1, 2$) and the large-l ($l \geq 3$) cases.

Considering the electron capture into the low-l Rydberg states [18, 19, 23], it is plausible to assume that the mixed flux through the central part of the Firsov plane give the main contribution to the transition amplitude. That is, in the low-l case the ionic Rydberg states have large eccentricities so that only a narrow cylindrical region around the ionic trajectory becomes relevant. The functions $\Psi_1(\vec{r}, t)$ and $\Psi_2(\vec{r}, t)$ are obtained by an analytic continuation from the regions where they are mainly localized (the solid region and the ionic region, respectively) to the central part of the Firsov plane. A generalization of the low-l TVM to the large-l Rydberg systems has been performed in Ref. [20]. Taking into account that the large-l Rydberg states are characterized by very small eccentricities it was assumed that the relevant area of the Firsov plane extends far from the ionic trajectory. This fact has been taken into account in the calculation of the wave function $\Psi_1(\vec{r}, t)$; namely, the JWKB form of the function Φ_{MA,μ_M} (analytically continued from the metallic region into the ionic region) valid in the central part $\rho \approx 0$ of the S_F plane is "bifurcated" on the whole Firsov plane by the appropriate matching procedure.

2.3. Calculation of $\mathcal{E}_{\mu_A}$

From Eq. (1) we see that the problem of reionization characteristic for the population of the large-l Rydberg states is reduced to the evaluation of the decay factor $\mathcal{E}_{\mu_A}(t)$, i.e., the rate $\Gamma^I_{\mu_A}(t)$.

By considering the complex energy eigenvalue problem of the of the Hamiltonian $\hat{\overline{H}}_2(t)$

$$\hat{\overline{H}}_2(t)\Phi_{AM,\mu_A} = E_A\Phi_{AM,\mu_A}, \tag{6}$$

one get the energies

$$E_A(R) = \mathrm{Re}E_A(R) - i\Gamma^I_{\mu_A}(t)/2; \quad (7)$$

from the imaginary part of the energies E_A we obtain, directly, the parabolic ionization rates $\Gamma^I_{\mu_A}$. The Hamiltonian $\hat{\overline{H}}_2$ differs from the final-channel Hamiltonian $\hat{H}_2$ used in the absence of reionization by the term $-U_0\Theta(-z)$, where U_0 is the depth of the potential well of the solid. The appropriate method of solving the eigenvalue problem expressed by Eq. (1) is the EEM.

The parabolic symmetry of the eigenvalue problem (8) in the low-l case appeared as a consequence of the fact that the radial electronic coordinate ρ could be neglected in comparison to the ion-surface distance R [22]. In contrast to the low-l case, in the case of large-l Rydberg states the electron transitions are mainly localized far from the ionic trajectory. In that case the parabolic symmetry is preserved for the quantity $\rho = \bar{\rho}$ treated as a variational parameter. The problem of calculation of the corresponding rate $\Gamma^I_{\mu_A}(t)$ can be performed in two steps. In the first step we calculate the rates $\Gamma^I_{\mu_A}(\bar{\rho}, R)$ with $\bar{\rho}$ as a parameter. In the next step, we calculate the quantity $\bar{\rho} = \bar{\rho}_0(\mu_A; R)$ using the variational requirement: $\delta\Gamma^I_{\mu_A}(\bar{\rho}, R)/\delta\bar{\rho} = 0 \Leftrightarrow \bar{\rho} = \bar{\rho}_0(\mu_A; R)$. In that way we obtain the rates $\Gamma^I_{\mu_A}(\bar{\rho}_0(\mu_A; R), R) = \Gamma^I_{\mu_A}(t)$. The low-$l$ results follow as a limiting case $\bar{\rho} \to 0$ of large-l case.

With the parabolic symmetry preserved, one can separate the variables in Eq. (8) in parabolic coordinates $\xi = r_A + z_A$, $\eta = r_A - z_A$ and φ, where the index "A" is addressed to the quantities defined in the coordinate system located at the ionic core. Bearing in mind the relation $\rho^2 = \xi\eta$, we get two one-dimensional effective eigenvalue problems (along the ξ and along the η directions), which both depend on the variational parameter $\bar{\rho}$. The large-l intermediate Rydberg system (formed in the neutralization process) is almost spherical and large; thus, its reionization is related to electron transitions far from the ionic trajectory ($\bar{\rho}_0 \gg 1$). The effective one-dimensional problems of the low-l reionization model [22] follow in the limit $\bar{\rho} \to 0$ from the corresponding large-l expressions. In both cases we use the scaled parabolic coordinates $\tilde{\xi} = \xi/2R\alpha_\xi$, $\tilde{\eta} = \eta/2R\alpha_\eta$ and φ, in order to obtain the canonical forms of the one-dimensional eigenvalue problems suitable for application of the EEM. We note that the scaling parameters α_ξ and α_η are different in the large-l and the low-l cases.

2.4. The final population probability $\overline{P}_{nl}$ and comparison with experiment

The electron exchange during the intermediate stages of the ion-surface interaction results in the formation of the final Rydberg system (n, l) at $t_{fin} \to \infty$. The experimentally verifiable population probability $\overline{P}_{nl}$, Eq. (2), can be calculated explicitly for all relevant values of the ion-surface parameters. The theoretical predictions are be compared with available experimental data for the ions ClVII and ArVIII. In presenting the theoretical results we shall focus on the l distributions, using the same normalization of the experimental data as in Ref. [20].

In Figs. 2 and 3 we present the renormalized population probabilities $\overline{P}_{nl}$ obtained by inclusion of the reionization (full curves) and the probabilities P_{nl} (dashed curves) [19],[20] for the ions ClVII (core charge $Z = 7$) and ArVIII (core charge $Z = 8$) with $7 \leq n \leq 12$ escaping the solid surface at velocity $v = 2.50$ a.u. and $v = 1.42$ a.u., respectively. The pronounced peaks of the P_{nl} distributions in the large-l region are lowered due to the reionization; note the existence of the threshold value $n = n_{thr} = 9$ in the low-l region, i.e. $P_{nl} = 0$ for $n > n_{thr}$ in the case of ClVII. The properly normalized beam foil experimental data from Ref. [1, 2] and Ref. [3] for ClVII and ArVIII ions, respectively, are presented by dots.

The experimental data for neutralization of the Ar^{7+} ion ($Z = 7$), and Ar^{8+} ion ($Z = 8$) obtained in the presence of microcapillary foil [4] are presented by circles. The available experimental data [4] are for $v \approx 0.2$ a.u., so that the experimental results are scaled to the intermediate velocity case (and properly normalized). We recognize that the proposed renormalized TVM predictions are in agreement with both sets of available experimental data.

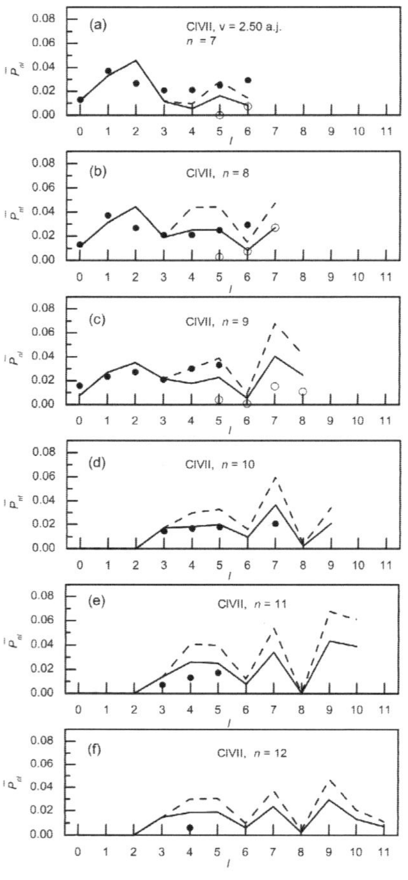

Figure 2. Renormalized population probabilities $\overline{P}_{nl}$ (full curves) and population probabilities P_{nl} (dashed curves) for the ClVII ion, via angular momentum quantum number l, for $7 \leq n \leq 12$, and $v = 2.50$ a.u. The low-l probabilities P_{nl} are from Ref. [19] and the large-l values P_{nl} are from Ref. [20]. Dots and circles are the beam-foil [1, 2] and the beam-microcapillary foil experimental data [4], respectively.

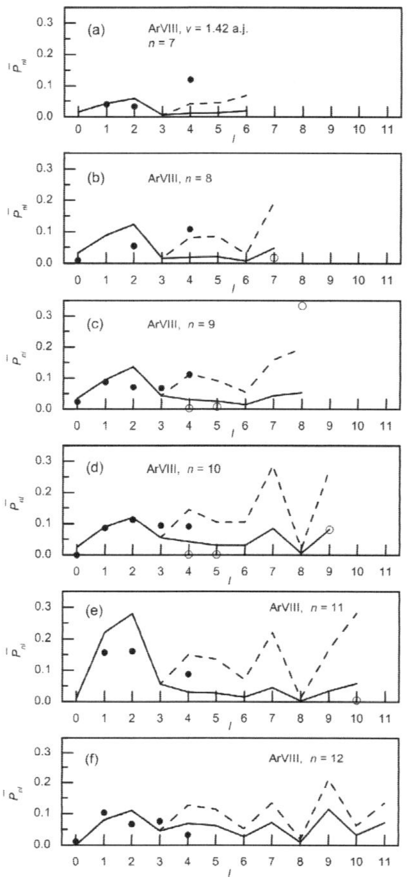

Figure 3. Renormalized population probabilities $\overline{P}_{nl}$ (full curves) and population probabilities P_{nl} (dashed curves) for the ArVIII ion, via angular momentum quantum number l, for $7 \leq n \leq 12$, and $v = 1.42$ a.u. The low-l probabilities P_{nl} are from Ref. [19] and the large-l values P_{nl} are from Ref. [20]. Dots and circles are the beam-foil [3] and the beam-microcapillary foil experimental data [4], respectively.

3. Concluding remarks

By the analysis presented in this article we complete our discussion of the l population distributions for the Rydberg states of multiply charged ions, for $v \approx 1$ a.u. The low-l [18, 19] and the large-l, but lower-n [20] studies have been performed within the framework of the TVM. The reionization mechanism [22] is adapted to the large-l case and also included in the model. The population distributions are obtained, which reflect the n- and l- selective population character, in agreement with both sets of available experiments: the beam-foil [1-3] and the beam-microcapillary foil [4] experiments.

Few additional concluding comments may be relevant for further extension of the presented theoretical approach.

First, in the presented model all ionic projectiles have been considered as point-like. The more realistic description of the ionic core, with core polarization effectively taken into account, can be included into the population/reionization model, using the modified wave function Ψ_2 [24, 25]. Second, the electron transitions in the very vicinity of the potential barrier top can be considered more accurately, like in Ref. [26]. Third, we found that the l distributions for the Rydberg states of multiply charged ions are very sensitive with respect to the ionic velocities. This circumstance can be used to test if the Rydberg states are formed inside the bulk of the foil, or over the surface, along the outgoing part of the ionic trajectory. It is possible to estimate upper velocity limits for the population-reionization process considered as a surface phenomenon.

Finally, we note that the experiments analyzed in this report were not designed for a direct observation of the quantum events at intermediate stages of ion-surface interaction. Therefore, comparing our theoretical data with the cited experiments, we only indirectly demonstrated the relevance of the reionization. The experimental evidence about the intermediate stages of the process will be of interest.

Acknowledgments
This work was supported in part by the Ministry of Science and Technological Development, Republic of Serbia (Project 14 1029).

References

[1] Veje E 1985 *Nucl. Instrum. Methods Phys. Res.* B **9** 586
[2] Veje E and Winter H *Z. Phys.* D **10** (1988) 457
[3] Bashkin S, Oona H and Veje E 1982 *Phys. Rev.* A **25** 417
[4] Morishita Y, Hutton R, Torri H, Komaki K, Brage T, Ando K, Ishii K, Kanai Y, Masuda H, Sekiguchi M, Rosmej F B and Yamazaki Y 2004 *Phys. Rev.* A **70** 012902.
[5] Burgdörfer J in *Review of Fundamental Processes and Applications of Atoms and Molecules* ed by C.D. Lin (World Scientific, Singapore,1993)
[6] Burgdörfer J, Lerner P and Meyer F W 1991 *Phys. Rev.* A **44** 5674
[7] Ducrée J J, Casali F and Thumm U 1998 *Phys. Rev.* A **57** 338
[8] Ducrée J J, Andrä H J and Thumm U 1999 *Phys. Rev.* A **60** 3029
[9] Wille U 1994 *Phys. Rev.* B **50** 1888
[10] Kürpick P and Thumm U 1996 *Phys. Rev.* A **54** 1487
[11] Borisov A G, Zimny R, Teillet-Billy D and Gauyacq J P 1996 *Phys. Rev.* A **53** 2457
[12] Nordlander P 1996 *Phys. Rev.* B **53** 4125
[13] Deutscher S A, Yang X and Burgdörfer J 1995 *Nucl. Instrum. Methods Phys. Res.* B **100** 336
[14] Kürpick P, Thumm U and Wille U 1998 *Phys. Rev.* A **57** 1923
[15] Sjakste J, Borisov A G and Gauyacq J P 2006 *Phys. Rev.* A **73** 042903
[16] So E, Bell M T and Softley T P 2009 *Phys. Rev.* A **79** 012901
[17] Kemmler J, Burgdörfer J and Reinhold C O 1991 *Phys. Rev.* A **44** 2993
[18] Nedeljković N N, Nedeljković Lj D, Vojvodić S B and Mirković M A 1994 *Phys. Rev.* B **49** 5621
[19] Nedeljković Lj D and Nedeljković N N 1998 *Phys. Rev.* B **58** 16455
[20] Nedeljković N N, Nedeljković Lj D and Mirković M A 2003 *Phys. Rev.* A **68** 012721
[21] Nedeljković N N, Nedeljković Lj D, Janev R K and Mišković Z L 1991 *Nucl. Instrum. Methods Phys. Res.* B **58** 519
[22] Nedeljković Lj D and Nedeljković N N 2003 *Phys. Rev.* A **67** 032709
[23] Nedeljković N N and Majkić M D 2007 *Phys. Rev.* A **76** 042902
[24] Nedeljković N N, Majkić M D, Galijaš S M D and Mitrović S B 2008 *App. Surf. Sci.* **254** 7000
[25] Nedeljković N N, Galijaš S M D and Majkić M D 2009 *Surf. Sci.* **603** 2403
[26] Nedeljković Lj D, Nedeljković N N and Božanić D K 2006 *Phys. Rev.* A **74** 032901

Interactions of ions with graphene

I Radović[1], N Bibić[1] and Z L Mišković[2]

[1]VINČA Institute of Nuclear Sciences, University of Belgrade, P.O. Box 522, 11001, Belgrade, Serbia

[2]Department of Applied Mathematics, University of Waterloo, Waterloo, Ontario, Canada N2L 3G1

E-mail: iradovic@vin.bg.ac.rs

Abstract. We evaluate the stopping and image forces on a charged particle moving parallel to a single sheet of graphene supported by an insulating substrate under the gating conditions. The forces are presented as functions of the particle speed and the particle distance for a broad range of charge-carrier densities in graphene. We also consider the effects of a finite gap between graphene and a supporting substrate, as well as the effects of a finite damping rate that is included through the use of Mermin's procedure. The damping rate is estimated from a tentative comparison of the Mermin loss function with a high-resolution reflection electron energy loss spectroscopy experiment.

1. Introduction

Graphene is a flat monolayer of carbon atoms tightly packed into a two-dimensional (2D) honeycomb lattice [1]. It is a basic building block for graphitic materials of all other dimensionalities: highly oriented pyrolytic graphite (HOPG, a stack of graphene layers), carbon nanotubes (rolled-up cylinders of graphene) and fullerene molecules (consisting of wrapped graphene by the introduction of pentagons on the hexagonal lattice) [2, 3]. Interactions of fast-moving charged particles with various carbon nanostructures have been investigated in recent years, e.g., in the electron energy loss spectroscopy (EELS) of carbon nanotubes [4] and isolated layers of free-standing graphene [5].

On the other hand, interactions of energetic heavy charged particles with graphene-based materials have been studied extensively for some time, e.g., in investigations of the directional effects in ion and molecule implantation into HOPG [6, 7], ion channelling through HOPG [8] and secondary electron emission from HOPG induced by fast ions [9] and clusters [10], as well as in ion channelling through carbon nanotubes [11, 12, 13, 14].

In order to better understand particle interactions with layered graphitic nanostructures and motivated by these developments, we explore here the effects of dynamic polarization of graphene on forces acting on ions moving parallel to it. Specifically, we calculate the dissipative force which opposes the ion's motion, called the stopping force, and the perpendicularly oriented conservative force which bends the ion's trajectory towards the sheet, called the image force.

Generally speaking, the easiest way to calculate these forces is based on a 2D hydrodynamic model [15] which proved itself to be a valuable theoretical tool for qualitative understanding of the high-frequency, collective excitations in carbon nanotubes [11, 16]. In the previous publications we have used a simpler, one-fluid hydrodynamic model, which treats all four carbon's valence electrons as a single fluid, to evaluate the dynamic image force on protons channelled through carbon nanotubes in the MeV energy range [13, 17, 18], as well as the stopping and image forces on fast ions moving parallel to a single sheet of graphene supported by an insulating substrate [19] and the stopping and image forces up to the second order on fast ions moving parallel to free-standing graphene [20]. We have also calculated, in the previous publications, the stopping and image forces on projectiles moving parallel to both carbon nanotubes [11, 12] and supported graphene [21] using the linearized two-fluid hydrodynamic model, which makes distinction between the contributions of carbon's σ and π electrons to plasmon excitations in these structures. We have also used the two-fluid hydrodynamic model to study the wake effect in interactions of fast ions with supported graphene [22].

In order to evaluate the stopping and image forces on slow ions moving parallel to a supported graphene under the gating conditions, the dielectric-response theory for surfaces and layered structures [23] is a convenient way to proceed, given that the dielectric function for graphene is available within the random phase approximation (RPA) based on a linear approximation for the π electron bands [24, 25, 26]. We also use a kinetic (Vlasov) equation approach which yields a relatively simple dielectric function that describes both the intraband single-particle excitations (SPEs) and plasmon excitations in graphene's π bands within the linearized electron energy dispersion approximation [27, 28].

Graphene usually appears in experimental situations as supported by a substrate [29, 30]. While the configuration of graphene on metal [29] opens interesting possibility of exciting novel modes of collective electron excitations due to plasmon hybridization [12, 31], we limit ourselves here to the insulating substrate, such as SiO_2 lying underneath a single graphene sheet [30]. Surprisingly, while Ishigami *et al.* [30] have found that the distance h between graphene and substrate is on the order of the distance between graphene layers in graphite or even larger, all theoretical models of graphene's dynamic response assume a zero gap between the graphene and a substrate [24, 25]. One of our principal goals is to demonstrate just how strong are the effects of finite h in the dynamic-polarization forces on moving ions, implying a need to include the gap size explicitly in modelling of other screening phenomena by graphene.

Although we consider the RPA dielectric function to be a basic, parameter-free model that provides an adequate description of both the interband and intraband SPEs, as well as plasmon excitations, in graphene, the model nevertheless has its shortcomings. For example, it ignores the local-field effects (LFE) due to electron-electron correlations [4, 32] and assigns an infinitely long lifetime to the electron excitations. The latter deficiency is often rectified in *ab initio* studies by applying a finite broadening, on the order of $0.5\,eV$, to the frequency domain for calculations of the loss function [5]. In a similar way, one can introduce a finite relaxation time, or decay (damping) rate, γ, to the RPA dielectric function for graphene using Mermin's procedure [33, 34]. Since there are many scattering processes that can give rise to a finite lifetime of the excited π electrons in graphene, an accurate determination of γ still presents a challenge [34, 35]. Therefore, we make one of our principal goals to estimate γ by comparing the RPA model with finite damping to the experimental data for the high-resolution reflection EELS (HREELS) spectra of graphene on a SiC substrate [36].

The parameters of primary interest in this study are the equilibrium density of charge carriers in graphene, n, the graphene-substrate gap height, h, and the damping rate, γ. The equilibrium density is particularly important because it determines the Fermi momentum of graphene's π-electron band, $k_F = \sqrt{\pi n}$, and the corresponding Fermi energy, $E_F = \hbar k_F v_F$, where $v_F \approx c/300$ is the Fermi speed of the linearized π band and c is the speed of light in free space. In this paper, we consider a wide range of densities $n \geq 0$, expressed as a multiple of the base value $n_0 = 10^{11}\ cm^{-2}$. Note that we use Gaussian electrostatic units.

2. Forces on moving charges

We use a Cartesian coordinate system with coordinates $\{\vec{R}, z\}$ and assume that graphene is located in the plane $z = 0$, where $\vec{R} = \{x, y\}$ is position in the plane and z distance from it. A substrate with dielectric constant ε_s is assumed to occupy the region $z \le -h$ underneath the graphene, whereas the region $z > -h$ is assumed to be vacuum or air. It can be shown that, for a point charge Ze moving parallel to graphene with speed v at a fixed distance $z_0 > 0$, the stopping and image forces are given by, respectively

$$F_s = \frac{2Z^2e^2}{\pi v}\int_0^\infty e^{-2qz_0}dq\int_0^{qv}\frac{\omega}{\sqrt{q^2v^2-\omega^2}}\,\mathrm{Im}\left[\frac{1}{\varepsilon(q,\omega)}\right]d\omega, \tag{1}$$

$$F_i = \frac{2Z^2e^2}{\pi}\int_0^\infty qe^{-2qz_0}dq\int_0^{qv}\frac{1}{\sqrt{q^2v^2-\omega^2}}\,\mathrm{Re}\left[\frac{1}{\varepsilon(q,\omega)}-1\right]d\omega, \tag{2}$$

where we have used the symmetry properties of the real and imaginary parts of the dielectric function of supported graphene

$$\varepsilon(q,\omega) = \varepsilon_0(q) + \frac{2\pi e^2}{q}\chi(q,\omega), \tag{3}$$

with $\chi(q,\omega)$ being the polarization function for free graphene and

$$\varepsilon_0(q) \equiv \frac{1}{1-\dfrac{\varepsilon_s-1}{\varepsilon_s+1}e^{-2qh}}, \tag{4}$$

is the background dielectric function which quantifies the effects of substrate on the response of graphene. Note that $\varepsilon_0(q)$ takes the values in the range between 1 and $(\varepsilon_s+1)/2$, characterizing, respectively, the case of a free-standing graphene ($h \to \infty$) and the case of a zero gap ($h = 0$) between graphene and a substrate.

We first present the stopping and image forces calculated with dielectric functions from the kinetic equation (KE) model [27] and the random phase approximation (RPA) model [24, 25, 37] for free graphene ($h \to \infty$) and a vanishing damping rate ($\gamma \to 0$). The results for both forces are normalized by $F_0 = Z^2e^2/(4z_0^2)$, the magnitude of the classical image force on a static point charge a distance z_0 from a perfect conductor, to better reveal differences between the two models.

The difference between the KE and RPA models at low particle speeds ($v < v_F$) is analyzed in figure 1. We compare the normalized stopping and image forces on a proton ($Z = 1$) moving at a speed $v = v_F/2$ above free graphene as a function of the particle distance z_0, for a broad range of densities. At such low speeds, one can see that the agreement between the KE and RPA models is better for the image force than it is for the stopping force. It follows from figure 1 that the condition $z_0k_F > 1$ may suffice as a rough criterion for the application of the KE model at low speeds. This condition is far less restrictive than $z_0k_F >> 1$, which is required for the application of the KE model at high speeds ($v > v_F$).

We now use the RPA dielectric function with a vanishing damping rate ($\gamma \to 0$) to evaluate the stopping and image forces, and investigate the effects of a finite graphene-substrate gap. We assume that graphene is supported by a SiO_2 substrate ($\varepsilon_s \approx 3.9$) and consider the gap heights $h = 0$ for the zero gap commonly considered in the literature, $h = 4$ Å for a realistic value [30] and $h \to \infty$ for free graphene.

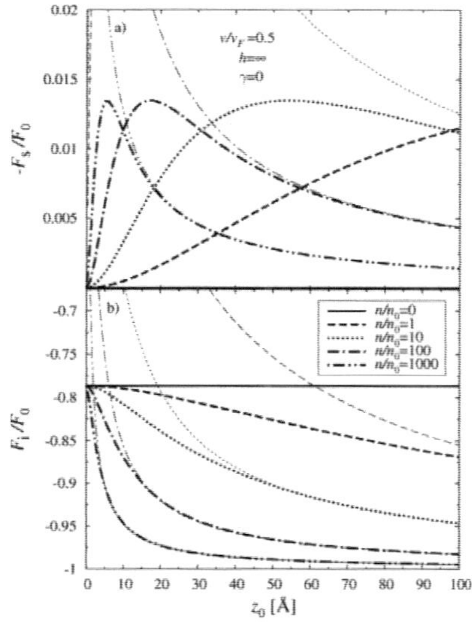

Figure 1. The stopping force (a) and image force (b) normalized by $F_0 = Z^2 e^2/(4z_0^2)$ and shown as a function of the distance z_0 of a proton ($Z=1$) moving at a reduced speed $v/v_F = 0.5$ above free graphene ($h \to \infty$) for several values of the reduced charge-carrier density n/n_0, where $n_0 = 10^{11}\ cm^{-2}$. The thick and thin lines represent the results from the RPA and KE models with vanishing damping ($\gamma \to 0$), respectively.

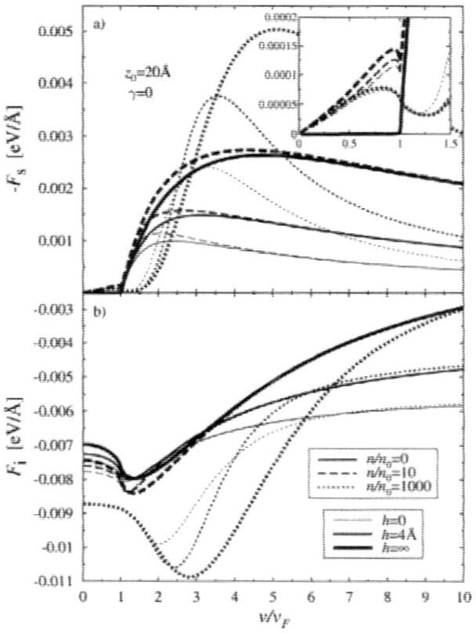

Figure 2. The stopping force (a) and image force (b) from the RPA model with vanishing damping ($\gamma \to 0$) shown as a function of the reduced speed v/v_F of a proton ($Z=1$) moving at a distance $z_0 = 20$ Å above graphene on a SiO_2 substrate ($\varepsilon_s \approx 3.9$). Results are shown for several values of the gap height h and several values of the reduced charge-carrier density n/n_0, where $n_0 = 10^{11}\ cm^{-2}$.

In figure 2, we compare the velocity dependence of the stopping and image forces on a proton moving at a distance $z_0 = 20\,\text{Å}$ above graphene for several gap heights and densities. For low particle speeds ($v < v_F$), the gap height has a relatively small influence on the stopping and image forces that diminishes as the charge-carrier density increases and effectively screens out the graphene-substrate gap. The density n is therefore the most important parameter in the low-speed behavior of both forces. For higher particle speeds, the charge carriers in graphene are not as effective in screening out the graphene-substrate gap, and hence the gap height has a much stronger effect on the stopping and image forces. In particular, figure 2 shows that for sufficiently high speeds ($v >> v_F$) an increase in the gap height tends to increase the strength of the stopping force and decrease the strength of the image force, but there is a range of moderate speeds for which this trend is reversed. One may conclude that in the RPA model, as in the KE model [27], any uncertainty or local variations in the gap height across graphene can lead to large fluctuations in the stopping and image forces, particularly for high particle speeds.

3. Comparison with HREELS experiment

In this section, we use the RPA model with a finite damping rate $\gamma > 0$ by means of the Mermin dielectric function $\varepsilon_M(q,\omega,\gamma)$ (see appendix in reference [37]). To obtain a reasonable estimate for γ, we compare the RPA model with finite damping to the experimental data for the HREELS spectra of graphene on a SiC substrate [36].

In figure 3, we display a tentative comparison between the HREELS data [36] and the Mermin loss function $\text{Im}[-1/\varepsilon_M(q,\omega,\gamma)]$ with $\hbar\gamma = 0$, 200, and $400\, meV$ and a gap height of $1\,\text{Å}$ for wavenumbers ranging from 0.008 to $0.102\,\text{Å}^{-1}$. The SiC substrate is treated in the static mode with dielectric constant $\varepsilon_s = 9.7$, and the equilibrium density in graphene is set at $n = 2\cdot 10^{13}\, cm^{-2}$ (hence $E_F \approx 570\, meV$ and $k_F \approx 0.08\,\text{Å}^{-1}$) to match experimental conditions. Note that since the HREELS data is scaled arbitrarily, the Mermin loss functions for $\hbar\gamma = 200$ and $400\, meV$ are scaled so that the maximum peak heights coincide with those from the experiment. For $\hbar\gamma = 0$, however, the singular plasmon peak prevents such a scaling, and so the Mermin loss function is scaled by the same factor as for the $\hbar\gamma = 400\, meV$ loss function.

In figure 3, the range $q \le k_F$ is particularly interesting because the Mermin loss function with $\hbar\gamma = 0$ exhibits three distinct features for these wavenumbers: a continuous spectrum of intraband SPEs for $0 < \omega < qv_F$, a continuous spectrum of interband SPEs for $\omega > v_F(2k_F - q)$, and a narrow plasmon line at $\omega = \omega_p(q)$ in the otherwise void interval $qv_F < \omega < v_F(2k_F - q)$. The fact that these three features are not visible in the experimental HREELS spectra can be tentatively explained by assuming that a large enough damping rate γ exists, due to various scattering mechanisms, that a broadened plasmon line merges into the two regions of SPEs to form a single peak that follows approximately the original plasmon dispersion curve, $\omega = \omega_p(q)$. Note that a broadening of the plasmon line for the $\hbar\gamma = 0$ loss function does occur for the wavenumbers $q = 0.077$ and $0.102\,\text{Å}^{-1}$ as the plasmon line crosses the boundary $\omega = v_F(2k_F - q)$ and enters into the region of interband SPEs, in which collective plasma oscillations decay into SPEs in a way that can be described by a finite Landau damping rate, γ_L [38]. For the Mermin loss function with a phenomenological damping rate γ, however, a broadening of the plasmon line occurs for all q. Figure 3 shows that a reasonably good qualitative agreement with the experiment can be achieved by using a gap height $h = 1\,\text{Å}$ and a damping rate $\hbar\gamma = 400\, meV$.

It is also interesting to consider the low-energy features in the experimental data which are not well reproduced in figure 3 by the present model. Given that graphene's π electron plasmon dispersion scales as $\omega_p(q) \propto \sqrt{q}$ at long wavelengths, one may expect that it will be strongly coupled with the non-dispersing Fuchs-Kliever surface phonon mode in the SiC substrate at the frequency around $116\,meV$ [36]. Such coupling was discussed recently in reference [39] where the strong optical phonon mode in SiC was described by a simple frequency-dependent dielectric function $\varepsilon_s(\omega)$ to be used in equation (4) for the overall dielectric response of epitaxial graphene. As a result, a strong avoided crossing between the graphene's plasmon mode and the phonon mode in the substrate was predicted theoretically in reference [39] and confirmed experimentally in reference [40]. It is interesting that this plasmon-phonon coupling scenario continues to attract considerable interest [41, 42, 43].

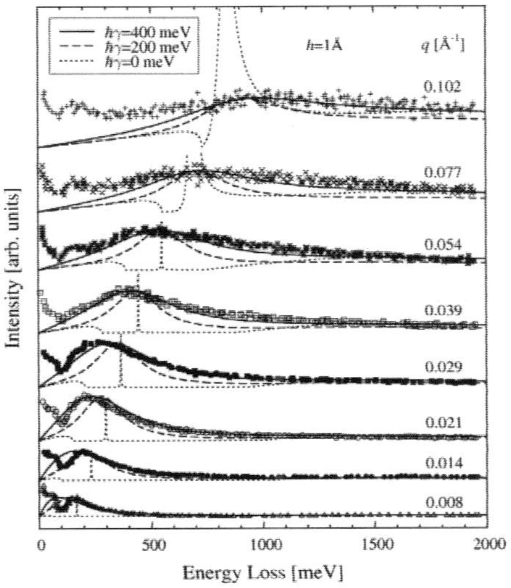

Figure 3. The Mermin loss function (in arbitrary units) versus the energy loss for graphene with a charge-carrier density $n = 2 \cdot 10^{13}\,cm^{-2}$ supported on a SiC substrate with static dielectric constant $\varepsilon_s = 9.7$ and a gap height $h = 1\,\text{Å}$. Model results are shown for damping rates $\hbar\gamma = 0$, 200, and $400\,meV$, while symbols show the HREELS experimental data from reference [36].

4. Conclusions

We have presented an extensive analysis of the stopping and image forces on an external point charge moving parallel to a single layer of supported graphene under the gating conditions. Calculations of the velocity and distance dependencies of the stopping and image forces were performed within the random phase approximation (RPA) for a broad range of charge-carrier densities with two major goals: to compare the results with a semiclassical kinetic equation (KE) model, and to examine of the effects of a finite graphene-substrate gap. The third goal was to estimate a finite damping rate introduced using Mermin's procedure.

With respect to the first goal, a comparison of the forces from the RPA and KE models in the regime of vanishing damping has revealed that the latter model may be justified for particle distances satisfying $z_0 k_F > 1$ for $v < v_F$ and satisfying $z_0 k_F >> 1$ for $v > v_F$. When combined through Bohr's adiabatic criterion, these conditions suggest that the KE model is valid only for heavily doped graphene with $E_F > \hbar v / z_0$, for which the effects of the interband single-particle excitations on the stopping and image forces are minimized. With respect to the second goal, the effects of a finite gap between graphene and a supporting substrate in the RPA model with vanishing damping have been found to be quite strong, particularly for medium to high particle speeds. These results have confirmed earlier findings from the KE model [27] and raise some concern over the common practice of treating graphene with a zero gap when dealing with the dynamic polarization forces on external moving charges. With respect to the third goal, we have made an effort to estimate the order of magnitude of the damping rate, γ, by providing a tentative fit of the RPA dielectric function modified by Mermin's procedure with experimental data for the HREELS spectra of graphene on a SiC substrate [36].

Acknowledgments
This work has been supported by the Ministry of Science and Technological Development of the Republic of Serbia (Project No. 141013). Z.L.M. acknowledges support by the Natural Sciences and Engineering Research Council of Canada. The authors are most grateful to Dr Ljupčo Hadžievski and Dr Duško Borka for many useful comments and continuing support.

References

[1] Geim A K and Novoselov K S 2007 *Nature Mater.* **6** 183
[2] Castro Neto A H, Guinea F, Peres N M R, Novoselov K S and Geim A K 2009 *Rev. Mod. Phys.* **81** 109
[3] Geim A K and Kim P 2008 *Sci. Am.* **298** 90
[4] Kramberger C, Hambach R, Giorgetti C, Rümmeli M H, Knupfer M, Fink J, Büchner B, Reining L, Einarsson E, Maruyama S, Sottile F, Hannewald K, Olevano V, Marinopoulos A G and Pichler T 2008 *Phys. Rev. Lett.* **100** 196803
[5] Eberlein T, Bangert U, Nair R R, Jones R, Gass M, Bleloch A L, Novoselov K S, Geim A K and Briddon P R 2008 *Phys. Rev.* B **77** 233406
[6] Ramos G and Scherzer B M U 1994 *Nucl. Instr. Meth.* B **85** 479
[7] Ramos G and Scherzer B M U 2001 *Nucl. Instr. Meth.* B **174** 329
[8] Yagi E, Iwata T, Urai T and Ogiwara K 2004 *J. Nucl. Mater.* **334** 9
[9] Cernusca S, Fürsatz M, Winter H P and Aumayr F 2005 *Europhys. Lett.* **70** 768
[10] Kaneko T, Kudo H, Tomita S and Uchiyama R 2006 *J. Phys. Soc. Jpn.* **75** 034717
[11] Mowbray D J, Mišković Z L, Goodman F O and Wang Y N 2004 *Phys. Rev.* B **70** 195418
[12] Mowbray D J, Mišković Z L and Goodman F O 2006 *Phys. Rev.* B **74** 195435
[13] Borka D, Petrović S, Nešković N, Mowbray D J and Mišković Z L 2006 *Phys. Rev.* A **73** 062902
[14] Borka D, Mowbray D J, Mišković Z L, Petrović S and Nešković N 2010 *New J. Phys.* **12** 043021
[15] Fetter A L 1973 *Ann. Phys.* **81** 367
[16] Stockli T, Bonard J M, Chatelain A, Wang Z L and Stadelmann P A 2001 *Phys. Rev.* B **64** 115424
[17] Borka D, Mowbray D J, Mišković Z L, Petrović S and Nešković N 2008 *Phys. Rev.* A **77** 032903
[18] Borka D, Mowbray D J, Mišković Z L, Petrović S and Nešković N 2008 *J. Phys.: Condens. Matter* **20** 474212
[19] Radović I and Borka D 2010 *Phys. Lett.* A **374** 1527
[20] Radović I, Hadžievski Lj, Bibić N and Mišković Z L 2009 *Mater. Chem. Phys.* **118** 293
[21] Radović I, Hadžievski Lj, Bibić N and Mišković Z L 2007 *Phys. Rev.* A **76** 042901
[22] Radović I and Borka D 2010 *Nucl. Instr. Meth.* B **268** 2649
[23] Pitarke J M, Silkin V M, Chulkov E V and Echenique P M 2007 *Rep. Prog. Phys.* **70** 1
[24] Wunsch B, Stauber T, Sols F and Guinea F 2006 *New J. Phys.* **8** 318
[25] Hwang E H and Das Sarma S 2007 *Phys. Rev.* B **75** 205418

[26] Barlas Y, Pereg-Barnea T, Polini M, Asgari R and Mac-Donald A H 2007 *Phys. Rev. Lett.* **98** 236601
[27] Radović I, Hadžievski Lj and Mišković Z L 2008 *Phys. Rev.* B **77** 075428
[28] Ryzhii V, Satou A and Otsuji T 2007 *J. Appl. Phys.* **101** 024509
[29] Bertoni G, Calmels L, Altibelli A and Serin V 2005 *Phys. Rev.* B **71** 075402
[30] Ishigami M, Chen J H, Cullen W G, Fuhrer M S and Williams E D 2007 *Nano Lett.* **7** 1643
[31] Prodan E, Radloff C, Halas N J and Nordlander P 2003 *Science* **302** 419
[32] Trevisanutto P E, Giorgetti C, Reining L, Ladisa M and Olevano V 2008 *Phys. Rev. Lett.* **101** 226405
[33] Mermin N D 1970 *Phys. Rev.* B **1** 2362
[34] Asgari R, Vazifeh M M, Ramezanali M R, Davoudi E and Tanatar B 2008 *Phys. Rev.* B **77** 125432
[35] Hwang E H and Das Sarma S 2008 *Phys. Rev.* B **77** 195412
[36] Liu Y, Willis R F, Emtsev K V and Seyller Th 2008 *Phys. Rev.* B **78** 201403(R)
[37] Allison K F, Borka D, Radović I, Hadžievski Lj and Mišković Z L 2009 *Phys. Rev.* B **80** 195405
[38] Pines D and Schrieffer J R 1962 *Phys. Rev.* **125** 804
[39] Allison K F and Mišković Z L 2010 *Nanotechnology* **21** 134017
[40] Liu Y and Willis R F 2010 *Phys. Rev.* B **81** 081406(R)
[41] Hwang E H, Sensarma R and Das Sarma S 2010 arXiv:1008.0862v1
[42] Tegenkamp C, Pfnür H, Langer T, Baringhaus J and Schumacher H W 2010 arXiv:1008.1013v1
[43] Koch R J, Seyller Th and Schaefer J A 2010 arXiv:1008.1130v1

Theory of below-threshold kinetic electron emission

P Tiwald, Ch Lemell, G Wachter, and J Burgdörfer

Institute for Theoretical Physics, Vienna University of Technology,Wiedner Hauptstrae 8-10, A-1040 Vienna, Austria, EU

E-mail: `paul@dollywood.itp.tuwien.ac.at`

Abstract. We investigate the electron emission from clean monotonically flat crystalline aluminum surfaces due to the impact of rare gas atoms at the surface under grazing angles of incidence. Recent improvements of experimental methods allowed for the first time the measurement of electron yields as small as $\gamma \approx 0.01$ electrons/projectile. Using these methods kinetic electron emission (KE) was found for projectile velocities well below the classical threshold v_{th} for KE, the minimum velocity a projectile has to have in order to transfer in a binary collision enough momentum to an electron at the Fermi edge (with $k = k_F$) to overcome the surface potential. Quantum mechanical effects smear out the Fermi edge creating electrons with momenta above k_F. We present a calculation of γ in the below threshold region within the framework of a classical trajectory Monte Carlo simulation in which special emphasis is put on the description of the projectile trajectory and on an accurate determination of the momentum distribution, i.e. the Compton profile of the surface electronic structure. We employ different methods for calculating the electronic structure and study its influence on. We find that realistic momentum distributions can account for kinetic electron emission in the below threshold region.

1. Introduction

Impact of atoms or ions on surfaces leads to a number of processes depending on the projectile type and energy transfer channels (kinetic and potential). For specific choices of the experimental parameters many of these channels are closed which allows for detailed investigation of selected processes. We study the emission of electrons from aluminum surfaces due to the impact of slow atomic projectiles. In general, electron emission processes can be divided by the "type" of energy transferred from the projectile to the electron. This is either potential or kinetic energy leading to potential (PE) or kinetic emission (KE), respectively. The use of rare gas atoms in their ground state as projectiles eliminates contributions of PE to the so-called secondary emission coefficient γ. For slow projectiles the two most important contributions to KE are quasi-molecular promotion (QMP) and direct collisions between projectile and target electrons, i.e. excitation of electron-hole pairs in the conduction band of the surface material by binary collisions. Electron emission by QMP occurs upon close contact of projectile and target atoms, when electronic levels are shifted giving rise to electronic excitations at avoided crossings or promotion of occupied states above the vacuum level. In recent experiments Winter et al. scattered rare gas atoms with kinetic energies between 2 and 28 keV off atomically flat aluminum surfaces (terrace lengths $\sim$ 2000 a.u.) under a grazing angle of incidence [1]. Within this setup projectile atoms move along well defined trajectories and are reflected from the surface with distances of closest approach large enough to exclude QMP. Therefore, binary collisions between

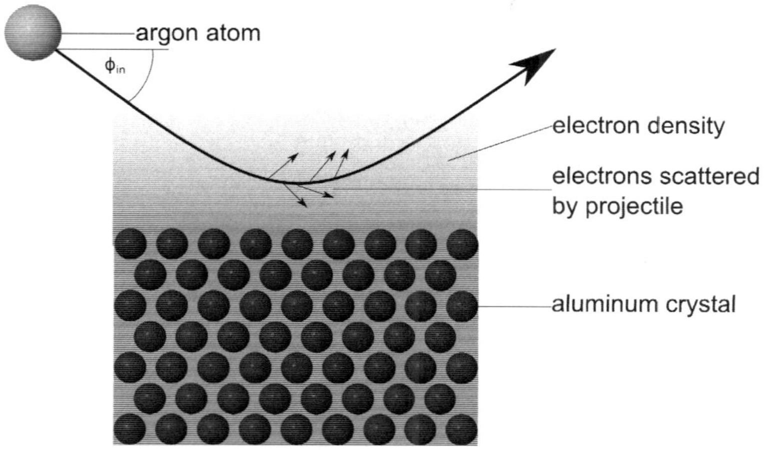

Figure 1. Schematic picture of a projectile trajectory in the experiments performed by the group of Winter.

projectiles and target electrons above the surface provide the most significant contribution to γ [2]. Fig. 1 shows a schematic picture of a projectile trajectory in the experiments performed by Winter et al.

Kinetic emission (KE) of target electrons induced by projectile atoms was already explored in the late 1970s. In 1979 Baragiola calculated the projectile velocity threshold v_{th} for KE due to collisions of projectile atoms with target electrons [3]. He considered a projectile entering a solid with a free electron gas of density n_e in which the highest available electron momentum is the Fermi momentum k_F. From these target properties a threshold for the onset of KE was derived treating a head-on collision within the binary-encounter approximation. As n_e is smaller above the surface ("spill-out") an even higher velocity threshold was expected due to a "local" Fermi momentum [4]. Measurements of Winter et al. (fig. 2) show, however, electron emission even below the threshold for bulk k_F. Since Winter and Winter derived a $\gamma \propto (v - v_{th})^2$ proportionality in the threshold region [5], $\sqrt{\gamma}$ is plotted versus the projectile velocity v_p.

Although several processes have been proposed to explain below threshold KE [5–8], definitive conclusions as to the origin of KE near the threshold were still outstanding. Our model is based on the notion that in difference to the free electron gas model (sharp Fermi edge) the momentum distribution of metal electrons is not bounded from above by k_F but, instead, contains off-shell components with $k > k_F$ due to correlation (interaction with other metal electrons) and corrugation (influence of the crystal structure) effects. Different methods of calculating such momentum distributions within the framework of density functional theory (DFT) are presented in section 3.

As the distance of closest approach strongly influences the expected KE yield ($\gamma \propto n(z) \propto n_{bulk} \cdot e^{-\alpha z}$), another prerequisite for the determination of γ at different projectile velocities is an accurate description of the projectile trajectory. Our method to determine projectile-surface potentials governing these trajectories is described in section 2. Finally, a classical trajectory Monte Carlo (CTMC) simulation of KE due to binary collisions of projectile atoms with target electrons including electron transport from the point of excitation to the detector is explained in section 4 with its results presented and discussed in section 5.

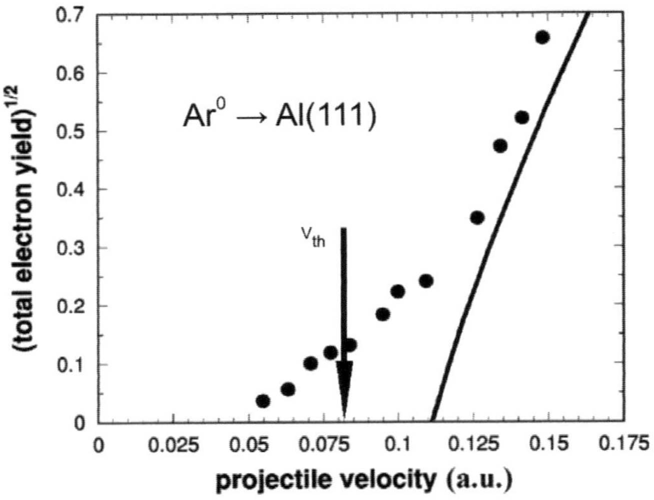

Figure 2. Results of a KE measurement for Ar atoms scattered off an Al(111) surface. The arrow indicates Baragiola's threshold. The line is a best fit to experimental data for larger projectile velocities according to a quasi-classical estimate [5].

2. The projectile surface potential

An appropriate test for the reliability of the projectile-surface potential is the comparison of classical trajectory simulations with rainbow scattering experiments. In these experiments projectiles are directed at the surface under a grazing angle of incidence and along a low indexed crystal direction. The small incidence angles lead to widely disparate kinetic energies parallel $E_{\|}$ and normal $E_{\perp}$ to the surface. The fast motion along low indexed crystal directions makes the projectile experience an effective two-dimensional atomic string potential of approximately sinusoidal shape. This shape leads to characteristic intensity distributions in the reflected atomic beam. Fig. 3 shows the schematic experimental setup with the measured intensity distribution of an 18 keV Ar beam directed onto a Al(111) surface ($E_{\perp} \approx 10$ eV) along the $\langle 1\bar{1}0 \rangle$ direction. The two intensity maxima at the edge of the distribution represent the rainbow peaks located at the rainbow angle θ_{rb}. For inert gas atoms θ_{rb} depends solely on the projectile's kinetic energy perpendicular to the surface. This dependence is very sensitive to the potential landscape above the aluminum surface and, hence, a good benchmark for the quality of projectile-surface potentials. We calculated the projectile-surface potential by considering the static limit of the collision complex consisting of the projectile and the target surface [9]. To this end we performed DFT slab calculations with periodic boundary conditions using the Abinit program package [10].

From classical trajectory simulations we determined the rainbow angles for Ar projectiles directed onto an Al(111) surface. Fig. 4(b) shows the near-perfect agreement of simulated and experimental results for $\theta_{rb}(E_{\perp})$. Especially in the low energy region, i.e. at large distances from the surface, there is a considerable improvement to earlier calculations in which the projectile-surface potential was approximated by a sum over two-body potentials, see fig. 4(a) [11].
In the kinetic electron emission experiments the inert gas projectiles are directed at the surface along a "random" (high indexed) direction. The projectiles experience a planar averaged surface potential, solely depending on the distance parallel to the surface normal. For our KE simulations we used the planar averaged DFT projectile-surface potentials.

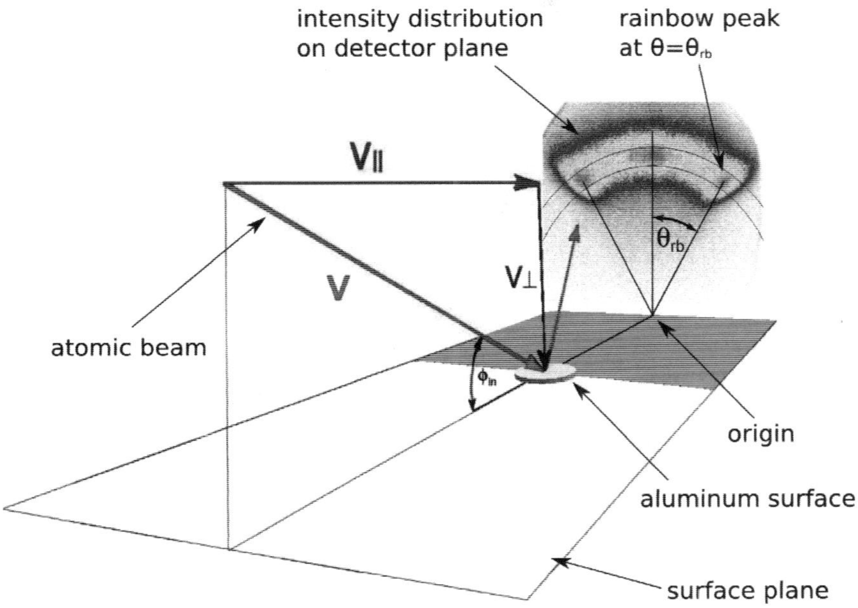

Figure 3. Schematic picture of the scattering geometry. The 2D intensity distribution on the detector plane arises from scattering of 10 keV Ar atoms from Al(111) along the $\langle 1\bar{1}0 \rangle$ direction at an angle of incidence $\phi_{in} = 2.35°$. This corresponds to an energy perpendicular to the surface of $E_{\perp} \approx 16.81$ eV.

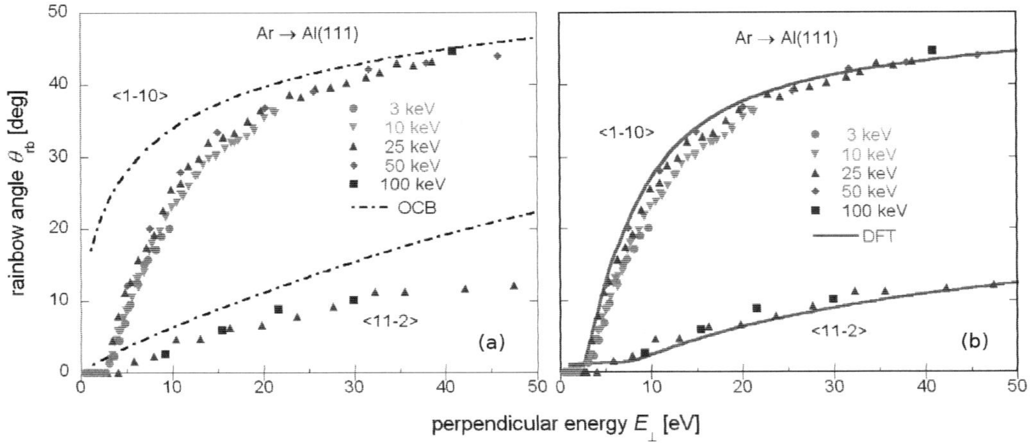

Figure 4. Rainbow angles as function of normal component of kinetic energy for Ar atoms scattered from Al(111) along low indexed directions. Symbols: experimental data for different total kinetic energies [11]. Curves: results from trajectory simulations based on (a) the O'Conner Biersack (OCB) potential [12] and (b) on Ar-Al(111) potentials from ab-initio DFT slab-calculations [9].

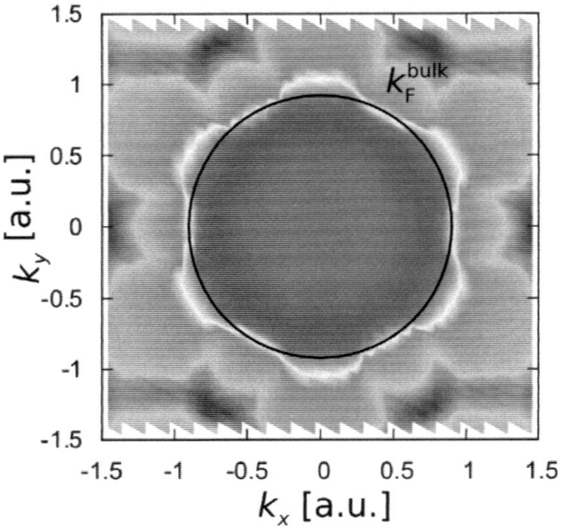

Figure 5. Lateral cut through the momentum distribution $\rho(\vec{k}_{\|}, z)$ at a distance of $z = 1$ a.u. in front of the topmost atomic layer of an Al(111) surface (logarithmic color code).

3. Momentum distributions above Al surfaces

Experiments showed that electron emission induced by projectile electron collisions occur well below Baragiola's threshold velocity for the projectile v_{th}. In his estimate for v_{th} he assumed the highest available electron momentum to be $k_{max} = k_F^{bulk}$. We conclude that the free-electron gas approximation is not sufficient to explain so-called sub threshold kinetic electron emission and suggest the use of more realistic momentum distributions taking into account many-particle effects (correlation) and a structured crystal potential featuring off-shell momentum components $k > k_F^{bulk}$. The momentum distributions above the Al(111) surface, i.e. the Compton profiles, are extracted from Kohn-Sham (KS) wave functions of ab-initio DFT slab calculations by corresponding Fourier-transforms [13]. In general, the KS pseudo-wave functions are not to be identified with the true wave functions and, therefore, it is a priori not obvious that the electron density $n_e(\vec{r})$ and the momentum density $\rho(\vec{k})$ are Fourier transforms of each other. In the following we postulate the applicability of the KS pseudo-wave functions which can be justified for the homogeneous electron gas. Its extension to realistic DFT calculation for surfaces can be viewed as an analogue of the local-density approximation to the exchange-correlation potential. Performing Fourier transforms of the KS wave function only for the coordinates parallel (x, y) to the surface yields the distance-dependent partial momentum distribution $\rho(\vec{k}_{\|}, z)$. Fig. 5 shows a lateral cut through $\rho(\vec{k}_{\|}, z)$ at a distance of $z = 1$ a.u. The hexagonal structure of the surface unit cell results in six humps at wave numbers of about $k_{\|} = 2\pi/a_{nn}$ with $a_{nn} \approx 5.4$ a.u., the nearest neighbor distance for the Al(111) surface.

$\rho(\vec{k}_{\|}, z)$ features small but finite momentum components above the surface larger than the Fermi momentum of bulk aluminum $k_F^{bulk} \sim 0.9$ a.u. Due to Heisenberg's uncertainty principle information about the z-dependence comes at the price of giving up any knowledge about the momentum component k_z. Since for grazingly incident particle beams high momentum components parallel to the surface are expected to be responsible for sub-threshold KE in head-on collisions with maximum momentum transfer $\rho(\vec{k}_{\|}, z)$ should be well suited to account for the onset of KE.

We also parameterized the momentum distribution above the surface by a z-dependent spherically symmetric Fermi distributions at high temperatures $\rho_{\mu,T}(\vec{k}, z)$. High temperatures lead to smearing of the Fermi edge and, therefore, to momentum components $k > k_F^{bulk}$. The z-dependence is introduced via an effective local temperature $T_{eff}(z)$ and an effective local chemical potential $\mu_{eff}(z)$. Note that $T_{eff}(z)$ and $\mu_{eff}(z)$ are not to be seen as physical temperature or chemical potential, respectively. They are fit parameters. In this sense $T_{eff}(z)$ can be interpreted as a measure for corrugation and correlation effects and $\mu_{eff}(z)$ plays the role of a z-dependent or local Fermi energy [5]. Both parameters are chosen such that $\rho_{\mu,T}(k_{\|}, z) = \int\int \rho_{\mu,T}(\vec{k}, z)\mathrm{d}k_z \mathrm{d}\varphi_k$ fits best to the partial momentum distribution $\rho(k_{\|}, z) = \int \rho(\vec{k}_{\|}, z)\mathrm{d}\varphi_k$, where φ_k is the azimuthal angle in momentum space.
As a third variant for the description of a locally varying momentum distribution we use the Husimi function $\rho_\sigma^H(\vec{k}, z)$ [14], the closest analog to the classical phase space distribution consistent with Heisenberg's uncertainty principle. It is calculated by averaging the Wigner function [15] over a Gaussian wave packet with minimum uncertainty.

4. Kinetic electron emission simulation

In the CTMC simulations scattering events between projectiles and surface electrons are treated within the binary-encounter approximation by neglecting, for the time of the collision, the influence of surrounding particles. This is justified in our simulation due to the large distance between single electrons. The Wigner-Seitz radius at a distance of $z = 1$ a.u. above the surface is $r_s \approx 2.3$ a.u. Initial electron momenta k are selected randomly from the Compton profiles discussed in sec. 3 and the scattering angle ϑ is randomly chosen from the differential elastic scattering cross section $\frac{\mathrm{d}\sigma^{el}}{\mathrm{d}\Omega}(\vartheta, E)$.
Electrons excited by the scattering event to energies above the vacuum level propagate close to the surface and may eventually leave the target. Within an electron propagation calculation [16] the trajectories of these particles are modeled individually to determine whether they leave the aluminum crystal or deposit their excess energy within the solid. We find that only a minor fraction (25%) of electrons excited to energies above the vacuum level leave the surface and contribute to the measured KE yield.
Note that all distribution in sec. 3 feature initial electron momenta with kinetic energies (but not total energies) above the vacuum level. This is not a surprise since even in a classical picture electrons may reach high velocities close to the nucleus but remain bound. A well-known quantum mechanical example is the electron momentum distribution $\rho(p) \propto (1 + p^2)^{-4}$ of the hydrogen atom with very large momenta which can be made visible ("put on shell") in ionization experiments [17]. Due to Heisenberg's uncertainty relation it is only possible to calculate partial, i.e. planar averaged (in real space), momentum distributions from KS wave functions. This, however, leaves high k-components in our distributions which are not balanced by the local potential $V_{el}(\vec{r})$ or its planar averaged version $V_{el}(z)$ which in turn leads to "auto-emission" (i.e. spontaneous emission in the absence of interactions) for electrons with initial energies above the vacuum level. In order to estimate the effect of "auto-emission" we performed calculations for the limiting cases in which electrons with initial energies above the vacuum level are fully taken into account or are completely neglected.

5. Results and Discussion

Fig. 6 show experimental data of $\sqrt{\gamma}$ as a function of the projectile velocity for argon atoms impinging on an aluminum (111) surface under an angle of incidence of $\phi_{in} = 2.2°$ and results of KE simulations in which the three momentum distributions of section 3 are used. In each figure two lines are plotted corresponding to the limiting cases in which "auto-emission" is allowed or suppressed.

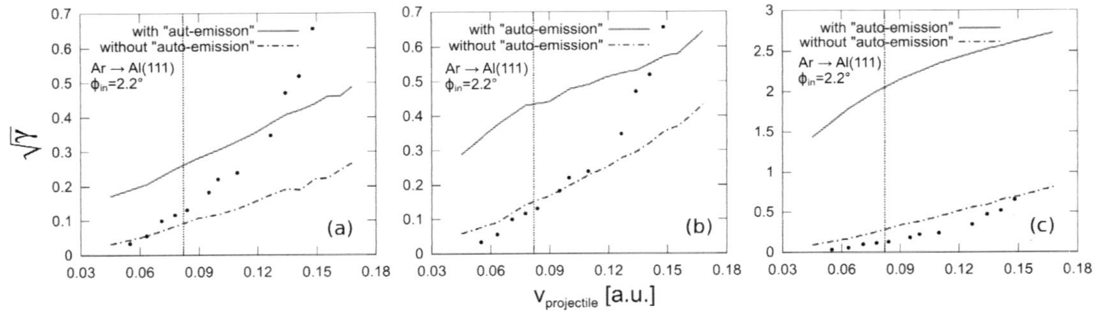

Figure 6. Lines: Secondary kinetic electron emission yields with respect to the projectile velocity calculated with the momentum distributions (a) $\rho(\vec{k}_{\|}, z)$, (b) $\rho_{\mu,T}(\vec{k}, z)$, and (c)$\rho_{\sigma}^{H}(\vec{k}, z)$. Points: experimental data. Dashed vertical lines at $v_p \approx 0.08$ a.u. indicate the classical threshold for KE.

Due to the high k-tails, all distributions feature sub threshold kinetic electron emission caused by binary-encounter processes with atomic projectiles. However, without a more elaborate method to treat electrons with initial energies above the vacuum level a quantitative analysis of sub threshold KE remains an open problem.

At low velocities $v_p < 0.12$ a.u. calculated yields compare well with experimental results when discriminating against auto-emission. At high velocities $v_p > 0.12$ a.u. the distributions lead to a quadratic rise of the yield $\gamma = \alpha v_p^2$ similar to experimental values. The proportionality α, however, differs by up to a factor 4. Tests showed that in the high velocity region even a z-independent momentum distribution equal to the bulk Fermi distribution cannot account for the increase of γ. The difference between measured and calculated yields has, therefore to be related to electron emission channels not treated in this work.

References

[1] Winter H P, Lederer S, Winter H, Lemell C and Burgdörfer J 2005 *Phys. Rev. B* **72** 161402
[2] Mix W, Tzanev S, Golichowski A, Snowdon K J, Bilic A and Gumhalter B 1995 *Surf. Sci.* **331-333** 332–336
[3] Baragiola R A, Alonso E V and Florio O A 1979 *Phys. Rev. B* **19** 121–129
[4] Winter H, Lederer S and Winter H P 2006 *Europhys. Lett.* **75** 964–970
[5] Winter H and Winter H P 2003 *Europhys. Lett.* **62** 739–745
[6] Sroubek Z, Chen X and Yarmoff J A 2006 *Phys. Rev. B* **73** 045427
[7] Lörincik J, Šroubek Z, Eder H, Aumayr F and Winter H 2000 *Phys. Rev. B* **62** 16116–16125
[8] Yarmoff J A, Than H T and Sroubek Z 2002 *Phys. Rev. B* **65** 205412
[9] Tiwald P, Schüller A, Winter H, Tökesi K, Aigner F, Gräfe S, Lemell C and Burgdörfer J 2010 *Phys. Rev. B* in press
[10] Gonze X *et al.* 2002 *Computational Materials Science* **25** 478–492
[11] Schüller A and Winter H 2007 *Nucl. Instrum. and Methods in Phys. Res. B* **256** 122–125
[12] O'Connor D J and Biersack J P 1986 *Nucl. Instrum. and Methods in Phys. Res. B* **15** 14–19
[13] Lemell C, Arnau A and Burgdörfer J 2007 *Phys. Rev. B* **75** 014303
[14] Husimi K 1940 *Proc. Phys. Math. Soc. Jpn.* **22** 264
[15] Wigner E 1932 *Phys. Rev.* **40** 749–759
[16] Ding Z J 1990 *Fundamental Studies on the Interactions of kV Electrons with Solids for Applications to Electron Spectroscopies* Ph.D. thesis Department of Applied Physics, Osaka University, Japan
[17] Lohmann B and Weigold E 1981 *Phys. Lett. A* **86** 139–141

Deposition and characterization of ultra thin diamond like carbon films

B Tomcik
Institute of Physics, Center for Atomic and Subatomic Physics, Laboratory for Atomic Collision Processes, Pregrevica 118, P.O.Box 68, Serbia

E mail: tomcik@ipb.ac.rs

Abstract. Amorphous hydrogenated and/or nitrogenated carbon films, a-C:H/a-C:N, in overall thickness up to 2 nm are materials of choice as a mechanical and corrosion protection layer of the magnetic media in modern hard disk drive disks. In order to obtain high density and void-free films the sputtering technology has been replaced by different plasma and ion beam deposition techniques. Hydrocarbon gas precursors, like C_2H_2 or CH_4 with H_2 and N_2 as reactive gases are commonly used in ion and plasma beam sources. Optimum incident energy of carbon ions, C^+, is up to 100 eV while the typical ion current densities during the film formation are in the mA/cm^2 range. Other carbon deposition techniques, like filtered cathodic arc, still suffer from co-deposition of fine nano-sized carbon clusters (nano dust) and their improvements are moving toward arc excitation in the kHz and MHz frequency range. Non-destructive film analysis like μ-Raman optical spectroscopy, spectroscopic ellipsometry, FTIR and optical surface analysis are mainly used in the carbon film characterization. Due to extreme low film thicknesses the surface enhanced Raman spectroscopy (SERS) with pre-deposited layer of Au can reduce the signal collection time and minimize photon-induced damage during the spectra acquisition.

1. Introduction

The main purpose of the carbon overcoat in the hard disk drive is to provide good corrosion and mechanical protection for the underlying magnetic-recording film under unfavorably environmental conditions and occasional contact with the read/write head[1,2]. To improve the disk corrosion protection properties over longer time, a thin flash layer of Cr is usually deposited over the magnetic layer prior the carbon overcoat. In principle, any carbide forming elements like Cr, Ti, Mo, Al or W can be used for such purpose. In industry the Cr is mainly used, and a technique of choice is the magnetron sputtering. In this work we tried to examine the capabilities of the Filtered Cathodic Arc (FCA) source in deposition of the Cr flash layer. With this source one may expect better surface coverage of the magnetic layer and more effectively filled micro-rough sites. An incoming particle flux is almost 100% ionized and possess the higher incident energy[3,4]. By physical and chemical sputtering/etching of the top 0.5 nm of the magnetic layer is also possible to achieve an ultra-smooth and polished surface, especially in combination with the substrate rotation.

In addition to the low thickness and favorable corrosion and wear protection properties of the overcoat there is a need for the low fly-height of the read/write head. This requires a small overall surface roughness of a disk. With easy control of ion energy of the FCA source one can meet most of these requirements, especially in the pre-conditioning of the magnetic layer with respect to the surface roughness and corrosion protection barrier. Up to now the carbon based implementation of the FCA source, e.g. with improved Pulsed Arc Deposition[5] and with better macroparticle filtration with double-band (twisted S-filter)[4,6], still suffers from the carbon nano-dust particles and a low film deposition rate. Therefore, in this work we are trying to exploit some beneficial FCA source capabilities for metal targets, other than carbon.

Protective corrosion properties of such a structure are usually evaluated by e.g.:a) corrosion spot density measurement[7]versus type of a flash layer element, flash layer thickness, eventual presence of macro/nanoparticles from the FCA source and overall surface roughness, b) implementing standard electrochemical tests[8] or some sort of the surface analytical study on the migrated cobalt compounds from the magnetic layer.

In this paper we opted for the analytical surface study of the overall disk structure, before and after the corrosion tests, using elastic collision effects of the highly energetic He^+ ion beams with disk depth profiling – Rutherford Backscattering Spectrometry (RBS).

2. Methodology and setup

All films were deposited on a silicon/glass disk in this sequence: a) magnetic layer by DC sputtering in Ar gas (1st magnetron), b) flash layer by FCA source at the base vacuum (10^{-5} mbar) and c) carbon overcoat, a-C or a-C:N film, in Ar+ N_2 by sputtering of carbon target (2nd magnetron), figure 1.

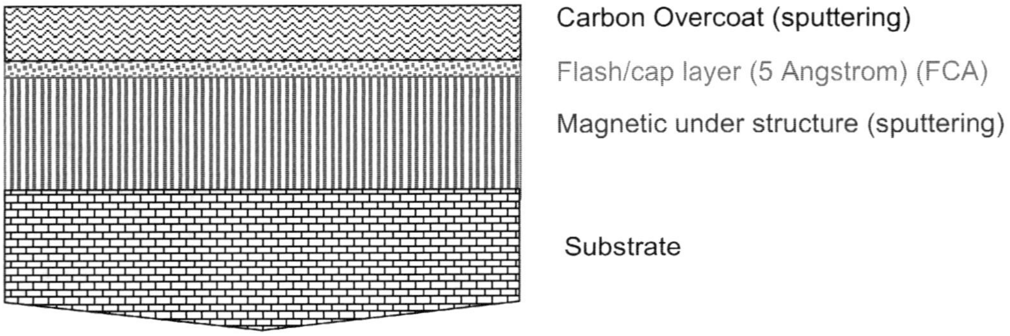

Figure 1. Disk structure and techniques employed in the film deposition over silicon and a glass disk.

The UHV chamber housed an FCA source and two magnetron sputtering sources with 3" targets. A substrate, either 2.5" glass disk or a Si wafer, was clamped on its holder mounted on the rotational stage and inserted into the load-lock chamber. The first step is sputter deposition of the 30 nm thin CoCrPtB magnetic layer. The substrate plane was parallel to the magnetron surface. After the first layer deposition the substrate holder was further positioned to normally face down the filtered cathodic arc source. During this step the base pressure in the chamber was ~$3x10^{-6}$ Torr and the Cr film was deposited in the residual gas environment, without Ar flow. The Cr target temperature increased with deposition time and reached T_{max}= 620 ^{0}C, while the arc voltage was constant at ~ U_{arc}=24.3 V, throughout the deposition. Deposition pressure after the formed equilibrium condition was ~ $1.5x10^{-4}$ Torr. The third layer, a-C film, was subsequently deposited exposing the substrate to the second magnetron located on the chamber top flange. The deposition conditions were summarized in Table 1.

During sputter deposition Ar flow in the chamber was kept constant at 30 sccm and the necessary deposition pressure was maintained by throttling the exhaust side of the turbomolecular pump. A target-to-substrate distance was ~ 60 mm while for the FCA source with the double bended 45-degree filter, a separation was around 115 cm.

Table 1. Overview of deposition parameters in the UHV chamber.

Layer	Deposition pressure (Torr)	Deposition rate (nm/s)	Ar flow sccm	Magnetr on power (W)	Arc current (A)	Typical thickness (nm)	Typical deposition times (s)
CoCrPtB	10×10^{-3}	0.26	30	80	/	30	120
Cr	1.5×10^{-4}	0.045	0	/	80	0.5	12
a-C	7×10^{-3}	0.043	30	100	/	2	46

2.1. Business Environment Corrosion Test

This test has been performed on the structure: a) Si substrate/CoCrPtB (30 nm)/Cr(0.5 nm)/a-C (2 nm) and b) Si substrate/CoCrPtB (30 nm)/a-C(2 nm), figure 2.

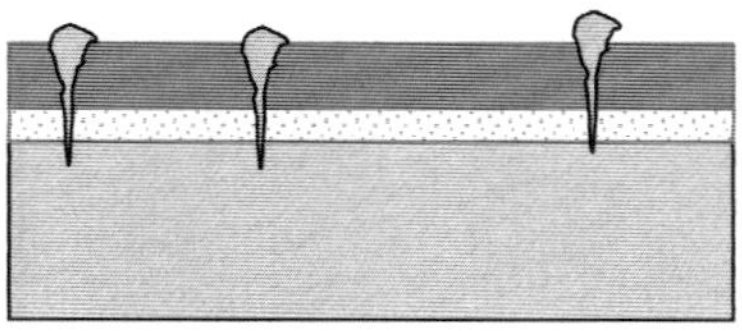

Figure 2. Model of the corrosion affected hard disk structure.

Half of the samples have been exposed to the business environment test (exposure to the 0.5 M HCl vapor for up to 284 hours) to promote corrosion and migration of cobalt to the top of the structure. Samples were analyzed by RBS technique.

2.2. Monte Carlo Simulation

In ultra-thin multilayer films, that can be deposited using energetic ion beams with filtered cathodic arc technology[6,9], is essential to predict interface widths and penetration depths of the ion forming films.

Main assumptions in the Monte Carlo simulations of the ion beam interaction during the film growth were: a) target is homogeneous and b) there are no thermal and vacancy induced diffusions. An ultra-thin, void-free, dense and smooth protective film can be produced by increasing the surface mobility of arriving film forming species. Optimization of the film structural properties, interface widths and a surface roughness can be accomplished by the proper selection of: substrate bias, energy of arriving ions and neutrals as well as their incident angle on the film surface. Ultra-smooth surface can be produced when ion beam polishing takes place along with a carbon film growth. The main process on the surface during ion beam deposition growth is a cascade process in the form of displacement collisions, vacancy production, replacement collision and interstitial atoms production.

3. Results and discussions

3.1. Corrosion study of the migrated Co using RBS technique

A corrosion behavior of the hard disk structure is usually investigated using some electrochemical tests, e.g. polarization resistance method[10,11]. Other extensively applied methods range from the most simplified ones like counting the number of corrosion spots with optical microscope to the more demanding methods that employ surface analysis, like XPS, AES or SIMS to measure the surface concentration of the migrated cobalt atoms. In this paper the corrosion related cobalt surface concentration was measured by RBS technique.

Surface area exposed to the beam was up to 1 mm in diameter. The incident He^{+} ion beam was 2 MeV and provided with a 15 keV energy resolution of silicon barrier detector, a depth resolution of ~ 10 nm for platinum. To improve a depth resolution across a thin scattering depth of Co on the top of hard disk structure, the *side* geometry has been applied, keeping always the same source-to-detector angle of 20 degree. In the *side* arrangement more target atoms were involved in the backscattering events and a better depth resolution of the Co and Pt peak has been achieved. The energy difference ΔE between ions scattered at the surface and ions emerging from the sample at the depth ΔZ is

$$\Delta E = KE_0 - E_1 = [\varepsilon]\, N\Delta Z \quad (1)$$

where $[\varepsilon]$ is the stopping cross-section factor and N is the atomic density of the sample. The kinematic factor K is known for every element. The incident ion with energy E_0 suffers first the energy loss on the migrated Co on the top on the surface, and additionally on the return trip to detector, after probing the magnetic layer with platinum content. Therefore, the broadening and energy position of the well resolved Pt peak can be used in evaluating the amount of cobalt on the top of the surface. The leading and trailing edge of Pt peak has been attenuated by passing the beam through the migrated cobalt. The total number of counts in the RBS spectrum is proportional to the beam current I, acceptance angle of the Si detector ($\Delta\Omega$), number of scattering centers per cm^2 ($N\Delta Z$) and the differential cross section of the scattered species ($d\sigma/d\Omega$)

$$C = I\, \Delta\Omega\, N\Delta Z\, d\sigma/d\Omega \quad (2)$$

To get initial insight on the RBS spectra, a corrosion measurement on the film CoCrPtB(30 nm)/a-C (2 nm) was first studied, as presented in figure 3.

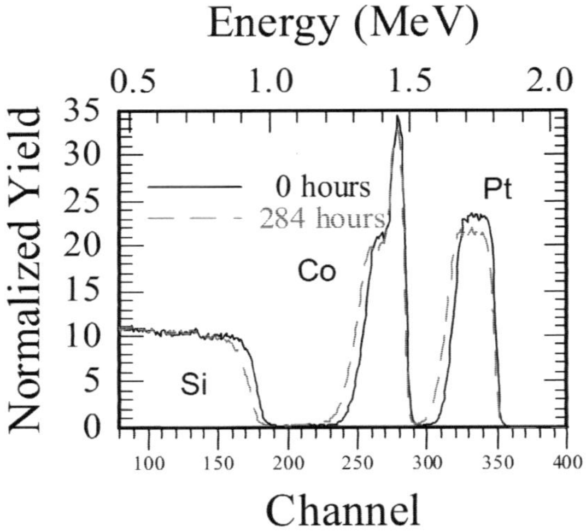

Figure 3. RBS spectrum of the simplified hard disk structure before and after 284 hours of exposure in 0.5 M of HCl vapor. The FWHM of the Co peak is 0.125 MeV after the exposure.

Exposure of the hard disk structure to the HCl corrosion vapor can be analyzed and quantified with RBS method by observing the Co signal: a) broadening of the Co peak takes place on the HCl treated sample (an increase of the FWHM of the Co peak from 0.106 MeV to 0.125 MeV), b) decrease in the normalized yield of Co-peak from 37 a.u. (for untreated sample) to 34.2 a.u. for the treated one. Also, on the overall spectrum an oxidization of the top of the surface has been observed.

A low-energy shift in the Pt signal, centered at ~ 1.75 MeV, for the corrosion treated sample indicates a loss of the He^+ ion energy on the migrated Co that started to surface a structure. After experiencing a collision, and a loss of incident energy with migrated Co, the He^+ ions start to sense the magnetic layer (Pt) on its original position (dashed line). Therefore, a shift in the leading/trailing edge of the Pt signal may be a good measure of the corrosion susceptibility of such a structure. The RBS corrosion quantification is a spot sensitive in nature. Other examined spots in area of 1mm on the same disk led to different conclusions. Therefore, it is necessary to measure disk corrosion parameters radially, e.g. from outer to inner disk diameter averaging contributions from all spots along the disk circumference. Also, in ultra thin films with thicknesses in the sub-nanometer range the influence of the local surface roughness can have a large impact on the energy position of Cr and Pt peaks and their respective widths. Optical microscopy combined with micro-Raman study proved also to be surface-spot sensitive in nature, implying that only integral evaluation techniques can lead to reliable conclusions.

3.2. Monte Carlo Simulation of the Film Growth

With increasing the ion incident angle with respect to the surface normal the number of backscattered ions goes up as well as the lateral and radial range of ions. The surface coverage of a film improves, with less imperfection in the film growth, mainly seen as pinhole defects formed after initial nucleation and coalescent/island film formation phase. To reduce the calculation time, 1000 ions with initial energy of 100 eV were followed in analysis.

In order to make more dense and pinhole-free film there is need for incident ions to impart energy to the target atoms more in the radial direction and improve the surface mobility of "thermalized" film forming species. The surface roughness on the microscope level can thus be also adjusted. In practice, all ion sources suffer from non-uniform flux distribution across its radius and exhibit pressure dependant "beam throw distances". In the FCA source the beam divergence is additionally influenced by the scanning magnetic field. Plasma electrons are closely guided by the magnetic field lines and ejected ions are tied to the electrons by electrostatic forces Therefore, the substrate rotation under to the beam slanted incident angle is essential for uniform film coverage and low surface micro-roughness.

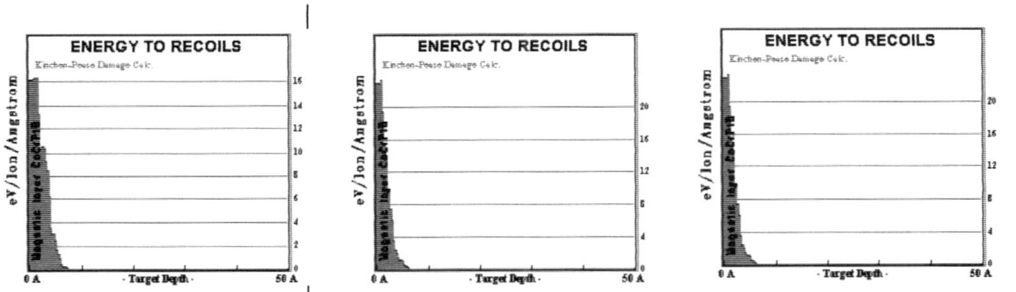

0 degree | 45 degree | 85 degree

Figure 4. A depth distribution of the energy of the substrate (target) atoms, CoCrPtB magnetic layer, acquired by the varying incident angle of Cr ions at 100 eV energy. The 1000 ions were taken in calculations. The calculated penetration depth of ions was always below 5 Angstrom.

The surface mobility of ad-ions can be substantially increased during the grazing incident angle film deposition. An average energy imparted to recoils (target atoms initially at rest) ranges from 16 eV/ion/Angstrom to 44 eV/ion/Angstrom with an ion incident angle change from 0 degree to 85 degree, respectively, figure 4. In the extremely thin nanometer and sub-nanometer thin films, that are now more common in the hard disk industry, especially when deposited by energetic ion beams, with energies over 100 eV per incoming ion, the Monte Carlo calculation of the stopping range of ions in material may be a good guidance in selecting the necessary deposition parameters and in evaluation of the interface widths.

3.3. SERS - Micro Raman study of the ultra thin carbon

The 1 nm and 8 nm thin ta-C overcoat films were deposited with filtered cathodic arc technique over the Si/CoCrPtB (20 nm) magnetic layer. To improve the low Raman signal-to-noise ratio for the 1 nm carbon film a flash layer of silver has been sputter deposited on the top of a structure.

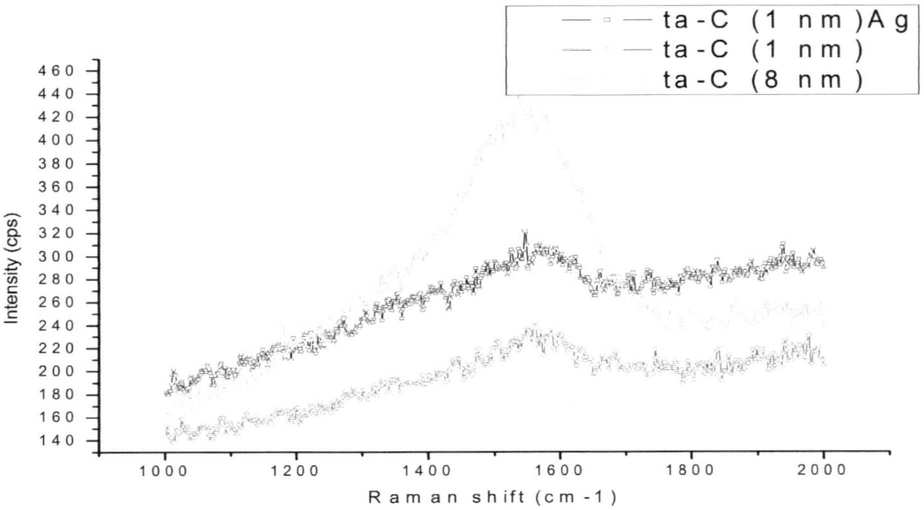

Figure 5. Micro Raman spectrum of the ta-C carbon film close to the detection limit. Lower curve, ta-C film,1 nm thin. Middle curve- SERS after the surface enhancement with sputter deposited silver film. The preliminary increase of signal-to-noise ratio of at least 30% has been observed.

Carbon film was analyzed in the 1000-2000 cm^{-1} range. Measurement has been done close to the detection limit of the micro-Raman setup. To improve the signal-to-noise ratio the 10 consecutive collections of the Ar 514 nm laser scattered lights has been acquired. The incident laser light had the output power of ~21 mW. For the reference purpose the micro Raman spectrum of 8 nm thick ta-C film has been presented, acquired under the same conditions, the top curve on figure 5.

Under constant measurement conditions, from run-to-run, and taking care of the same laser light focusing/collecting area, being ~3 micrometer in diameter, the area under the broad carbon band and especially under the deconvoluted G-peak (at around the shift at 1580 cm-1) is proportional to the carbon film thickness. Deposition of silver nanorods, instead of implemented sputtered silver film may further enhance the excitation of localized surface plasmons. Distribution of metal nanoparticles (Ag, Au, Pt) on top of the carbon film can be a method of choice in the micro Raman carbon film evaluation at extremely thin structures. Surface signal enhancement is especially strong for carbon double bonds, C=C. In the investigated film most of carbon bonds were in the tetrahedral, sp3 configuration, resulting in the lower signal enhancement. The SERS technique can be used in monitoring the carbon content under the G (graphite) peak and carbon corrosion protection study of the magnetic layer.

4. Summary

The filtered cathodic arc technology can be successfully used not only in the read/write head manufacturing but also on the hard disk surfaces. The low film deposition rate may restrict its implementation to only the extremely thin structures like the flash layer, prior to carbon overcoat. The beam scanning across the magnetic layer at incident angles other than 0^0 with respect to the surface normal may be of advantage for improved surface coverage and reduction of the surface roughness on the microscopic level.

In the corrosion study of the hard disk structure the RBS technique can be successfully implemented. The energy position and broadening of the clearly distinctive Pt peak in the RBS spectra of the magnetic structure containing Pt can be used in the evaluation of the corrosion protective properties of the overcoat material. To improve the depth resolution of the Co and Pt spectrum the incident He^+ beam should be oriented at the grazing incident angle, geometry offering more backscattering events with the target material.

Upon assumption of the target elemental composition, its density and thickness from one side, and energy, incident angle and type of the incident ion on another side, a Monte Carlo simulation can give a clue to the expected ranges-penetrations depths of the elements/ions on the boundaries, interface widths and help in selecting the appropriate layer thicknesses for the barrier-flash layer.

Acknowledgment
Author would like to thank Prof. Thomas Osipowicz, National University of Singapore, for the RBS measurements and A*STAR, Singapore's Agency for Science, Technology and Research.

References

[1] Bhushan B, "Chemical, mechanical and tribological characterization of ultra-thin and hard amorphous carbon coatings as thin as 3.5 nm: recent developments" 1999 *Diamond and related materials* **8** 1985-2015
[2] Grundy P J 1998 "Thin film magnetic recording media" *J. Phys. D: Appl. Phys.* **31** 2975-90
[3] Baldwin D A and Falabella S 1995 "Deposition processes utilizing a new filtered cathodic arc source" *38th Annual Technical Conference Proceedings, Society of Vacuum Coaters* **505/856-7188** 309-316
[4] Shi X, Tay B K, Tan H S, Zhong L, Tu Y Q, Silva S R P and Milne W I 1996 "Properties of carbon ion deposited tetrahedral amorphous carbon films as a function of ion energy" *J. Appl. Phys.* **79(9)** 7234-40
[5] Witke T, Schuelke T, Schultrich B, Siemroth P and Vetter J, 2000 "Comparison of filtered high-current arc deposition (Φ-HCA) with conventional vacuum arc methods" *Surf. and Coat. Technol.* **126** 81-88
[6] Anders A and Ryan F W 2000 " Ultrathin ta-C films on heads deposited by twist-filtered cathodic arc carbon plasma", TRIB-Vol. 10 *Symposium on Interface Tribology Towards 100 Gbit/in^2 and Beyond* ASME p.43-50
[7] Guruz M U, David V P, Chung Y W, Lacerda M M, Bhatia C S, Yu Y H and Lee S C 2001 "Corrosion performance of ultrathin carbon nitride overcoat synthesized by magnetron sputtering" *Thin Solid Films* **381** 6-9
[8] Tomcik B, Seng S C, Balakrisnan B and Lee J Y 2002 "Electrochemical tests on the carbon protective layer of a hard disk" *Diamond and related materials* **11** 1409-15
[9] Chun S Y and Chayahara A, 2000 " Pulsed vacuum arc deposition of multilayers in the nanometer range" *Surf. and Coat. Technol.* **132** 217-221
[10] Novotny V and Staud N 1988 *J. Electrchem. Soc. Electrochem. Sci. Technol.* **135**(2)2931-2938
[11] Novotny V, Itnyre G, Homola A and Franco L 1987 *IEEE Trans. Mag.* **23** (5) 3645-47

Generation of atmospheric pressure non-thermal plasma by diffusive and constricted discharges in air and nitrogen at the rest and flow

Yuri Akishev, Mikhail Grushin, Vladimir Karalnik, Igor Kochetov, Anatoly Napartovich, Nikolay Trushkin

SRC RF TRINITI, Pushkovykh St.-12, Troitsk, Moscow region, 142190, Russia

e-mail: akishev@triniti.ru

Abstract. Main subject of this paper is low current atmospheric pressure gas discharges powering with DC power supplies. These discharges are widely used for generation of non-thermal or non-equilibrium plasma in air and nitrogen which are much cheaper compared to rare gases like He or Ar. Molecular nitrogen as plasma forming gas has a unique capability to accumulate huge energy in vibration, electron (metastables) and dissociated (atomic) states. Besides, all active species have a long life-time, and they can be therefore transported for a long distance away from the place of their generation. Different current modes (diffusive and constricted) of these discharges are discussed. Experimental and numerical results on generation of chemically active species in the diffusive and constricted mode are presented.

1. Introduction

The weekly ionized non-thermal plasma (NTP) is of great interest for many applications because of its strong non-equilibrium state wherein an average electron energy T_e exceeds markedly gas temperature T_g, i.e. electrons in the NTP are strongly overheated compared to neutral gas. Energetic electrons due to frequent collisions with the neutrals excite and dissociate effectively atoms and molecules of the plasma-forming gas that results in a creation of physically-, and bio-chemically active gaseous medium in a practically cold background gas.

The NTP at atmospheric pressure is of considerable interest for practice. A reason is that sustaining the NTP at atmospheric pressure at first allows us to avoid the use of expensive vacuum equipment and second gives opportunity to use the NTP for treatment of the exhausted gases and polluted liquids. The second opportunity cannot be realized at all with use of the NTP at low pressure. Another advantage of atmospheric pressure NTP compared to plasma at low pressure is its ability to intensify plasma treatment that is well suited for fast activation of the subjects to be treated.

At present there are many kinds of plasma sources working at low and atmospheric pressure and using MW, RF, low frequency, pulsed and DC power supplies for NTP generation [1-10]. The subject of this paper is low current atmospheric pressure gas discharges powering with DC power supplies. These discharges are widely used for generation of non-thermal or non-equilibrium plasma in air and nitrogen which are much cheaper compared to rare gases like He or Ar. Molecular nitrogen as plasma forming gas has a unique capability to accumulate huge energy in vibration, electron (metastables) and dissociated (atomic) states. Besides, all N_2-active species have a long life-time, and they can be therefore transported for a long distance away from the place of their generation.

Low current DC discharges at atmospheric pressure exhibit different current modes (diffusive and constricted). Physical reasons that are responsible for sustention of diffusive and constricted mode of these discharges are discussed. Experimental and numerical results on generation of chemically active species in the diffusive and constricted mode are presented.

2. The most important parameters of atmospheric pressure NTP

The NTP at atmospheric pressure is very complicated subject and characterized by numerous parameters. We would like to point to the most important of them. The measurement and/or calculation of these parameters is to be main concern for the experiment and numerical modeling the NTP.

The most important parameter is the reduced electric field strength E/N which determines a value of the average electron energy Te (E is the electric field strength, N is the gas density). Besides the E/N ratio influences the electron energy distribution function (EEDF) which governs a distribution of the energy released by NTP in between different channels like gas heating, vibration and electronic state excitation, dissociation, ionization, etc.

Other important parameters characterizing the NTP are the number densities of the charged particles: electrons Ne, negative ions Nn and positive ions Np. These densities vary over a wide range (10^9 – 10^{13} cm^{-3}) and determine the intensity of electron and ion collisions providing a transfer of electrical energy to the NTP.

The ratio between electron temperature Te of plasma and translation Tg and vibration Tv temperatures of the process gas shows the degree of non-equilibricity of the NTP. As a rule, $Tg \ll Tv \ll Te$. In such a case, energetic electrons generate intensively numerous bio-chemical active species. High magnitude of Tv provides effectively the promotion for non-equilibrium plasma-chemical reactions under cold ambient. A magnitude of Tg determines NTP-applicability for treatment of thermal-sensitive materials.

Bio-chemical reactivity of the NTP is determined by sort and number density of active species generated by the NTP. Because of this, solid data on number densities of the radicals and metastables like O, OH, CH, He*, N_2*, etc formed by particular NTP are needed, and they allow us to judge the applicability and efficiency of this plasma to the solution of any specific problem.

3. On problems referred to sustaining the diffusive atmospheric pressure NTP

The atmospheric pressure NTP can exist in two forms: diffusive (or glow) and constricted. As a rule, diffusive NTP is characterized by lower gas temperature and current density than those in the constricted NTP. Gas discharge with diffusive NTP has a narrow current region for its existence. Development of ionization instability in the diffusive NTP results in its transition to the constricted mode. Let us point to several physical reasons provoking the constriction of diffusive discharge.

In the absence of vortex electric fields, each gas discharge can be roughly divided into three main parts: two near-electrode layers and plasma column. In the course of discharge forming, both cathode and anode layers are going to the states characterized by so-called normal current densities. Cathode/anode normal current density grows drastically (quadratic) with an increase in gas pressure and equals to 100-200 A/cm^2 for discharge in N_2 and air at atmospheric pressure (note a diffusive plasma column is characterized by much lower current density). A tendency of near-electrode layers to come to the states with high current density leads to formation of the cathode/anode current spots which provoke a constriction of diffusive plasma column.

Normal current density in He is lower than that in molecular gases by a factor 100. Due to this, time τ needed for normal near-electrode layer formation is much longer compared to that in the most of molecular gases. So, formation of normal spots can be avoid, if period T of the alternating applied voltage $T \ll \tau$. This condition can be satisfied in DBD with He (or mixture with He) that enables us to generate a diffusive (or homogeneous) NTP at atmospheric pressure [11, 12]. Diffusive DBD can be obtained also in N_2 at atmospheric pressure but within very restricted region of the experimental parameters [13].

4. Influence of an electrode geometry and gas flow on a discharge stability

Steady-state normal cathode layer is characterized by strong electric field. It is therefore difficult to destroy cathode spot or diminish its current density by any external action. To extend/spread DC gas discharge in transverse to the current direction, cathode is fabricated as sectioned, i.e. the cathode surface has been broken mechanically into numerous small pieces each of them can be ballasted (or not) by individual resistor. In the sectioned discharge, there are many small cathode spots regularly dispersed over the cathode surface. Fortunately, small cathode spots do not provoke a constriction of plasma column. The reason is that the Dark Faraday Space which shields diffusive plasma column from cathode layer with high electric field.

Figure 1a demonstrates the sketch of the electrode system with the sectioned cathode. Every cathode element is ballasted by a resistor of 1MOhm. This system works in atmospheric pressure gas flow and generates diffusive (glow) discharge (Fig.1b). Installation can be easily scale up and enables us to generate diffusive plasma inside big volume. For details see [14-17].

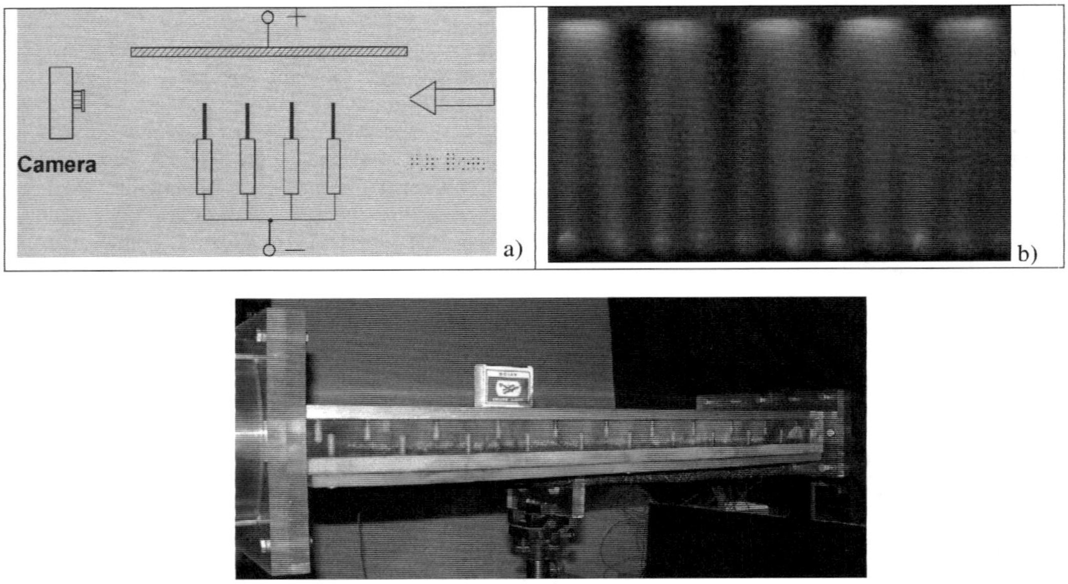

c)

Fig.1. a) Sketch of the electrode system with the sectioned cathode; b) Image (fragment) of DC diffusive (glow) discharge with the sectioned cathode (below) in airflow at atmospheric pressure. The photo is taken by camera in upwind direction; c) Image of DC gas discharge chamber with the sectioned cathode. Lilac color inside discharge chamber corresponds to the diffusive plasma. Gas flow rate of this installation is 300 m^3/hour.

Steady-state normal anode layer is characterized by much lower electric field compared to that in cathode layer. However anode spot is much more dangerous for stability of a diffusive plasma column because the anode spot actively initiates plasma column constriction. Fortunately, owing to low electric field, normal anode spot can be easily destroyed, i.e. transformed into anode layer with low current density which equals close to that in diffusive plasma column. One can possibly avoid a normal anode spot formation by two approaches: with changing the shape of anode and blowing the anode. This idea is illustrated with figure 2.

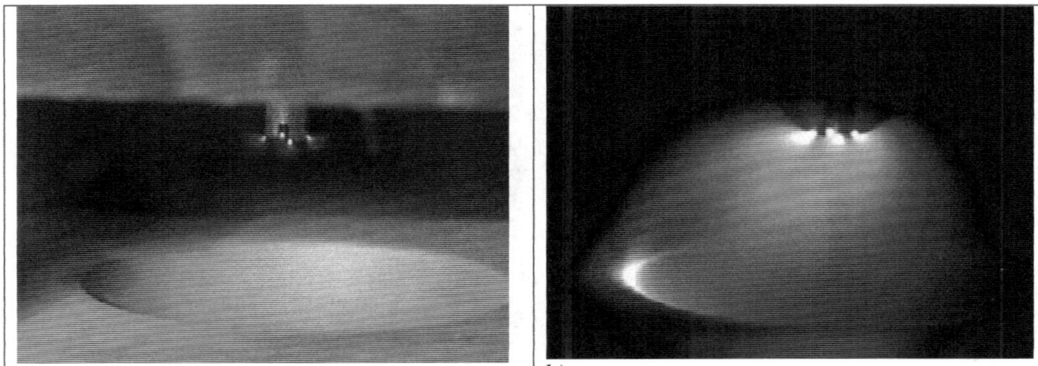

a) b)

Fig.2. a) Image of the electrode geometry providing the stabilization of both the anode layer against its transition to normal anode spot and the diffusive plasma column against its constriction (from [18]). Cathode (upper) and anode (bottom) are fabricated with several sharpened spikes and spherical crater. b) Side view of steady-state glow discharge with diffusive NTP in airflow at atmospheric pressure, I=1mA.

5. NTP created by constricted discharges at atmospheric pressure

Diffusive state of the NTP at atmospheric pressure is unstable in regards to development of different ionization instabilities. In the most cases atmospheric pressure NTP exists therefore in a constricted form like filaments or microdischarges (MDs).

Advantages of the constricted NTP: constricted discharges at atmospheric pressure can be realized much easier compared to high pressure diffusive discharges; the constricted NTP is also non-equilibrium plasma and can generate effectively and abundantly different sorts of active species.

Disadvantages of the constricted NTP: filamentation (constriction) of a discharge leads to the reduction of a gas volume activated by NTP; mixing active species of the constricted NTP with non-activated surrounding gas diminishes non-equilibricity of the afterglow; MDs provide intensive local energy deposition that can result in non-homogeneous surface treatment and even cause a local damage in the case of thin films.

5.1. Constricted (filamentary) discharges in transverse gas flow

Filamentary discharge in transverse gas flow is always non-stationary and exists only as pulsing self-repeated discharge. In general, this discharge is similar to well-known gliding arc between two divergent electrodes but not the same to that. The key features of filamentary discharge in transverse gas flow can be illustrated by example of the constricted discharge in ambient airflow between needle and knife (Fig.3). Gas velocity is 50 m/s. Average current is 2.5 mA.

Figs. 3a) and 3b) show front (in upwind direction) and side view of DC constricted (filamentary) discharge in transverse airflow. With the naked eye this discharge looks more or less homogeneous as it seen in Fig.3b. In fact, the constricted discharge exists in the pulsing self-repeated regime the period of which includes a formation of filament, its blowing and decay. Repetition frequency of the current pulses is not constant but varies within 1-10 kHz. A set of the side view shots (##1-4) elucidates a spatial-time behavior of the self-repeated filamentary discharge: #1 – gas gap breakdown or formation of the filament; #2-3 – extension of the filament due to its blowing from discharge gap; #4 – break and decay of the filament. Exposure time of each shot is 5 μs. Time interval between shots #1 and #4 equals to 150 μs.

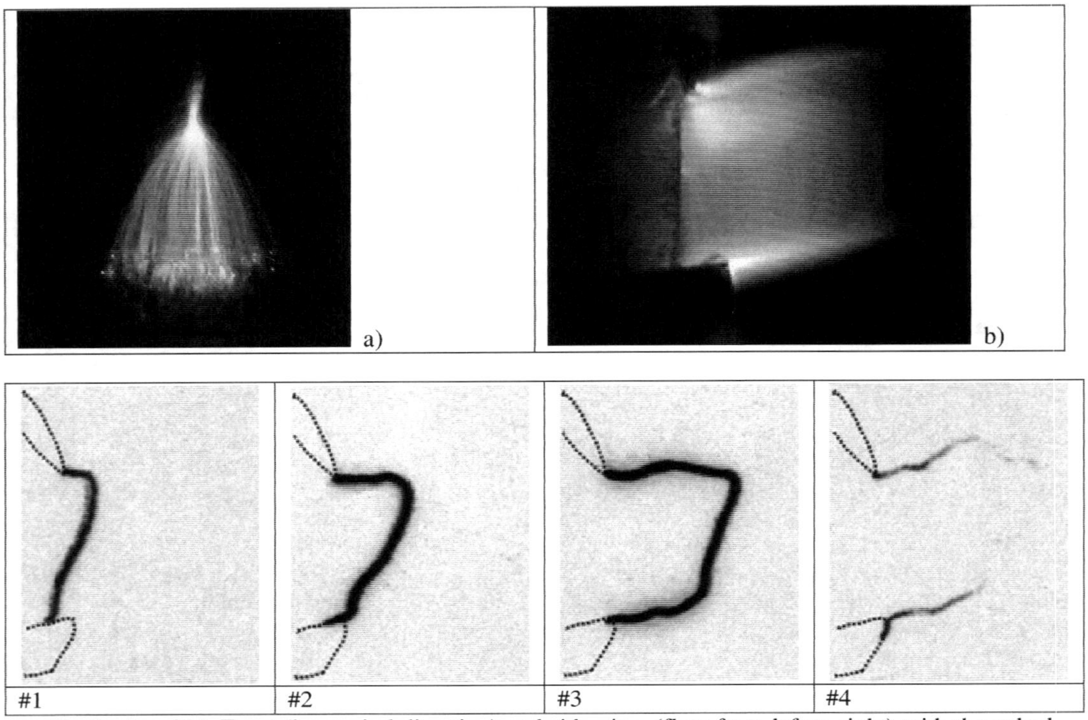

Fig.3. a) and b): Front (in upwind direction) and side view (flow from left to right) with the naked eye of DC constricted (filamentary) discharge in transverse airflow. Electrodes are needle (upper) and knife (bottom). Exposure time is 0.1s. Gas velocity is 50 m/s. Average current is 2.5 mA. Constricted discharge exists in the pulsing self-repeated regime. Repetition frequency of the current pulses is not constant but varies within 1-10 kHz. A set of the side view shots (##1-4) elucidates a spatial-time behavior of self-repeated filamentary discharge: #1 – gas gap breakdown or formation of the filament; #2-3 – extension of the filament due to its blowing from discharge gap; #4 – break and decay of the filament. Exposure time of each shot is 5 μs. Time interval between shots #1 and #4 equals to 150 μs.

Fig.4 presents experimental evidences of non-equilibricity and high efficiency of the constricted NTP generated by pulsing self-repeated DC discharge in transverse airflow. Parameters of discharge are the same as those in Fig.3. The coordinate "zero" in the horizontal axis corresponds to the outlet of the gas discharge gap. Spectroscopic measurements allow us to prove the generation of such active species as electronic excited nitrogen molecules $N_2(C)$, oxygen atoms O and radicals OH. Besides, one can see also a big difference between vibration and translation temperatures.

However spatial behavior of the species formed by high-energetic electrons is opposite to that for vibration temperature determined by collisions of low-energetic electrons with N_2-molecules. The reason is that increase in the filament length due to its blowing results in diminishing the reduced electric field inside the constricted plasma. These changes in the E/N value influence strongly the EEDF high-energetic "tail" that is responsible for electronic excitation, dissociation and ionization but not so much the EEDF low-energetic part (the EEDF "core") responsible for vibration excitation.

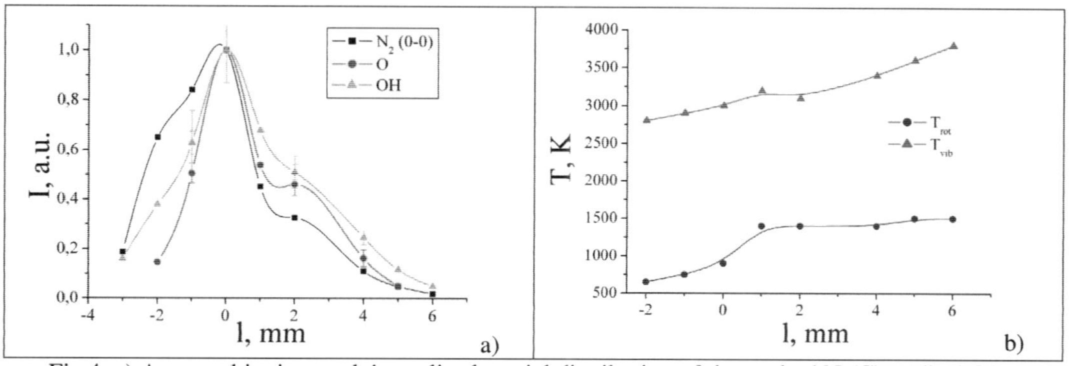

Fig.4. a) Averaged in time and normalized spatial distribution of the excited $N_2(C)$, radical OH and atom O **in** and **outside** the gap for pulsing self-repeated filamentary discharge in transverse airflow. b) Averaged in time spatial distribution of the vibration and rotation temperatures **in** and **outside** the gap. Experimental parameters of the constricted discharge are the same as those in Fig.3. The coordinate "zero" in the horizontal axis corresponds to the outlet of the gas discharge gap.

5.2. **Constricted (filamentary) discharges in longitudinal gas flow**

Longitudinal discharge was activated inside a glass tube blown out with N_2 flow at atmospheric pressure [19]. The diameter of glass tube and inter-electrode gap were the subjects to variation from 5 to 20 and 10 to 40 mm, respectively. Images of short constricted discharge and its long and wide afterglow are shown in Fig.5a. Volt-Ampere Characteristics of this discharge is presented in Fig.5b.

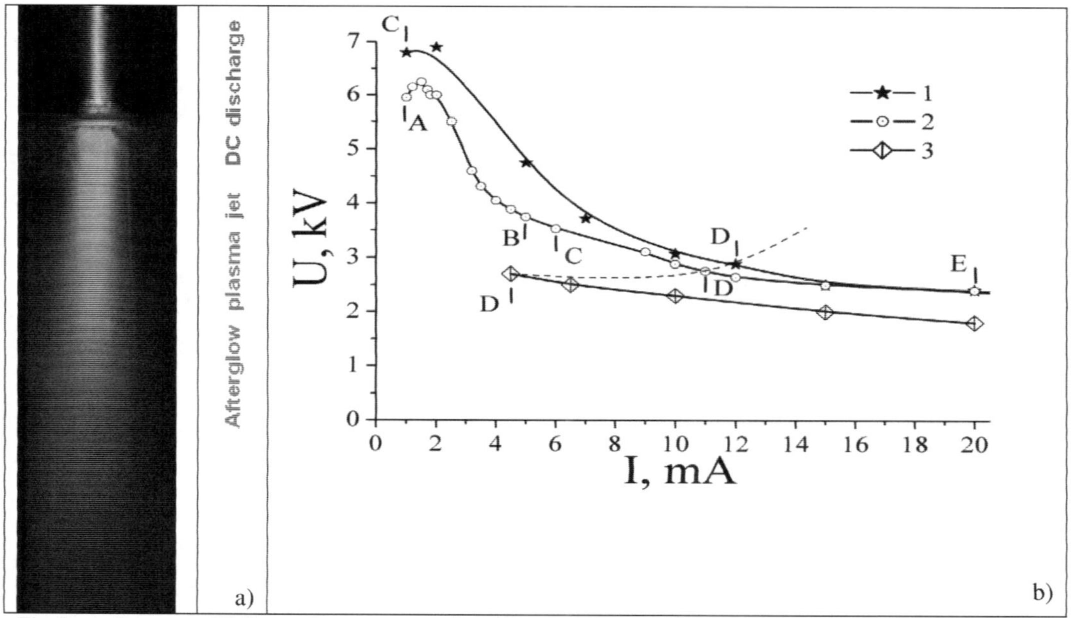

Fig.5. a): Images of DC steady-state constricted discharge (upper) and its afterglow in longitudinal N_2 flow at atmospheric pressure. Discharge is activated in glass tube with inner diameter of 5 mm. Inter-electrode distance is 15 mm. Radius of the constricted plasma column is about 1mm. b) Averaged in time Volt-Ampere Characteristics of the constricted DC discharge. Gas flow rate is variable parameter: 1 – 320 cm^3/s; 2 - 190 cm^3/s; 3 - 0 cm^3/s (gas at the rest). Purity of N_2 is 99.999%.

Figure 6a shows a dependence of rotational Trot (or Tgas) and vibration Tvib temperatures inside the constricted discharge in longitudinal N_2 flow vs average discharge current under different gas flow rate of N_2. For comparison, the temperatures T_r and T_v measured in the CGD at rest gas are shown as well. One can see, both temperatures depend on the current, however their dependences are not similar to each other: rotation temperature T_r grows monotonically with an increase in current but behavior of the vibration temperature T_v is non-monotonic having maximum about 6000K at the current I≈7-8mA. Maximum deviation from thermodynamic equilibrium in the NTP (T_v>>T_r) there is in the current region 2mA<I<20mA. With increase in current both temperatures, T_r and T_v, tend to approach each to other. It corresponds to gradual transition of the constricted discharge to the state with thermodynamic equilibrium like to that in arc discharge.

Figure 6b presents the distribution of Tgas and Tvib along N_2 afterglow formed by plasma jet ejected from the tube with the constricted discharge. Cross-section of afterglow is bigger compared to that of the constricted discharge. Both the expansion of plasma jet on outlet the tube and mixing with ambient non-activated gas result in diminishing of temperatures Tgas and Tvib that is seen in Fig.6b.

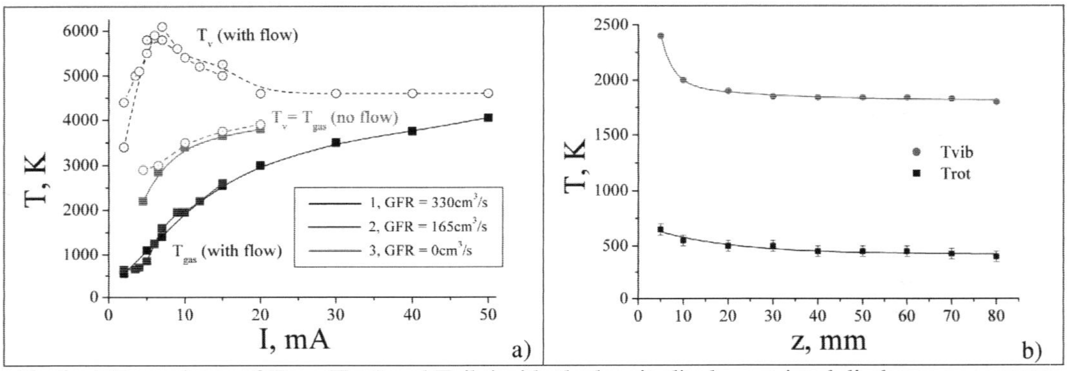

Fig.6. a) Dependence of Tgas (Trot) and Tvib inside the longitudinal constricted discharge vs average discharge current. Gas flow rate (GFR) of N_2 is variable parameter (see the legend). b) Distribution of Tgas and Tvib along N_2 afterglow (plasma jet ejected from the tube with the constricted discharge). Gas flow rate is 320 cm^3/s.

Optical spectra of the constricted N_2 flow discharge and its afterglow are presented in Fig.7 One can see (Fig.7a) the emission of the first positive system 1^+ ($N_2(B^3\Pi^+ \rightarrow A^3\Sigma)$), second positive system 2^+ ($N_2(C^3\Pi^+ \rightarrow B^3\Pi^+)$) and first negative system 1^- ($N^+_2(B^2\Sigma^+ \rightarrow X^2\Sigma^+)$) give the most input in total emission from the constricted plasma column. Besides there are weak bands of CN radical ($CN(B^2\Sigma \rightarrow X^2\Sigma)$). An existence of CN emission from the constricted discharge is attribute to trace level impurities containing inevitably in nitrogen of high purity (99.999%).

Spectrum of emission from afterglow is distinct from that of the constricted plasma – with the naked eye the constricted plasma column looks as red colour but afterglow has purple or lilaceous colour. The emission from afterglow does not contain spectral lines of nitrogen but there is intensive emission of CN, NO and NH. These particles are formed due to secondary reactions of N_2 active species with trace admixtures of ppm-level like O_2, H_2O, CxHy.

Note the rate of primary active species transformation in reactions with the impurities increases with an increase in pressure of "pure" gas: the higher gas pressure, the stronger influence of the impurities on the composition of active species in afterglow and in the discharge as well.

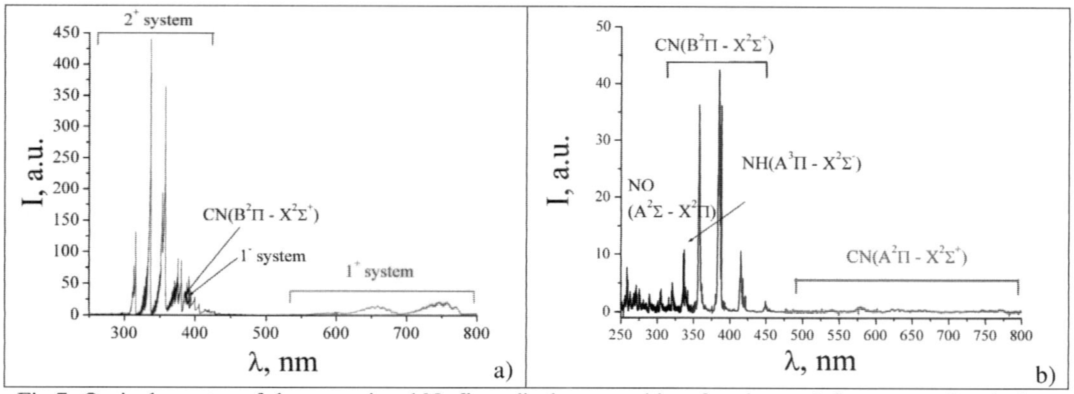

Fig.7. Optical spectra of the constricted N_2 flow discharge and its afterglow. a) Spectrum of emission from the constricted plasma column. b) Spectrum of emission from afterglow jet. Discharge current is 40 mA.

6. Numerical calculations on the composition of reactive species in the NTP

Vast diversity of the collision processes forming the NTP provides, from one side, a variety of plasma properties, but from other side, the complexity in modeling the plasma itself and plasma chemistry as well. The present state in both the designing/invention of plasma sources and NTP-processing is characterized, as a rule, by a "trial and error" strategy. Besides, some of the important plasma parameters not always can be determined by experimentally. In such a case, important role belongs to numerical modelling of both the reactive species generation by NTP and their behaviour in afterglow. Numerical simulation of the diffusive NTP in airflow was done in [15, 16]. Here main attention will be paid to simulation of the constricted NTP.

Structure of the program module that is normally used for simulation of the NTP is presented in Fig.8. In general, such module includes the blocks with electron scattering cross-sections data, plasma chemistry kinetic data and several solvers: Boltzmann equation solver, plasma chemistry solver and gas dynamic and electric circuit solver. The input for calculations are experimental conditions such as gas mixture, temperature and pressure, gas flow rate and velocity, electrode and chamber geometry, voltage amplitude, frequency, parameters of electric circuit, etc. The output of simulations is a composition of the NTP and reactive species.

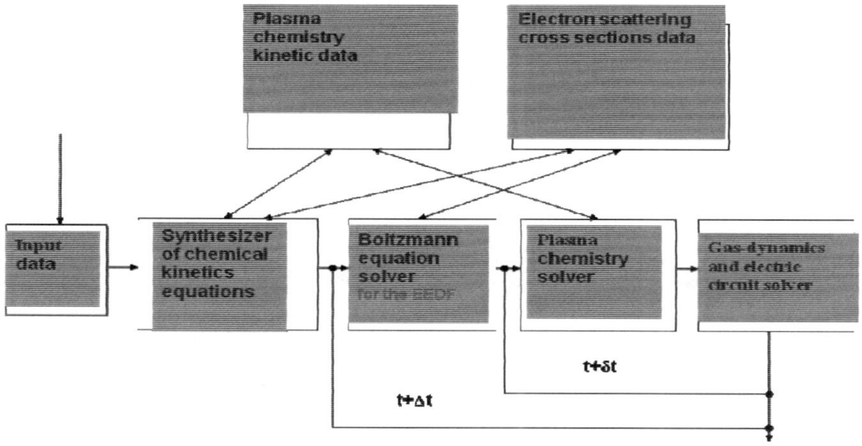

Fig.8. Structure of program module that is used normally for simulation of the NTP.

Journal of Physics: Conference Series **257** (2010) 012014 doi:10.1088/1742-6596/257/1/012014

6.1. Modeling of the constricted NTP in transverse airflow at atmospheric pressure

The processes taken into account in the simulation of the NTP in airflow are listed below. The results obtained are presented in Fig. 9.

$O_2 + e \rightarrow O_2(vibr) + e$	$H_2O + e \rightarrow H_2O(rot) + e$
$O_2 + e \rightarrow O_2(0.98) + e$	$H_2O + e \rightarrow H_2O(vibr1) + e$
$O_2 + e \rightarrow O_2(1.63) + e$	$H_2O + e \rightarrow H_2O(vibr2) + e$
$O_2 + e \rightarrow O^- + O$	$H_2O + e \rightarrow H + OH + e$
$O_2 + e \rightarrow O_2(4.5) + e$	$H_2O + e \rightarrow H_2O^+ + e + e$
$O_2 + e \rightarrow O + O + e$	$H_2O + e \rightarrow OH + H^-$
$O_2 + e \rightarrow O + O(^1D) + e$	$H_2O + e \rightarrow H_2 + O^-$
$O_2 + e \rightarrow O_2^+ + e + e$	$N_2(A3) + H_2O \rightarrow N_2 + H + OH$
$O_2 + e \rightarrow O + O^+ + e + e$	$N_2(A3) + O_2 \rightarrow N_2 + O + O$
$O_2 + e + (O_2) \rightarrow O_2^- + (O_2)$	$N_2(A3) + O_2 \rightarrow N_2O + O$
$N_2 + e \rightarrow N_2(vibr) + e$	$N_2(A3) + N_2O \rightarrow N_2 + N_2 + O$
$N_2 + e \rightarrow N_2(E{=}1) + e$	$N_2(A3) + NO_2 \rightarrow NO + O + N_2$
$N_2 + e \rightarrow N_2(esum) + e$	$N + NO \rightarrow N_2 + O$
$N_2 + e \rightarrow N + N + e$	$N + NO_2 \rightarrow N_2O + O$
$N_2 + e \rightarrow N_2^+ + e + e$	$N + NO_2 \rightarrow NO + NO$
$H_2O + e \rightarrow H_2O(rot) + e$	$N + O_2 \rightarrow NO + O$
$H_2O + e \rightarrow H_2O(vibr1) + e$	$NO + O_3 \rightarrow NO_2 + O_2$
$H_2O + e \rightarrow H_2O(vibr2) + e$	$O + O_2 + O_2 \rightarrow O_3 + O_2$
$H_2O + e \rightarrow H + OH + e$	$H_2O + e \rightarrow OH + H$
$H_2O + e \rightarrow H_2O^+ + e + e$	$H_2O + e \rightarrow H_2 + O$

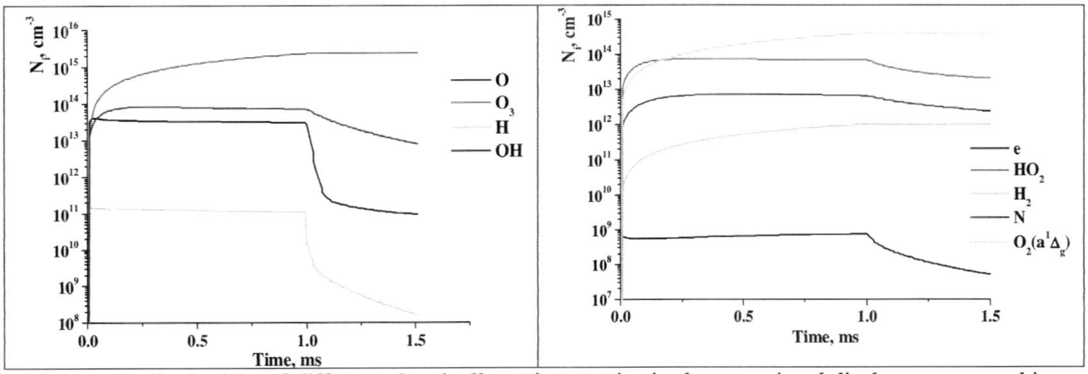

Fig.9. Evolution in time of different chemically active species in the constricted discharge zone and in post discharge region. DC constricted (filamentary) discharge in transverse airflow. Experimental conditions correspond to those in Fig.3.

6.2. **1-D modeling of steady-state constricted N_2 glow discharge at atmospheric pressure**

Numerical model takes into account the radial transfer of the particles and heat due to diffusion and heat conduction and includes the set of appropriate balance equations for charged particles: electrons Ne and positive ions, N^+, N_2^+, N_3^+, N_4^+. The charged particle kinetics considers the ion conversion and several types of ionization processes: direct ionization, stepwise ionization, associative and pooling ionization and electron-ion recombination and ambipolar diffusion as well. Electronically excited states $N_2(A)$, $N_2(B)$, $N_2(C)$, $N_2(a')$; vibration energy for the ground state of $N_2(X,v)$ and gas heating were included also. The needed rate coefficients were taken from literature and calculated with use of Boltzman equation solver developed by I.Kochetov and A.Napartovich. For details see [20].

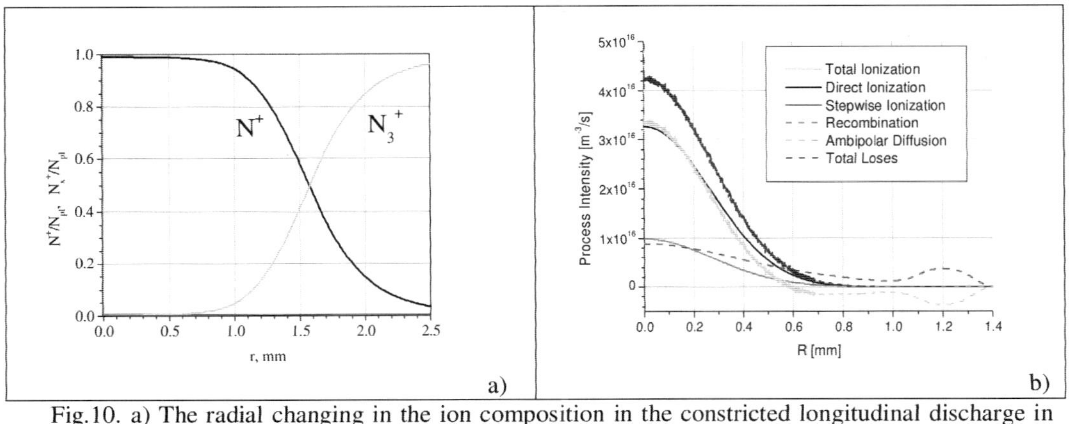

Fig.10. a) The radial changing in the ion composition in the constricted longitudinal discharge in N_2 flow. I=40 mA. Gas flow rate is 114cm^3/s. b) Ionization balance in hot core of the constricted plasma.

In the constricted plasma column, gas temperature changes in a radial direction from high to low magnitude (3500 K at the axis and 520 K at the wall). Under such conditions, the impact of ion conversion on discharge radial structure reveals itself the most obviously: main sort of positive ions is atomic N^+ ion inside hot plasma, and molecular N_3^+ ion in cold periphery region of discharge (see

Fig.10). The strongest changing in the ion composition occurs within a thin cylindrical layer at the nearest periphery of plasma column ($0.75 \leq r \leq 1.5$ mm) where 14% of the total current is transferred.

Based on information in Fig.10b one can conclude that ionization balance in hot core of the constricted N_2-plasma column is non-local because it is controlled by ambipolar diffusion but not recombination. From this it amazingly follows that constricted plasma column in N_2 at atmospheric pressure is similar to diffusive glow discharge at low pressure under Schottky regime!

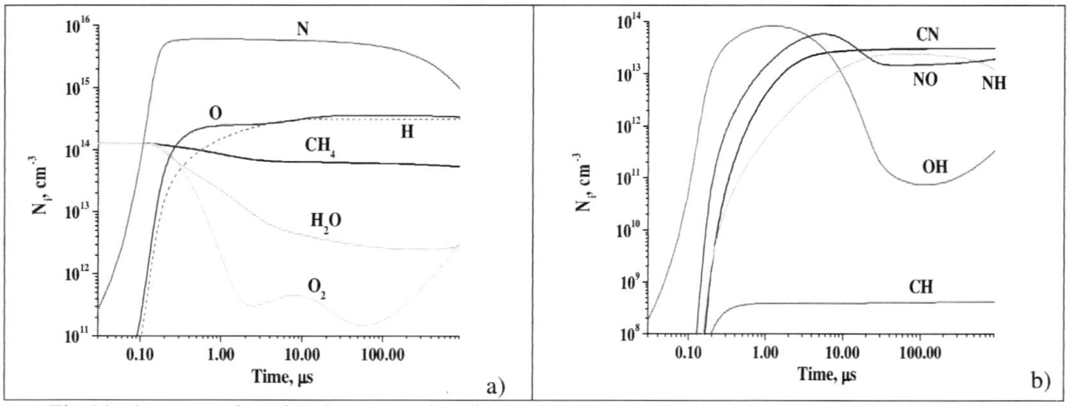

Fig.11. Atoms and molecules (a) and radicals (b) number density distribution along afterglow of N_2-plasma jet containing impurities: 5 ppm CH_4, 5 ppm O_2 and 5 ppm H_2O. Time = distance from the nozzle/gas velocity. The reduced energy deposition in the constricted discharge is 600 J/liter.

Figure 11 presents numerical results on composition of non-N_2 active species in N_2-afterglow due to influence of gaseous impurities of ppm-level. The impurities of such level there are in the balloons with "pure" nitrogen (99.999%) delivered by the manufacturer. Non-N_2 active species with number densities about 10^{14} cm^{-3} can dramatically change both the paths of plasma chemical reactions and final results of the NTP treatment.

7. Conclusion

Atmospheric pressure NTP can be generated in two forms: diffusive and constricted. Diffusive NTP can be created by AC discharges (for instance, by DBD) and DC discharges with sectioned electrodes in transverse gas flow. Constricted NTP can be generated by AC and DC discharges in transverse and longitudinal gas flow.

In the case of transverse gas flow the constricted NTP is non-stationary (discharge is pulsing and self-repeated), in the most cases of longitudinal gas flow the constricted NTP is steady-state. As a rule, constricted NTP characterized by higher gas temperature nevertheless strong non-equilibricity of plasma can be kept.

In "pure" gases from balloons always there are trace impurities. Plasma chemical reactions of these impurities with primary active species generated by NTP change drastically the composition of active species. This effect has to be taken into account in the use of atmospheric pressure plasma treatment and interpretation of the final results.

Acknowledgement.

Authors thank to the RFBR for partial support of this work (grant No 08-02-00601a).

References

[1] Ishijima T, Sugiura H, Saito R, Toyota H, Sugai H 2010 *Plasma Sources Sci Technol.* **19** 015010
[2] Sakiyama Y and Graves D 2009 *Plasma Sources Sci Technol.* **18** 025020
[3] Waskocnig J, Niemi K, Knake N, Graham L, Reuter S, Schulz-von der Gathen V and Gans T 2010 *Plasma Sources Sci Technol.* **19** 045018
[4] Roth J, Rahel J, Dai X and Sherman D 2005 *J. Phys. D: Appl. Phys.* **38** 555-567
[5] Buntat Z, Smith I R and Razali N A M 2009 *J. Phys. D: Appl. Phys.* **42** 235202
[6] Kogelschatz U 2003 *Plasma Chem. Plasma Process.* **23** 1-46
[7] Briels T M, Veldhuizen E M and Ebert U 2008 *J. Phys. D: Appl. Phys.* **41** 234008
[8] Heesch E J M, Winands G J J and Pemen A J M 2008 *J. Phys. D: Appl. Phys.* **41** 234015
[9] Akishev Yu, Grushin M, Dyatko N, Kochetov I, Napartovich A, Trushkin N, Minh Tran Duc and Descours S 2008 *J. Phys. D: Appl. Phys.* **41** 235203
[10] Bibinov N, Dudek D, Awakowicz P, Engemann J 2007 *J. Phys. D: Appl. Phys.* **40** 7372-7378
[11] Laroussi M, Alexeff I, Richardson J P and Dyer F F 2002 *IEEE Trans. Plasma Sci.* **30** 158
[12] Akishev Yu, Demyanov A, Karalnik V, Pankin M, Trushkin N. 2001 *Plasma Phys. Rep.* **27** 164-171
[13] Gherardi N, Naude N, Cambronne J-P and Massines F 2005 *J. Phys. D: Appl. Phys.* **38** 530-538
[14] Akishev Yu, Deryugin A, Kochetov I, Napartovich A and Trushkin N 1993 *J. Phys. D: Appl. Phys.* **26** 1630-1637
[15] Akishev Yu.S., Deryugin A.A., Elkin N.N., Kochetov I.V., Napartovich A.P., and Trushkin N.I. 1994 *Plasma Phys. Rep.*, **20** (5), 437-441.
[16] Akishev Yu.S., Deryugin A.A., Karal'nik V.B., Kochetov I.V., Napartovich A.P., and Trushkin N.I. 1994 *Plasma Phys. Rep.*, **20** (6), 511-524.
[17] Akishev Yu.S., Deryugin A.A., Kochetov I.V., Napartovich A.P., and Trushkin N.I. 1994 *Plasma Phys. Rep.*, **20** (6), 525-532.
[18] Callebaut T., Kochetov I., Akishev Yu., Napartovich A., Leys C. 2004 *Plasma Sources Science and Technology.* **13** (2), 245-250.
[19] Akishev Yu, Grushin M, Karalnik V, Petryakov A and Trushkin N 2010 *J. Phys. D: Appl. Phys.* **43** 075202
[20] Akishev Yu, Grushin M, Karalnik V, Petryakov A and Trushkin N 2010 *J. Phys. D: Appl. Phys.* **43** 215202

Journal of Physics: Conference Series **257** (2010) 012015 doi:10.1088/1742-6596/257/1/012015

Collective phenomena in volume and surface barrier discharges

U Kogelschatz[1]

Retired from ABB Corporate Research, Segelhof 1, 5405 Baden, Switzerland

e-mail: u.kogelschatz@bluewin.ch

Abstract. Barrier discharges are increasingly used as a cost-effective configuration to produce non-equilibrium plasmas at atmospheric pressure. This way, copious amounts of electrons, ions, free radicals and excited species can be generated without significant heating of the background gas. In most applications the barrier is made of dielectric material. Major applications utilizing mainly dielectric barriers include ozone generation, surface cleaning and modification, polymer and textile treatment, sterilization, pollution control, CO_2 lasers, excimer lamps, plasma display panels (flat TV screens). More recent research efforts are devoted to biomedical applications and to plasma actuators for flow control. Sinusoidal feeding voltages at various frequencies as well as pulsed excitation schemes are used. Volume as well as surface barrier discharges can exist in the form of filamentary, regularly patterned or diffuse, laterally homogeneous discharges. The physical effects leading to collective phenomena in volume and surface barrier discharges are discussed in detail. Special attention is paid to self-organization of current filaments and pattern formation. Major similarities of the two types of barrier discharges are elaborated.

1. Introduction

Gas discharges, predominantly used at atmospheric pressure and formed in narrow gaps between closely spaced electrodes, at least one of which is covered by a special coating are called barrier discharges (BDs). Typical barrier materials used are thin glass or ceramic plates or tubes. In laboratory experiments also the use of resistive, ferroelectric and semiconducting materials has been investigated. Also the use of porous ceramic layers and dielectric barriers with controlled surface conductivity has been reported. By far the most important BDs are those using dielectric barriers: dielectric barrier discharges or DBDs, in the literature also often referred to as silent discharges. Due to better diagnostics, advanced computer simulations and novel applications DBDs, known since 1857, have experienced a genuine renaissance during the past two decades. Reviews of the subject and of the older literature on barrier discharges were published by Kogelschatz *et al* [1], by Wagner *et al* [2] and by Fridman *et al* [3]. A detailed discussion of various properties of barrier discharges and their important industrial applications can also be found in the book *Non-Equilibrium Air Plasmas at Atmospheric Pressure* by Becker *et al* [4]. More recent research activities focus on DBD torches [5], on flow control [6] and on biomedical applications [7]. In this contribution the focus is on the discharge physics, mainly on collective phenomena such as filamentation, pattern formation and self-organization, and on the conditions for obtaining diffuse, laterally homogeneous discharges.

[1] To whom any correspondence should be addressed.

2. Volume barrier discharges (VBDs)

2.1 *Filamentary VBDs*

In many atmospheric-pressure VBDs a multitude of thin current filaments referred to as microdischarges (MDs) is observed. The duration of a single microdischarge is a few ns, the transferred charge is of the order 0.1 nC for a 1 mm air gap, increasing with gap spacing and specific capacitance of the dielectric barrier(s). Although previously it had been assumed that these MDs are randomly distributed in space and time we now know that that they occur only during certain phases of the discharge cycle and that there is quite a bit of spatial coordination and self-organization of MDs [8]. The charge deposited by a MD at the dielectric surface determines its duration by reducing the electric field in the gap at this location. It also has an influence on adjacent or subsequent MDs. As long as the voltage is rising new MDs occur in new locations where the electric field is still high. When the voltage reverses the next MDs preferentially seek locations of previous MDs where the field has collapsed. This memory effect caused by the surface charges deposited on the dielectric is extensively used in plasma displays [9]. At elevated frequencies additional volume memory effects are observed when the time between succeeding half cycles becomes too short for the ions or electrons to decay or be swept out of the gap. Also long-lived metastables can carry information from the previous half cycle.

Since all MDs are initiated at about the same breakdown voltage and the overall gap voltage stays close to the breakdown value during the active discharge phase a rather simple equivalent circuit for such an ensemble of a large number of MDs can be given (figure 1a). The discharge gap is represented by two anti-parallel Zener diodes, clamping the gap voltage at the breakdown value. In this case the voltage-charge Lissajous figure of the DBD resembles an ideal parallelogram (figure 1b) in which the slopes of the straight lines are related to the total capacitance of the configuration during the phases without discharge and to the capacitance of the dielectric barrier(s) during the active phases. The voltage change due to the charge transfer of a single MD is less than the line width in the diagram. Under certain conditions pronounced voltage collapses are observed that are indicative of simultaneous discharge of a large number of MDs covering a considerable fraction of the electrode area (figure 1c). The diagrams were obtained with a laboratory ozone generator with a 1 mm wide annular discharge gap formed by an outer steel tube and an inner glass tube of 1 m length (sinusoidal feeding voltage, frequency: 1 kHz, electrode area: 1200 cm^2).

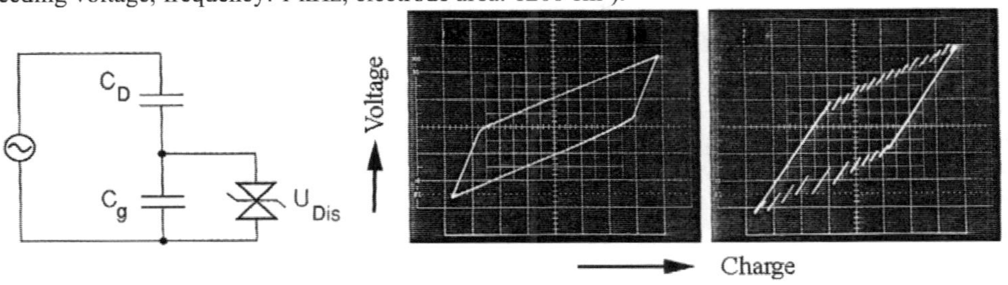

Figure 1. Simple equivalent circuit and voltage-charge Lissajous figures of a filamentary DBD in O_2 without and with self-pulsing.

This self-pulsing of the discharge has also been observed in mixtures CO_2 and CH_4 [10]. It depends on the slope dU/dt of the feeding voltage when the breakdown voltage in the gap is reached and on the surface conditions. They are influenced by humidity (surface conductivity) and also by deposition on or etching of the surface. Pulsing effects can occur only during the positive half cycle, only during the negative half cycle, or in both. The reason for this behaviour is that during the delay phase of breakdown the radiation of one initial MD can trigger adjacent MDs through the action of UV photons releasing electrons from the electrode surfaces. In the case of fast rising voltages (pulsed DBDs) the MDs can be synchronized in such a way that the entire gap voltage collapses in a short time.

2.2 *Single filament measurements*

The last decade, by the use of advanced diagnostic techniques, brought much more detailed information about the plasma parameters inside the MD columns. Different groups succeeded in stabilization the location of an individual current filament so that repetitive measurement techniques could be applied. Probably the most powerful tool is cross-correlation spectroscopy (CCS), first introduced to DBD investigations by Kozlov and co-workers at Moscow State University in 1995, described in detail in 2001 [11] and more recently reviewed in a book chapter by Wagner *et al* [12]. CCS allows sensitive emission spectroscopy with high spatial and temporal resolution. Such measurements provided important data about the local and temporal evolution of the electric field and the relative electron density inside a MD channel and led to a better understanding of the electrical breakdown processes leading to MD formation, the discharge physics as well as the involved plasma chemistry. Subsequent investigations were extended to rare gases and diffuse barrier discharges. Other groups focused on the kinetics of excimer formation and on the quantitative evaluation of VUV excimer radiation in Ar and Xe [13]. Perhaps a remark of caution should be added. An isolated single MD filament may not have exactly the same properties as a MD squeezed in between adjacent MDs and forced to share the available surface area and dielectric capacitance with its neighbours.

2.3 *Single filament modelling*

Early 2D numerical simulations of MD formation [14] showed that after an initial electron avalanche phase an ionisation wave is formed that upon arrival at the momentary cathode results in the formation of a thin cathode fall region of high electric field strength. In air, O_2 or N_2 at atmospheric pressure this fall region extends only a few μm into the discharge gap. The current peak of a single MD follows the formation of the cathode fall. Immediately thereafter charge accumulation on the dielectric leads to a reduction of the electric field at this location until the MD is extinguished. Consequently a MD can be regarded as a self-arresting discharge of a few ns duration. More advanced numerical models were published for atmospheric pressure air [15] and N_2 [16]. Computed emission intensities of the second positive and the first negative systems of nitrogen agree with recent experimental data. Experimental results for a relatively long (up to 1 μs) Townsend pre-phase and for the velocity of the cathode-directed ionizing wave are satisfactorily reproduced in the models. Secondary electron emission from the dielectric surface caused by impinging ions on the momentary cathode is an important process. Inside the MD channel electron densities of the order 10^{14} cm^{-3} are reached. At the dielectric the MD channel spreads into a wider surface discharge. With its thin high field region (reduced field E/n about 1000 Td) at the momentary cathode and a channel of quasi-neutral plasma (positive column) at relatively low reduced field (E/n about 100 Td) each MD in its high current phase can be characterized as a transient high pressure glow discharge. In these models the cathode layer on the dielectric is formed as an extremely thin (5 - 10 μm thick) space charge layer. Both modelling efforts come to the conclusion that photon-induced volume processes (photoionisation) must be considered to adequately describe the expansion of the MD channel across the surface of the cathode dielectric during the MD quenching phase.

2.4 *Pattern formation in VBDs*

Under certain conditions the weakly ionized plasma in a DBD can also show regular self-organized luminous patterns. Many authors described the formation of luminous spots and regular discharge patterns in barrier discharges between narrowly spaced planar electrodes [17]. Boyers and Tiller described hexagonal arrays of "plasma bubble domains" which they identified as regular patterns of localized glow discharges. They already distinguished between dim uniform Townsend discharges, assemblies of glow discharge domains and bright uniform normal or abnormal glow discharges. Kogelschatz showed a hexagonal DBD discharge pattern resembling Rayleigh-Bénard convection cells (figure 2c). Breazeal and co-workers talked about static and dynamic two-dimensional current-density patterns of "self-extinguishing discharge avalanches" [17].

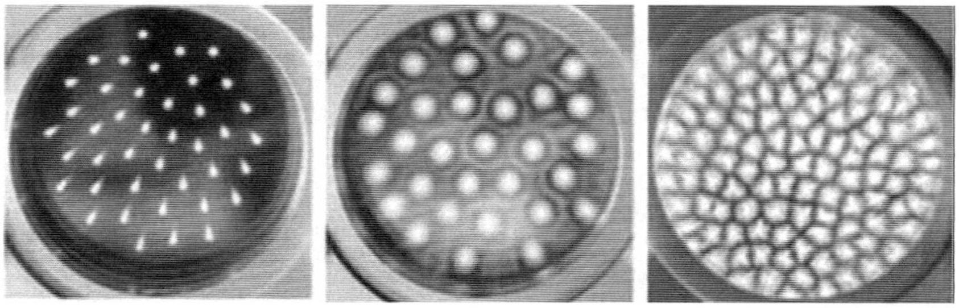

Figure 2. Two-dimensional discharge patterns obtained in 5-mm wide cell with planar walls filled with xenon/chlorine mixtures at different pressures (visible diameter: 8 cm, 1 kHz sine voltage).

Figure 2 shows photographs of self-organized discharge structures taken in 1991. A round flat silica glass cell of 5 mm width with planar walls was filled with Xe/Cl_2 mixtures at different pressures. The walls served as dielectric barriers. These photographs were taken with a normal camera and relatively long exposure times (20 ms) indicating fairly stationary conditions. The picture in the middle clearly shows discharge areas of different luminosity and presumably different current density with intermediate dark regions. Very beautiful discharge patterns including spirals, hexagonal and square lattice and interlacing superlattice patterns have been published by Lifang Dong and co-workers [18]. They used Ar and Ar/air mixtures in a relatively simple DBD arrangement with planar glass plates as dielectric barriers and water electrodes. Most experiments were performed close to atmospheric pressure with 1.4 mm or 1.5 mm discharge gaps and a sinusoidal feeding voltage with frequencies ranging from 49 to 62 kHz. Figure 3 is taken from their 2008 publication.

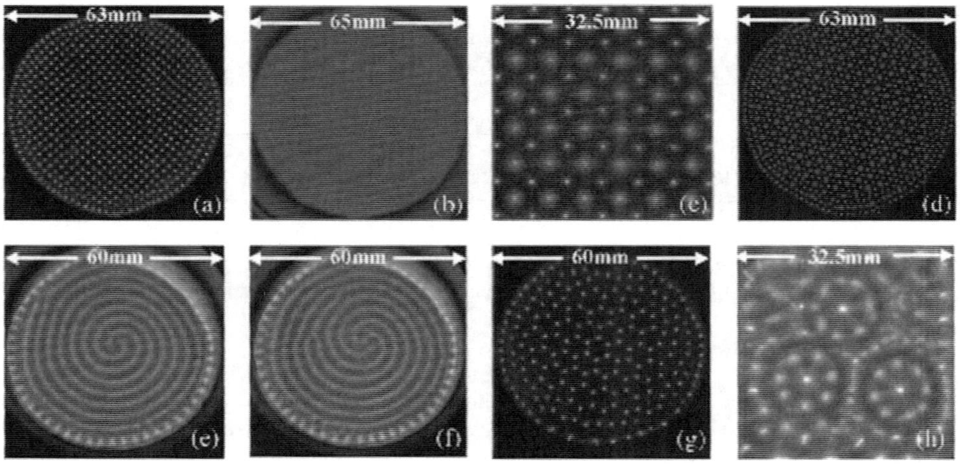

Figure 3. Regular patterns in a DBD operating in Ar/air mixtures. Taken from [18c] with permission. (© 2008 IEEE)

Progress in the understanding of these DBD patterns was made by synchronizing short exposure image converter pictures with current traces and spectroscopic diagnostics. Miralaï *et al* [19] presented a systematic investigation of the optical appearance of DBDs in air, He, Ar and N_2. In air they always obtained a filamentary discharge, in He and pure N_2 patterns of luminous columns or totally diffuse discharges, in Ar bright luminous columns which were narrower and more densely packed. At low driving frequency (4 kHz) also in Ar diffuse discharges could be obtained, using a MgO coating on the steel electrode. Gurevich *et al* [20] reported on the spatio-temporal evolution of luminous patterns in the form of concentric rings (figure 4). In N_2 at a pressure of 2.5 hPa after ignition they first observed a concentric ring pattern with a bright spot at the centre in the positive half cycle of the operating voltage. When the voltage was further raised they got an inverse pattern with a dark spot at the centre and dark regions at the position of the previously bright rings. Changing to the negative half cycle they first obtained the ring pattern with the dark centre followed by the pattern with the bright spot at the centre. The two ring patterns per half cycle corresponded to two current peaks per half cycle superimposed on the sinusoidal displacement current. This is a clear indication that the spatiotemporal behaviour of the ring pattern is determined by the charges previously deposited on the glass plates.

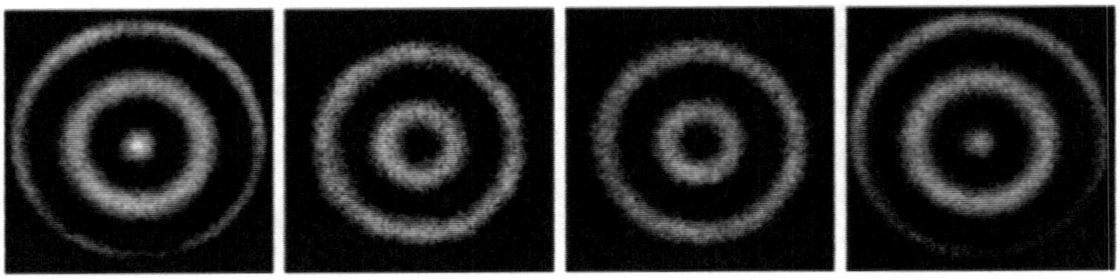

a b c d

Figure 4. Concentric ring patterns in a DBD in N_2. Reprinted with permission from Gurevich E L, Zanin A L, Moskalenko A S and Purwins H-G *Phys. Rev. Lett.* 91, 154501, 2003. Copyright (2003) by the American Physical Society. http://prl.aps.org/abstract/PRL/v91/i15/e154501

Bernecker *et al* [21] observed a similar spatiotemporal behaviour with hexagonal luminous structures in Ne. At the beginning of a cycle first a bright luminous hexagonal structure appeared, followed by a much fainter honeycomb structure in the space between. These two interlacing structures are related to a first high current peak and to a much smaller current peak during one half cycle of the driving voltage. The authors suggest that the bright regions can be characterized as glow-discharge domains while the darker regions in between are diffuse Townsend discharges. The type of pattern obtained in a DBD depends on the gas properties, the width of the discharge gap and the operating pressure, but also on the barrier thickness and material and on the driving frequency. Also the shape of the electrodes has a strong influence by setting strong boundary conditions [22].

In recent years it became possible to measure localized remnant surface charges in DBDs and to demonstrate that they correspond to observed self-organized luminous glow patterns [23]. Moreover, the experimental results of the surface-charge distribution measurements are in good agreement with those calculated in a full 3D discharge simulation solving the corresponding transport equations [24].

2.5 *Modelling of pattern formation in VBDs*

Purwins and co-workers presented a number of papers on pattern formation in gas discharges and used phenomenological two or three component activator-inhibitor models (reaction diffusion models) to reproduce the observed luminous patterns [25]. A variety of self-organized patterns could be generated without using details about gas discharge physics. In this approach the gas discharge is treated as a homogeneous medium with S-shaped current-voltage characteristic that contains a section with negative differential conductivity. Ebert and Arrayás [26] published a more general approach to pattern formation in low current density electrical discharges based on a minimal model of reaction, drift and diffusion of two charged species coupled nonlinearly to an externally imposed electric field. Further details on the transition from Townsend to glow discharge were published by Šijačič and Ebert [27] and by Raizer *et al* [28].

Brauer *et al* [29] used a self-consistent two-dimensional fluid model of a He DBD ($2.4 \cdot 10^4$ Pa, 0.5 mm gap). The authors could show how an initially homogeneous plasma abruptly turned into regularly spaced glow discharge regions after about 30 μs, which corresponded to 6 cycles of the applied sinusoidal voltage. Zhang and Kortshagen [30], using a more complex kinetic model for He including

20 ppm N_2, showed that also the opposite can occur. A patterned He DBD can transit to a transversely homogeneous discharge. In conclusion, it can be stated that the self-organized behaviour of DBDs can be simulated by phenomenological activator-inhibitor models or, more precisely, by 2D or 3D models treating the discharge physics and chemical kinetics. The latter are also able to predict the size and the spacing of the glow discharge regions found in experiments. Also the time it takes to reach stationary conditions is reproduced in the simulations. An increase in pressure leads to smaller glow domains and closer spacing, in experiment as well as in simulation.

2.6 *Diffuse volume barrier discharges*
Early experimental work on diffuse or laterally homogeneous DBDs at atmospheric pressure was initiated by Okazaki and co-workers at Sophia University in Tokyo about 1987 [31] and followed up by Massines and her group in Toulouse [32]. This work started a world wide activity in the field. Recent advances have been reviewed by Massines *et al* [33]. In the beginning, all diffuse DBDs were referred to as atmospheric pressure glow discharges (APGs, APGDs). Later one learned to distinguish between glow discharges and more Townsend-like discharges (APTDs). It is relatively easy to obtain homogeneous DBDs in He, Ne or N_2. With additives or a special sandwich structure placing a wire mesh between the dielectric and the metal electrode also in air, O_2 and Ar stable glows were obtained in atmospheric pressure DBDs, even when using a 50 Hz source [34]. These experiments were repeated and confirmed by many groups [35]. The function of the wire mesh at the back side of the dielectric remained a mystery for a long time. Through the work of Tepper [36] and that of Fang [35] it was clarified that the mesh initiates a corona discharge that evenly charges the PET (polyethyleneterephthalate) dielectric, which is an electret material. When a certain threshold voltage is reached charge carriers are spontaneously expelled into the main gas gap (polarization switching).

More detailed information about homogeneous DBDs and transitions between diffuse and filamentary DBDs were obtained by advanced diagnostics using short exposure ICCD images and spectroscopy (CCS, LIF) linked to electrical measurements [37]. The existence and extension of a high field region at the cathode in the case of He discharges was confirmed by Stark effect measurements [38]. The measured values of the electric field near the cathode (about 10 kV/cm) and the extension of the cathode fall region (about 0.6 mm) indicate that these DBD glow discharges do not reach the state of a fully developed glow discharge. From atmospheric pressure DC glow discharges in He between metal electrodes the following experimental results were reported: current density at cathode 2.3 A/cm^2, electric field at cathode: 62 kV/cm, width of cathode layer: 0.06 mm [39]. A reduction of the current density reduces the field strength and increases the extension of the fall region [40]. It appears that in most DBDs the dielectric barrier quenches the discharge at an early stage of cathode fall development already at a peak current density of a few mA/cm^2.

From the large number of publications on homogeneous DBDs one must conclude that many parameters including the properties of the power supply, the dielectric barriers, the used gas and pressure, the geometrical configuration and, most strikingly, the operating frequency have an influence. Memory effects including surface charges and long living species from the previous half cycle play a dominant role. For this reason the first few cycles after ignition differ significantly from the final quasi-stationary state, which normally is achieved within less than 10 cycles. In the first cycles filaments and moving current fronts are observed. Nevertheless it is possible to specify working domains in which homogeneous DBDs can be reproducibly obtained. Most homogeneous DBDs are characterized by one short current pulse per half cycle of the operating voltage. An interesting alternative is the multi pulse mode in which a number of regularly spaced short current pulsed of descending peak height are obtained per half cycle [41]. Very recently it was demonstrated that a fast gas flow can influence the number of pulses and also the shape of the short current pulses [42].

2.7 *Numerical modelling of diffuse VBDs*
A few remarks concerning the terms used in the literature should be made. The diffuse or "homogeneous" DBD described in many papers is at best transversely or laterally homogeneous. In

the direction of the electric field it is never homogeneous. Many authors distinguish between glow discharge DBDs and Townsend-like DBDs. These expressions are borrowed from low pressure dc discharges. A glow discharge is characterized by a high field region at the cathode (cathode fall region), a region of negative glow and a column of quasi-neutral plasma (positive column) in addition to intermediate dark spaces. A classical Townsend discharge is characterized by electron and ion concentrations small enough not to noticeably distort the applied electric field. Both, over a wide range, maintain a voltage independent of the current. In the extensive literature on DBDs the terms are used in a less stringent form. Apparently homogeneous DBDs with a high field region near the momentary cathode, indicative of a cathode fall region, are called glow or glow-like DBDs. Uniform barrier discharges in the Townsend regime are frequently characterized by much lower current densities and by the absence of a quasi-neutral plasma region. In diffuse DBDs we often observe operating modes in which the cathode fall is not fully developed due to the current limitation by the dielectric barriers but space charge concentrations are too high to qualify as classical Townsend discharges. For this intermediate range the qualification subnormal glow discharge is sometimes used.

A large number of papers deal with the numerical modelling of homogeneous He DBDs at atmospheric pressure. Contrary to low pressure discharges, molecular species in the form of the excited dimer He_2^* and the ion He_2^+ now gain importance [43]. The influence of impurities in rare gas plasmas is stressed [44, 45]. Even a minute 1 ppm N_2 admixture to pure He will make N_2^+ ions as abundant as He_2^+ ions. Above 17 ppm admixture nitrogen ions govern the positive charge carriers and N_4^+ takes a leading role [45]. Excited He species in connection with impurities play an important role in sustaining a diffuse discharge. A detailed numerical model of such a discharge in a 5mm gap between glass plates operated at 35 kHz shows a short current pulse of sub-microsecond duration per half cycle, a high field region at the cathode indicating the structure of a glow discharge at maximum current density (2.3 mA cm^{-2}) and resembling the structure of a subnormal glow discharge with residual electrons at the momentary anode just before the next current peak [46]. These results are in excellent agreement with measurements by Luo *et al* [47]. Also the appearance of a multitude of short current pulses per half cycle observed when the voltage is raised could be explained by modelling [48]. By varying the thickness of the dielectric Mangolini *et al* could demonstrate its current limiting properties [49]. A thicker dielectric reduces the observed current peak and shifts the operating point along the subnormal branch of the current voltage characteristic from a glow-like towards a more Townsend-like appearance. Detailed 2D models [30] could also reproduce observed radially moving current fronts during the ignition phase [50] and transitions from an initially filamentary to a diffuse discharge. This can be achieved by increasing the operating frequency or the applied voltage or by decreasing the permittivity of the dielectric. In the first case more luminous regions appear and coalesce to form a transversely uniform glow discharge. In the second case, lowering the permittivity, the number density decreases and finally results in a transversely uniform Townsend discharge. Penning ionisation by metastables, secondary electron emission induced by ions and metastables and remnant surface charges have a large influence.

Most authors treating homogeneous DBDs in nitrogen at atmospheric pressure come to the conclusion that this discharge in fact has many similarities with a Townsend discharge [51]. The observed current pulse is much longer (about 30-90 μs) than in He and the peak current density is lower by at least an order of magnitude (typically 0.1 – 0.3 mA cm^{-2}). The evolution of a N_2 DBD at 130 kHz was investigated by Panousis *et al* in a coupled electro-dynamic and kinetic model [52]. Most modellers succeed in approximating measured current and voltage traces. The long living metastable triplet state $N_2(A^3\Sigma_u)$ plays an important role. When efficient quenchers of this species like O_2 or NO are added the discharge behaviour changes dramatically [53]. Also Townsend-like DBDs can exhibit multipulse behaviour with several short current pulses per half cycle for which also an analytical model was proposed [54]. Stability analysis suggests that Townsend discharges are more stable against lateral perturbations than glow discharges [27, 55]. In their numerical simulation of the short high-current pulses obtained by polarization switching Golubovskii and co-workers come to the conclusion that under these conditions the nitrogen Townsend discharge changes into a glow mode [56].

3. Surface barrier discharges (SBDs)

Like volume barrier discharges also surface barrier discharges (SBDs) have a long history. First applications for boxer chargers in electrostatic precipitators and for ozone generators were proposed by Masuda *et al* [57]. They used rows of parallel open HV electrodes separated from the ground electrode by a thin dielectric layer (figure 5b).

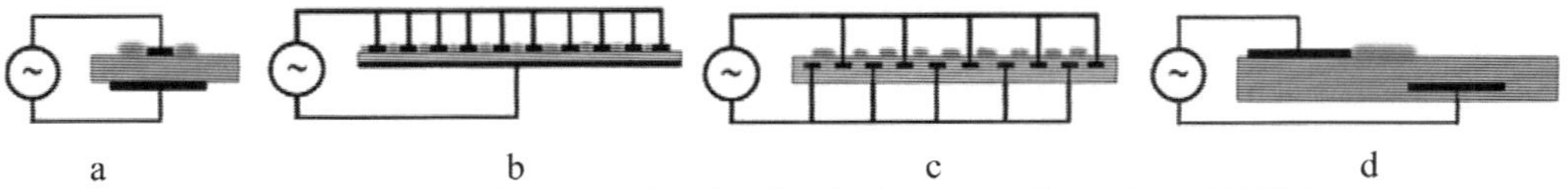

Figure 5. Examples of surface barrier discharge configurations (SBDs).

Based on Masuda's ideas different electrode configurations have been investigated (figure 5). The most prominent is the coplanar barrier discharge (CBD) used in plasma displays with rows of parallel linear electrodes buried in or covered by dielectric material (figure 5c) [9]. This large-volume application and the more recent prospect of using SBD plasma actuators for flow control have stimulated substantial research activities. Devices intended for flow control normally use one open electrode and another one slightly displaced and buried inside or below dielectric material (figure 5d).

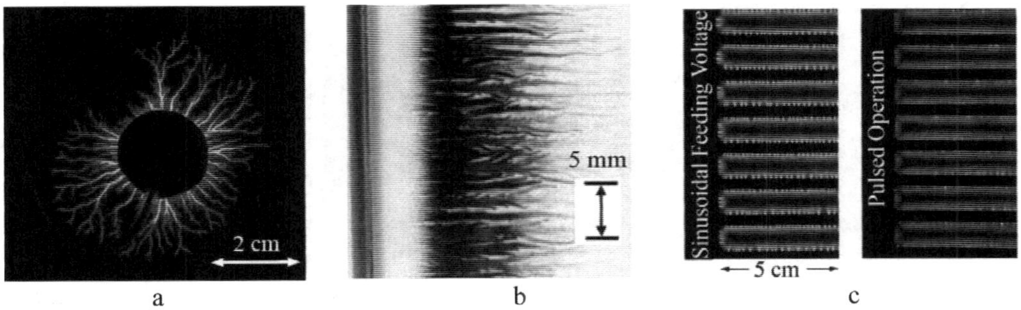

Figure 6. Manifestations of surface discharges (SBDs) taken from [58, 62a, 59] with permission.

Figure 6a, taken from the dissertation of Katia Allégraud [58], shows an ICCD image of a surface discharge in Ar (exposure time: 100 ns, taken 200 ns after streamer ignition). Figure 6b is a dust figure of a surface discharge in air of a linear electrode and figure 6c shows photographs of a section of a large-area electrode configuration (22 cm by 14 cm) [59]. The gas discharge is always initiated in the inhomogeneous field at or close to the edge of one electrode and moves along the dielectric surface as a surface discharge (creeping discharge, sliding discharge). In atmospheric pressure air the discharge consists of bright filaments during the positive half-period of the applied voltage [60]. During the negative half-period the emission is weaker and more diffuse. Many experiments were conducted in ambient air. Traces of the current filaments can be documented by dust figures, originally proposed by Lichtenberg in 1777 [61], by photographic Lichtenberg figures, by charge measurements and, more recently, by fast imaging techniques, now reaching gating times of sub-nanosecond duration. The structured charge arrangement covers an area that increases with time and voltage amplitude. A stepwise propagation is observed caused by a repeated re-ignition of the current channels, sometimes referred to as surface streamers. The thin filaments have currents and transported charges similar to those found in microdischarges in VBDs [62]. At peak current a quasi-neutral plasma with electron densities of the order 10^{14} cm^{-3} is established after formation of a thin cathode layer near the negative electrode, even in the case it is covered by a dielectric layer. The charge deposited on the dielectric surface reduces the electric field in the filament and finally chokes the current flow. In addition to the barrier thickness and permittivity also the barrier structure and its chemical composition define the discharge characteristics and its structure [63]. Charge-voltage Lissajous figures resemble those of volume DBDs [60]. Some authors using CBDs reported that in nitrogen and ambient air at high power densities (100 W/cm^{-3}) the discharge turns into a "visually almost uniform plasma" [64]. They called it a diffuse coplanar surface dielectric barrier discharge (DCSD, DCSBD). From the diagnostics presented it is not clear whether this discharge is really homogeneous or a very close spacing of filaments. Earlier investigations on SBDs in He came to the conclusion that under certain conditions diffuse discharges can be obtained [65]. The conclusion was that, similar to the experience with volume DBDs, remaining metastables and free electrons from the previous half cycle play a crucial role.

In some experiments a striking synchronization of current filaments in SDs was reported [60, 66]. Kashiwagi and Itoh investigated this effect in detail and showed that UV radiation from the first filament was responsible for triggering adjacent filaments. In air at atmospheric pressure the most effective spectral range was about 115 nm in the vacuum UV and reached for a distance of 120 mm. These photons of 10 eV energy fall into an absorption window of oxygen and are capable of releasing secondary electrons from dielectric surfaces or photoionising NO, which is rapidly formed in N_2/O_2 mixtures. Numerical simulations of the dynamic behaviour of surface streamers by Tanaka et al [67] showed that the experimental results can only be reproduced when photoelectron emission and volume photoionisation are taken into consideration.

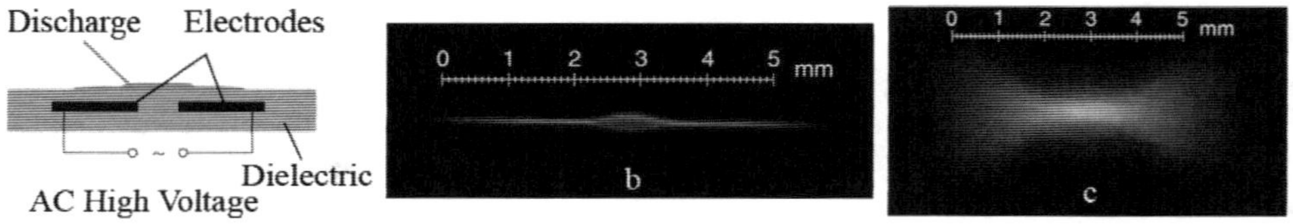

Figure 7. Coplanar surface DBD with embedded needle electrodes. b) side view, c) top view. Taken from [69] with permission.

More recent simulations of the surface current filaments in O_2 and in air present more details about the propagation and the thickness of the surface discharges [68]. Experimental values for the length of the filaments and the charge density on the dielectric can now be adequately simulated. An interesting detail is the prediction that the streamer head starting from a positive electrode travels at a height of 0.05 cm above the dielectric surface. This is in agreement with recent short exposure CCD images by Hoder *et al* [69] showing that the surface discharge is located slightly above the surface. It maybe of interest to note that well arranged equally spaced SBDs reminiscent of pattern formation in VBDs have been observed [70]. In certain electrode configurations also phases of diffuse SBDs have been theoretically predicted and recently also found experimentally [71].

4. Conclusions

It can be concluded that the plasma properties obtained in VBDs and SBS are similar. Filamentation, pattern formation and diffuse appearance have been observed in both. Collective effects caused by UV photons and remnant surface charges play a crucial role. The overall discharge behaviour is dominated by the current limiting properties of the dielectric barrier. Current densities can vary by several orders of magnitude and cover the full range of Townsend-like discharges to atmospheric pressure glow discharges. The VBD type that comes closest to the values expected from similarity laws is the filamentary discharge observed in atmospheric pressure in O_2, N_2 or air. In these filaments the current density in the quasi-neutral column and the properties of the cathode fall region correspond to those expected for atmospheric pressure glow discharges.

Numerical modelling has been very successful in providing new insight into the chemical kinetics, the influence of impurities and the distribution of charges in the discharge gap. One of the major results is that plasma chemistry substantially influences electrical discharge behaviour. Nevertheless there is still considerable debate on individual processes involved, especially at the surfaces. It is still an open question how previously stored electrons or ions are released from or neutralized at a dielectric surface. Spontaneous or thermal electron desorption and secondary electron emission induced by impinging ions, metastables and UV photons are among the processes suggested. In many models rather arbitrary assumptions about the coefficients for secondary electron release from dielectric surfaces had to be introduced to match experimental data. Further research is needed about the physisorption kinetics of charged particles at dielectric boundaries to find out how electron-ion recombination, transverse charge motion and secondary electron emission is effected.

References

[1] Kogelschatz U, Eliasson B and Egli W 1999 *Pure Appl. Chem.* **71** 1819-28; Kogelschatz U 2002 *IEEE Trans. Plasma Sci.* **30** 1400-8; 2003 *Plasma Chem. Plasma Process.* **23** 1-46

[2] Wagner H-E, Brandenburg R, Kozlov K V, Sonnenfeld A, Michel P and Behnke J F 2003 *Vacuum* **71** 417-36

[3] Fridman A, Chirokov A and Gutsol A 2005 *J. Phys. D: Appl. Phys.* **38** R1-24

[4] Becker K H, Kogelschatz U, Schoenbach K H and Barker R J eds 2004 *Non-Equilibrium Air Plasmas at Atmospheric Pressure* (Bristol and Philadelphia: IOP)

[5] Laroussi M and Akan T 2007 *Plasma Proc. Polym.* **4** 777-88

[6] Moreau E 2007 *J. Phys. D: Appl. Phys.* **40** 605-36; Corke T C, Post M L and Orlov D M 2009 *Exp. Fluids* **46** 1-26; Miles R B, Opaits D F, Shneider M N, Zaidi S H and Macheret S O 2009 *Eur. Phys. J. Appl. Phys.* **47** 22802

[7] Fridman G, Friedman G, Gutsol A, Shekhter A B, Vasilets V N and Fridman A 2008 *Plasma Process. Polym.* **5** 503-33; Dobrynin D, Fridman G, Friedman G and Fridman A 2009 *New J. Phys.* **11** 115020; Laroussi M 2009 *IEEE Trans. Plasma Sci.* **37** 714-25

[8] Chirokov A, Gutsol A, Fridman A, Sieber K D, Grace J M and Robinson K S 2004 *Plasma Sources Sci. Technol.* **13** 623-35; 2005 *IEEE Trans. Plasma Sci.* **23** 300-1

[9] Boeuf J-P 2003 *J. Phys. D: Appl Phys.* **36** R53-79

[10] Kozlov K V, Michel P and Wagner H-E 2000 *Plasmas Polym.* **5** 129-49

[11] Kozlov K V, Wagner H-E, Brandenburg R and Michel P 2001 *J. Phys. D: Appl. Phys.* **34** 3164-76

[12] Wagner H-E, Kozlov K V and Brandenburg R 2008 *Cross-correlation emission spectroscopy* in *Low Temperature Plasmas* vol 1 ed R Hippler *et al* (Weinheim: WILEY-VCH) chapter 10 pp 271-306

[13] Merbahi N, Sewraj N, Marchal F, Salamero Y and Millet P 2004 *J. Phys. D: Appl. Phys.* **37** 1664–78; Merbahi N, Ledru G, Sewraj N and Marchal F 2007 *J. Appl. Phys.* **101** 123309; Sewraj N, Merbahi N, Marchal F, Ledru G and Gardou J P 2009 *J. Phys. D: Appl. Phys.* **42** 045206

[14] Eliasson B and Kogelschatz U 1991 *IEEE Trans. Plasma Sci.* **19** 309-22; Braun D, Gibalov V and Pietsch G 1992 *Plasma Sources Sci. Technol.* **1** 166-72; Li J and Dhali S K 1997 *J. Appl. Phys.* **82** 4205–10; Steinle G, Neundorf D, Hiller W and Pietralla M 1999 *J. Phys. D: Appl. Phys.* **32** 1350-6; Gibalov V I and Pietsch G J 2000 *J. Phys. D: Appl. Phys.* **33** 2618–36

[15] YurgelenasYu V and Wagner H-E 2006 *J. Phys. D: Appl. Phys.* **39** 4031-43; Yurgelenas Yu V and Leeva M A 2009 *IEEE Trans. Plasma Sci.* **37** 809-15

[16] Papageorghiou L, Panousis E, Loiseau J F, Spyrou N and Held B 2009 *J. Phys. D: Appl. Phys.* **42** 105201

[17] Boyers D G and Tiller W A 1982 *Appl. Phys. Lett.* **41** 28–30; Kogelschatz U 1992 *Silent discharges and their applications* in *Proc. 10th Int. Conf. on Gas Discharges and their Applications* (Swansea, UK, 13-18 September 1992) vol 2 ed W T Williams pp 972-80; Breazeal W, Flynn K M and Gwinn E G 1995 *Phys. Rev. E* **52** 1503-15

[18] Dong L, Yin Z , Wang L, Fu G, He Y, Chai Z and Li X 2003 *Thin Solid Films* **435** 120-3; Dong L, Yin Z, Li X, Chai Z and He Y 2006 *Plasma Sources Sci. Technol.* **15** 840-4; Dong, L, Fan W, He Y and Liu F 2008 *IEEE Trans. Plasma Sci.* **36** 1356-7

[19] Miralaï S F, Czeremuszkin G and Wertheimer M R 2000 *High speed camera imaging of dielectric barrier discharges in He, Ar, N_2 and air* in *Proc. 7th Int. Symp. on High Pressure Low Temperature Plasma Chemistry* (Greifswald, Germany, 10-13 September 2000) vol 1 ed H-E Wagner *et al* pp 33-7

[20] Gurevich E L, Zanin A L, Moskalenko A S and Purwins H-G 2003 *Phys. Rev. Lett.* **91** 154501

[21] Bernecker B, Callegari T, Blanco S, Fournier R and Boeuf J P 2009 *Eur. Phys. J. Appl. Phys.* **47** 22808

[22] Duan X, Ouyang J, Zhao X and He F 2009 *Phys. Rev.* E **80** 016202.1-5

[23] Stollenwerk L, Laven J G and Purwins H-G 2007 *Phys. Rev. Lett.* **98** 255001; Stollenwerk L 2009 *New J.*

Phys. **11** 103034

[24] Stollenwerk L, Amiranashvili S, Boeuf J-P and Purwins H-G 2006 *Phys. Rev. Lett.* **96** 255001

[25] Astrov Y, Ammelt E, Teperick S and Purwins H-G 1996 *Phys. Lett.* A **211**, 184-90; Purwins H-G, Radehaus C, Dirksmeyer T, Dohmen R, Schmeling R and Willebrand H 1989 *Phys. Lett.* A **136** 480-4; Astrov Yu A and Purwins H-G 2001 *Phys. Lett. A* **283** 349-54;

[26] Ebert U and Arrayas M 2001 *Pattern formation in electric discharges* in *Coherent Structures in Complex Systems* ed D Reguera *et al* (Berlin: Springer) pp 270-82

[27] Šijačič D D and Ebert U 2002 *Phys. Rev.* E **66** 066410

[28] Raizer Yu P, Ebert U and Šijačič D D 2004 *Phys. Rev.* E **70** 017401

[29] Brauer I, Punset C, Purwins H-G and Boeuf J P 1999 *J. Appl. Phys.* **85** 7569-72; Müller I, Punset C, Ammelt E, Purwins H-G and Boeuf J P 1999 *IEEE Trans. Plasma Sci.* **27** 20-1

[30] Zhang P and Kortshagen U 2006 *J. Phys. D: Appl. Phys.* **39** 153-63

[31] Kanazawa S, Kogoma M, Moriwaki T and Okazaki S 1988 *J. Phys. D: Appl. Phys.* **21** 838-40; Yokoyama T, Kogoma M, Moriwaki T and Okazaki S 1990 *J. Phys. D: Appl. Phys.* **23** 1125-8

[32] Massines F, Rabehi A, Decomps P, Gadri R B, Ségur P and Mayoux C 1998 *J. Appl. Phys.* **83** 2950-7; Massines F, Gherardi N, Naudé N and Ségur P 2005 *Plasma Phys. Control. Fusion* **47** B577-88; Enache I N, Naudé N, Cambronne, J P, Gherardi N and Massines F 2006 *Europ. Phys. J. Appl. Phys.* **33** 15-21

[33] Massines F, Gherardi N, Naudé N and Ségur P 2009 *Eur. Phys. J. Appl. Phys.* **47** 22805

[34] Okazaki S, Kogoma M, Uehara M and Kimura Y 1993 *J. Phys. D: Appl. Phys.* **26** 889-92

[35] Trunec D, Brablec A, Šťastný F and Bucha J 1998 *Contrib. Plasma Phys.* **38** 435-45; Wang X, Luo H, Liang Z, Mao T and Ma R 2006 *Plasma Sources Sci. Technol.* **15** 845-8; Fang Z, Qiu Y, Zhang C and Kuffel E 2007 *J. Phys. D: Appl. Phys.* **40** 1401-7

[36] Tepper J and Lindmayer M 2000 *Investigations on two different kinds of homogeneous barrier discharges at atmospheric pressure* in *Proc. 7th Int. Symp. on High Pressure Low Temperature Plasma Chemistry* (Greifswald, Germany, 10-13 September 2000) vol 1 ed H-E Wagner *et al pp* 38–43

[37] Gherardi N and Massines F 2001 *IEEE Trans. Plasma Sci.* **29** 536-44; Naudé N, Cambronne J-P, Gherardi N and Massines F 2005 *Eur. Phys. J. Appl. Phys.* **29** 173-80; Choi J H, Han M H, Baik H K and Song K M 2007 *J. Appl. Phys.* **101** 033304; Dilecce G, Ambrico P F and De Benedictis S 2007 *Plasma Sources Sci. Technol.* **16** 511-22; Brandenburg R, Navrátil Z, Jánsky J, Stahel P, Trunec D and Wagner H-E 2008 *J. Phys. D: Appl. Phys.* **42** 085208;

[38] Obradović B M, Ivković S S and Kuraica M M 2008 *Appl. Phys. Lett.* **92** 191501; Ivković S S, Obradović B M, Cvetanović N, Kuraica M M and Purić J 2009 *J. Phys. D: Appl. Phys.* **42** 225206

[39] Wang Q, Koleva I, Donnelly V M and Economou D J 2005 *J. Phys. D: Appl. Phys.* **38** 1690-7; Arkhipenko V I, Kirillov A A, Safronau Ya A, Simonchik L V and Zgirouski S M 2009 *Plasma Sources Sci. Technol.* **18** 045013

[40] Shi J J and Kong M G 2003 *J. Appl. Phys.* **94** 5504-13

[41] Visentin G, Mangolini L, Orlov K, Kortshagen U and Heberlein J *Experimental study of multiple current pulses in atmospheric pressure glow discharges* in *Proc. 15th Int. Symp. on Plasma Chemistry* (Orléans, France, 9-13 July 2001) vol 8 ed A Bouchoule *et al* pp 3251-6, Radu I, Bartnikas R, Czeremuszkin G and Wertheimer M R 2003 *IEEE Trans. Plasma Sci.* **31** 411-421; Wang D, Wang Y and Liu C 2006 *Thin Solid Films* **506-507** 384-88; Sublet A, Ding C, Dorier J-L, Hollenstein Ch, Fayet P and Coursimault F 2006 *Plasma Sources Sci. Technol.* **15** 627-34

[42] Reichen P, Sonnenfeld A and Rudolf von Rohr Ph 2010 *J. Phys. D: Appl. Phys.* **43** 025207

[43] Kutasi K, Hartmann P and Donko Z 2001 *J. Phys. D: Appl. Phys.* **34** 3368-77; Wang Y and Wang D 2005 *Phys. Plasmas* **12** 023503; Martens T, Bogaerts, A, Brok W and van Dijk J 2007 *Anal. Bioanal. Chem.* **388** 1583-94; Martens T, Brok W J M, van Dijk J and Bogaerts A 2010 *Appl. Phys. Lett.* **96** 091501

[44] Yuan X and Raja L L 2002 *Appl. Phys. Lett.* **81** 814-6

[45] Martens T, Bogaerts A, Brok W J M and van Dijk J 2008 *Appl. Phys. Lett.* **92** 041504

[46] Martens T, Brok W J M, van Dijk J and Bogaerts A 2009 *J. Phys. D: Appl. Phys.* **42** 122002

[47] Luo H, Liang Z, Bo L, Wang X, Guan Z and Wang L 2007 *Appl. Phys. Lett.* **91** 221504

[48] Akishev Yu S, Demyanov A V, Karalnik V B, Pankin M V and Trushkin N L 2001 *Plasma Phys. Rep.* **27** 164-71; Golubovskii Yu B, Maiorov V A, Behnke J and Behnke J F 2003 *J. Phys. D: Appl. Phys.* **36** 39-49; Yuan X, Shin J and Raja L L 2006 *Vacuum* **80** 1199-205; Maiorov V A and Golubovskii Yu B

2007 *Plasma Sources Sci. Technol.* **16** S67-75; Lü B, Wang X , Luo H-Y and Liang Z 2009 *Chinese Phys.* B **18** 646-51; Zhang Y, Gu B, Wang W, Wang D and Peng X 2009 *J. Appl. Phys.* **106** 023307

[49] Mangolini L, Anderson C, Heberlein J and Kortshagen U 2004 *J. Phys. D: Appl. Phys.* **37** 1021-30

[50] Mangolini L, Orlov K, Kortshagen U, Heberlein J and Kogelschatz U 2002 *Appl. Phys. Lett.* **80** 1722-4; Malik D A, Orlov K E, Miroshnikov I V and Smirnov A S 2005 *Tech. Phys. Lett.* **31** 500-2; 2008 *J. Appl. Phys.* **103** 033303

[51] Massines F, Gherardi N, Naudé N and Ségur P 2005 *Plasma Phys. Control. Fusion* **47** B577-88; Naudé N, Cambronne J-P, Gherardi N and Massines F 2005 *Eur. Phys. J. Appl. Phys.* **29** 173-80; Brandenburg R, Maiorov V A, Golubovskii Yu B, Wagner H-E, Kozlov K V, Behnke J F, Behnke J 2005 *J. Phys. D: Appl. Phys.* **38** 2187-98; Golubovskii, Yu B, Maiorov V A, Behnke J and Behnke J F 2002 *J. Phys. D: Appl. Phys.* **35** 751-61; 2003 *J. Phys. D: Appl. Phys.* **36** 39-49; Golubovskii, Yu B, Maiorov VA, Li P and Lindmayer M 2006 *J. Phys. D: Appl. Phys.* **39** 1574-83; Luo H, Liang Z, Wang X, Guan Z and Wang L 2010 *J. Phys. D: Appl. Phys.* **43** 155201

[52] Panousis E, Papageorghiou L, Spyrou N, Loiseau J-F, Held B and Clément F 2007 *J. Phys. D: Appl. Phys.* **40** 4168-80

[53] Brandenburg R, Maiorov V A, Golubovskii Yu B, Wagner H-E, Kozlov K V, Behnke J F, Behnke J 2005 *J. Phys. D: Appl. Phys.* **38** 2187-98

[54] Nikandrov D S and Tsendin L D 2005 *Tech. Phys.* **50** 1282-94

[55] Kudryavtsev A A and Tsendin L D 2002 *Tech. Phys. Lett.* **28** 1036-42; Golubovskii Yu B, Maiorov V A, Behnke J and Behnke J F 2003 *J. Phys. D: Appl. Phys.* **36** 975-81; Naudé N and Massines F 2008 *IEEE Trans. Plasma Sci.* **36** 1322-3

[56] Golubovskii Yu B, Maiorov V A, Behnke J F, Tepper J and Lindmayer M 2004 *J. Phys. D: Appl. Phys.* **37** 1346-56

[57] Masuda S, Washizu M, Mizuno A and Akutsu K 1978 *Boxer charger - a novel charging device for high resistivity powders* in *Proc. IEEE/IAS Annual Meeting* (Toronto, Canada, 1-5 October 1978) vol 1B 16-22; Masuda S, Akutsu K, Kuroda M, Awatsu Y and Shibuya Y 1988 *IEEE Trans. Ind. Appl.* **24** 223-31

[58] Allegraud K 2008 *Décharge à barrière diélectrique de surface: physique et procédé* Ph D Thesis (Palaiseau, France: Ecole Polytechnique) http://pastel.paristech.org/4783/

[59] Williamson J D, Trump D D, Bletzinger P and Ganguly B N 2006 Comparison of high-voltage ac and pulsed operation of a surface dielectric barrier discharge *J. Phys. D: Appl. Phys.* **39** 4400–6

[60] Allegraud K, Guaitella O and Rousseau A 2007 *J. Phys. D: Appl. Phys.* **40** 7698-706

[61] Lichtenberg G C 1777 *Novi Comment. Goett.* **8** 168-80

[62] Gibalov V I and Pietsch G J 2000 The development of dielectric barrier discharges in gas gaps and on surfaces *J. Phys. D: Appl. Phys.* **33** 2618-36; 2004 *J. Phys. D: Appl. Phys.* **37** 2082-92

[63] Kozlov M V, Sokolova M V, Temnikov A G, Timakov V V and Vereshchagin I P 2003 *Plasmas Polym.* **8** 179-97

[64] Šimor M, Ráhel J, Vojtek P, Černak M and Brablec A 2002 *Appl. Phys. Lett.* **81** 2716-8; Černák M, Černáková L, Hudec I, Kováčik D and Zahoranová A 2009 *Eur. Phys. J. Appl. Phys.* **47** 22806

[65] Štefečka M, Korzec D, Širý M, Imahori Y and Kando M 2001 *Sci. Technol. Adv. Mat.* **12** 587-93

[66] Kashiwagi Y and Itoh H 2006 *J. Phys. D: Appl. Phys.* **39** 113-8; Allegraud K and Rousseau A 2009 *IEEE Trans. Diel. Electr. Insul.* **16** 435-9

[67] Tanaka M, Murooka Y and Hidaka K 1987 *J. Appl. Phys.* **61** 4471-8

[68] Gibalov V I, Tkachenko I S and Lunin V V 2008 Russ. *J. Phys. Chem.* A **82** 1020-23; Solov'ev V R, Konchakov A M, Krivtsov V M and Aleksandrov N L 2008 *Plasma Phys. Rep.* **34** 594-608

[69] Hoder T, Šíra M, Kozlov K V and Wagner H-E 2008 Investigation of the coplanar barrier discharge in synthetic air at atmospheric pressure by cross-correlation spectroscopy *J. Phys. D: Appl. Phys.* **41** 035212

[70] Srivastava A K and Prasad G 2008 *Phys. Lett.* A **372** 6101-6

[71] Boeuf J P, Lagmich Y, Unfer T, Callegari T and Pitchford L C 2007 *J. Phys. D: Appl. Phys.* **40** 652-62; Hoskinson A R and Hershkowitz N 2010 *J. Phys. D: Appl. Phys.* **43** 065205

Microwave discharges: generation and diagnostics

Yu.A. Lebedev

Topchiev Institute of Petrochemical Synthesis RAS, Leninsky Prospect, 29, Moscow, 119991, Russia

E-mail: lebedev@ips.ac.ru

Abstract. Microwave discharges are widely used for generation of quasi-equilibrium and nonequilibrium plasma for different applications. Microwave plasma can be generated at pressures from 10^{-5} Torr up to atmospheric pressure in the pulse and continuum wave regimes at incident powers ranged between several Watts and hundreds of kW. The plasma absorbed power can be high enough and runs up to 90% of the incident power. This paper reviews the methods of microwave plasma generation, general features of microwave plasma, and selected aspects of microwave plasma diagnostics.

1. Introduction

Microwave discharges (MD) are the electrical discharges generated by the electromagnetic waves with frequencies exceeding 300 MHz (wave length λ[cm]=30/f [GHz]). The used wavelengths of microwaves are in the range from millimeters up to several tens of centimeters and should correspond to permitted microwave frequencies for industrial, medical and scientific applications. The frequency 2.45 GHz is the most commonly used.

The starting stage in development of MD and microwave induced plasma (MIP) was strongly coupled with successes in radar technique. For example the antenna switches are the microwave gas discharge plasma devices which use the high power microwave pulse for plasma generation to prevent the damage of low signal microwave receiver in the moment of passing this pulse through the microwave circuit. Further development of microwave technique created the necessary prerequisites for application microwave devices in different areas of science and technique and in particular for generation of gas discharge plasma. Now MDs are widely used for generation of quasi-equilibrium and nonequilibrium plasma for different applications: for generation of the active medium in the gas discharge lasers, light sources, in plasma chemistry, analytical chemistry, for creation of artificial ionized areas in the Earth's atmosphere, recovering of the Earth's ozone layer, etc. MD is the nonelectrode discharge and has some advantages over other types of electrical discharges [1-14]:

- the simplicity of plasma generation both with high (>100W/cm^3) and low (< 1W/cm^3) levels of absorbed powers;
- the wide region of operating pressures (from 10^{-5} Torr up to atmospheric pressure);
- the simplicity of control the plasma internal structure by means of changes in electrodynamic characteristics of microwave-to-plasma applicator;
- the possibility of plasma generation both in small and large chambers, including the free space;
- the possibility of plasma generation in the electrode discharge systems without any contamination of gas phase or plasma treated samples by products of the electrode erosion;

- the possibility to treat a large gas volumes (including a cleaning process for the reactor walls) by scanning the plasma region (which is small as compared with chamber dimensions) over the chamber;
- the possibility to realize of joint action of the plasma and electromagnetic field on the treated substances (e.g., powders) to increase the energy efficiency of plasma chemical process;
- the developed family of the highly effective microwave-to-plasma applicators permits to choose the necessary design for any application.

MD can be generated in the pulse and continuum wave regimes at incident microwave powers ranged between several Watts and hundreds of kW. The plasma absorbed power can be high enough and runs up to 90% of the incident power. Plasma density usually exceeds the critical density n_{ec} (n_{ec} [cm^{-3}]$\approx 1,24\bullet 10^{10} f^2$[GHz]). Problems of comparison of MIP properties and plasma of other discharges are analyzed in ([10] Pr7-369).

Information of MD is presented in numerous books, journal papers and in the Proceedings of many plasma conferences. Two of conferences should be specially mentioned because they contain the comprehensive information on MD. It is the International Workshop "Strong Microwaves in Plasmas" ("Strong Microwaves: Sources and Applications" since 2008) [6, 7]. The second one is the International Workshop "Microwave Discharges: Fundamentals and Applications" [8-14]. This review is mainly based on these Proceedings and it permits to shorten the list of references.

2. Microwave plasma generators

2.1. Typical structure of microwave plasma stand (microwave plasma generator)

The microwave plasma generators typically include several elements (figure 1). It consists of the microwave power source (usually the magnetron generator), element protecting the magnetron from the reflected power (any nonreciprocal device, e.g., circulator), standing wave ratio meter, matching circuit and microwave-to-plasma applicator. The last one is the main element of plasma generator because it realizes the input of microwave energy to plasma. It defines the efficiency of plasma generator (the portion of the incident power absorbed in plasma), levels of minimal and maximal plasma powers, bandwidth of the system, and structure of electromagnetic field in plasma, uniformity and size of plasma. Depending on the properties of the applicator some of elements can be removed from typical structure of microwave generator (figure 1). For example it can realize the matching functions and in that case the microwave generator can be directly connected with the applicator. The applicator defines also the correlation of external (pressure, power, frequency, etc) and internal (mean energy and energy distributions of plasma particles, electron density, etc) plasma parameters. Therefore the term "MIP" unites the plasma objects with different structures and internal parameters and the field frequency is the only common quantity. It is not enough for plasma characterization especially for comparison with the plasma of other types of discharges. So the most correct approach is considering the MIP in the particular applicator (it is the MD).

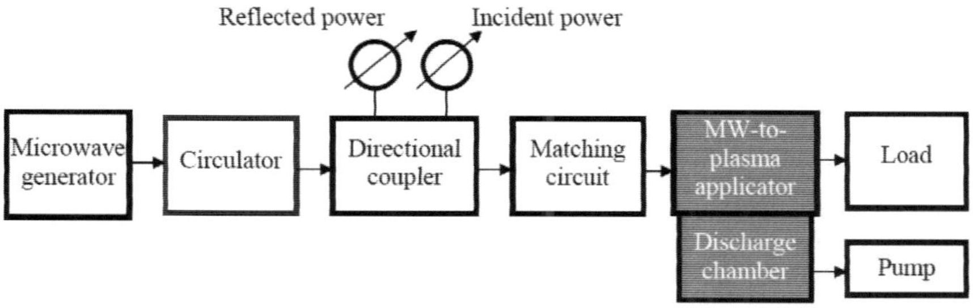

Figure 1. Typical structure of the microwave plasma stand

Engineering methods used in designing the microwave-to-plasma applicators are defined by requirements to microwave devices and elements with electrically distributed parameters (waveguides, resonators, etc). All existing applicators can be conditionally divided into several classes:

- cavity microwave plasma generators;
- waveguide microwave plasma generators;
- surface wave plasma generators;
- slow wave plasma generators;
- wave beam microwave plasma generators;
- microwave plasma generators with distributed energy input;
- initiated microwave discharges;
- electrode microwave plasma generators;
- microwave plasma generators with external magnetic fields;
- plasma generators with combinations of microwave and other fields.

It is non-unique principle of applicators classification. For example almost all low pressure microwave plasma generators on the base of physics of microwave power absorption can be classified as the generators with "plasma resonance".

Several examples from numerous microwave plasma devices will be presented below.

2.2. Cavity microwave plasma generators (figure 2)

First two classes of plasma generators (#2.2, 2.3) can be referred to so-called plasma generators with the "localized discharge zone" as the discharge exists inside the applicator and the discharge sizes are limited by it dimension (characteristic dimensions of plasma are less then the wavelength).

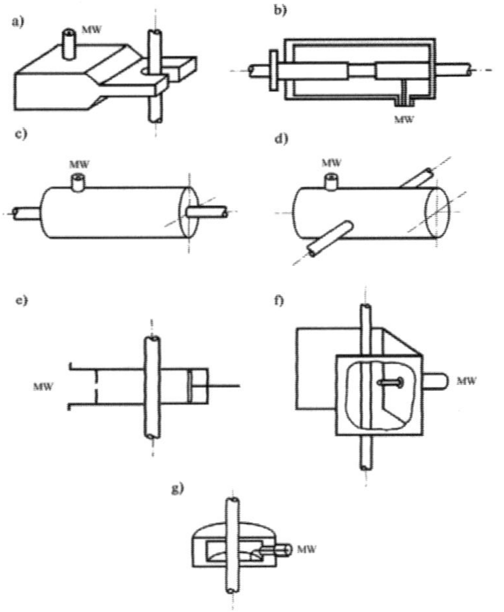

Figure 2. Cavity microwave plasma generators

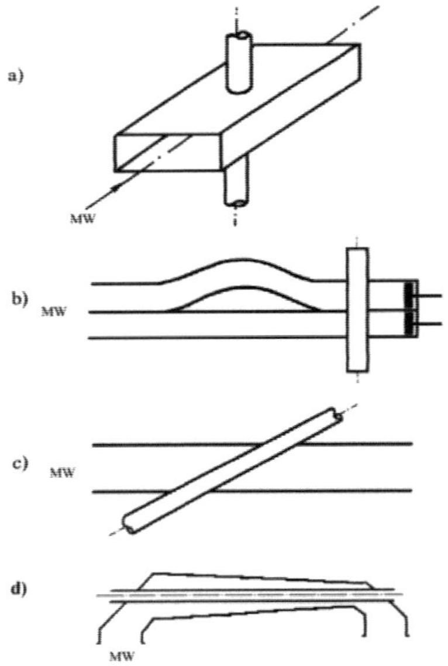

Figure 3. Waveguide microwave plasma generators

Cavity plasma generators are among the first microwave plasma generators. Plasma exists in the low losses dielectric tubes (quartz, ceramics) placed in the prismatic (a, e), coaxial (b), cylindrical (c, d, g), opened (f) resonators. High and low quality resonators are used for generation of plasma with density exceeded n_{ec} but the density less than n_{ec} can be obtained in the high quality resonator.

2.3. Waveguide microwave plasma generators (figure3)

Electromagnetic energy is introduced in plasma in the discharge tubes by rectangular, cylindrical or coaxial waveguides. Discharge tubes orientation respectively to the waveguide axis can be different. Plasmatron on the mode H_{10} (a) is one of the simplest and widespread plasma generator from this class. To improve the matching of the applicator the length of one part of the parted waveguide can be increased by $\lambda/4$ respectively to another part (b) or the discharge tube can be placed at a certain angle to the waveguide axis (c). Long slab of plasma can be generated in the tubes placed along the waveguide axis. To increase the uniformity of plasma slab the waveguide dimensions can be changed along the tube (c) or one can feed the applicator from the opposite ending of the waveguide.

2.4. Surface wave microwave plasma generators (figure4)

Special classes of microwave applicators were designed to increase the plasma sizes. In contrast to first two classes they are based on the "propagating waves" or 'radiating structures'. It the first case plasma exists outside the applicator and sustained by the surface waves (SW).

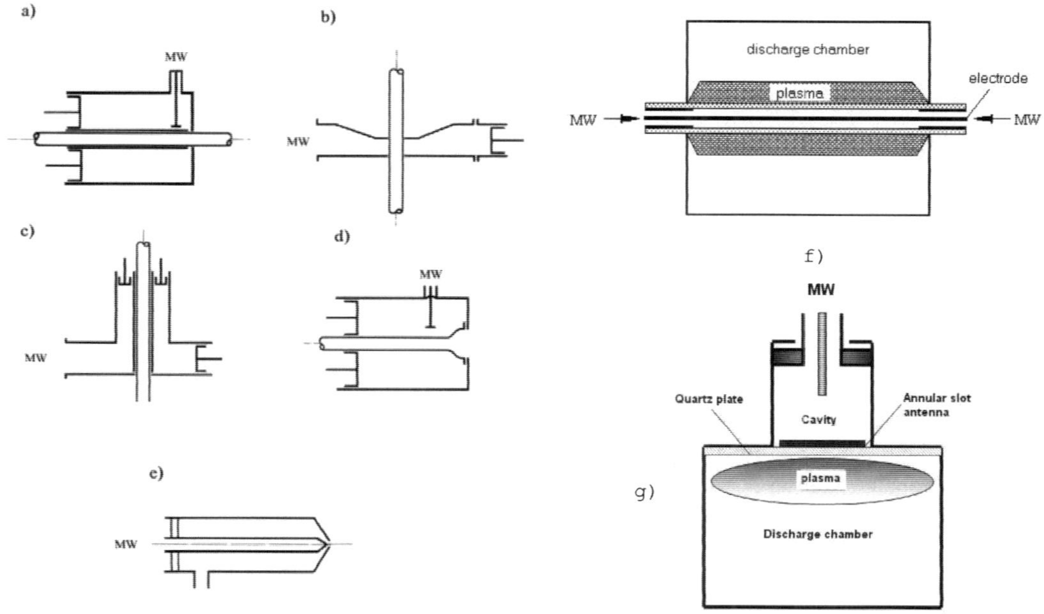

Figure 4. Different designs of surface wave devices on the base of cylindrical (a, d) rectangular (b, c) and coaxial structures (e), duo-plasmaline ([10] Pr7-99) (f), large area surface wave plasma source (g) ([11] p. 163)

These types of applicators have large bandwidth, can generate microwave plasma inside the dielectric tube (figure 4, a, b, c, d), in free space (e) and outside the tube (f)

2.5. Microwave plasma generators with radiating structure: slow wave plasma generators, wave beam microwave plasma generators

Slow wave systems are the periodic structures served for reducing the phase velocity of electromagnetic wave. Electromagnetic field existing near such structures can be used for large volume plasma generation. As the field structure is non-uniform, special methods should be applied to increase the plasma uniformity. For example it is possible to use two similar slow wave structure having inputs from opposite ends.

Unique possibilities of microwaves for plasma generation give the wave beams. Plasma can be generated in the region of maximal electric field in the focused wave beam. Crossed beams can be used to stabilize the plasma position. This is the source of freely localized discharge which can be initiated in the Earth's atmosphere

2.6. Microwave plasma generators with distributed energy input (figure 5)

Generators of this class can produce the large area/volume plasma and give possibility to change the internal plasma spatial structure. The powered waveguide is connected with the discharge chamber by adjusted couplers (rods, slots, orifices) to control the necessary level of plasma uniformity.

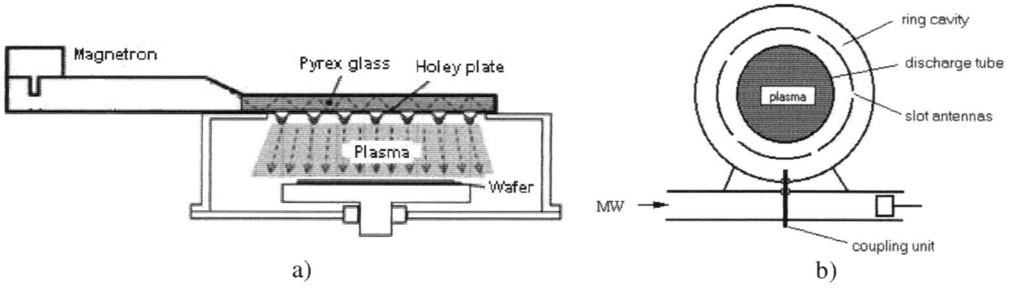

Figure 5. (a) Waveguide holey-plate plasma generator ([11] p.175)) and (b) SLAN-system (slotted antennas) ([10] Pr7-1)

2.7. Initiated microwave discharges, electrode microwave plasma generators (figure 6)

Initiated microwave discharges are the discharges which exist at low levels of incident microwave power and can not be ignited without any initiator (ionizing radiation, auxiliary electrode, solid particles, etc). Their peculiarities are:

- low maintaining power;
- localization of the plasma region;
- large volume/area plasma can be produced by the specially adjusted set of initiators;
- possibility of the electric field strength control in a plasma;
- strongly nonequilibrium plasma at atmospheric pressure.

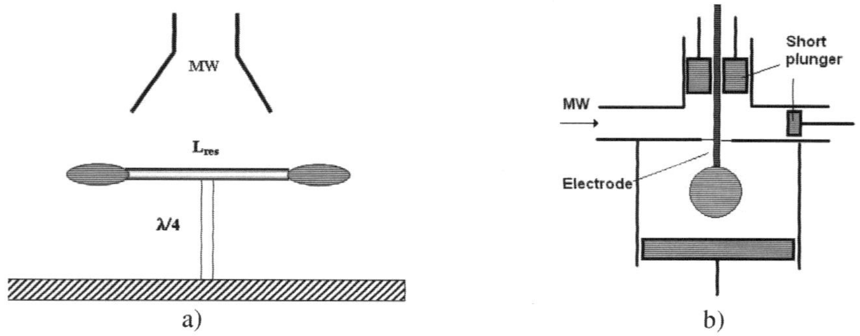

Figure 6. Resonance antenna initiator (a) [15] and electrode microwave discharge (b) [16]

2.8. Microwave plasma generators with external magnetic fields (figure 7)

Any microwave plasma system can be placed in the magnetic field to provide:

- the transport of plasma particles to the necessary direction (a);
- the matching of plasma device with the microwave power transmitted line;

- the limitation of plasma zone (protection of the walls, dielectric windows from the action of plasma, decrease of the charged particles losses which leads to decrease of microwave maintenance field strength, etc) (b, c);
- the ECR-regime ($f(Hz)=f_{ce}=2.8 \cdot 10^{10}B_0(T)$) (d);
- the plasma anisotropy.

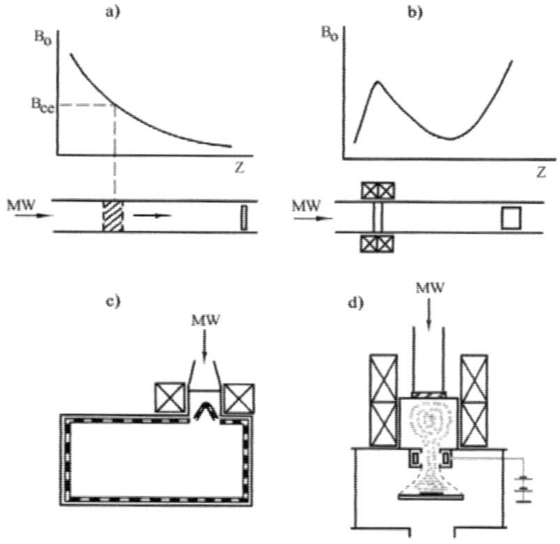

Figure 7. Microwave plasma generators with external magnetic fields

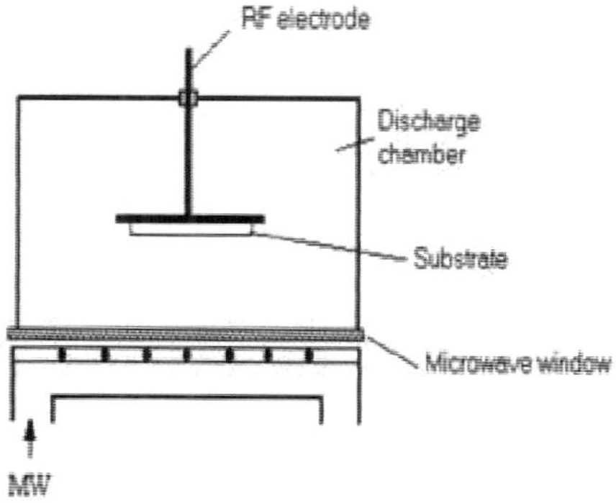

Figure 8. Plasma generator with combination of MW and RF fields.

2.9. Plasma generators with combinations of microwave and other fields (figure8)

Combined discharges are used if it is necessary to:

- use the positive properties of each of discharges;
- generate the plasma at power levels insufficient for existence of self-sustained discharge;
- produce plasma with desirable properties (additional possibility of control the plasma parameters; widening of range of variation of parameters; increase the active particle densities, homogeneity, etc);
- increase the stability of plasma system.

3. Diagnostics of microwave discharges

All methods of plasma diagnostics which are successfully used in other types of discharges can be applied for diagnostics of microwave discharges taking into consideration the peculiarities of microwave plasma. Besides that it is necessary to mention that almost all modern methods of diagnostics are the experimental-computational methods and require either, one or another plasma model to process the experimental data for calculation the plasma parameters. These models should also correspond to the plasma peculiarities.

First of all when choose and apply methods of diagnostics it is necessary to take into account the inherent non-uniformity of microwave plasma (if special methods do not used to suppress it) and level of non-uniformity is often unknown *a priory*. Reasons of this non-uniformity are:

- skin-effect produces non-uniform microwave field;
- plasma resonance can produce the non-uniform and non-Ohmic power absorption;
- regions of plasma resonance can generate the electron beams and produce the anisotropy of plasma (this effect was observed at pressures below 50 mTorr);

- electrons in non-uniform microwave filed experience a force proportional to ∇E_m^2. These force can causes the spatial redistribution of electrons in plasma. Within the range of fields strength used in low power low temperature plasma devices (e.g., in plasma chemistry) the electron spatial distribution is not disturbed by non-uniformity of microwave field.

Level of non-uniformity of microwave plasma depends on the power consumption, pressure, plasma gas, electric field frequency, discharge chamber design (e.g., the shape of electrodes, or microwave-to-plasma applicator, etc.). Thus the integral methods of plasma diagnostics give serious error not only in the averaged plasma parameters but also in the dependences of parameters on plasma conditions. This means that the methods of diagnostics with spatial resolution should be used in general for diagnostics of microwave plasma.

Another source of measuring error is related with direct influence of microwaves on measuring device (e.g., electrostatic probe, thermocouple).

Several methods of microwave plasma diagnostics will be considered below. The fundamentals of the methods are described in numerous books (see e.g. [17-19]). Here we emphasize the difficulties and possibilities of methods in measurements of the key microwave plasma parameters.

3.1. Electrodynamic measurements

Electrodynamic measurements give information both on microwave-to-plasma applicator with plasma (MD) and MIP, e.g. it is possible to define the plasma impedance and then calculate the averaged plasma parameters (electron density and temperature). Technique and approaches are similar to that used in microwave plasma diagnostics (see e.g. [20]). Here we consider measurements of one of the most important plasma characteristics, namely the plasma absorbed power P_{abs}. This parameter is used in plasma physics and applications.

If the plasma is the only source of microwave power losses in the system, P_{abs} can be defined from the power balance equation using the measured incident P_{in} and reflected P_{ref} powers:

$$P_{abs}=P_{in}-P_{ref}.$$

The real situation is more complicated and there are additional sources of microwave power losses:

- losses in the walls of the discharge chamber and transmitted lines and connectors (these losses grow with increasing standing wave ratio);
- radiation from the windows and leakage from different microwave connectors.

It is difficult to separate these sources of losses from the absorption of microwaves by plasma.

It is also necessary to take into account the sources of errors in the power measurements:

- Error in the measured incident power in the case of high standing wave ratio in the point of measurement (the mode of pure running wave should be realized).
- Error in the measured powers can be caused by the fact that the powered microwaves can contain not one frequency only but the set of frequencies of comparable amplitudes and their correlation depends on microwave power and matching of the plasma load. This source of error is important if the resonant measuring systems are used in the power or the standing wave ratio meters.

These factors complicate the measurements of plasma absorbed power and cause the uncontrollable error in the measured P_{abs}. To get over the difficulties it is possible to use several methods.

The direct measurement of the plasma absorbed power by calorimeter is one of the methods.

Energy deposited in plasma is lost through the walls of plasma chamber and/or carried away with the gas flow. In the case of low flow rate or in low pressure plasma the first channel is the main and to measure P_{abs} it is necessary to calorimetry the walls of discharge tube (figure 9). The calorimeter liquid should be low-loss dielectric at microwave frequencies because it operates under the action of electromagnetic field. Organic silicon oil or liquids can be used for this purpose. In the case then the gas flow is the main channel of energy losses, the calorimetry of gas flow can be done outside of the applicator where the electromagnetic field is absent and requirements to the dielectric is softer.

If the electromagnetic field in the applicator is slightly disturbed by the presence of plasma, P_{abs} can be defined by subtraction of power absorbed in the system without plasma (blank measurement) from

Journal of Physics: Conference Series **257** (2010) 012016 doi:10.1088/1742-6596/257/1/012016

the power absorbed in the system with plasma. Results are shown in figure 10. It is seen that the corrected power curve differs markedly from uncorrected curve.

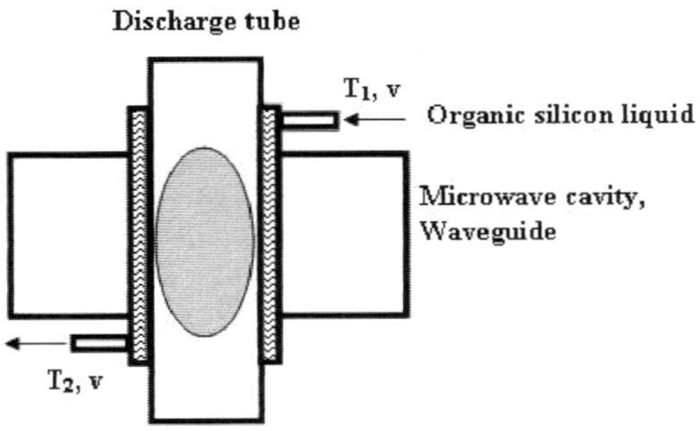

Figure 9. Calorimetric method of measurement of plasma absorbed power.[21]

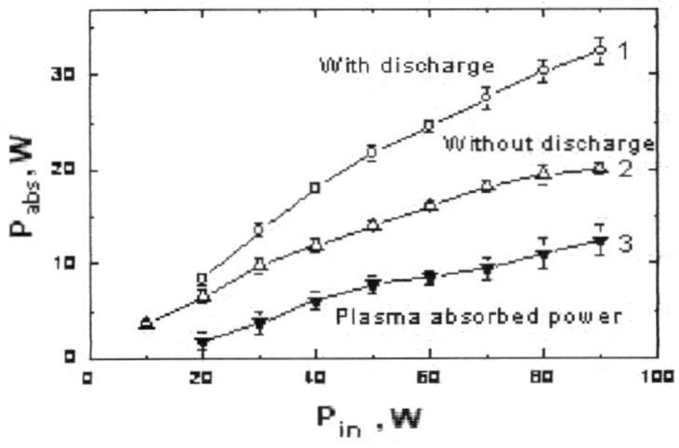

Figure 10. Measurement of plasma absorbed power by subtraction of blank absorption.[22]

3.2. Probe measurements in microwave plasma.

Electrostatic probes are the effective tool for local diagnostic of parameters of charged components of plasma [23-27]. The same as any invasive methods the probes can cause the perturbation of plasma. In microwave (and RF) plasma diagnostics there are additional disadvantages: perturbation of electromagnetic field, perturbation of probe voltage-current characteristic (VCC) by electromagnetic fields. Here we will consider the role of the last factor [26].

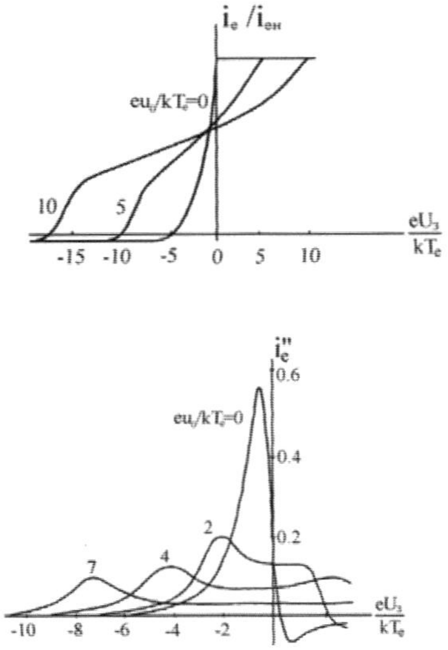

Figure. 11. Influence of alternating field on VCC (a) and its second-order derivation (b) (theory).

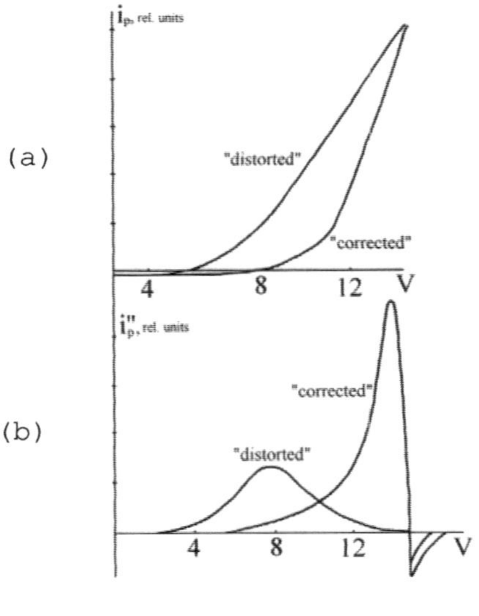

Figure. 12. Influence of alternating field on VCC (a) and its second derivation (b) (experiment).

Nature of VCC perturbation by electromagnetic fields is related with nonlinear voltage-current probe (probe sheath) characteristics and frequency transformation of alternating signal v_0 applied to the sheath. Rectification of alternating sheath voltage leads to appearance of DC voltage in addition to the external DC probe voltage (so-called "self-biasing of the probe") which moves VCC along the voltage axis. Since the nonlinearity of VCC depends on the external DC voltage (the operating point), the self-biasing is also the function of it. This effect leads not only to parallel move of VCC along the axis but also to deformation of it. This deformation in one's turn causes the deformation of the voltage second-order derivation of VCC which defines the electron energy distribution function. Methods of minimization of this error are clear from equivalent probe circuits (figure 13).

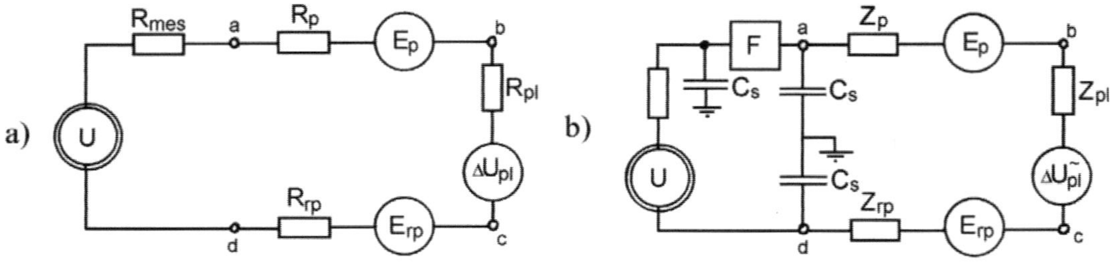

Figure 13. DC (a) and rf (b) equivalent probe circuits: U is the probe voltage; R_{mes} is the measuring resistance; R_p, R_{rp}, R_{pl} and Z_p, Z_{rp}, Z_{pl} are the resistances and impedances of the probe, reference probe and plasma, resp.; E_p and E_{rp} are the emf of the probe and reference probe sheaths; C_s is the stray capacitance; F is the high-impedance filter; b and c are the positions of probe and reference probe in plasma; ΔU_{pl} is the difference in the space potentials in the places of probe and reference probe; $\Delta U_{\tilde{pl}} = \Delta U_{pl} + u_0 \sin(\omega t)$.

To decrease the influence of microwaves on the probe results it is desirable to:

- place the probe and reference probe on the equipotential surface for the alternating field. This approach is generally realized in the double probe method;
- increase the probe surface (decrease both the probe impedance Z_p and the internal alternating signal u~ applied to the probe sheath when other circuit components being equal). It is necessary to take into account the increase of negative influence of geometric plasma perturbation and electron depletion in the vicinity of the probe due to increased probe current;
- increase the impedance of the external part of the probe circuit (application of high-impedance resonant filters, e.g. cavities ("F" in figure 13b)). Good results for the double probe method give introducing of resistors (20-30 kOhm) in series with the probes. Filters should be placed at the minimal distance from the collecting surface of the probes to suppress the role of the stray capacitors C_s;
- use the external alternating signal of necessary frequency, amplitude and phase in the probe circuit to compensate the alternating signal from plasma (active method).

Application of all above mentioned methods can minimize the influence of microwave field on VCC. Absolute criteria of residual influence of microwaves are absent. To prove the real probe accuracy it is necessary attract the additional information. Minimal VCC distortion can be indicated by minimal probe floating potential which depends on u_0

$$\Delta U_{fl} = -\frac{kT_e}{e} \ln I_0\left(\frac{eu_0}{kT_e}\right).$$

Correct use of probe method give information on the density and mean energy of electrons, electron energy distribution function. Use of the probe technique together with the Boltzmann equation gives information of the electric field strength in plasma (see e.g. [28]).

3.3. Measurement of plasma emission integrated over spectrum.

Spatially resolved plasma emission integrated over spectrum gives information on the structure of MIP. Since the plasma emission is proportional to the specific absorbed power these measurements show the distribution of the power consumption over the plasma volume (e.g. [29]).

3.4. Optical emission spectroscopy

Method of optical emission spectroscopy is described in detail in numerous monographs (e.g. [17-19]).

3.4.1. Measurements of gas temperature.

The gas temperature T_g is an important parameter that determines both the interaction of the electromagnetic field with the plasma (via the ratio ν/ω, where ν is the effective collision frequency of electrons with heavy particles) and kinetics of the plasma processes. Here we will consider two methods which are widely used in studying plasmas with spectral devices having a low and moderate spectral resolution. Measurements with spatial resolution give information on gas heating process.

The gas temperature in nitrogen plasma can be determined from the unresolved rotational structure of the molecular rotational–vibrational–electronic transitions, assuming that the rotational and gas temperatures $T_{\rm rot}$ and T_g, are equal to one another [30-37]. The temperature $T_{\rm rot}$ is determined by comparing the measured and calculated spectra of the second positive system of nitrogen (figure 14).

For the measured rotational temperature $T_{\rm rot}$ to be equal to the actual gas temperature T_g, it is necessary that the following conditions are satisfied: the populations of the rotational levels of an electronically excited state should obey a Boltzmann distribution; in the electronic state from which the excitation occurs, there should be equilibrium between the rotational and translational degrees of freedom; the mechanism for a change in the rotational distribution after excitation should be known; the kinetic processes responsible for the population of the excited state under study should not disturb the rotational distribution; the rotational relaxation time should be shorter than the residence time of molecules in the discharge; the lifetime of the electronically excited state under analysis should be much longer than the rotational relaxation time; and the emission in the bands under study should not be reabsorbed.

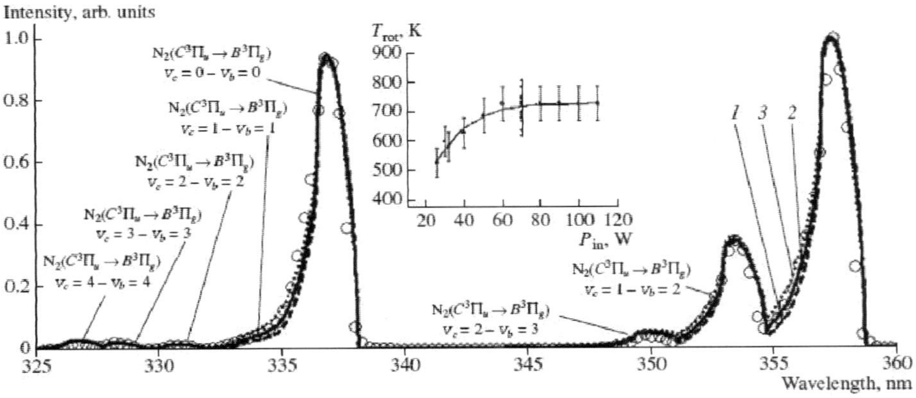

Figure 14. Calculated (curves *1–3*) and measured (circles) emission spectra of the second positive system of molecular nitrogen in the spectral range 325–360 nm from electrode microwave discharge in nitrogen at a pressure of p = 1 Torr; gas flow rate of 20 sccm; $P_{\rm in}$ = 70 W; and $T_{\rm rot}$ =(*1*) 900, (*2*) 700, and (*3*) 550 K. The inset shows $T_{\rm rot}$ as a function of $P_{\rm in}$.[37]

The gas temperature in hydrogen plasma can be determined from the relative intensities of rotational lines of the electron excited molecules $H_2(d^3\Pi_u)$ (bands of Fulcher α-system $H_2(d^3\Pi_u \rightarrow a^3\Sigma_g)$) [38, 39]. Rotational temperature of the ground state of H_2 is calculated taking into account the ratio of rotational constants of the ground (B_0) and excited states (B') ($T_{\rm rot0} = T_{\rm rot*}\ B_0/B'$). This method can be used only if the $H_2(d^3\Pi_u)$ state is depopulated via a radiative transition to the ground state and the collisions with atoms and molecules can be neglected.

3.5. Measurements of electron density and electric field strength

Method of relative intensities of lines or bands emission in combination with measurements of absolute intensities of emission (absolute populations of radiating states) is used for non-invasive measurements of electron density and electric field strength measurements [40-45]. As usual the assumption that the electronic states are excited by electron impact and quenched via radiative decay without reabsorbtion (coronal model) applied for processing of experimental data. The correctness of the coronal model should be proved for conditions of each experiment. It is the simplest case of application of the method but more complicated models of excitation of plasma emission can be used if necessary for realization of this method.

For example the method applied to microwave nitrogen discharge included several successive steps [44]. The experimental ratios of 2^+ and 1^- nitrogen plasma intensities were modeled on the base of kinetic model which includes the calculation of the electron energy distribution function. Electron densities n_e were defined from the absolute intensities of nitrogen bands emission. This iteration procedure was continued up to full agreement of experimental and calculation results giving the values of electric field E and n_e.

In hydrogen microwave plasma small quantity of Ar was added for diagnostics and intensities of H_α, H_β and Ar lines were measured [45]. If Ar lines are excited by electron impact and H_α, H_β lines are excited by electron impact in the processes of dissociative excitation of H_2, processing of ratios of H_α/Ar H_β/Ar intensities and absolute values of lines give the values of E and n_e. This method contains the internal criteria of correctness of the procedure: the values of E and n_e calculated using the lines H_α and H_β should be equal.

4. Summary

- Microwave plasma generators can produce the quasi-equilibrium or non-equilibrium plasma in wide region of pressures, powers in different gases providing manifold internal plasma parameters.
- Any plasma process can be realized on the base of devices from the already designed family of different effective microwave-to-plasma applicators. New microwave plasma devices are continually developed. Efficiency of microwave plasma application in comparison with other discharges should be analyzed in each case.
- Diagnostics of microwave plasma is complicated by presence of microwave field and non-uniformity of plasma.
- These facts should be taken into account in the realization of discharge diagnostics and modeling. Spatially resolved diagnostic methods should be used for microwave plasma diagnostics.
- Competent use of probe and optical methods for microwave plasma diagnostics gives information on gas temperature, spatial distributions of plasma absorbed power, microwave field strength, electron density, and densities of plasma particles.

References

[1] *The Applications of Plasmas to Chemical Processing* 1967 ed R F Baddour and R S Timmins (Cambridge, Mass.: MIT Press)

[2] *Microwave Power Engineering* 1968 ed E. Okress (New York: Acad. Press)

[3] Batenin V M, Klimovskii I I, Lusov G V, Troitskii V N 1988 *Generators of Microwave Plasma* (Moscow: Energoatomizdat)

[4] Lebedev Yu A 1998 *Chemistry of Nonequilibrium Microwave Plasma /Plasma Chemistry* ed L S Polak and Yu A Lebedev (Campbridge: Cambridge Interscience Publishing)

[5] *Microwave Excited Plasmas 1992* ed M Moisan and J Pelletier (Amsterdam: Elsevier)

[6] *Strong Microwaves in Plasma* 1996, 1999, 2002, 2005 ed A G Litvak (Nizhny Novgorod: Institute of Applied Physics RAS)

[7] *Strong Microwaves: Sources and Applications* 2008 ed A G Litvak (Nizhny Novgorod: Institute of Applied Physics RAS)

[8] *Microwave Discharges: Fundamentals and Applications Proc. NATO ARW (Vimeiro, Portugal, 11-14 May 1992) (NATO ASI, Series B: Physics* 1993 *vol 302)* ed C Ferreira and M Moisan (New York and London: Plenum)

[9] *Microwave Plasma and its Applications* 1994 *Proc. Int Workshop (Zvenigorod, Russia, 5-8 September, 1994)* ed Yu A Lebedev (Moscow: the Moscow Phys Soc)

[10] *Microwave Discharges: Fundamentals and Applications 1997 Proc.3 Int Workshop (Fontevraud, France, 20-25 April 1997)* ed C Boisse-Laporte and J Marec (J de Phys IV 1998 vol 8 Pr7)

[11] *Microwave Discharges: Fundamentals and Applications Proc. 4 Int Workshop (Zvenigorod, Russia, 18-22 September, 2000)* ed Yu A Lebedev (2001 Moscow: Yanus-K)

[12] *Microwave Discharges: Fundamentals and Applications Proc. 5 Int Workshop (Greifswald, Germany, 8-12 July, 2003)* ed A Ohl (Greifswald: INP)

[13] *Microwave Plasma and its Applications Proc. 6 Int Workshop (Zvenigorod, Russia, 11-15 September, 2006)* ed Yu A Lebedev (2006 Moscow: Yanus-K)

[14] *Microwave Discharges: Fundamentals and Applications Proc. 7 Int Workshop (Hamamatsu, Japan, 22-27 September, 2009)* ed M Kando and M Nagatsu (2006 Moscow: Yanus-K)

[15] Bychkov D V, Grachov L P, Esakov I I 2009 *Techical Physics* **79** N 3 39

[16] Lebedev Yu A, Mokeev M V, Solomakhin P V et al. 2008 *J. Phys. D: Appl. Phys.*, **41**, 194001.

[17] *Plasma Diagnostic Techniques* 1965 ed R H Huddlestone and S L Leonard (New York, London: Academic Press)

[18] *Plasma Diagnistics* 1968 ed W Lochte-Holtgreven (Amsterdam: North-Holland Publ Comp)

[19] Ochkin V N 2006 *Spectroscopy of Low Temperature Plasma* (Moscow: Nauka)

[20] Heald M A, Wharton C B 1965 *Plasma Diagnostics with Microwaves* (New-York, London, Sydney: John Willey & Sons Inc.)

[21] Lebedev Yu A 1995 *Plasma Souces Sci & Technol* **4** 474

[22] Lebedev Yu A, Mokeev M V 2001 *Plasma Phys. Reports* **27** 418

[23] Mott-Smith H, Langmuir I 1926 *Phys. Rev.* **28** 727

[24] Swift J D Schwar M J R 1970 *Electrical probes for plasma diagnostics* (London.: Iliffe books)

[25] Kozlov O V 1969 *Electric Probe in Plasma* (Moscow: Atomizdat)

[26] Ivanov Yu A, Lebedev Yu.A, Polak L S 1981 *Methods of Contact Diagnistics in Nonequilibrium Plasma Chemistry* (Moscow: Nauka)

[27] Demidov V I, Kolokolov N B, Kudryavtsev A A 1996 *Probe Method of Diagnostics of Low Temperature Plasma* (Moscow: Energoatomizdat)

[28] Bardos L, Lebedev Yu A 1998 *Technical Physics* **43** 1428

[29] Lebedev Yu A, Mokeev M V, Tanatinov A V 2000 *Plasma Phys. Reports* **26** 272

[30] Gubanov A M 1970 *J Applied Spectroscopy* (Rus) **XII** 794

[31] Gubanov A M 1971 *Optics and Spectroscopy* (Rus) **XXX** 211

[32] Phillips D.M. 1975 *J. Phys. D: Appl. Phys.* **8** 507

[33] Sahli K, Hochard L, Pointu A M 1993 *Meas. Sci. Technol.* **4** 685

[34] Chelouah A, Marode E, Hartmann G et. al. 1994 *J Phys D: Appl.Phys.* **27** 940

[35] Chelouah A, Marode E, Hartmann G 1994 *J Phys D: Appl. Phys.* **27** 770

[36] Laux C O, Spence T G, Kruger C H et. al. 2003 *Plasma Sources Sci. & Technol.* **12** 125

[37] Lebedev Yu A, Solomakhin P V, Shakhatov V A 2007 *Plasma Phys. Reports* **33** 157

[38] Qing Zhou, Otorbaev D K, Brussaard G J H et al 1996 *J. Appl. Phys.* **80** 1312

[39] Lebedev Yu A, Mokeev M V 2003 *Plasma Phys. Reports* **29** 226

[40] Berdichevskii M G and Marusin V V 1973 *Zh. Prikl. Spektrosk.* **18** 1055

[41] Yamagishi M, Mizuno K, De Groot J 1988 *J. Quant. Spectrosc. Radiat. Transfer* **40** 519

[42] Spyrou N, Manassis C 1989 *J. Phys. D: Appl. Phys.* **22**, 120

[43] Gallimberti I, Hepworth J K, Klewe R C, 1974 *J. Phys. D: Appl. Phys.* **7**, 880

[44] Lebedev Yu A, Solomakhin P V, Shakhatov V A 2008 *Plasma Phys. Reports* **34** 614

[45] Lebedev Yu A, Mokeev M 2003 *Plasma Phys. Reports* **29** 983

25th Summer School and International Symposium on the Physics of Ionized Gases—SPIG 2010 IOP Publishing
Journal of Physics: Conference Series **257** (2010) 012017 doi:10.1088/1742-6596/257/1/012017

Parametric decay instability control by non-monochromatic pumps

V I Arkhipenko[1], E Z Gusakov[2] and L V Simonchik[1]
[1] *Stepanov Institute of Physics, National Academy of Sciences of Belarus, pr. Nezavisimostsi 68, 220072 Minsk, Belarus*
[2] *Ioffe Institute, Russian Academy of Sciences, ul. Politechnicheskaya 26, 194021 St-Petersburg, Russia*

E-mail: simon@imaph.bas-net.by

Abstract. The wide variety of physical effects accompanying the parametric decay instability driven by the frequency modulated pump is studied experimentally and theoretically. It is shown that pump frequency modulation does not influence the instability when the modulation frequency is much faster than the decay wave transient time in the interaction region. However in the case of slower modulation it is demonstrated that the resonant enhancement and suppression may take place depending on the modulation rate. A scheme of active parametric decay instability feed-back control is proposed. A possibility of deep instability suppression by launching of an additional (small power) pump wave possessing a frequency shifted by the value equal to the frequency separation of ion acoustic eigen modes is demonstrated as well. The recovery of microwave power absorption at the parametric decay instability suppression is shown using measurements of the plasma luminosity and fluxes of accelerated electrons.

Introduction

The parametric decay instability (PDI) excitation is a reason for anomalous reflection and absorption of electromagnetic waves in experiments on laser fusion and RF heating in magnetic confinement devices. In laser fusion experiments excitation of stimulated Raman and Brillouin scattering instabilities and two plasmon decay can lead to substantial reflection of light and generation of fast electrons changing the power deposition in the target [1]. In RF heating experiments PDIs, especially excited at the plasma edge, cause variations of power absorption profiles which are not always favorable as in lower hybrid heating [2]. Anomalous reflection is also accompanying RF heating experiments as in the case of electron Bernstein wave heating where the induced backscattering PDI [3] due to relatively low power thresholds [4] is often observed. Moreover, as it was shown recently, the PDI is observed at ECR heating on TEXTOR tokamak as well [5, 6].

According to the homogeneous plasma PDI theory the frequency width of a PDI is determined by its growth rate γ_0 [7]. Therefore the frequency broadening of the pump wave $\Delta\omega$ exceeding the PDI growth rate ($\Delta\omega/\gamma_0>1$) should lead to the PDI threshold growth. Practically the pump wave spectral broadening can be associated with different mechanisms among which the finite frequency width of the generation of powerful oscillators and multi-mode pump composition. The pump frequency broadening can be caused by the pump phase and frequency modulation as well. Based on results of homogeneous plasma theory, the random pump phase modulation was proposed in [8] for the PDI control. However, a more detailed theoretical analysis performed in [9] in the framework of inhomogeneous plasma theory demonstrated a weak sensitivity of the PDI to the random pump phase modulation. It should be mentioned that theoretical conclusions [9] were obtained within the model utilizing extremely fast (delta-correlated) pump phase modulation, in particular, much quicker than the transient time of the slow daughter wave in the decay region. In the present paper we demonstrate in the case of slower modulation a wide variety of physical effects accompanying the parametric decay instability driven by the frequency modulated pump in inhomogeneous plasma, in particular the PDI resonant enhancement and suppression.

2. The experimental approach

2.1. Experimental installation "Granit"

The experiment was carried out at the linear magnetized plasma device [10] (see figure 1*a*) with magnetic field of about 0.35 T, in which plasma was produced by 200 W ECR discharge at frequency 9.8 GHz in a tube 2 cm in diameter and 100 cm long filled by argon at pressure 1–2 Pa. Plasma diffuses from ECR region along discharge tube and magnetic field, its density exponentially decreases with scale $a = 3 - 5$ cm. Inhomogeneity scale across magnetic field is $b \approx 0.4$ cm. The electron density estimation provided by the cavity method in the middle part of installation results in value of $n_e = 10^{12}$ cm^{-3}, electron temperature was measured by the multi-grid analyzer at the level of $T_e = 2\ eV$. The axial and radial distribution of plasma luminosity was controlled by the optical diagnostics

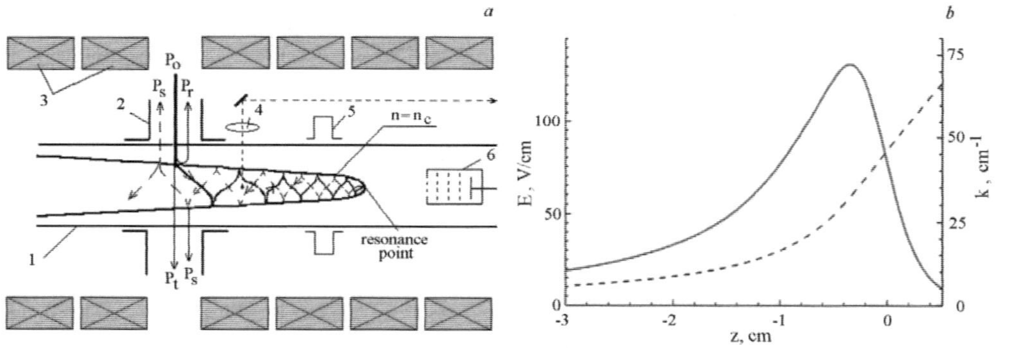

Figure 1. The experimental device and wave propagation scheme – (*a*). P_0, P_r, P_s, P_t – incident, reflected, scattered and transmitted waves; $n = n_c$ – the critical density surface; 1 – quartz tube; 2 – waveguide; 3 – magnetic field coils; 4 – optical system; 5 – microwave cavity; 6 – the fast electron multi grid analyzer.
EPW field (solid) and parallel wave number (dashed) calculated distributions, z = 0 – the resonance point position – (*b*).

utilizing the fiber light guide. The microwave radiation backscattered in plasma was studied in the waveguide using microwave homodyne detectors and spectral analyzers.

2.2. The Trivelpiece-Gould mode pump wave

The electron plasma pump wave (Trivelpiece-Gould (TG) mode) at frequency $f_0 = \frac{\omega_0}{2\pi} = 2480\ MHz$ was excited in this plasma using a waveguide system possessing electric directed along the balloon axis. The dispersion relation for this wave is given by $k_\perp^2 = \left[\frac{\omega_{pe}^2}{\omega_0^2} - 1\right] k_\parallel^2$, where $k_\parallel^2$ and $k_\perp^2$ are the components of the wave vector parallel and transverse to the magnetic field, ω_{pe} is the electron plasma frequency.

The high density plasma ($n_e\ r,z\ > n_c$, where n_c - critical electron density) creates a plasma waveguide weakly inhomogeneous in axial direction for the electron plasma wave (EPW). According to the results of analytical theory [10] accounting for the realistic 2D plasma geometry in the vicinity of the hybrid resonance, where $\omega = \omega_{pe}(0, z)$, EPW field there takes a form

$$E_o = \left(\frac{2P_o'}{\omega_o}\right)^{1/2} \frac{k^{3/2}}{(3r_d^2 bk^3 + 1)^{1/2}} \exp\left[-\int_{-\infty}^{z}(k' + ik'')dz' - \frac{k}{2b}r^2 - i\omega_o t\right] + cc, \quad (1)$$

where $P_0' = \kappa P_0$ - is the part of power P_0, which goes into excitation of fundamental radial TG mode ($\kappa \approx 0.2$); k - is the component of wave vector **k** along the exterior magnetic field direction. It is defined in the neighbourhood of the resonance point from the equation:

$$3r_d^2(k + ik'')^2 - \frac{z}{a} - \frac{2}{(k' + ik'')b} + i\eta'' = 0, \quad (2)$$

where η'' – imaginary part of the longitudinal dielectric plasma permeability.

The calculated dependencies of electric field amplitude and wave vector for TG mode are shown in figure 1*b*. The calculation parameters measured or calculated using experimental data were as follows: P_0 = 20 mW, κ = 0.2, f_0 = 2350 GHz, T_e = 1.8 eV, a = 5 cm, b = 0.43 cm, ν_{ea} = 4.7 10^7 c^{-1}, δ = 0.003 – the fraction of super thermal electrons in the resonance poin, T_h = 8 eV – effective temperature of the super thermal electron tail.

2.3. Accelerated electron generation

Propagating, as a Gaussian wave beam (the fundamental plasma waveguide mode), towards decreasing electron density to a point of a plasma resonance, where $\omega_0 = \omega_{pe}(0, z)$, the wave slows down. Its wave number parallel to magnetic field increases, the beam width decreases and,

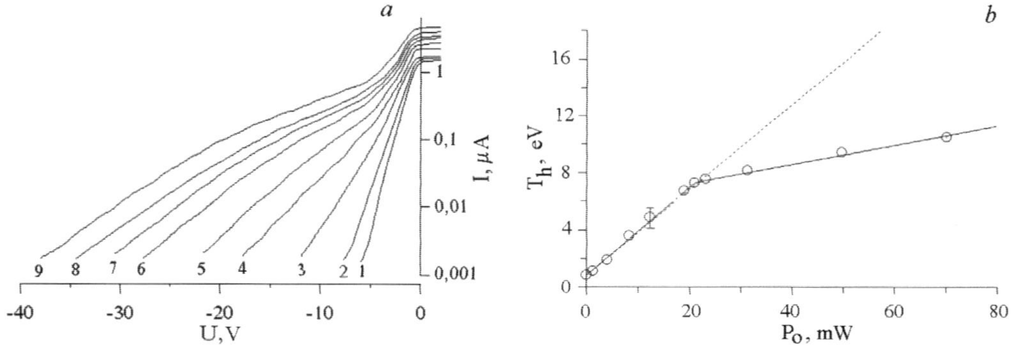

Figure 2. The multi-grid analyzer characteristics (*a*) and the fast electron tail effective temperature for growing pump power (*b*). 1 – initial plasma, 2 – P_0 = 1 mW, 3 – 4, 4 – 9, 5 – 12, 6 – 18, 7 – 23, 8 – 35, 9 – 50 mW.

consequently, its electric field grows drastically. After that the collisional absorption of wave as well as its interaction with electrons via the Landau mechanism should take place in the vicinity of the resonance leading to the wave absorption. Accordingly the electron distribution function differs substantially from the Maxwellian dependence for energy exceeding the thermal one $E >> T_e$ (figure 2*a*) and may be well approximated by the bi-Maxwellian dependence

$$f_e(v) = \left(\frac{m_e}{2\pi T_e}\right)^{1/2} \exp\left(-\frac{m_e v^2}{2T_e}\right) + \delta\left(\frac{m_e}{2\pi T_h}\right)^{1/2} \exp\left(-\frac{m_e v^2}{2T_h}\right), \quad (3)$$

As it is seen in figure 2*b*, the effective temperature of the super thermal electron tail T_h grows with EPW power when it is small. However this growth saturates at the power level of 20 mW, thus indicating reduction of the power fraction absorbed due to the Landau damping [11].

2.4. Experimental manifestation of absolute parametric decay instability at monochromatic pump
The saturation of the effective sub-thermal electron tail temperature growth shown in figure 2*b* at $P_0 > 20$ mW was always accompanied by excitation of stimulated backscattering (BS) parametric decay instability $l_0 \rightarrow l_0' + s$ occurring in the vicinity of resonance point [10]. It was found that at very low power level $P_0 \sim 10$ mW the convective decay instability accompanied by the generation of a reflected fundamental TG mode l_0' and an ion-acoustic wave s propagating along the magnetic field

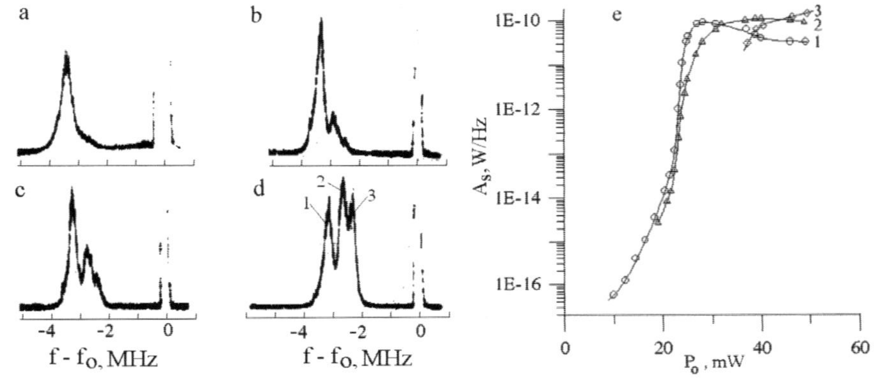

Figure 3. The induced BS spectrum modification (*a* - *d*) and evolution of amplitude shifts of BS satellites (1, 2 and 3) at growing pump power (*e*). (*a*) P_0=17 mW, (*b*) 23 mW, (*c*) 27 mW and (*d*) 40 mW. Numerals 1–3 in (*d*) correspond to amplitude dependences in (*e*).

toward the lower density is excited. In this case, a satellite arises in the spectrum of the scattered signal down shifted from the pump frequency by about 3 MHz (see figure 3*a*). At small instability threshold excess (10 mW < P_0 < 20 mW) the dependence of the BS signal on the pump power (Figure 3*e*) was exponential, typical for convective decay instability [12, 13]. However at higher pump power $P_0 > 20$ mW, even steeper dependence was observed in [14] indicating the absolute instability onset (see figure 3*e*, curve 1). The absolute instability mechanism, according to [14], is related to the complicated spatial structure of pump wave, namely, to the small fraction of the first radial mode present in the pump along with the dominant fundamental radial TG mode ($P_1 \leq 0.1P_0$) This small fraction interacting with the BS wave leads to generation of the ion acoustic (IA) wave in spatial point shifted by $\delta z \approx 0.3$–0.5 cm from the main decay region. This IA wave propagates back to the decay region where it experiences amplification and generates the BS fundamental TG mode l_0', thus leading to formation of the feedback loop and onset of the absolute PDI. The absolute decay instability is a coherent process with the limited number of oscillatory modes excited, which close to the threshold manifest themselves by narrow lines in the BS spectrum (Figure 3 *a-d*). According to [15], the instability growth rate and an unstable spectrum structure are determined by the time of the IA wave circulation in the feedback loop $\tau = \frac{\delta z}{c_s} \sim \frac{a}{c_s} \approx 1.5$–$2.5\ 10^{-6}$ s, where c_s is the IA velocity.

3. Absolute PDI at non-monochromatic pump

The frequency modulated microwave pump was provided by a voltage controlled oscillator possessing minimal and maximal frequencies 2200 and 3200 MHz correspondingly. The output of the generator

Journal of Physics: Conference Series **257** (2010) 012017 doi:10.1088/1742-6596/257/1/012017

in the present experiment took a form $U(t) = U_m \cos\left\{2\pi \int_0^t \left[f_0 \pm \Delta f \left[U_C \ t' \ \right] \right] dt' \right\}$ where $\Delta f(U_c)$ – frequency deviation as a function of control voltage U_c. For this generator maximal deviation Δf_{max} ~ ±500 МГц at the voltage variation in range from 3 V up to 20 V, a band-pass of the control input was up to 50 MHz. It allows controlling the pump frequency at high rate and using different frequency variation waveforms. The pump could be amplified using travelling-wave tube amplifiers up to the level of 1 W.

The amplitude of parametric satellite was measured using enhanced scattering technique [10]. For this purpose the probing wave at frequency f_p = 2.35 GHz < f_0 and small power (< 5 mW) was launched into plasma, so that $f_0 - f_p < \_f_{max}$. In vicinity of own resonance point the probing wave is effectively scattered on excited in plasma ion-sound wave and scattered wave amplitude is proportional to ion–sound one.

3.1 Linear pump frequency modulation

To obtain the pump frequency modulation close to linear the saw-tooth voltage from special generator was used as control voltage for sweep-generator. In this case the frequency of pump was modulated as $f = f_0 - \Delta f\, t/T$, where f_o = 2650 MHz; Δf = 300 MHz; $T \geq 1$ μs – modulation period. Then the frequency variation rate is given by expression $v_{sw} = \Delta f / T$ resulting in drift of the three wave resonance point where the PDI takes place in inhomogeneous plasma. The corresponding decay point velocity is equal to $v_d = \dfrac{2a}{T}\dfrac{\Delta f}{f_0}$.

In figure 4*a*, the spectra of BS signal at different frequency rates are shown. As it can be seen, scattered signal has maximal value at T = 5.5 μs. This corresponds to the frequency variation rate of 60 MHz/μs. The corresponding velocity of the decay point is $v_d \sim 2\ 10^5$ cm/s. It is close to the ion-

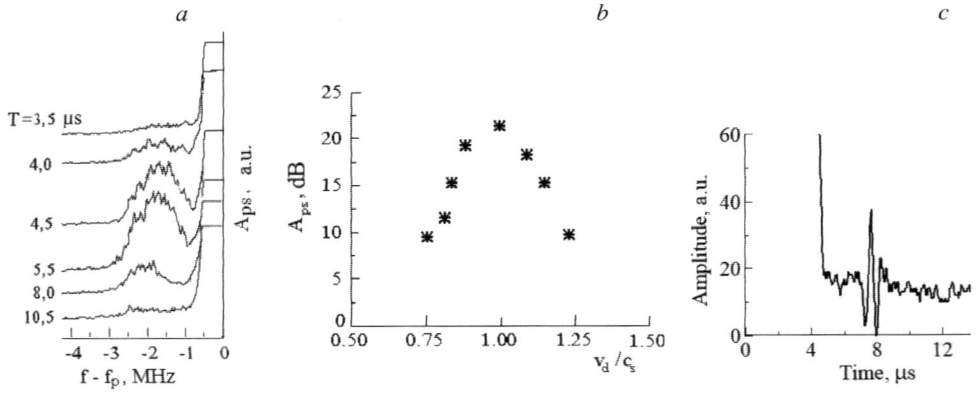

Figure 4. The PDI spectra at different sweep speed (*a*), satellite amplitude A_{ps} against the normalized velocity of decay point displacement (*b*) and a burst of ion-acoustic wave in decay region at $v_d/c_s \sim 1$ (*c*).

acoustic velocity calculated for the experimental plasma parameters $c_s = (T_e/M_{Ar})^{1/2} \sim 2.2\ 10^5$ cm/s. At higher and smaller frequency modulation rate the scattering signal decreases substantially. Its dependence on the decay point velocity normalize by the ion acoustic speed is shown in Figure 4b. It possesses a well pronounced maximum at $v_d \cong c_s$ indicating the resonant nature of the enhancement effect. In this case the direction and velocity of decay point drifting in plasma are close to ion–acoustic wave ones. It results in significant decreasing of the convective energy losses in decay region and, as consequence, leads to PDI stimulation, which is revealed in scattered signal amplitude growth. According to [16], the spatial amplification coefficient expression which describes the slow decay wave amplification in the presence of linear frequency modulated pump wave in the inhomogeneous plasma can be written as the following

$$S = \frac{\pi \gamma_0^2 \ell_*^2}{|v_2 - v_d||v_1 - v_d|},$$

where ℓ is the phase mismatch length, v_1 and v_2 are the velocities of decay waves.

Since the PDI stimulation is resonant and its duration is short in time, so the excited ion-acoustic wave in decay region looks likes a soliton (Figure 4c). It should be noted that during time of reverse variation of the frequency the scattered signal was suppressed to the level less than -20 dB.

3.2 Harmonic pump modulation

To obtain the harmonic frequency modulation the controlling voltage was formed using the standard sinusoidal RF generator.

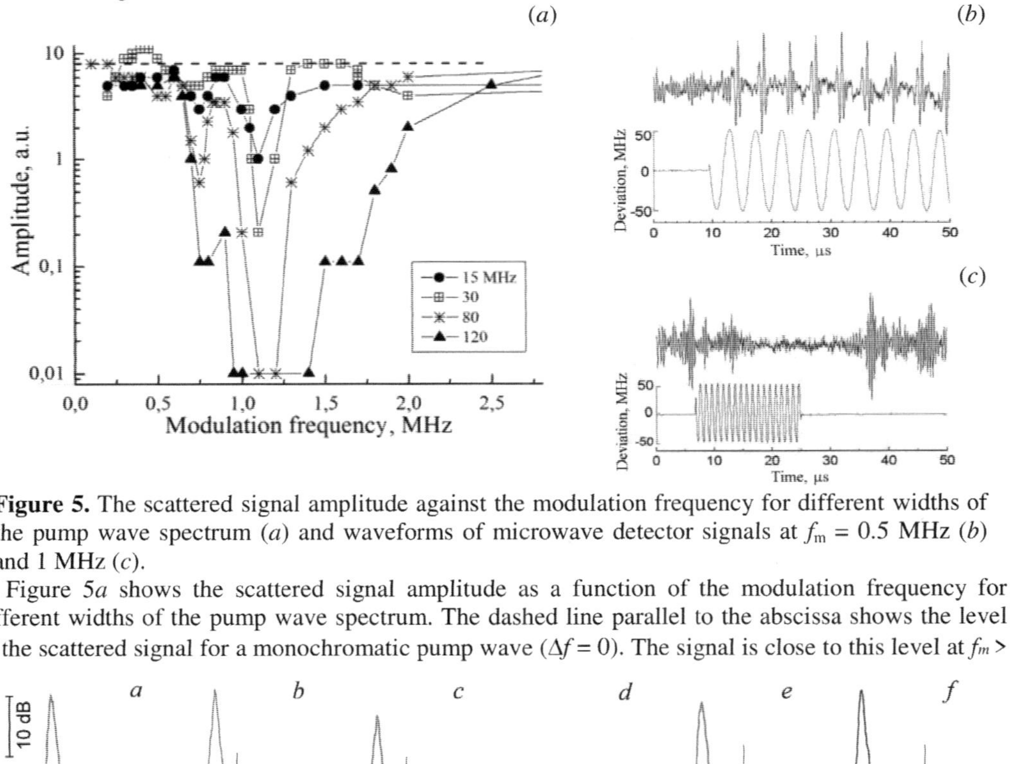

Figure 5. The scattered signal amplitude against the modulation frequency for different widths of the pump wave spectrum (*a*) and waveforms of microwave detector signals at f_m = 0.5 MHz (*b*) and 1 MHz (*c*).

Figure 5*a* shows the scattered signal amplitude as a function of the modulation frequency for different widths of the pump wave spectrum. The dashed line parallel to the abscissa shows the level of the scattered signal for a monochromatic pump wave ($\Delta f = 0$). The signal is close to this level at f_m >

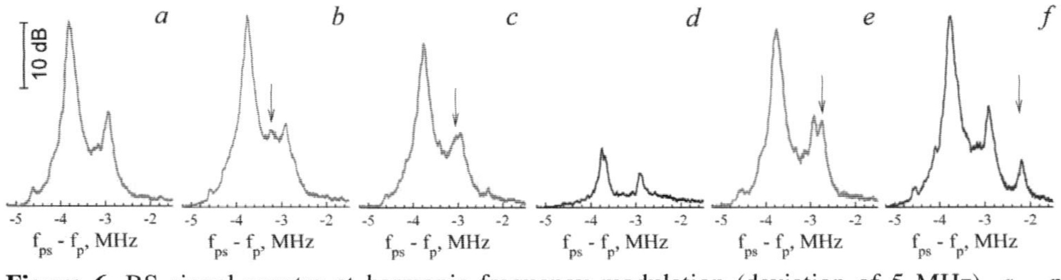

Figure 6. BS signal spectra at harmonic frequency modulation (deviation of 5 MHz). *a* – no modulation, *b* – modulation frequency is 0.5 MHz, *c* – 0.7 MHz, *d* – 0.85 MHz, *e* – 1.0 MHz and *f* – 1.7 MHz. The arrow indicates the additional line caused by the pump frequency modulation.

1.5 MHz and f_m < 0.6 MHz. The certain increase in the scattered signal is observed for Δf = 120 MHz and f_m = 0.5 MHz. It is related to the enhancement of decay instability occurring if the velocity of propagation of the decay point coincides with that of the ion-acoustic wave as in case of linear frequency modulation. Therefore the corresponding microwave detector signal looks like a sequence of bursts (Figure 5c).

Drastically resonant suppression of the parametrically scattered signal is observed around modulation frequency f_m = 1 MHz within a wide range of deviation values. To clarify the nature of this suppression the experiment was performed for the coherent regime of the $l \to l' + s$ absolute instability when scattering spectrum at pump power of 40 mW consist of two stable narrow lines at frequency $f_{ps}^{(k)} = f_p - f_{IA}^{(k)}$ (Figure 6*a*) corresponding to the instability eigen modes. The lines spectral separation varies depending on experimental conditions in the range 0.4 – 0.9 MHz in agreement with the estimation $f_{IA}^{(k)} - f_{IA}^{(k-1)} \approx \tau^{-1}$ derived in [14]. Application of the harmonic pump frequency modulation results in additional lines appearing in the BS spectrum at frequencies $f_{IA}^{(k)} \pm f_m$. The variation of the BS spectrum at growing f_m is shown in Figures 6*b* – 6*f*. Arrows in figures 6*b, c, e and f* are indicating additional lines caused by the pump frequency modulation. As it is seen in Figure 6*d*, suppression of the BS signal occurs at the coincidence of the modulation frequency and the eigen mode frequency difference $f_{IA}^{(k)} - f_{IA}^{(l)} \sim \frac{c_s}{a}$.

At small frequency deviation δf and close to the PDI threshold the suppression region is very narrow, confirming the resonant nature of this effect (see Figure 7*a*), whereas at larger deviation and the threshold excess the BS suppression become deeper and broader in modulation frequency.

It is important to note that frequency modulation leads not only to suppression of the probing wave BS, the IA wave generation and accordingly anomalous reflection, but also to enhancement of the pump wave absorption in the vicinity of the resonance point. The last statement is confirmed by observations of the plasma luminosity growth (Figure 7*b*) and super thermal electron production at energy higher than 20 eV (Figure 7*c*) in the decay region initiated by the pump frequency modulation switching on. We used the retarding potential about 20 V on the analyzer in order to exclude the contribution of thermal electrons with average energy of about 2 eV.

3.2.1 The PDI feed-back control experiment. Based on the observations of a very high sensitivity of PDI to the pump frequency modulation at frequency close to the difference of the PDI eigen frequencies it was proposed to use a signal obtained as a result of the backscattering signal double frequency down-conversion for the pump frequency modulation [17]. The developed feed-back scheme provides the control signal at the absolute PDI eigen frequency difference $f_{IA}^{(k)} - f_{IA}^{(l)}$. It filters the backscattering signal suppressing the signal at pump frequency and at the probing frequency. At the microwave detector it mixes the IA frequencies amplifying the signal at frequency $f_{IA}^{(k)} - f_{IA}^{(l)}$ and suppressing simultaneously the signal at the ion acoustic wave frequencies $f_{IA}^{(k)}$ and $f_{IA}^{(l)}$.

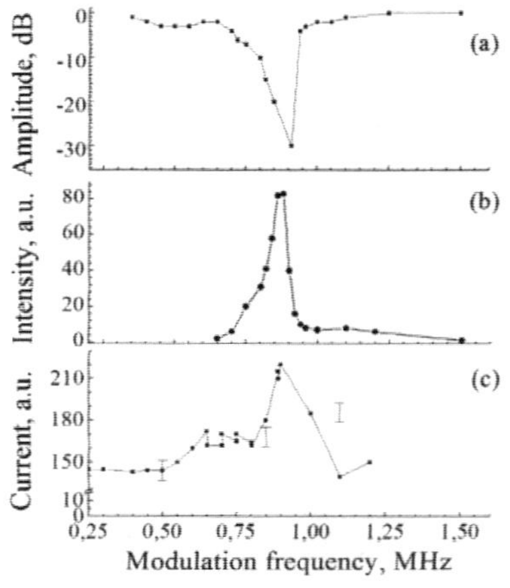

Figure 7. Dependences of the BS signal (*a*), plasma luminosity growth (*b*) and the super thermal (>20 eV) electron current (*c*) on the modulation frequency (δf = 7.5–12

The possibility of substantial (a factor of 5) suppression of the PDI ion-acoustic wave and anomalous pump wave reflection using the feed-back scheme was shown for frequency deviation of 7.5 and 10 MHz. It is important to note that the PDI suppression was not possible with a control signal at IA frequency $f_{IA}^{(k)}$ proportional to the IA wave amplitude provided by the scheme in the case when the pump wave frequency was not filtered out.

3.3 Stochastic pump modulation

It should be stressed that resonant enhancement of the convective amplification coefficient is possible not only for linear or harmonic pump frequency modulation, but also for stochastic modulation with finite correlation time as was shown in the numerical modeling [16].

To produce the stochastic frequency modulation of the output signal the controlling voltage U_c was formed using a photomultiplier current in the photon–counting mode, when a photocathode was illuminated by a filament lamp. In this case, the spectrum of the photocurrent approximates the white–noise spectrum. To change the correlation time of controlling voltage signal from 0.01 up to 2 μs the limitation of photomultiplier frequency band by RC–filter was used in experiments.

In Figure 8*a* the obtained dependencies of the PDI induced scattering signal amplitude on pump frequency modulation band $2\Delta f$ are shown at different correlation time. All curves have the growth part with obviously expressed maximum and followed decreasing part. It testify to enhanced plasma

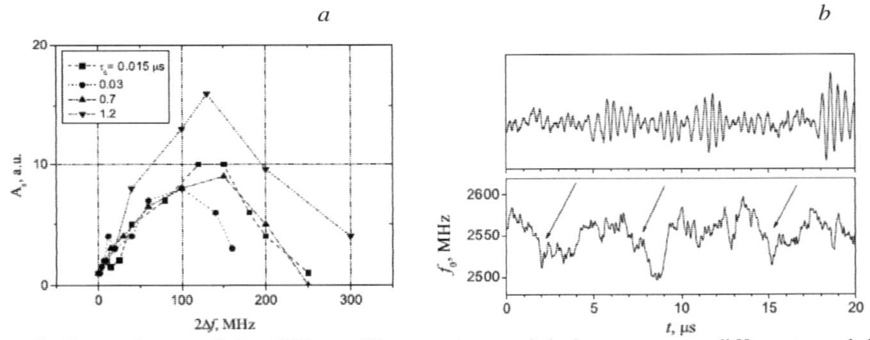

Figure 8. Dependence of the PDI satellite on the modulation range at different modulation correlation time (*a*) and waveforms of microwave detector signal (above) and frequency variation (*b*).

wave amplification and PDI stimulating at small pump frequency deviation and PDI suppression at modulation band > 300 MHz. One can see that maximal satellite amplitude grows with correlation time.

In Figure 8*b* the detected homodyne signal (upper waveform) and time dependence of pump frequency are shown for the case of stochastic modulation with deviation ±50MHz. The splashes of oscillations at frequency of about 3MHz corresponding to daughter ion–sound wave frequency seen in the upper picture coincide with the periods of linear frequency modulation, as shown in the bottom waveform in Figure 8b.

3.4 PDI suppression by complementary pump.

PDI suppression method developed in previous sections, unfortunately, is difficult to realize at high (lasing) frequencies or with powerful RF generators because the fast modulation technique is needed. However a possibility of the deep PDI suppression by launching of the additional (small power) pump wave possessing a frequency shifted by the value equal to the frequency separation of ion acoustic eigen modes exists in these full-scale experiments. To demonstrate it in our model experiment, the parametric decay instability $l \rightarrow l' + s$ was excited by the EPW pump at frequency $f_0 = 2335$ MHz and power of about 40 mW resulting in red-shifted satellite observed in BS spectrum (Figure 9(*a*)). When additional EPW at frequency $f_p = f_0 + 5$ MHz and small power ($P_p \sim 15$ mW) was launched into plasma by the same waveguide system a complicated BS spectrum consisting of two pump lines and two down shifted satellites was observed due to BS of the additional pump off the parametrically driven small scale ion-sound wave (Figure 9(*b*)).

At high enough pump frequency difference $|f_p - f_0| \geq 2$ MHz (Figure 9(b) and (*f*)) the complementary pump do not influence the pump BS spectrum. However, as it is seen in Figure 9(c) and (e), at frequency difference $|f_p - f_0| \approx 1$ MHz the main pump BS decreases by several orders of magnitude, thus indicating the next to total absolute PDI suppression. At $|f_p - f_0| < 1$ MHz the BS signal recovers, however possessing different spectrum shown in Figure 9(d). At the variation of complementary pump power from 40 mW to 4 mW the optimal suppressing frequency changes from 1.2 MHz to 0.8 MHz. The physical reason for this change is related to the variation of ion acoustic eigen mode frequency due to density profile flattening caused by growth of absorbed microwave power. A more detailed analysis of suppression efficiency dependence on frequency reveals several suppression maxima corresponding to easier and deeper suppression of the absolute instability. The pump frequency difference corresponding to these maxima is equal to the frequency difference of ion acoustic eigen modes excited by the instability and observable as separate lines in BS spectrum in Figure 9(*a*). It should be also mentioned that the strongest suppression occurs at the complementary pump power comparable to the main pump power.

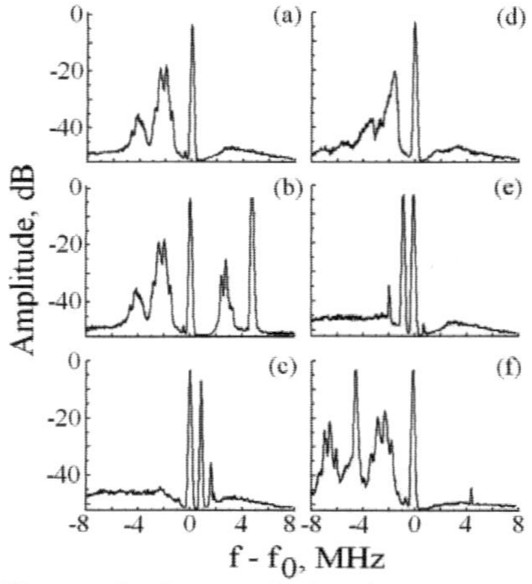

Figure 9. Scattered signal spectra at different frequencies of additional pump wave.

It is natural to assume that the anomalous backscattering suppression due to the complementary pump influence should result in enhancement of power absorbed in plasma and accordingly in growth of fast electron production and plasma luminosity. The measured luminosity distribution is shown in Figure 10(*a*) for the cases of $f_p - f_0 = 5$ MHz and $f_p - f_0 = 1$ MHz corresponding to spectra (*b*) and (*c*) in Figure 9. The hybrid resonance position is given in Figure 10 by arrow. As it is seen in Figure 10(*a*), suppression of the instability at $f_p - f_0 = 1$ MHz is accompanied by plasma luminosity growth compared to the case $f_p - f_0 = 5$ MHz and by shift of the luminosity maximum further from the EPW

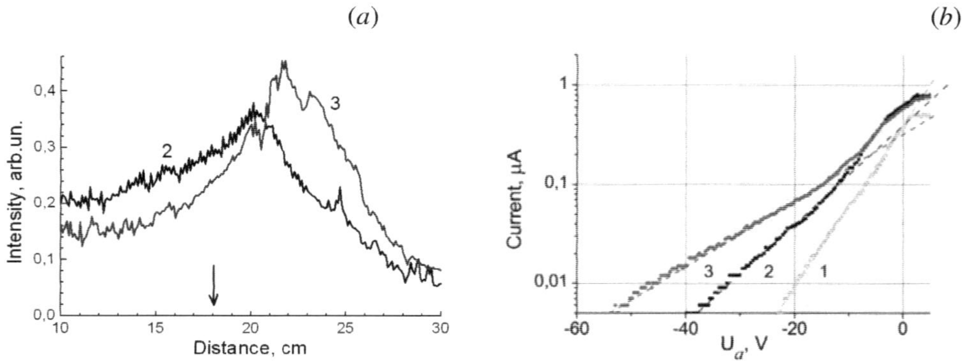

Figure 10. Axial plasma luminosity distribution (a) and analyzer voltage current characteristics (b) registered at pump frequency difference of 5 MHz (2) and 1 MHz (3), 1 – background plasma. Arrow corresponds to the hybrid resonance position.

excitation region indicating suppression of the anomalous reflection and absorption growth.

The fast electron analyzer situated in the low density plasma at 5 cm from the hybrid resonance (see Figure 1) also registered evidences of the RF power absorption growth at the PDI suppression. As it is seen in Figure 10(*b*), the effective temperature of the fast electron tail, which was 4.2 eV in the unperturbed plasma, enhanced up to 7.1 eV at application of the two-frequency pump in the case $f_p - f_0 = 5$ MHz and further increased up to 9.3 MHz when the pump frequency difference take the resonance value $f_p - f_0 = 1$ MHz providing the deepest instability suppression.

4. Conclusions

The wide variety of physical effects is accompanying the parametric decay instability driven by the frequency modulated pump. As it is confirmed experimentally, the pump frequency modulation does not influence the instability when the modulation frequency is much faster than the decay wave transient time in the interaction region. In the case of slower modulation, the resonant enhancement and suppression may take place instead. Coming to the physical reason of the observed effects it is necessary to stress that the physical reason for the observed PDI resonant enhancement is provided by suppression of convective losses of the daughter wave from the decay region, drifting due to the slow pump frequency modulation at the ion acoustic speed. The strong resonant suppression of the most dangerous absolute inhomogeneous plasma PDI is observed at a minimal frequency deviation (less than 1%) when the modulation frequency is equal to frequency separation of the stable lines observed in the backscattering spectrum which correspond to ion acoustic wave eigen modes excited in plasma by the absolute PDI. Based on this effect a scheme of active PDI feed-back control was proposed.

A possibility of deep instability suppression by launching of an additional (small power) pump wave possessing a frequency shifted by the value equal to the frequency separation of ion acoustic eigen modes is demonstrated as well. The recovery of microwave power absorption at the parametric decay instability suppression is shown using measurements of the plasma luminosity and fluxes of accelerated electrons.

Keeping in mind that excitation of discrete frequency spectrum due to creation of feedback loop, when the part of energy taken away by a daughter wave from the decay region is retuned back is the common feature of absolute PDIs, both the harmonic modulation and the feed back control methods may be recommended for the PDI suppression. In the case of laser fusion experiments due to the short interaction time only the first method based on harmonic pump frequency modulation is applicable, whereas in RF heating experiments both methods can be useful.

Acknowledgements

Financial support of BRFBR-RFBR Grants (F10R-010, 10-02-90003-Bel_a), RFBR Grant 10-02-00887 and scientific school grant 6214.2010.2. is acknowledged.

References

[1] H.A. Baldis, D.S. Montgomery, J.D. Moody, C. Labaune, S.H. Batha, K.G. Estabrook, R.L. Berger and W. L. Kruer 1992 *Plasma Phys. Control. Fusion* **34**, 2077
[2] V.N. Budnikov and M.A. Irzak 1996 *Plasma Phys. Control. Fusion* **38**, A135
[3] H.P. Laqua 2007 *Plasma Phys. Control. Fusion* **49**, R1
[4] E.Z. Gusakov and A.V. Surkov 2007 *Plasma Phys. Control. Fusion* **49,** 631
[5] E. Westerhof et al. 2008 *Reveiw Scientific Instruments* **79**, 093503
[6] E.Z. Gusakov and A.Yu. Popov *2010, JETP Letters*, **91**, *656*
[7] V. P. Silin, 1973 Parametric Influence of High Power Radiation on Plasma (Nauka, Moscow).
[8] E. Valeo and C. Oberman 1973 *Phys.Rev.Letters* **30**, 1035
[9] G. Laval, R. Pellat, and D. Pesme 1976 Phys. Rev. Lett. **36**, 192
[10] V. I. Arkhipenko, V.N. Budnikov, E.Z. Gusakov, I.A. Romanchuk, L.V. Simonchik 1987 *Plasma Phys. Report* **13**, 693

[11] V. I. Arkhipenko, V.N. Budnikov, E.Z. Gusakov, I.A. Romanchuk, L.V. Simonchik 1986 *Pis'ma ZhTF* **12**, 1190
[12] A.D. Piliya, 1971 *Proceedings of the Tenth International Conference on Phenomena in Ionized Gases,* Oxford **1**, 320
[13] M.N. Rosenbluth 1972 *Phys. Rev. Lett.* **29**, 565
[14] V. I. Arkhipenko, V.N. Budnikov, E.Z. Gusakov, L.V. Simonchik 1987 *Zh.Exp.Theor.Phys.* **93,** 1221.
[15] V. I. Arkhipenko, V.N. Budnikov, E.Z. Gusakov, L.V. Simonchik 1987 *JETP Letters* **46**, 17
[16] V. I. Arkhipenko, E. Z. Gusakov, V.A. Pisarev, L.V. Simonchik and B.O. Yakovlev 2004 *Physics of Plasmas* **11,** 71
[17] V. I. Arkhipenko, E. Z. Gusakov, L. V. Simonchik, F. M. Truhachev Phys. Rev. Lett. **101**, 175004 (2008)

Journal of Physics: Conference Series **257** (2010) 012018 doi:10.1088/1742-6596/257/1/012018

Effect of gas mixture composition on tar removal process in a pulsed corona discharge reactor

E A Filimonova and G V Naidis

Joint Institute for High Temperatures, Russian Academy of Sciences, Izhorskaya 13, bld.2, 125412, Moscow, Russia

e-mail: helen@oivtran.ru

Abstract. Results of simulation of naphthalene removal from nitrogen and its mixture with water vapour by a sequence of corona discharge pulses are presented. A comparison of the data calculated on the basis of two approaches is presented: with approximate and detailed account of reactions involving charged species. It is shown that approximate account gives the efficiency of naphthalene removal close to that obtained basing on more detailed approach.

1. Introduction

The use of pulsed corona discharges is considered as a perspective method for gas cleaning from harmful admixtures [1]. In particular, it has been shown [2] that such discharges are effective tools for removal of heavy hydrocarbons (tars) from products of biomass gasification. In these works naphthalene removal by a sequence, including up to several thousands, of corona discharge pulses has been experimentally studied in pure nitrogen and its mixtures with H_2, CO, CO_2 and H_2O, at the initial tar content around 1000 ppm. The remaining tar molar fraction has been obtained as a function of the specific energy input.

In [3] the results of numerical modeling of naphthalene removal by pulsed corona discharges, for conditions close to those of experiment [2], have been presented. The simulation of the cleaning process included a description of two stages. The first, fast stage is generation of primary active species (radicals, ions, excited molecules and atoms) during the corona discharge pulses, the discharge having a structure of thin plasma filaments – streamers. The densities of primary active species obtained at streamer simulation were used as the input parameters for the modeling of the second, slow stage: the chain of chemical transformations triggered by the primary species.

For account of the role of ions as active species in the cleaning process an approximate approach was used in [3]: the ion-molecular reactions and electron-ion recombination processes were assumed to be infinitely fast. In this approach the reactions involving ions were accounted for as additional sources of primary neutral active species (atoms, radicals, excited molecules) and were not included to the kinetic scheme describing the chemical transformations at the second stage. In the given work for simulation of naphthalene removal in pure nitrogen and its mixture with water vapor a more rigorous approach is used, with explicit account of ion-molecule reactions and electron-ion recombination processes during the second, postdischarge stage of gas treatment. The modeling results based on these two approaches are compared with each other and with experimental data [2].

The consideration of more detail kinetics containing ions is aimed at more accurate account of reactions involving ions, as additional (to reactions between neutral species) effective channels of the removal process, especially in strong electric field of pulsed corona discharge. The increased interest to the kinetic processes involving ions is noted also in other fields such as plasma assisted combustion and plasma medicine. The other aspect related with ion kinetics is associated with the role of electronegative gas admixtures in streamer start, propagation and branching phenomena [4].

2. Formation of primary active species during streamer propagation

The number densities of primary active species produced by the streamers are evaluated via corresponding *G*-values - the numbers of particles produced per 100 eV of input energy. For calculation of the *G*-values the method, based on streamer simulation, is used analogous to that described in [3]. At evaluation of the *G*-values for dissociation of nitrogen and water molecules the dissociation cross sections are taken from [5] and [6], respectively. The values G_j for the production of active species of sort *j* for a given mixture composition are estimated by weighting of partial *G*-values as

$$G_j = \sum_i X_i G_{ij} \qquad (1)$$

where X_i is the molar fraction of species of sort *i*, G_{ij} is the partial *G*-value for production of species *j* in the reaction of electrons with particles of sort *i* (normalized to the molar fraction of species *i* equal to unity). The partial *G*-values are assumed to be independent of the mixture composition, gas pressure, temperature, applied voltage and geometry of the discharge gap.

Obtained estimates for the partial *G*-values in nitrogen and its mixtures with water vapor are $G_{OH} = G_H = 1.4$, $G_{N(4S)} = G_{N(2D)} = 0.5$, $G_{N2(A3\Sigma)} = 2.2$, G_e=0.7. The *G*-value for production of $N_2(A^3\Sigma)$ is estimated with account of cascade transitions from higher triplet and singlet states of nitrogen molecules. The dominating channel of nitrogen dissociation is assumed to be $N_2 + e \rightarrow N(^4S) + N(^2D) + e$ [7]. Note that some uncertainty exists concerning the *G*-values for N_2 and H_2O dissociation. The effective dissociation cross sections are governed by competition of predissociation and collisional quenching and, hence, may decrease with growth of the gas density [8]. The cross section presented in [5] corresponds to low gas pressures, so that the evaluated, basing on this cross section, *G*-value for N_2 dissociation is an above estimate (in [3] twice lower *G*-value for this process was used). As for H_2O dissociation, we assume, basing on the data given in [6], that its dominating channel is $H_2O + e \rightarrow OH + H + e$. Note that in [9] another dissociation channel, $H_2O + e \rightarrow O(^1D) + H_2 + e$, was taken as dominating.

The number densities n_j of active species vary with the relative gas number density δ (the ratio of the gas number density to its normal value, 2.5×10^{19} cm^{-3}, at the gas pressure of 1 bar and temperature of 300 K) as δ^2. They are estimated as

$$n_j = \frac{G_j}{G_e} n_e \qquad (2)$$

where the evaluated *G*-value for generation of electrons is $G_e = 0.7$ and the number density of electrons n_e is $5\times10^{14}\,\delta^2$ cm^{-3}.

Table 1. G-values in pure N_2 (I, I a) and in 17.5%H_2O+82.5% N_2 mixture (II, II a).

	I	I a	II	II a
H	–	–	0.945	0.245
OH	–	–	0.945	0.245
$N_2(A^3\Sigma)$	2.94	2.24	1.85	1.85
N	0.5	0.5	0.42	0.42
$N(^2D)$	0.5	0.5	0.42	0.42
N_2^+	–	0.7	–	0.55
H_2O^+	–	–	–	0.15
E	–	0.7	–	0.7

In table 1 the *G*-values are given in pure nitrogen and its mixture with H_2O for two approaches: the first approach, described in [3], (columns I and II) estimates contribution of ions approximately, the second one (columns Ia and IIa) is based on inclusion, to the kinetics scheme, of reactions involving ions. These data are used at the modeling of the postdischarge stage of the cleaning process.

3. Results and discussions

The simulations have been performed for the conditions of experiments [2]: at pressure of 1 bar and temperature of 200° C, other parameters are indicated in [3]. The results in [3] have been obtained for pure N_2 and mixtures 90%N_2+10%CO_2, 80%N_2+10%CO_2+10%CO and 50%N_2+20%CO+12%CO_2 +17%H_2+1%CH_4. Here the modeling results for the mixture of nitrogen with water vapor 17.5%H_2O+82.5%N_2 are discussed and compared with results for pure nitrogen. The initial (just after streamer stage) gas composition for two mixtures is given in table 2, evaluated in accordance with G-values from table 1.

Table 2. Initial concentration of components in pure N_2 (I, I a) and in 17.5%H_2O+82.5% N_2 mixture (II, II a).

	I	I a	II	II a
Components	cm^{-3}	cm^{-3}	cm^{-3}	cm^{-3}
N_2	$1.50\cdot10^{19}$	$1.50\cdot10^{19}$	$1.24\cdot10^{19}$	$1.24\cdot10^{19}$
H_2O	–	–	$2.63\cdot10^{18}$	$2.63\cdot10^{18}$
$C_{10}H_8$	$1.33\cdot10^{16}$	$1.33\cdot10^{16}$	$1.24\cdot10^{16}$	$1.24\cdot10^{16}$
H	–	–	$2.52\cdot10^{14}$	$6.53\cdot10^{13}$
OH	–	–	$2.52\cdot10^{14}$	$6.53\cdot10^{13}$
$N_2(A^3\Sigma)$	$7.83\cdot10^{14}$	$5.97\cdot10^{14}$	$4.92\cdot10^{14}$	$4.92\cdot10^{14}$
N	$1.33\cdot10^{14}$	$1.33\cdot10^{14}$	$1.10\cdot10^{14}$	$1.10\cdot10^{14}$
$N(^2D)$	$1.33\cdot10^{14}$	$1.33\cdot10^{14}$	$1.10\cdot10^{14}$	$1.10\cdot10^{14}$
N_2^+	–	$1.86\cdot10^{14}$	–	$1.46\cdot10^{14}$
H_2O^+	–	–	–	$4.00\cdot10^{13}$
E	–	$1.86\cdot10^{14}$	–	$1.86\cdot10^{14}$

In figure 1 the remaining fraction of naphthalene is shown versus the specific energy input in pure N_2 and in the mixture of 82.5% N_2 +17.5% H_2O. Two cases are considered: with approximate account of contribution of ions (I and II) and with inclusion of ion reactions in the kinetics scheme (I A and II A). One can see that the results are almost identical. Note that addition of H_2O increases the calculated cleaning efficiency, in agreement with experimental data, whereas the addition to nitrogen of such gases as CO, CO_2 and H_2 decreases the removal efficiency [3].

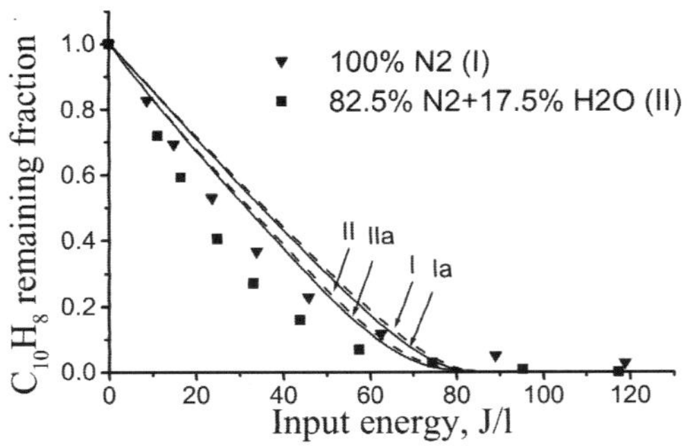

Figure 1. Naphthalene decomposition in mixtures versus the specific energy input in the gas. Symbols are experiments, lines are calculations; dash lines (I and II) are first approach, solid lines (Ia and IIa) are second one.

The leading reactions in pure nitrogen are presented and discussed in [3]. In this case the main reactions for $C_{10}H_8$ destruction are:

$N_2(A^3\Sigma) + C_{10}H_8 \rightarrow$ products, $k = 1.6 \cdot 10^{-10}$ cm^{3}s^{-1}, [10]
$N(^2D) + C_{10}H_8 \rightarrow$ products, $k = 3 \cdot 10^{-11}$ cm^{3}s^{-1} , [10]
$N_2(A^3\Sigma) + N_2(A^3\Sigma) \rightarrow N_2 + N_2(B^3\Pi, C^3\Pi)$, $k = 4 \cdot 10^{-10}$ cm^{3}s^{-1}, [10].

In the mixture of 82.5%N_2+17.5%H_2O other channels are added to the reactions mentioned above:

$OH + C_{10}H_8 \rightarrow H_2O + C_{10}H_7$ $k = 1.1 \cdot 10^{-17} T^2 \exp(-969/T)$ cm^{3}s^{-1},
$H + C_{10}H_8 \rightarrow$ products, $k = 1.4 \cdot 10^{-9} \exp(-2566/T)$ cm^{3}s^{-1},
$H + C_{10}H_7O \rightarrow C_{10}H_7OH$, $k = 1.7 \cdot 10^{-10}$ cm^3 s^{-1},
$O + C_{10}H_8 \rightarrow H + C_{10}H_7O$, $k = 2.3 \cdot 10^{-11} \exp(-902/T)$ cm^{3}s^{-1}.

The rate constants of these reactions were taken from [11] (see the discussion in [3]).

Below the processes involving charged species (electrons and ions) are considered. The chain of fast ion-molecule reactions transforms primary positive ions N_2^+ and H_2O^+ to more stable positive ions N_4^+ and H_3O^+.

$N_2 + N_2^+ + M \rightarrow N_4^+ + M$ $k = 1.5 \cdot 10^{-26} T^{-1}$ cm^{6}s^{-1}, [12]
$H_2O + H_2O^+ \rightarrow OH + H3O^+$ $k = 1.7 \cdot 10^{-9}$ cm^{3}s^{-1}, [13]
$H_2O + N_2^+ \rightarrow N_2 + H_2O^+$ $k = 2 \cdot 10^{-9}$ cm^{3}s^{-1}, [14]
$H_2O + N_4^+ \rightarrow N_2 + N_2 + H_2O^+$ $k = 3 \cdot 10^{-10}$ cm^{3}s^{-1}, [14]

The latter recombine with electrons with production of radicals:

$e + N_4^+ \rightarrow N_2 + N_2(A^3\Sigma)$ $k = 6.0 \cdot 10^{-4} T^{-1}$ cm^{3}s^{-1} , [14, 15] (*)
$e + N_4^+ + M \rightarrow N_2 + N_2(A^3\Sigma) + M$ $k = 9.35 \cdot 10^{-21} T^{-2.5}$ cm^{6}s^{-1}. [14]
$H_2O + e + H_2O^+ \rightarrow OH + H + H_2$ $k = 1.24 \cdot 10^{-14} T^{-3.5}$ cm^{6}s^{-1}, [13]
$e + e + H_2O^+ \rightarrow H_2O + e$ $k = 7.81 \cdot 10^{-9} T^{-4.5}$ cm^{6}s^{-1}, [13]
$e + H_2O^+ \rightarrow OH + H$ $k = 6.6 \cdot 10^{-6} T^{-0.5}$ cm^{3}s^{-1}, [14]
$e + H_2O^+ \rightarrow O + H + H$ $k = 3.0 \cdot 10^{-6} T^{-0.5}$ cm^{3}s^{-1}, [14]
$e + H_2O^+ \rightarrow O + H_2$ $k = 2.4 \cdot 10^{-6} T^{-0.5}$ cm^{3}s^{-1}, [14]
$H_2O + e + H_3O^+ \rightarrow OH + H + H + H2O$ $k = 1.24 \cdot 10^{-14} T^{-3.5}$ cm^{6}s^{-1}, [13]
$e + H_3O^+ \rightarrow OH + H + H$ $k = 6.0 \cdot 10^{-7}$ cm^{3}s^{-1}, [16]
$e + H_3O^+ \rightarrow H + H_2O$ $k = 4.0 \cdot 10^{-7}$ cm^{3}s^{-1}, [16]

The evolution of concentrations of these ions in streamer trace is demonstrated in figure 2; the time of electron-ion recombination is about 10 ns. Note that in the previous approach [3] the influence of ions was estimated only through reaction (*).

The evolution of produced radicals is shown in figure 3; the lifetime of O, OH and H radicals is around one millisecond, in contrast to excited nitrogen molecules, with lifetime about 10 μs. Due to the presence of water an additional quenching of the radicals most effective for removal, namely $N_2(A^3\Sigma)$ and $N(^2D)$, takes place in the reactions [10]:

$H_2O + N_2(A^3\Sigma) \rightarrow OH + H + N_2$ $k = 5.0 \cdot 10^{-14}$ cm^{3}s^{-1}.
$H + N_2(A^3\Sigma) \rightarrow H + N_2$ $k = 2.1 \cdot 10^{-10}$ cm^{3}s^{-1}.
$N(^2D) + H_2O \rightarrow NH + OH$ $k = 1.33 \cdot 10^{-11}$ cm^{3}s^{-1}.

The decomposition of formed complex ions also gives the additional H and OH radicals:

$H_2O + H_3O^+ + M \rightarrow H_3O^+(H_2O) + M$ $k = 9.61 \cdot 10^{-20} T^{-3}$ cm^{6}s^{-1} , [13]
$H_2O + e + H_3O^+ \rightarrow OH + H + H + H_2O$ $k = 1.24 \cdot 10^{-14} T^{-3.5}$ cm^{6}s^{-1} , [13]
$H_3O^+(H_2O)_2 + M \rightarrow H_2O + H_3O^+(H_2O) + M$ $k = 2.0 \cdot 10^{-13}$ cm^{3}s^{-1}, [17]
$H_2O + H_3O^+(H_2O) + M \rightarrow H_3O^+(H_2O)_2 + M$ $k = 4.0 \cdot 10^{-29}$ cm^{6}s^{-1} , [17]
$e + H_3O^+(H_2O)_2 \rightarrow H + 3H_2O$ $k = 1.56 \cdot 10^{-5} T^{-0.2}$ cm^{3}s^{-1}, [14]
$e + H_3O^+(H_2O)_2 + M \rightarrow H + 3H_2O + M$ $k = 7.79 \cdot 10^{-20} T^{-2.5}$ cm^{6}s^{-1}, [14]

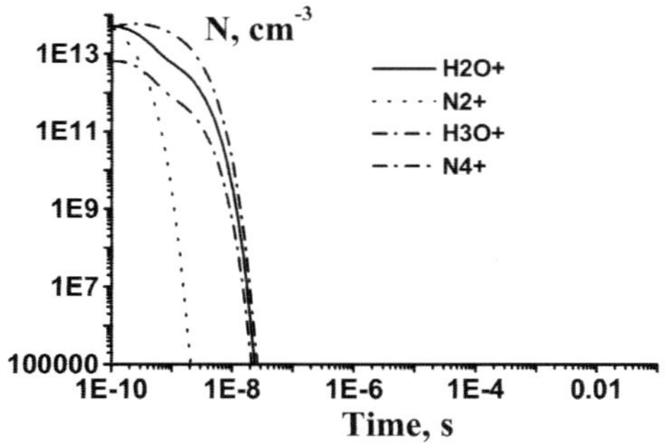

Figure 2. Evolution of ions in the mixture of 17.5% H_2O+82.5% N_2 in streamer trace.

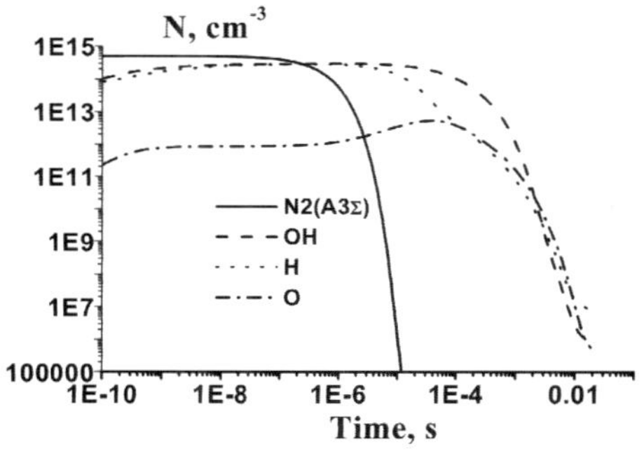

Figure 3.Evolution of radicals in the mixture of 17.5% H_2O+82.5% N_2 in streamer trace.

The time dependencies of the densities of complex ions and electrons in the streamer trace in first tens of discharge pulses are presented in figure 4, where the data inside the streamer channels are shown. After averaging the concentrations inside and outside the channels (see [3]), the remaining in the reactor the ions $OH^-(H_2O)$ and $H_3O^+(H_2O)_2$ have concentrations about $1\cdot10^9$ cm^{-3} With the growth of NO_2 concentration the contribution of ion $NO_2^-(H_2O)$ in neutralization of positive complex ion $H_3O^+(H_2O)_2$ increases:

$H_2O + OH^- + M \rightarrow OH^-(H_2O) + M$ $\quad k = 2.5\cdot10^{-28}$ cm^{6}s^{-1}

$NO_2 + OH^-(H_2O) \rightarrow OH + NO_2^-(H_2O)$ $\quad k = 1.5\cdot10^{-11}$ cm^{3}s^{-1}.

$NO_2^-(H_2O) + H_3O^+(H_2O)_2 \rightarrow 4H_2O + HNO_2$ $\quad k = 5.0\cdot10^{-15}$ cm^{3}s^{-1}.

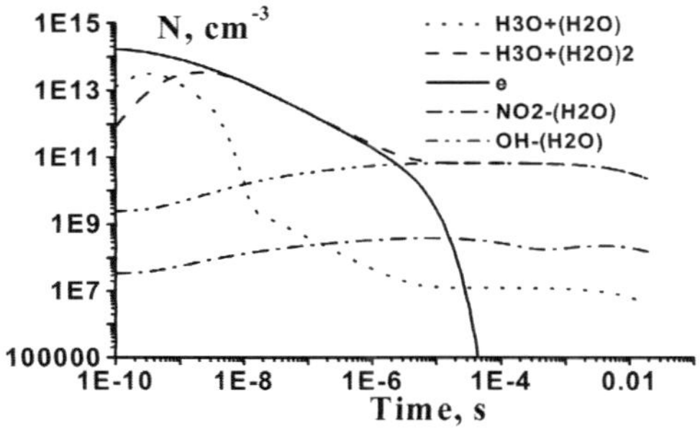

Figure 4. Evolution of complex ions and electrons in the mixture of 17.5% H_2O+82.5% N_2 in streamer trace.

Besides chemical transformations in reactions with active species, the naphthalene experiences the thermal decomposition, both in the discharge reactor and in the inactive volume [2,3]. The more is the treatment time, the more amount of naphthalene is thermally decomposed. In the considered nitrogen-water mixture, at the initial $C_{10}H_8$ molar fraction of 825 ppm, during 18 min of treatment 10 ppm of $C_{10}H_8$ is thermally decomposed.

Note that the effect of formation of tar ions in ion-molecular reactions with participation of tar molecules may be neglected, because the time of this process is (at number densities of tar molecules about 10^{16} cm^{-3} and rate constants ~ 10^{-9} cm^3/s [18]) about 10^{-7} s, much larger than the time of dissociative recombination.

4. Conclusions

The comparison of two approaches to the modeling of tar removal in the pulsed corona reactor has been performed. In the framework of the first approach it is assumed that the ion-molecule reactions are infinitely fast and may be taken into account approximately at estimation of the G-values for primary neutral active species. The second one is based on the consideration of plasma-chemical kinetics with participation of ions and electrons. The results of calculations have shown that the difference between the values of removed $C_{10}H_8$ concentration obtained using these two approaches, both in pure N_2 and its mixture with H_2O, is weak. It follows that at simulation of cleaning process the effects related with charged species may be estimated approximately, as additional contributions to the G-values for primary neutral active species.

References

[1] van Veldhuizen E M (ed) 2000 *Electrical Discharges for Environmental Purposes: Fundamentals and Applications* (New York: Nova Science)

[2] Nair S A, Pemen A J M, Yan K, van Heesch E J M, Ptasinski K J and Drinkenburg A A H 2003 *Plasma Chem. Plasma Process.* **23** 665

Pemen A J M, Nair S A, Yan K, van Heesch E J M, Ptasinski K J and Drinkenburg A A H 2003 *Plasmas and Polymers* **8** 209

Nair S A, Yan K, Pemen A J M, van Heesch E J M, Ptasinski K J and Drinkenburg A A H 2005 *Ind. Eng. Chem. Res.* **44** 1734

[3] Bityurin V A, Filimonova E A and Naidis G V 2009 *IEEE Trans. Plasma Sci.* **37** 911

[4] Pancheshnyi S 2005 *Plasma Sources Sci.Technol.* **14** 645

[5] Itikawa Y 2006 *J. Phys. Chem. Ref. Data* **35** 31

[6] Itikawa Y and Mason N 2005 *J. Phys. Chem. Ref. Data* **34** 1

[7] Zipf E C and McLaughlin R W 1978 *Planet. Space Sci.* **26** 449

[8] Slovetsky D I 1980 *Mechanisms of Chemical Reactions in Nonequilibrium Plasmas.* (Moscow: Nauka)

[9] Yousfi M Bekstein A, Merbahi N, Eichwald O, Ducasse O, Benhenni M and Gardou J P 2010 *Plasma Sources Sci. Technol.* **19** 034004

[10] Herron J T 1999 *J. Phys. Chem. Ref. Data* **28** 1453

[11] NIST Chemical Kinetics Database on the Web, http://kinetics.nist.gov

[12] Gordiets B, Ferreira C M , Pinheiro M J and Ricard A 1998 *Plasma Sources Sci.Technol.* **7** 363

[13] Bychkov V L and Yurovskii V A 1993 *High Temp.* **31** 5 (Translated from *Teplofiz. Vys. Temp.* **31** 8)

[14] Matzing H Tech. *Report Kernforschungszentrum Karlsruhe,* KfK 4494, 1989

[15] Cao Y S and Johnsen R 1991 *J. Chem. Phys.* **95** 1991

[16] Viggiano A A, Ehlerding A, Hellberg F, Thomas R D, Zhaunerchyk V, Geppert W D, Montaigne H, Larsson M, Kaminska M and Osterdahl F 2005 *J. Chem. Phys.* **122** 226101

[17] Popov N A 2010 *Plasma Phys. Rep.* **36** (Translated from *Fiz. Plasmy* **36** 867)

[18] Midey A 1999 *Sci. Report AFRL-VS-TR-2002- 610-1590*

On the Measurement of Subsonic Flow in a Capacitively Coupled Helicon Plasma Source

Åshild Fredriksen, Lekha Nath Mishra, Njål Gulbrandsen, and Wojciech J. Miloch
Department of Physics and Technology, Faculty of Science, University of Tromsø, N-9037 Tromsø, Norway

E-mail: ashild.fredriksen@uit.no

Abstract. Plasma parameters and the subsonic flow from a capacitively coupled, cylindrical plasma source of the Njord helicon device are investigated by means of a Mach probe and a retarding field energy analyzer (RFEA). 13.56 MHz and 600 W RF power is inserted into the argon working gas under low-pressure conditions and moderate magnetic field. By means of a downstream field coil, the magnetic field is shaped from a purely expanding field to a configuration with more parallel field lines. It is shown that the downstream plasma density along the outer rim of the source increases significantly and there is a sudden increase by nearly 20 V in the plasma potential already after a moderate increase in the downstream magnetic field. The investigation of the flow indicates that current ratios derived from the Mach probe result in an apparent flow in the direction towards the source, while the current bratios derived from the RFEA indicate a flow in the direction away from the source. PIC simulations demonstrate that the acceptance angle of the probes, being nearly 180° for the Mach probe, and about 45° for the RFEA, can critically affect the current ratios and hence the subsonic flow measured by the probes in the weakly magnetized plasma in our device. The first section in your paper

1. Introduction.

It is now well established that the plasma expanding from inductively-coupled helicon source into a larger downstream vacuum chamber, can sustain an electrical double-layer (DL) structure near the entrance to the vacuum chamber [1]. Evidence that this type of DL exists also in the Njord helicon device has been presented earlier [2]. An extensive number of recent studies have not yet found a decisive explanation of the mechanism behind these, so-called current-free DLs, as no external electrical circuit is needed to form them [3-5]. In particular, the role played by the magnetic field in the formation of the DL has been debated. It is well established that it disappears when the field in the source decreases below a certain value [6]. Recent findings show evidence the high density in the source as well as a DL can form also for very low magnetic field [7, 8]. The magnetic field is an important factor in the formation of the DL, but the details of the role it plays are not yet clear. It has been shown that the addition of a downstream magnetic field applied in the expansion chamber, extinguishes the potential drop at the source entrance and subsequently, the DL and the signature beam disappear [9]

In the present paper, we investigate in more detail the measurements of a subsonic flow. We demonstrate how the Mach probe measurements result in reversed flows. The application of the retarding-field energy analyzer (RFEA) probe in the measurement of subsonic flow is investigated and compared to analysis of simulated data. Finally, the dependence of acceptance angle on the current ratio for Mach number analysis is demonstrated and shown to be in agreement with the actual probe data. In Section 2, we explain the experimental setup and diagnostics. Measurements are reported in Section 3, and the PIC simulations are discussed and compared to our data sets in Section 4.Other paragraphs are indented.

2. Experiment and diagnostics.

The Njord experiment [2] consists of a 0.6m diameter and 1.5m long chamber with a spherical dome at one end. The chamber is pumped to a base pressure of typically 0.05–0.1 mPa. A helicon plasma source with the same specifications as the source of the Chi Kung device at ANU [10] is attached axially on the dome, as shown in Fig. 1. The source is constructed from a 30 cm long Pyrex glass tube with inner diameter 13.8 cm. The outer edge of the tube is chosen as origo of the z-axis for measurements along the centre of the column. A Boswell-type saddle antenna is formed around its outer diameter. The tube and antenna are enclosed by a cylindrical aluminium former, holding a pair of magnetic field coils of diameter 24 cm and width 9.5 cm. The centre of the first coil is placed 18 mm downstream of the outer edge of the Pyrex tube, and the centre of the second is displaced by 20 cm with respect to the centre of the first. Currents of 5 and 7 A in the source coils produce maxima of the magnetic fields of $B \approx 25$ mT. At $z = 55$ cm downstream from origo, a third (guide) coil is placed, close to the intersection between the dome and the main chamber. It produces $B \approx 8$ mT at 20A current. 13.56 MHz CW RF power, typically 300–800W forward and less than 50W reflected, is fed from a Henry 8K Ultra amplifier through an air-cooled □-matching network to the antenna Three radially and one axially oriented ports are placed on the dome 50 cm downstream from origo, as indicated in figure 1. Probes can be inserted through these ports for measurements along the axial and radial directions.

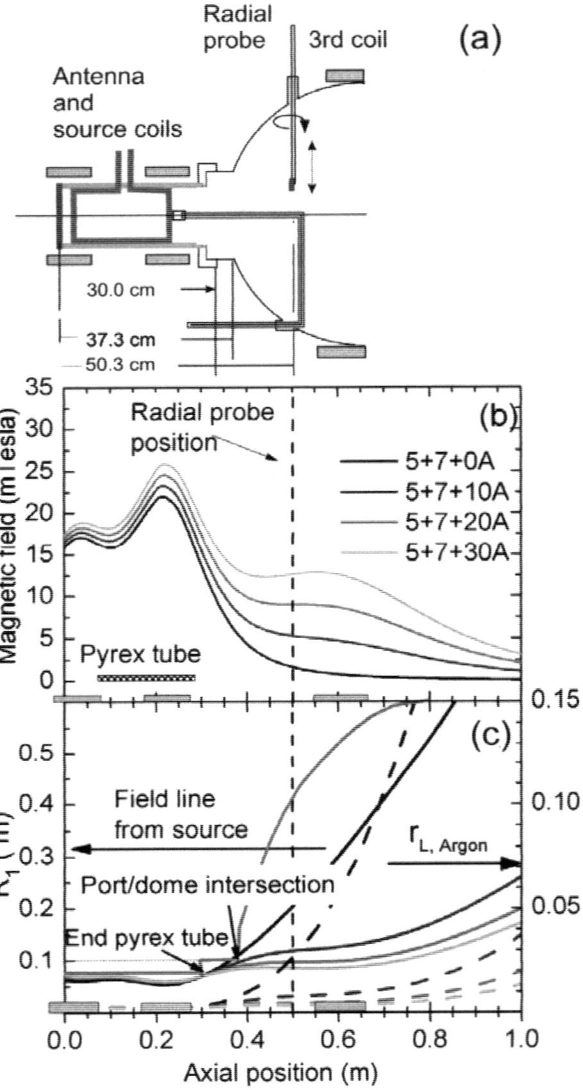

Figure 1. Sketch of the experiment layout (a), axial magnetic field (b), and field line expansion and Larmor radii of ions with axial position (c). Field lines are drawn as full lines with scale at left axis. Dashed lines represent r_L, with scale at right axis.

For the data presented here, a retarding field energy analyzer (RFEA) of outer dimensions WidthxLengthxHeight $\approx$ 20x40x10 mm was used to obtain plasma potential, beam density and beam energy. The RFEA placed in the radial port (figure 1) could be rotated 360° around its own axis, and background plasma parameters were obtained with the RFEA looking 90° to the flow direction. As the plasma is steady state, the total current to the collector could be obtained in both up- and down-stream direction and Mach probe analysis performed (see below). The grids and plates of the RFEAs [11] were organized 0.5mm apart as follows: an aperture plate with a gridded 2mm diameter hole, an electron repeller grid biased at typically −90V, a discriminator grid swept from −100 to +100V at zero offset, a secondary electron repeller grid biased at −18Vand a collector plate at −9V. To capture the energy distribution at the plasma potential, additional batteries between the sweep and the grid produced positive offsets up to 90V in steps of 9V. The discriminator was biased in 400 steps per sweep. At each step, the collector current, measured over a 35.7 kΩ resistor, was digitized into 200 samples which were then averaged into one single value, and stored to the file for further processing. The derivative of the current I(Vd) to the collector as a function of the discriminator bias, is proportional to the ion energy distribution through the relation [11]:

$$f(V_d) = \frac{1}{q} f(E) = \frac{m}{T^4 A q} \frac{1}{v} \frac{dI}{dV_d} \tag{1}$$

where V_d is the bias potential of the discriminator grid, E is the ion energy, q the elementary charge, m the ion mass. T is the transparency of one grid, A the area of the aperture and v is the ion thermal velocity.

Furthermore, a Mach probe, consisting of two parallel 2x4 mm rectangular electrodes separated by ceramic insulation, was used to test direct flow measurements in accordance with standard Mach probe theory [12]. With each electrode biased to -130 V, the ion saturation current I_{sat}^+ was collected in the upstream direction towards the source (viewing angle $\theta = 0^\circ$) and in the downstream direction looking away from the source ($\theta = 180^\circ$). According to the theory, the Mach number M can be obtained as

$$\frac{I_{sat}^+(\theta)}{I_{sat}^+(\theta+\pi)} = \frac{\int_{\theta-\Delta\theta}^{\theta+\Delta\theta} j(M,\theta)d\theta}{\int_{\theta-\Delta\theta}^{\theta+\Delta\theta} j(M,\theta+\pi)d\theta} = \exp\left(KM \cos\theta \frac{\sin\Delta\theta}{\Delta\theta} \right) \tag{2}$$

Here, the angle surface normal of the probe, θ, is zero with the probe looking towards the source, and $\Delta\theta$ is the field of view of the probe. I is the total current collected at the probe surface, and j is the current density. K is a constant depending on model and magnetization, with values typically in the range 1.34 to 1.45 [13, 14]. The Mach probe analysis has been most successfully used in magnetized plasmas [13, 15]. Originally proposed for flows in non-magnetized plasma [16], the Mach probe method has been less successful in these plasmas, due to the problem pointed out by Hutchinson [17], and confirmed experimentally by [18]. It was shown that ions from the upstream direction can be deflected and collected at the downstream surface of a Mach probe, resulting in a much lower flow measurement or even a flow reversal. In the experiments presented here, we will show evidence that this effect is important in our plasma, but that ion currents collected by the RFEA are less sensitive to this wake effect.

In the experiments reported here, for the controlled parameters we used RF power of 400–600W, and the argon gas fill pressure was 0.024 Pa (1.5 sccm). The currents in the first and second source coils were set to 5A and 7A, respectively, where the first is at the outer end of the source. The current in the third coil was varied between 0 and 30 A. Some resulting B-fields are shown in figure 1 b).

As the current in the 3rd coil is increased from 0 A to 30 A, the magnetic field at the axial position of the radial ports increases from about 2 mT to about 12 mT, resulting in the argon Larmor radius to decrease from about 5 cm to 5 mm. Within the same B-field range, the radial position of the field lines emerging from the edge of the source tube decreases from 20 cm to 8 cm at the radial ports, as shown in Fig. 1 c).

It was shown earlier [9] that for lower downstream magnetic fields, these field lines cross the walls of the port supporting the source tube, but as the 3rd coil current increases above about 6-8 A, they are restricted to within the chamber. The confinement of the electrons significantly improves, and the downstream plasma potential increases, eventually destroying the double layer creating the ion beam. From I = 8 A and upwards, a beam can no longer be discerned, and the entire plasma population flows at subsonic speed.

It should be noted that the double layer can be sustained for an inductive tuning of the matching network. In the present investigating of subsonic flows, the antenna was capacitively matched to the RF-power. Thus no beam was generated neither at the lowest downstream magnetic fields.

3. Measurements

Radial profiles of potential and total ion current measured by the RFEA pointed in the normal direction to the flow are shown in Fig. 2. The plot representing density profiles (Fig. 2 a) are obtained from the total areas of Gaussian fits to the derivatives of the 'I vs V'-curves. This method has the advantage that instrumental offsets in the ion current are removed in the derivative. Comparisons with the more commonly used technique of direct reading of the ion saturation current with offsets removed [2, 19], show good agreement, provided a good Gaussian fit to the actual distribution is possible. In the Figure, the capacitive coupling is evident from the hollow density profiles with maxima centred at the field lines arriving from around the edge of the source. The density increases as the magnetic field improves the confinement. The plasma potential profiles, derived from the maximum of the Gaussian fit, have only a small gradient towards the edge of the column. A large increase in potential between 0 and 6 A in the 3rd coil is evident, in agreement with what is found with the RF coupled inductively to the plasma source. For the flow measurements, the RFEA was pointed successively in the up- and down-stream direction to collect the ion energy distribution (IED) in both directions, yielding the total Isat+ by integration of the distribution.

Typical IEDs from two different radial positions at moderate magnetic field of $I_3 = 10$A, i.e. $B = 5$ mT and $r_L = 1$ cm at the probe, are shown in Fig. 3. At the radial position $r = 9$ cm (Fig. 3 a) the integrated area of the IED at $\theta = 180^\circ$ is smaller than the area at $\theta = 0^\circ$, indicating a flow in the positive z-direction away from the source. On the other hand, at r = 4 cm (Fig. 3 b) the integrated area of the IED at $\theta = 180^\circ$ is slightly larger than that at $\theta = 0^\circ$, indicating a flow reversal. Also, the Mach probe was used to obtain I_{sat}^+ simultaneously in the up- and downstream direction to deduce flow from the current ratios as given in Eq. 2.

The ratios of the ion saturation currents from the upstream and downstream looking Mach probe are shown in Fig. 4 a) for a set of different downstream magnetic fields, as represented by the 3rd coil current. Apart from the case of the lowest magnetic field which shows a ratio > 1, for all higher magnetic fields, the ratio is less than 1 in most except the outermost of the radial positions, indicating a flow towards the source. The profile is also qualitatively very different from the current ratios obtained by the RFEA, shown in Fig. 4 b). While the current ratios of the RFEA are largest where the largest densities are observed and increasing with density and magnetic field, the Mach probe ratio is decreasing with increasing magnetic field to values down to 0.7. As such values indicate the unphysical situation of a flow towards the source, it is likely that this phenomenon is due to upstream ions being collected by the downstream looking surface of the probe, as demonstrated earlier [17, 18]. On the other hand, the effect is becoming more severe by increasing magnetic field, which is somewhat counter-intuitive, as the Larmor radius is down to about 5 mm (about the size of the Mach probe) at the highest magnetic field. One explanation might be that the plasma potential is increasing rapidly with increasing magnetic field, so that the probe bias with respect to the plasma is increasing, and hence the deflection of the ion trajectories might be stronger.

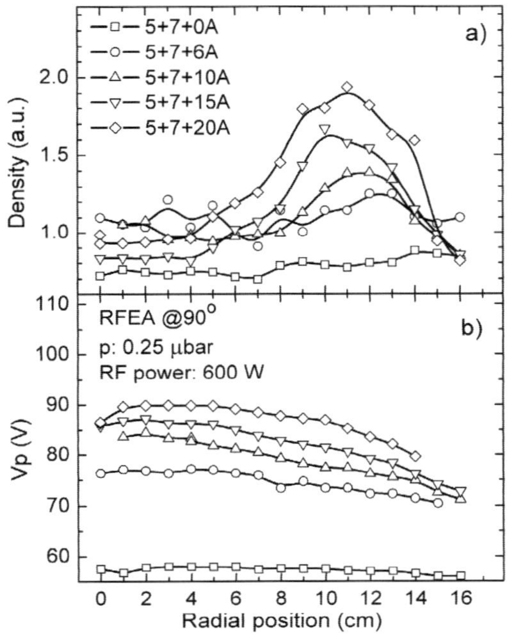

Figure 2. Density profiles (a), and V_p obtained from RFEA looking in the direction normal to the flow, for a set of four different currents in the 3rd coil.

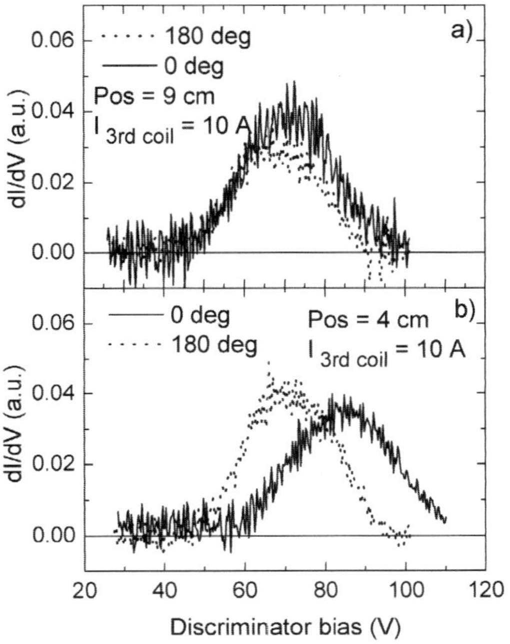

Figure 3 Typical IEDs at a) r = 9 cm and at b) r = 4 cm, with B = 5 mT at the RFEA probe. Full lines are IEDs at $\theta = 0^{\circ}$ and dotted lines are IEDs at $\theta = 180^{\circ}$

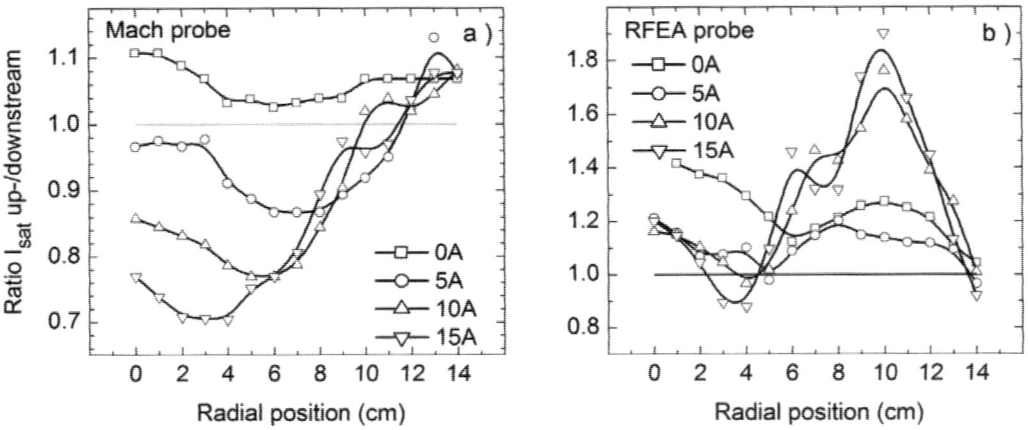

Figure 4 Current ratios from Mach-probe (a), and from up- and down-stream looking RFEA (b)

On the other hand, we also have seen strong indications of an electron beam from the source in Langmuir probe measurements, and this could also represent a possible source of erroneous ion current ratios as well. The ratios of the ion current to the RFEA looking towards the source over the current to the RFEA looking downstream, indicate, quite contrary, flows in the direction away from the source, except from a small region between 3 -4 cm from the axis for 5 A in the 3rd coil. The indicated flow is highly inhomogeneous over the plasma cross-section, which nevertheless is in agreement with the density profile, providing an inhomogeneous plasma pressure which in turn is more confined along the magnetic field as it increases. Thus, it is reasonable that the flow fields are restricted to a more narrow range in the radial positions.

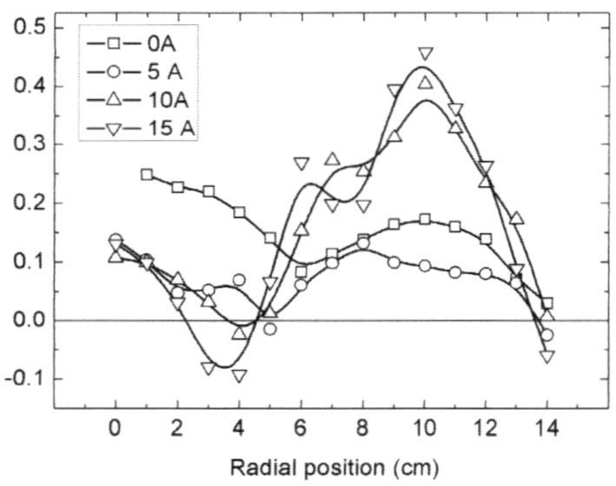

Figure 5. Mach numbers calculated from the current ratios taken form RFEA measurements in Fig. 4 b)

Another point in favour of the RFEA ratios, is the fact that the collecting surfaces of the Mach probe is in direct contact with the plasma, while the probe surface of the RFEA as seen from the plasma, is at ground. This means that the potential difference between the RFEA and the plasma is 130 V less than between the Mach probe and the plasma. Moreover, the RFEA is bigger (2x4 cm), with a smaller collecting aperture of 2 mm. The fact that the aperture at the front creates a total acceptance angle of about 45o prevents the ions with large velocity components parallel to the probe surface from entering. Most of the upstream ions collected at the downstream surface of the Mach probe would have such velocity components, as their trajectories have been bent from a large velocity parallel to the flow. These would be collected by the Mach probe due to its more open-faced geometry, with an acceptance angle of nearly 180 °. Thus, there are indications that the current ratios measured by the RFEA are more reliable than of the Mach probe for our plasma parameters.

In Fig. 5, the resulting radial profiles Mach numbers obtained from the RFEA data are shown. The value taken for the constant K is taken to be 1.4. The Mach numbers computed from this data set are within comparable range of those measured by others in helicon sources [20]. As the estimated ion acoustic velocity $c_s = (T_e/M_i)^{1/2}$ is 3.5 km/s at an estimated electron temperature T_e of 5 eV and ion mass M_i of argon, the resulting flow speeds would range from about 350 m/s to a maximum of 1500 m/s at the highest magnetic field. To what point it is accurate, is still not yet fully investigated.

4. Simulations

To improve our understanding of flow measurements with the RFEA, we have performed PIC simulations applying the DiP3D code, which has been designed for simulations of objects in complex plasma environments [21, 22], to a set of plasma parameters for some simple cases of subsonic flows, as indicated from the present measurements. For the present study, the code has been upgraded to

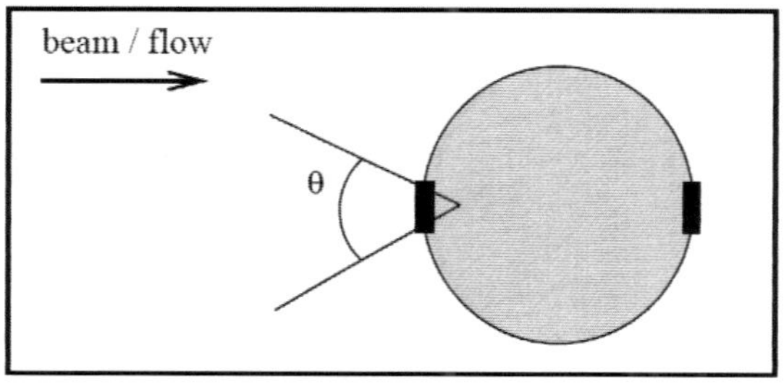

Figure 6. Geometry of the numerical simulations

account also for external uniform magnetic field [23], and collisions [24]. We simulate argon plasma with parameters closely representing typical conditions in the helicon plasma device Njord [9]. The plasma density is n = 2.5 • 10^{16} m^{-3}, and the neutral gas density nn = 2.0 • 10^{19} m^{-3}. The electron and ion temperatures are T_e = 4 eV and T_i = 0.4 eV, respectively. A spherical analyzer of radius $r \approx 4\times10^{-4}$ • 10–4 m with the aperture radius a/r = 0.15 and varied acceptance angle θ is placed in the centre of the simulation box of length L = 4 • 10–3 m in each direction, as shown in Fig. 6. The analyzer is biased at Φ = −45 V with respect to the plasma potential. The size of the analyzer is much smaller than in real experiments, which we compensate by a strong magnetic field (B = 0.6 T), ensuring that the ratios of characteristic lengths in the system are maintained. We consider the case of plasma with subsonic flow.

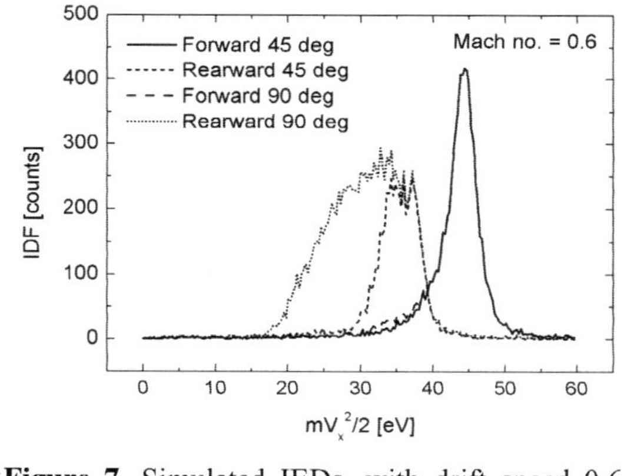

Figure 7. Simulated IEDs, with drift speed 0.6 cs., obtained by a probe with acceptance angles 90° and 45°, in forward and rearward directions.

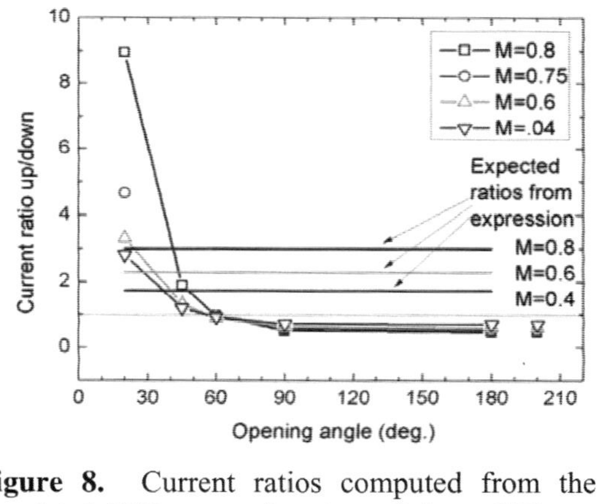

Figure 8. Current ratios computed from the simulated IEDs, as a function of three different subsonic acceptance angles

In Fig. 7 is shown a set of simulated ion energy distributions (IEDs) at 0.6 times the ion acoustic speed c_s, which reach the probe surface through two acceptance angles of 45o and 90o, respectively. On the IEDs observed in the forward direction (facing the source), no effect can be observed from the variation of the acceptance angle. The two distributions overlap almost completely, apart from a very small increase in the low-energy tail of the 90° acceptance angle. With the rearward looking probe, a smaller acceptance angle of 45° leads to a large decrease in the low-energy part of the IED compared to the IED with 90° acceptance angle. This indicates that at the wake side the IED contains a large amount of ions with low energy in the *x* direction. As this low-energy contribution is removed with smaller acceptance angle, it indicates that this part of the distribution has a large y-component, so they will be ions with a trajectory nearly parallel to the probe surface. They can thus be interpreted as ions originating from the up-stream plasma, with trajectories bent towards the probe and collected at the back side, as reported by others [17, 18, 25].

Ratios representing the up- and downshifted saturation currents were computed from the ratio of the integrated distributions (total counts) between forward and rearward looking probe, according to Eq. 2, for a number of acceptance angles (referred to as opening angle in the plot). The result is shown in Fig. 8 for a set of four different plasma flow speeds, at Mach numbers 0.4, 0.6, 0.8, and a single value of 0.75 at 20°. The horizontal lines indicate the current ratios as they would be expected from the known speed of the input IED from Eq. 2. It is evident that the larger acceptance angles will yield apparently reversed flows with respect to the true speeds of the IEDs, with a turning point from reversed to forward (ratio=1) flow at an acceptance angle of about 60°. The turning point is quite independent of the actual flow speed, which indicates that it may depend on the geometry only. However, the sensitivity of the ratio to the acceptance angle is increasing rapidly as the angle is decreasing, and the correct Mach number is derived in a range of the acceptance angle between about 35° and 32°. This result imposes a quite severe restriction on the aperture of the RFEA with respect to its distance between the front grid and the collector. In the case of our RFEA, with a 2.0 mm diameter aperture and distance between front grid and collector of 2.4 mm, the acceptance angle is 45°. We note that the current ratios in the simulated data at 45° are in the range between 1 and 2, which agrees quite well with the measured ratios of the Fig. 4 b). Thus, we may conclude that a standard analysis of our RFEA data will generally be measuring the right direction of the flow, but will tend to underestimate the speed. On the other hand, the Mach probe, with a much larger acceptance angle of nearly 180°, will measure reversed flow (ratio < 1), as in the measurements shown in Fig. 4a). It should be noted that the simulations were made for a case of V_p = 45 V, that is, for a situation with a low magnetic field. On the other hand, in our data sets, V_p is in the range between 70-90 V, and hence the data cannot be directly compared. A more extensive study of the effect of magnetic field and plasma potential yet needs to be carried out in order to create a working knowledge about the capabilities of the RFEA as a probe for flow measurements.

5. Conclusions

In this paper, we have explored the feasibility of the Mach probe and the RFEA to be applied to flow measurement in a weakly magnetized, expanding plasma. The analysis of the Mach probe is shown to indicate a flow in the opposite direction to that of the RFEA. The importance of the acceptance angle of the RFEA has been demonstrated with PIC simulation of IEDs for a set of different angles. For plasma potential V_p = 45 V we find that the simulations result in reversed flow at acceptance angles larger than 60°, and that flows of the correct order would be inferred from the collected IEDs at acceptance angles between 30 to 35 degrees. This result is in good agreement with data from both the Mach and RFEA probes. Further investigations of the role of V_p and magnetization are in process.

References

[1] Charles C and Boswell R W 2004 *Phys. Plasmas* **11** 1706.
[2] Byhring H S, Charles C, Fredriksen A and Boswell R W 2008 *Phys. Plasmas* **15** 102113
[3] Fruchtman A 2006 *Phys. Rev. Lett.* **96** 065002
[4] Lieberman M A, Charles C and Boswell R W 2006 *J. Phys. D* **39** 3294
[5] Meige A, Boswell R W, Charles C and Turner M M 2005 *Phys. Plasmas* **12** 052317
[6] Charles C 2007 Plasma *Sources Sci. Technol.* **16** R1
[7] Lafleur T, Charles C and Boswell R W 2010 *Phys. Plasmas* **17** 043505
[8] Corr C S, Zanger J, Boswell R W and Charles C 2007 *Appl. Phys. Lett.* **91** 241501
[9] Fredriksen A, Mishra L N and Byhring H S 2010 *Plasma Sources Sci. Technol.* **19** 034009
[10] Charles C and Boswell R W 2003 *Appl. Phys. Lett.* **82** 1356
[11] Charles C, Boswell R W and Porteous R K 1992 *J. Vac.Sci. Technol.* A **10** 398

[12] Shikama T, Kado S, Okamoto A, Kajita S and Tanaka S 2005 *Phys. Plasmas* **12** 044504
[13] Hutchinson I H 1988 *Phys. Rev.* A **37** 4358
[14] Hutchinson I H 2003 *Plasma Phys. Contr. Fusion* **45** 1477
[15] Riccardi C, Barni R and Fredriksen A 2004 *Rev. Sci. Instrum.* **75** 4341
[16] Hudis M and Lidsky L M 1970 *J. Appl. Phys.* **41** 5011
[17] Hutchinson I H 2002 *Phys. Plasmas* **9** 1832
[18] Ko E and Hershkowitz N 2006 *Plasma Phys. Contr. Fusion* **48** 621
[19] Charles C 2005 *IEEE Trans. Plasma Sci.* **33** 336
[20] Sun X, Biloiu C, Hardin R and Scime E E 2004 *Plasma Sources Sci. Technol.* **13** 359
[21] Miloch W J 2010 *Plasma Phys. Contr. Fusion in press*
[22] Miloch W J, Kroll M and Block A D 2010 *Phys. Plasmas* **17** 103703
[23] Langdon C K B 1992 *Plasma Physics via Computer Simulation* (Bristol Adam Hilger)
[24] Vahedi V and Surendra M 1995 *Comp. Phys. Comm.* **87** 179
[25] Miloch W J, Gulbrandssen N, Mishra L N \and Fredriksen Å 2010 *Appl. Phys.Lett. in review*

Numerical experiments on 2D strongly coupled complex plasmas

Lu-Jing Hou*, **A. V. Ivlev, H. M. Thomas and G. E. Morfill**
Max-Planck-Institut for extraterrestric Physics, Giessenbachstrae 85748 Garching Germany
E-mail: *hou@mpe.mpg.de

Abstract. The Brownian Dynamics simulation method is briefly reviewed at first and then applied to study some non-equilibrium phenomena in strongly coupled complex plasmas, such as heat transfer processes, shock wave excitation/propagation and particle trapping, by directly mimicking the real experiments.

1. Introduction

Laboratory generated complex (dusty) plasma is a suspension of micron-sized charged dust particles in a weakly ionized plasma with electrons, ions, and neutral atoms or molecules. These dust particles usually acquire a few thousand electron charges by absorbing surrounding electrons and ions and/or by secondary electron/photoelectron emission, float over the lower electrode in the discharge chamber due to the plasma sheath electric field, and interact with each other via a dynamically screened Coulomb potential while undergoing Brownian motion due primarily to fluctuations of local field and frequent collisions with the neutral molecules. When the interaction potential energy between charged dust particles significantly exceeds their kinetic energy, they become strongly coupled and can form ordered structures comprising liquid and solid states that are observable even with naked eyes. Since the motion of charged dust particles in complex plasmas can be directly observed in real time by using a video camera, such systems have been generally regarded as a promising model system to study many phenomena occurring in solids, liquids and other strongly-coupled systems at the kinetic level, such as phase transitions, transport processes, and collective dynamics [1, 2].

The complex plasma physics has now grown into a mature research field with a very broad range of interdisciplinary facets [2]. In addition to usual experimental and theoretical study, computer simulation in complex plasma has been playing an important role in bridging experimental observations and theories and in understanding many interesting phenomena observed in laboratory. The present contribution is intended to demonstrated the basic ideas of numerical experiments on strong coupled complex plasmas (SCCPs) and will focus on a class of non-equilibrium computer simulations that mimic the real complex plasma experiments in laboratory (i. e., numerical experiment). The simulation method, i. e., the so-called Brownian Dynamics methods, is reviewed in Sec. 2 and then examples, such as simulations of heat transfer and shock wave propagation in two-dimensional (2D) dust solids and liquids, are presented in Sec. 3 and Sec. 4, respectively. A short summary is given in Sec. 5

Journal of Physics: Conference Series **257** (2010) 012020 doi:10.1088/1742-6596/257/1/012020

2. Brownian dynamics method

In a typical experiment, charged dust particles are immersed in very weakly ionized plasmas, where absolute majority of components is the neutral gas atoms/molecules. Therefore it is not surprising to see that many (if not all) dynamic processes are heavily influenced by neutral gas, for example, melting of plasma crystal, collective dynamics and transport processes. On one hand the neutral gas provides a viscous/dissipative media, which takes away extra kinetic energy of dust particles and cools down the system, while on the other hand neutral gas molecules frequently collide with dust particles and set them into Brownian motion. Behaviors of dust particle are therefore better simulated by Brownian dynamics method, which may be regarded as a generalization of the usual Molecular Dynamics (MD) method and is based the Langevin equation and its integral (or Kramers equation written in Hamilton form) [3, 4, 5],

$$\begin{aligned} \frac{d}{dt}\mathbf{r} &= \mathbf{v} \\ \frac{d}{dt}\mathbf{v} &= -\gamma\mathbf{v} + \frac{1}{m}\mathbf{F} + \mathbf{A}(t). \end{aligned} \tag{1}$$

Here, as usual, m, $\mathbf{v}$ and $\mathbf{r}$ are respectively the mass, velocity and the position of a Brownian particle, and $\mathbf{F}$ is the systematic (deterministic) force resulting from external sources and/or from the inter-particle interactions in a many-particle system of Brownian particles interacting with the ambient gas of light particles. What is different from Newton's equations is the appearance of dynamic friction, $-\gamma\mathbf{v}$, and the stochastic (Brownian) acceleration, $\mathbf{A}(t)$. It should be stressed here that the Brownian acceleration may include either internal random kicks from surrounding gas molecules or any external heating sources that can be represented or approximated by a Gaussian white noise , such as fluctuations of local field, a scanning laser beam or other instabilities in the system. When the system is in thermal equilibrium, the friction and the stochastic acceleration are related to the ambient temperature via fluctuation-dissipation theorem. Note that the friction coefficient γ is usually regarded as constant [3].

The Langevin equation (1) may be integrated numerically in a manner similar to that used for Newton's equations in MD simulations. The most popular algorithms are the so-called Euler-like (EL) method, and Beeman-like (BL) method [3]. However, the above two methods has first-order and second-order accuracy, respectively, and very small time-steps is needed, in order to maintain a necessary accuracy and to conserve the system energy, especially when the damping is very small (say $\gamma < 0.05$, which is a common parameter range in complex plasma experiments). Therefore a Gear-like (GL) method [4, 5], which has higher orders of accuracy, excellent stability, and a negligibly small energy drift on the long time scales, is developed recently to circulate this issue. The Gear-like BD method will therefore be used in all the following simulations.

3. Heat transfer

The first example [6] is to study the heat conduction in 2D dust solids and liquids, motivated by recent experiments on thermal conduction in 2D SCCPs [7, 8] in both crystalline and solid/liquid mixture states. Typically, $N = 10000$ particles are simulated in a rectangular area with periodical boundary condition in y direction and confining boundary condition in x direction. Particles interact with each other via pairwise Yukawa potential [9]:

$$\Phi(r) = \frac{Q^2}{r}\exp(-r/\lambda_D), \tag{2}$$

with Q, r and λ_D being the particle charge, interparticle-distance and screening length, respectively. Such a system can be fully characterized by three parameters, i. e., the Coulomb coupling parameter, $\Gamma = Q^2/(aT)$, the screening parameter, $\kappa = a/\lambda_D$, and the damping

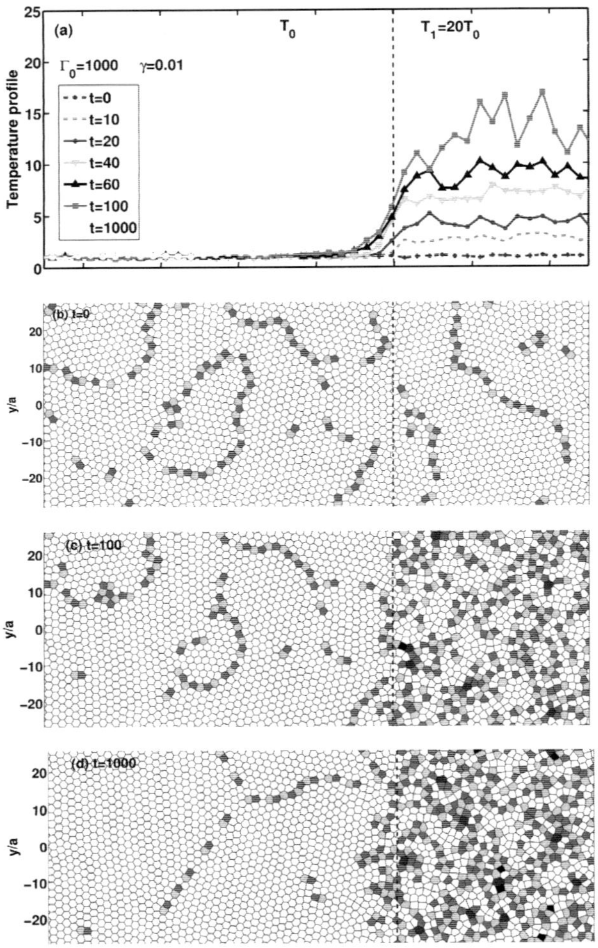

Figure 1: Time-evolution of temperature profile (a) and system structures (b)-(d), for initial coupling strength of $\Gamma_0 = 1000$, $\gamma = 0.01\omega_0$ and high temperature $T_1 = T_0$ ($T_{0/1} = 1/\Gamma_{0/1}$). The time unit is $\tau_0 = 1/\omega_0$. Black, red, white, green, and blue pixels denote respectively, four-, five-, six-, seven- and eight-folder structures. Vertical dash-line marks the border between heated and unheated region.

rate, ν/ω_0, due to the neutral gas, where T is the system temperature (in energy units), $a = (\pi\sigma_0)^{-1/2}$ the Wigner-Seitz radius, and σ_{d0} the equilibrium dust number density. The dust plasma frequency is defined by $\omega_0 = \left[2Q^2/(ma^3)\right]^{1/2}$. Γ and γ are varied to realize different equilibrium states and different damping rates, while $\kappa = 1$ is kept constant throughout our simulation and in subsequent discussions, because $\kappa \sim 1$ is the most typical value found in experiments and, moreover, no substantial effects of varying κ are expected in the features to be discussed in the following.

Our simulation is directly reproducing recent experiments [7, 8], and is different from the usual method of non-equilibrium simulation for heat conduction [10]. The system is firstly brought to an equilibrium with desired temperature (T_0) in either liquid or solid state. The

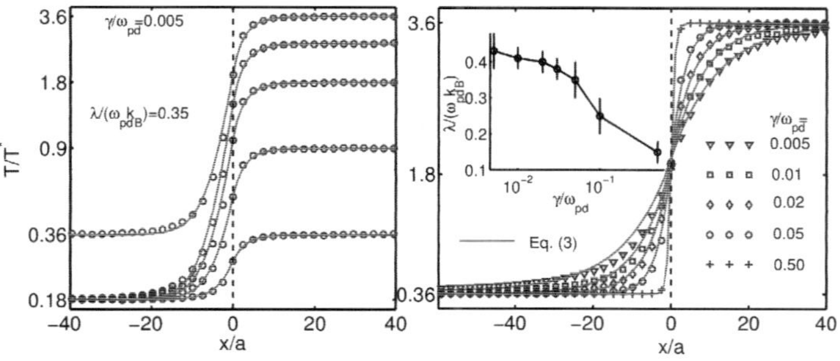

Figure 2: Stationary temperature profiles for: (a) $\gamma = 0.05\omega_0$ but different system states and temperature gradients and; (b) $T_0 = 0.36T^*$ and $T_1 = 3.6T^*$ but different damping rate. In both figures, symbols are simulation results while solid lines are fits according to Eq. (3). For (a), $\lambda = 0.35\omega_0 k_B$, and for (b) different λ values(given in the inserted plot) are obtained. Copyright (2009) by Institute of Physics Publishing [6].

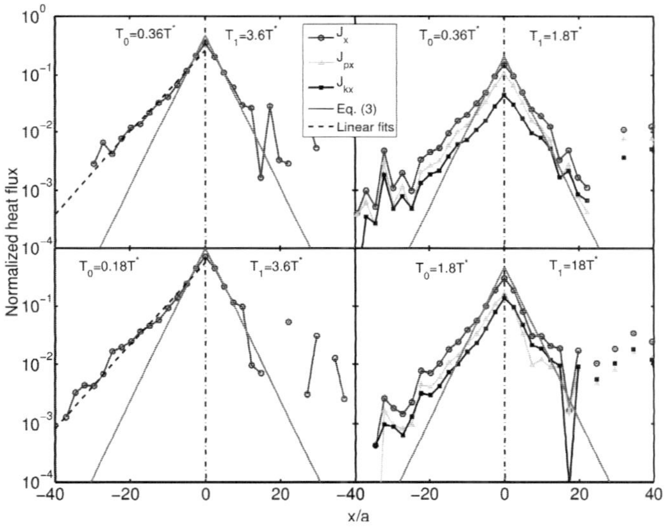

Figure 3: Heat flux for different system states and different temperature gradients with $\gamma = 0.05\omega_0$. Symbols are simulation results, solid lines are fits according to Eq. (3), using $\lambda = 0.35\omega_0 k_B$, and dash lines in the two panels on the left are direct linear fits of the heat flux with $\lambda = 1.2\omega_0 k_B$. Copyright (2009) by Institute of Physics Publishing [6].

melting point for $\kappa = 1$ is at $\Gamma^* \approx 180$ [11], and we'll denote the corresponding temperature as T^*. Then the right half of the system ($x > 0$), is heated to a higher temperature (T_1) by applying a Gaussian white noise with desired strength. The evolution of the, system structures, temperature profile and also the heat flux are recorded. In particular, Fig. 1 depicts the time evolution of temperature profile and system structure in the first 1000 time units of heating. A steady state is approached after a substantially long period.

The microscopic heat flux for ith particle is defined as:
$\mathbf{J}_i(t) = \mathbf{v}_i E_i + \frac{1}{2}\sum_{j=1,j\neq i}^{N} \mathbf{r}_{ij}(\mathbf{F}_{ij}\cdot\mathbf{v}_{ij}) - \mathbf{r}_i(\mathbf{F}_{ext}\cdot\mathbf{v}_i)$ where $E_i = (1/2)(mv_i^2 + \sum_{j=1,j\neq i}^{N}\phi_{ij}) + \phi_{ext}$ is the particle energy. The total flux in a region is then a summation of the microscopic flux

of all particles therein divided by its area A, i. e., $\mathbf{J}(t) = (1/A)\sum_{i\in A}\mathbf{J}_i(t)$. We are mainly interested in the x-component of the heat flux J_x. The three terms on the right-hand-side of above equation correspond to respectively contributions from: (1) the particle migration, which is believed to be the main mechanism of heat transport in gas and is denoted as J_{kx} hereafter, (2) particle interactions, i. e., phonon scattering, which is dominant in solids and is denoted as J_{px} hereafter, and (3) the external force. Since the external force acts on only a few rows of particles around the two confining boundaries, its *direct* influence on the heat flux is localized. Neglecting the external contribution will bring it back to the standard one [12, 13].

Analytical model: The heat transfer in our specific case can be described by Fourier's law: $\mathbf{J} = -\lambda\nabla T$ together with energy balance between heat conduction and energy dissipation due to damping: $\nabla(\lambda\nabla T) = 2\gamma n(T - T_0)k_B$ [8], where λ is the heat conductivity. According to [8], the heat conductivity λ is almost a constant in a wide range of coupling strength that covers both liquid and solids states. We therefore assume that λ a constant in the following derivation of analytical model and one has,

$$T(x) - T_0 = \frac{T_1 - T_0}{2}e^{\sqrt{\frac{2n\gamma k_B}{\lambda}}x}, (x < 0); T(x) - T_1 = \frac{T_0 - T_1}{2}e^{-\sqrt{\frac{2n\gamma k_B}{\lambda}}x}, (x > 0). \quad (3)$$

Expressions for heat flux may be obtained in a straightforward way, and we omit the results here. Since λ is the only unknown parameter in Eq. (3), it may be measured by fitting the stationary temperature profile (STP) to Eq. (3). It should also be mentioned that the Fourier's law could break down for low dimensional crystalline systems [13], largely due to a slow decay of equilibrium correlations of the heat current and a divergence of the finite-size conductivity. However, both of them may be avoided in complex plasmas because of the finite damping effect.

Figure 2 (a) shows examples of STPs for different system states and temperature gradients with $\gamma = 0.05\omega_0$, which is close to the experimental condition of [8]. Symbols are measurements from simulation, while solid lines are analytical fits according to Eq. (3). These fits give a constant heat conductivity, $\lambda = (0.35 \pm 0.05)\omega_0 k_B$, or in terms of thermal diffusivity D_T: $D_T \approx 22\mathrm{mm}^2/s$ using parameters from Ref. [8]. This value is between the experimental measurement for crystalline state ($30\mathrm{mm}^2/s$) [7] and that for solid/liquid mixture phase ($9\mathrm{mm}^2/s$) [8]. Fits for high temperature (e. g. the two upper-most curves) suggest a slightly smaller λ. Nevertheless, the value is in the range of the error bar for the present measurement.

Figure 2 (b) shows STPs for different damping rate with other parameters fixed. Fits with Eq. (3) give a damping-dependent heat conductivity, as is shown in the inserted plot that λ rises slightly with the decrease of γ. Note that this tendency is contradictory with that given by the analytical model in [14], which predicts an increase of λ with increase of damping rate and was confirmed by their experiment [14]. However, this model is based on an empirical relation between diffusion and heat conduction coefficients obtained by fitting simulation results for three-dimensional (3D) simple liquid without damping, and their experiment was also performed in a 3D complex plasma liquid [14]. Whereas in our simulation we study 2D systems covering both liquid and solid states and/or with a liquid-solid mixture phase. So the discrepancy could have been caused by the different dimensionality and system states, as is known that transport processes depends much on these two factors. In our simulation, the damping effect is taken into account self-consistently and increase of λ with decrease of γ may be intuitively understood as follows. It has two effects on heat conduction: direct energy dissipation and indirect suppression of phonon propagation. The first one is only related to kinetic energy of the system, and had been explicitly taken into account in Eq. (3), whereas the second one affects the collective modes and is not included in Eq. (3). Therefore decrease of γ means less damping of phonon propagation, more efficient heat transfer through phonon scattering and consequently a higher heat conductivity.

Figure 3 shows distributions of J_x, J_{px} and J_{kx} for different system states and temperature gradients, together with the analytical result derived from Eq. (3). Firstly, it may be seen that the kinetic part J_{kx} and the phonon part J_{px} have different weights in different system states. As expected, J_{px} is clearly dominant for solid state and low temperature liquid state, while J_{kx} dominates for high temperature liquid. The critical temperature where the two parts become equal is about $6T^*$. Secondly, the heat flux is not symmetrical about the heating interface. The decay of heat flux on the low-temperature side is slower and one needs a smaller slope, consequently a larger λ to fit J_x on this side, indicating a higher heat conductivity for lower temperature. Thirdly, the agreement between analytical results and simulation depends closely on system states. One generally observes a better agreement on the high temperature side and for higher temperature. These features suggest that λ becomes temperature-dependent.

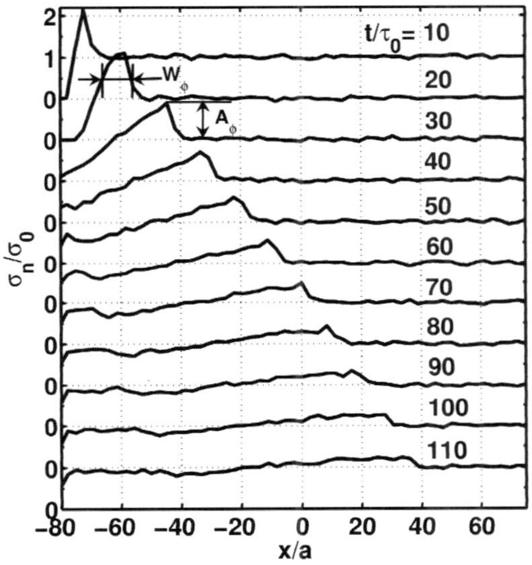

Figure 4: Density profiles of the wave for subsequent $100\tau_0$, when $\Gamma = 20$. Copyright (2008) by the American Physical Society [15].

4. Simulation of dust-density wave propagation and particle trapping

The second example [15] is to study the propagation of large amplitude dust-density waves (DDWs) [16] in 2D SCCPs and particle trapping therein, motivated by recent experimental observations of particle-wave interactions in SCCPs [17, 18]. In our simulation, we use a rectangular geometry with confining boundary conditions in the x direction (also the wave propagation direction) and periodic boundary conditions in the y direction, and consider 3000 dust particles interacting via Yukawa potential Eq.(2). We use again $\kappa = 1$ throughout our simulation. Under these conditions, we have a rectangular system with a length of $160a$ (approximately) ($x \in [-80a, 80a]$) and width $80a$ ($y \in [-40a, 40a]$). Such a geometry could prevent a perfect triangular crystal to form at very large Γ, and consequently introduce some defects around the confining boundary in deep solid state. To reduce influence from these effects, we perform most of our simulations in liquid state or high-temperature solid state ($\Gamma = 800$ is the highest coupling strength). To further reduce the boundary effect, we make our analysis of trapping (which could be affected by defects) in a much smaller region (approximately $x \in [-45a, 45a]$). So the margin between the geometrical boundaries ($x = -80a$ and $80a$) and the

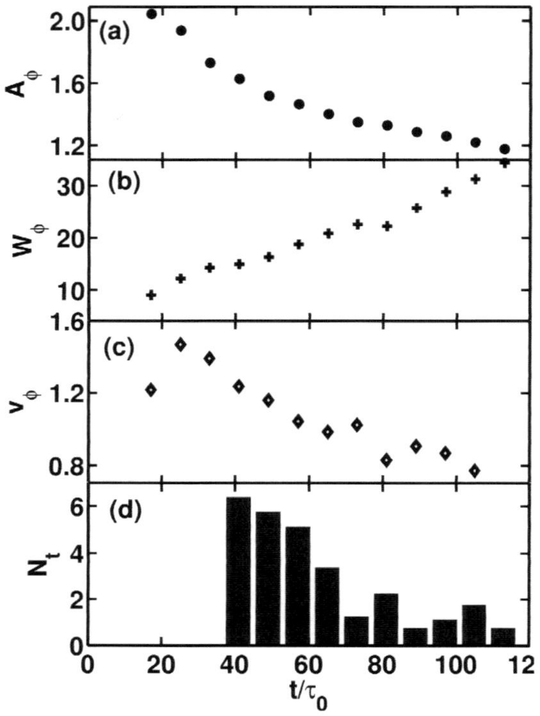

Figure 5: Time dependent (a) wave amplitude ($A_\phi = \sigma_n/\sigma_0$), (b) wave width W_ϕ [(defined by $0.5(\sigma_n - \sigma_0)$, and normalized by a], (c) wave speed v_ϕ (normalized by v_s), and (d) number of trapped particles, when $\Gamma = 20$. Note here that the record for particle trapping starts from $t = 35\tau_0$ in order to avoid any possible boundary effect and effect from initial excitation (for example numerical heating during the wave excitation). Copyright (2008) by the American Physical Society [15].

data analysis boundaries ($x \approx -45a$ and $45a$) is at least $35a$, which is large enough to diminish the spatial correlation in our simulation. For convenience of later discussion, we introduce here τ_0 (the reciprocal of ω_0), and the sound speed $v_s = \omega_0 a$, respectively.

The simulation starts by generating an equilibrium system with an expected temperature, then pulse waves are excited at one confining end. As for the wave-excitation, one may imagine that there is a piston moving back and forth periodically about the left boundary at $x = -80a$, and the motion of the piston follows a symmetrical triangular function with a constant speed ($v_p = 0.8v_s$) and period ($\tau_p = 110\tau_0$). When the piston is moving forth, it compresses particles in front it and so launches the pulse wave. We have observed that the so-excited wave reaches its maximum amplitude in a few τ_0 of the excitation, and detaches from the piston once the wave speed exceeds that of the piston. After that, the wave propagates freely along the x direction as a soliton [19], as is shown in Fig. 4: the evolution of density profiles σ_n (normalized by σ_0) for subsequent $100\tau_0$.

We see that particles in the wave region is highly compressed, as the compressional rate is always large than unity, i. e., $\sigma_n/\sigma_0 > 1$. The highest compression rate is reached at the very beginning and is around $\sigma_n/\sigma_0 = 2.2$, which is comparable to the recent experimental realization [19] in 2D plasma crystals. The wave amplitude A_ϕ (defined by the peak height of

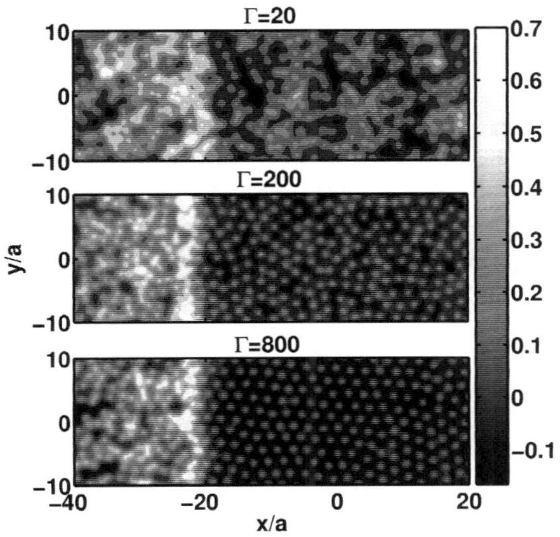

Figure 6: Snapshots of potential energy contour (normalized by $\phi_0 = (Ze)^2/a$) in a plane parallel to the dust layer with a distance $h = 0.8a$ for different Γ (from top to bottom $\Gamma = 20$, $\Gamma = 200$ and $\Gamma = 800$, respectively). Copyright (2008) by the American Physical Society [15].

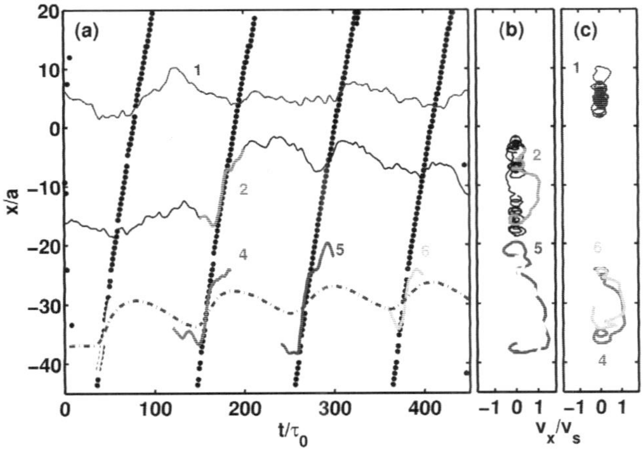

Figure 7: (a) Periodogram of 4 successive pulse waves, and samples of trajectories for trapped and untrapped particles, when $\Gamma = 20$. Here dots show the time-dependent wavefronts (defined by the peak positions in Fig. 4); the (blue) dash-dot line shows the averaged trajectory of particles, whose initial positions fall into the range $-32a > x > -42a$; other numbered lines are trajectories for 6 different particles: the line 1 [(blue) thin dash-line] is an example of trajectories for untrapped particles; and lines 2-6 are examples of trapped-particle trajectories. Note here that, for clarity, only one of the trapped-particle trajectories, line 2, is shown for whole range, with the (black) thin solid-line for untrapped time and (red) heavy solid-line(s) for the time around trapped period. Copyright (2008) by the American Physical Society [15].

the compressional rate) continuously decays, with the time marching on. According to another experimental observation in 2D plasma crystal [20], the nonlinearity becomes pronounced when $(\sigma_n - \sigma_0)/\sigma_0 > 0.1$. So one can immediately conclude from Fig. 4 that the wave here is strongly nonlinear in most of the time. And the dissipation of the pulse wave may be attributed to the Epstein damping [19] as well as the viscous heating of the equilibrium particles, and the latter is then responsible for the continuous broadening of the wave-width W_ϕ [defined by the width at half-amplitude i. e., $0.5(\sigma_n - \sigma_0)$], as is clearly shown in Fig. 4. Less obvious is that the wave speed v_ϕ also decreases continuously during the wave propagation and is found proportional to the wave amplitude or local density. More quantitative illustrations of these wave properties are given in Fig. 5. Similar tendencies had also been observed in previous experiments [19, 20].

In the following, we illustrate the microscopic process of wave propagation and wave-particle interaction by analyzing the dynamics of individual particles. A microscopic view of wave propagation in the dust layer is given in Fig. 6: snapshots of the potential energy (including the Hartree part of the energy) [11] over the plane. Note here that $0.8a$ is the root of $mv_s^2/2 = \phi(r)$, i. e., the interparticle distance at which the interaction potential energy between two particles equals their kinetic energy when they have relative speed v_s. Such a treatment is mainly to remove the singularity of potential distribution in the dust plane, and consequently to obtain a better visualization.

The three figures in Fig. 6, from top to bottom, give wave propagation in three different equilibrium states, $\Gamma = 20$: a high temperature liquid state, $\Gamma = 200$: a state close to the liquid-solid phase transition and, $\Gamma = 800$: a solid state, respectively. Particle positions can be readily identified from those small circles or spots. All the three figures are divided into two distinct zones with sharp boundaries indicated by the enhanced brightness around $x/a = 18$, which actually give roughly locations of the wavefront (defined by the peak positions of density profiles shown in Fig. 4). Obviously, the waves are propagating from left to right of the plane. On the right hand side, the system is not affected yet and are in their respective equilibria, whereas on the left, where the waves just swept by, the systems are in a great disorder.

A small number of particles are observed traveling along with the wave for certain periods of time. Fig. 7 shows the periodogram (wavefront position v.s. time) of 4 successive pulse waves, together with trajectories (only component in x direction) of typical trapped and untrapped particles, and averaged trajectory as well. The curvature of the wavefront with time indicates the variation of the wave speed, as shown above in Fig. 5 (b). As in a classical wave-particle interaction picture, most of the particles just bounce forward and backward when the wavefront is approaching and leaving this area, such as the trajectory #1 in Fig. 7. Such a tendency may be seen more clearly in the averaged trajectory (dash-dot line), which oscillates back and forth during the passage of the wave front. Note here that there is a net shift in the averaged trajectory toward the wave propagation direction. We found that this shift is proportional to the amplitude of the wave and that it might be related to those particles which are carried along by the wave for a considerable distance: the trapped particles. Lines $2-6$ in Fig. 7 are examples of trapped trajectories. Comparing with an untrapped trajectory (Line #1) and averaged trajectory (dash-dot line), the trapped ones (Lines $2-6$) are characterized by extended periods during which the trajectory lines are parallel to the wavefront curves. This shows that these particles have a speed very close to the wave speed and travel together with the wave. The corresponding trajectories in $v_x - x$ phase space are also shown in Fig. 7 (b) and (c), from which one can see that the trapping picture is different from the classical one in a sinusoidal potential well. Since the wave here is in the form of a pulse-wave or soliton, the trapping occurs in front of a single moving potential barrier. The trapped particles usually have extended trajectories in phase space, while untrapped ones making local circles. The similar tendency had also been observed in recent experiment [18].

It should be pointed out that in the simulation, all particle trajectories are recorded, but only those particles, which lie in a very narrow region ($5\lambda_D$ wide in x direction) around the moving wavefront and at the same time have an x-component of velocity above 90% of local wave speed, i. e., when $v_x > 0.9v_\phi$, are defined as trapped particles. Typically, there are a few hundreds (about 300 when $\Gamma = 20$ and less for higher Γ) of particles are trapped during the 4 successive pulse-wave propagations shown in Fig. 7. Trapped particles have a finite lifetime (*trapping time*), and will lose their energy and become detrapped very soon (If v_x of a trapped particle decreases to below $0.9v_\phi$, we say the particle is detrapped.), mainly due to the collisions with surrounding particles. The time interval, during which $v_x > 0.9v_\phi$, is then defined as the *trapping time*, for the convenience of later discussion. Typical *trapping time* is around a couple of τ_0 and fewer particles can be trapped for longer period and vice versa. In our observation, the longest trapping time is about $12\tau_0$.

As is shown above that the trapping picture is quite different from a classical one that we are familiar with in ordinary ion-electron plasmas. The differences mainly come from nonlinear effect associated with large amplitude waves and the pronounced interparticle interaction due to strong-coupling effect in our system. The former makes it possible to trap a large number of particles in the main part of the distribution (near $v = 0$), instead of only "resonant particles" with speed very close to the wave speed, and it may still be explained by the classical trapping condition, while the latter poses new condition for trapping to happen. The short-range interaction or correlation between particles must also be included by introducing the concept of "caging" potential formed by all surrounding particles. Now trapping of a particle needs not only enough energy to accelerate it to wave speed, but also enough energy to overcome the "caging" potential barrier. In other words, the strong coupling effect makes the trapping harder to happen in SCCPs than in ordinary weakly coupled plasmas, while on the other hand, makes the detrapping more significant and frequent.

5. Summary

Thus we have demonstrated the basic ideas of numerical experiments on strongly coupled complex plasmas through two examples, i. e., the heat transfer and shock wave propagation and particle trapping, by directly mimicking the real experiments. This kind of numerical simulation will be helpful to explain many phenomena observed in real experiments and to explore more details of physical pictures that an real experiment cannot easily obtain.

References

[1] Shukla P K and Eliasson B 2009 *Rev. Mod. Phys.* **81** 25
[2] Morfill G E and Ivlev A V 2009 *Rev. Mod. Phys.* **81** 1353
[3] Allen M P and Tildesley D J 1989 *Computer Simulation of Liquids* (New York, Oxford University Press)
[4] Hou L J and Mišković Z L 2008 `arXiv:0806.3912`
[5] Hou L J, Mišković Z L, Piel A and Shukla P K 2009 *Phys. Plasmas* **16** 053705
[6] Hou L J and Piel A 2009 *J. Phys. A: Math. Theor.* **42** 214025
[7] Nunomura S, Samsonov D, Zhdanov S and Morfill G 2005 *Phys. Rev. Lett.* **95** 025003
[8] Nosenko V, Zhdanov S, Ivlev A V, Morfill G, Goree J and Piel A 2008 *Phys. Rev. Lett.* **100** 025003
[9] Konopka U, Morfill G E and Ratke L 2000 *Phys. Rev. Lett.* **84**, 891
[10] Donkó Z and Hartmann P 2004 *Phys. Rev. E* **69** 016405
[11] Hartmann P, Kalman G J, Donkó Z and Kutasi K 2005 *Phys. Rev. E* **72** 026409
[12] Volz S G and Chen G 2000 *Phys. Rev. B* **61** 2651
[13] epri S, Livi R and Politi A 2003 *Phys. Rep.* **377** 1
[14] Fortov V E, Vaulina O S and Petrov O F *et al* 2007 *Phys. Rev. E* **75** 026403
[15] Hou L J and Piel A 2008 *Phys. Plasmas* **15** 073707
[16] Piel A, Klindworth M, Arp O, Melzer A and Wolter M 2006 *Phys. Rev. Lett.* **97** 205009
[17] Schwabe M, Rubin-Zuzic M and Zhdanov S *et al*, *Phys. Rev. Lett.* **99** 095002
[18] Liao C T, Teng L W, Tsai C Y and I Lin 2008 *Phys. Rev. Lett.* **100** 185004
[19] Samsonov D, Ivlev A V, Quinn R A, Morfill G E and Zhdanov S K 2002 *Phys. Rev. Lett.* **88** 095004
[20] Nosenko V, Nunomura S and Goree J 2002 *Phys. Rev. Lett.* **88** 215002

Regularities and irregularities of the Stark broadening parameters for singly ionized noble gases

R. J. Peláez[1,3], S. Djurović[2], M. Ćirišan[2], J. A. Aparicio[1] and S. Mar[1]

[1]Departamento de Física Teórica, Atómica y Óptica, Universidad de Valladolid, P. Prado de la Magdalena s/n, 47071 Valladolid, Spain
[2]Faculty of Sciences, Department of Physics, Trg Dositeja Obradovića 4, 21000 Novi Sad, Serbia
[3]Laser Processing Group, Inst.Óptica, CSIC, Serrano 121, 28006 Madrid, Spain

E-mail: ramon@opt.uva.es djurovic@uns.ac.rs

Abstract. In the last years, different trends and regularities of Stark broadening parameters (halfwidths and shifts of spectral lines) have been analyzed. Conditions related to atomic structure of the element, as well as plasma conditions are responsible for regular or irregular behavior of the Stark broadening parameters . The absence of very close perturbing levels makes Ne II a good candidate for the analysis of the Stark broadening parameters regularities. The other two elements considered in this work, Kr II and Xe II, have complex spectra and present strong perturbations, leading to the appearance of Stark broadening parameters irregularities in some cases. In this work, we analyze the influence of perturbations on Stark broadening parameters within the multiplets.

1. Introduction

Spectroscopy of ionized noble gases has a great importance for both laboratory and astrophysical plasmas. Generally, spectra of inert gases are important for many physics areas, like laser physics, fusion diagnostics, photoelectron spectroscopy, collision physics, astrophysics etc. Stark halfwidths and shifts of spectral lines are frequently employed for plasma diagnostic purposes. For example, atomic data of argon, krypton and xenon will be useful for the spectral diagnostics of ITER [1]. In addition, the software used for stellar atmosphere simulation, like TMAP [2] and SMART [3], require a large amount of atomic and spectroscopic data. Availability of these parameters will be useful for a further development of stellar atmosphere and evolution models [4].

Stark broadening parameters data can be useful for verification of theoretical calculations, as well as for investigation of regularities and systematic trends of these parameters within a multiplet, supermultiplet or transition array [5,6]. This regular or nearly regular behavior has been experimentally found in many different works, for ionized atoms see Ref. [7-11]. The existence of Stark broadening parameters regularities within a multiplet, supermultiplet or transition array may be useful in order to make data interpolation. In this work we analyze these similarities of Stark broadening parameters of Ne II, Kr II and Xe II spectral lines within the chosen multiplets. According to Ref. [5,6], differences in the Stark halfwidth and shift values of the lines belonging to the same multiplet should be within a few percents and possibly lower than 1% and 10%, respectively.

2. Plasma Stark broadening parameters measurements and data treatment

Stark broadening parameters analyzed in this work have been measured in the laboratory of plasma spectroscopy of the University of Valladolid. The experimental set-up is described in detail in previous work [12]. Here, only short description is given.

Measurements were performed in a pulsed plasma. Pulses were created by discharging a capacitor bank, charged up to 9 kV, through a cylindrical Pyrex tube. A mixture of helium, as carrying gas, and several percents of the investigated gas was continuously flowing through the tube at a pressure of 3 kPa. The percentage of the noble gas and the flow of the gas mixture was adjusted in order to obtain minimal self-absorption, which was checked for every line in the recorded spectrum. Self-absorption was negligible for all lines considered in this work.

Electron density was determined by a two wavelength interferometric method and the highest density was about $2 \cdot 10^{23}$ m^{-3} with the estimated experimental error lower than 10%. Electron temperature was determined by a Boltzmann-plot and it was in the range 15000 – 27000 K for Kr II and Xe II, and 25000 – 45000 K for Ne II. The error of temperature measurements was estimated to be lower than 15%. An example of the typical evolution of electron density and temperature along the plasma life is shown in figure 1. Line spectra were recorded during the first 200 μs of the plasma life and the typical exposure time was 5 μs.

In order to obtain the Stark halfwidth from the total halfwidth other broadening mechanisms, like Doppler or instrumental broadening, have been taken into account. On the other hand, van der Waals and resonance broadening were negligible for the plasma conditions in this experiment. Stark halfwidths were plotted as a function of the electron density and linear trends were obtained. The value of the slope at 10^{23} m^{-3} corresponds to the Stark halfwidth of the line. The Stark shift was obtained by plotting the relative position of the line centre as a function of electron density. The procedure is described in detail in Ref. [13].

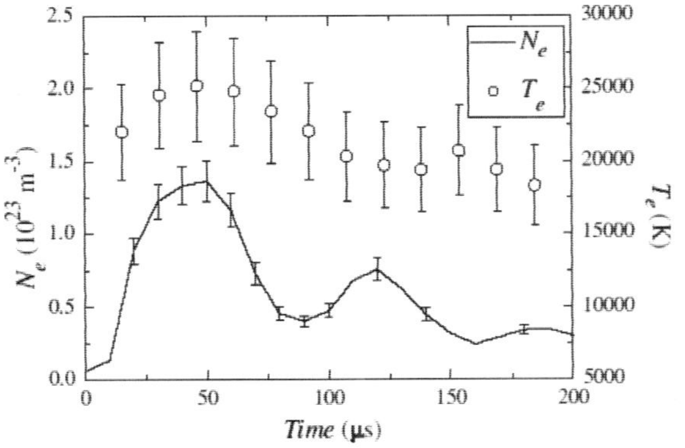

Figure 1. Typical evolution of electron density, N_e, obtained by two wavelength interferometry, and temperature T_e, obtained from Boltzmann-plot.

Stark broadening parameters are normalized to the temperature of 35000 K in case of Ne II, 21000 K in case of Kr II and 22000 K in case of Xe II using $T^{1/2}$ dependence. Detailed description of experimental data treatment is given in Ref. [12]

3. Results and discussion

Measured Stark halfwidths and shifts of ionized noble gases Ne II, Kr II and Xe II are given in Tables 1, 2 and 3, respectively. In the first four columns Tables contain transitions, terms, degenerations and wavelengths of the observed spectral lines. In the next two columns the measured parameters Stark halfwidths, w_m, and Stark shifts, d_m are shown. Data are expressed in frequency units in order to avoid the influence of the wavelength, for the comparison purpose. The obtained results are compared with available results of other authors.

Experimental errors of the measured Stark halfwidths of Ne II lines are estimated to be lower than ±20%, while in case of shifts these errors can be even higher due to very small values of the measured shifts [13].

Two multiplets of the transition array 3p – 3d have been selected in case of Ne II. As can be seen in Table 1, Stark broadening parameters of the spectral lines belonging to the same multiplet have very similar values. The measured halfwidths of the lines belonging to the multiplet $^4D^o$ – ^4F vary ±7% from the average value, while in case of multiplet $^4P^o$ – ^4D the variation is ±3%, corresponding to differences lower than 2 pm. This value is close to the experimental precision. All the lines of the 3p – 3d transition array show similar red shifts. Differences between the measured shifts are within ±7% for multiplet $^4D^o$ – ^4F and ±11% for multiplet $^4P^o$ – ^4D, corresponding to less than 1 pm. In order to illustrate these agreements, experimental profiles of seven Ne II lines belonging to the multiplet (^3P)3p $^4P^o$ – (^3P)3d ^4D are shown in figure 2. The lines have been normalized in area and a clear overlapping can be observed.

Table 1. Experimental Stark broadening parameters of Ne II spectral lines normalized to electron density $N_e = 1 \times 10^{23}$ m^{-3} and electron temperature T_e = 35000 K. Results from this work [14] are marked as TW, from Platiša et al. [15] as PL, from Konjević and Pittman [16] as KP, from Purić et al. [17] as PU, from Milosavljević et al. [18] as MI and from Djeniže et al. [19] as DJE.

Transition	Term.	J	λ (nm)	w_m (10^{11}s^{-1})	d_m (10^{11}s^{-1})
(^{3}P)3p – (^{3}P)3d	$^4D^o$ – ^4F	5/2-5/2	318.874	3.10 TW	-0.63 TW
		5/2-7/2	319.859	3.13 TW	-0.62 TW
				3.08 KP	
		1/2-3/2	321.374	3.43 TW	-0.68 TW
		3/2-5/2	321.433	3.25 TW	-0.60 TW
		7/2-9/2	321.819	3.03 TW	-0.65 TW
				2.63 PL	-0.22 PU
	$^4P^o$ – ^4D	5/2-3/2	301.731	3.01 TW	-0.60 TW
				2.77 KP	
		5/2-5/2	302.702	3.07 TW	-0.65 TW
				2.59 KP	-0.05 DJE
		5/2-7/2	303.446	3.13 TW	-0.72 TW
				2.21 MI	0.00 DJE
		3/2-3/2	303.772	2.99 TW	-0.75 TW
				2.33 MI	0.02 DJE
		1/2-1/2	304.556	3.02 TW	-0.61 TW
		3/2-5/2	304.756	3.00 TW	-0.76 TW
		1/2-3/2	305.467	3.02 TW	-0.62 TW

Finally, a reasonable explanation for the Stark broadening parameters regularities observed in this work may come from the analysis of the Ne II energy levels scheme. Ne II energy levels are well described by LS coupling scheme and all perturbing levels are too far from the transition levels to be able to produce significant perturbations of the latter ones. Under these conditions, the predicted regularities of Stark broadening parameters within the same multiplet have been obtained experimentally.

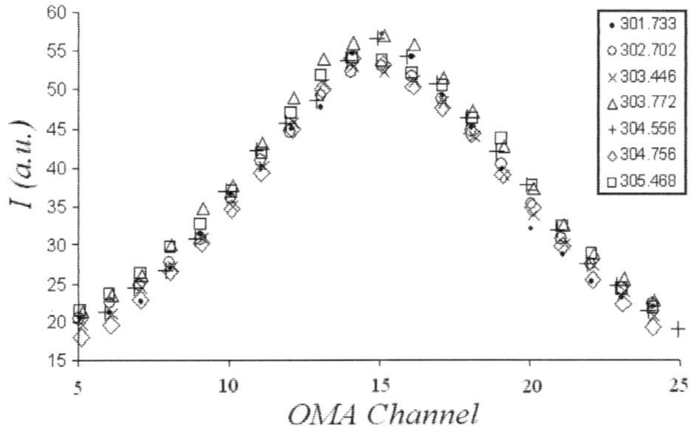

Figure 2. Profiles of 7 Ne II lines belonging to the multiplet $^4P^o - {}^4D$.

Comparison between the present halfwidth results and the results of other authors (Table 1) shows a satisfactory agreement and supports the above conclusions. There is some disagreement between the present results and the results from Ref. [19] in case of measured shifts, for example, for the 303.446 nm and 303.772 nm line. Still, one should keep in mind that the difference is only about 3 pm.

The ratio of experimental and theoretical halfwidth values, w_m/w_{MSE}, is about 1.5 for all considered spectral lines. For the theoretical calculations modified semiempirical formula MSE [20] was used. In case of shifts, the ratio d_m/d_{MSE} was between 2 and 2.5. Here, modified semiempirical formula [21] was used for theoretical calculations. It is well known that line shifts are less accurate than the line halfwidths, both measured and calculated.

Behavior of the Ne II lines is not a typical situation. Kr II has more complex spectra and its energy levels are not well described by the LS coupling scheme, especially the levels having high *l*. In addition, the average distance between the transition and perturbing levels is lower. Therefore, in some situations, there are strong perturbations that break the regularities in transition arrays and multiplets. Stark broadening parameters of Kr II lines from 6 multipletes in 2 transition arrays are shown in Table 2 [22]. Experimental errors for the measured halfwidths are estimated to be lower than ±6%, while in case of shifts are within ±12%.

Table 2. Experimental Stark broadening parameters of Kr II spectral lines normalized to electron density $N_e = 10^{23}$ m^{-3} and electron temperature $T_e = 21000$ K. Results from this work [22] are marked as TW.

Transition	Term.	*J*	λ (nm)	w_m (10^{11}s^{-1})	d_m (10^{11}s^{-1})
4d – (^{1}D)5p	$^2D - {}^2D^o$	3/2-5/2	516.680	4.45 TW	
		3/2-3/2	520.022	9.90 TW	1.81 TW
	$^2D - {}^2F^o$	3/2-5/2	616.880	3.66 TW	0.28 TW
(^{3}P)4d – (^{1}D)5p	$^2F - {}^2F^o$	5/2-7/2	660.501	3.39 TW	
(^{1}D)5p – (^{1}D)6s	$^2F^o - {}^2D$	7/2-5/2	448.988	11.8 TW	-4.03 TW
	$^2P^o - {}^2D$	3/2-5/2	459.280	14.2 TW	-6.76 TW
	$^2D^o - {}^2D$	3/2-3/2	503.385	17.4 TW	-5.08 TW
		5/2-5/2	508.652	10.6 TW	-4.71 TW

It is obvious that, in case of 516.680 nm and 520.022 nm line, halfwidths similarity within the multiplet $^2D - {}^2D^o$ is broken. $^2D_{5/2}{}^o$, the upper level of 516.680 nm line, has no close perturbing levels and the halfwidth of the line is $4.45 \cdot 10^{11}$ s^{-1}. On the other hand, $^2D_{3/2}{}^o$, the upper level of 520.022 nm line, has a very close perturbing level ($^2P_{1/2}$), so the line has a twice larger halfwidth. One part of the

Journal of Physics: Conference Series **257** (2010) 012021 doi:10.1088/1742-6596/257/1/012021

energy levels diagram, containing the considered levels, is shown in figure 3 [23]. Positions of the perturbing levels can also explain the difference between the shifts of the 520.022 nm and 616.880 nm line. The halfwidth of the 516.680 nm line and the halfwidths of other two spectral lines, 616.880 nm and 660.501 nm, from the multiplets belonging to the 4d – 5p transition array, show a reasonable regularity (differences < 15% from the average value of $3.9\cdot 10^{11}s^{-1}$).

Perturbations of the upper energy levels of the transitions are more evident. But, if the lower level has a very close perturbing level, perturbation in the broadening can also be detected. In the 5p – 6s transition array we can distinguish two lines (459.280 nm and 503.385 nm) whose lower levels $^2P^o_{3/2}$ and $^2D^o_{3/2}$ are perturbed, as shown in figure 3. In this case, the upper and lower levels approach each other; hence there is an additional red shift which is in good agreement with the measurements. In this particular case, the energy differences between the lower transition level and the perturbing level are only 23.6 cm^{-1} for $^2P^o_{3/2}$ and 5.1 cm^{-1} for $^2D^o_{3/2}$. This is the reason why these perturbing effects cause a bit higher shifts than the shifts of the other two lines in 5p – 6s transition array.

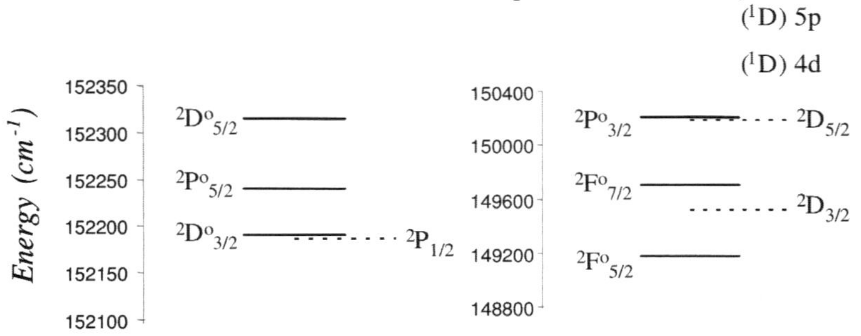

Figure 3. Energy scheme of some (^{1}D)5p Kr II levels (continuous lines), and the nearest perturbing 4d and (^{1}D)4d levels (dashed lines).

In case of Kr II there are no available experimental data of other authors for the comparison. Similar situation is with the theoretical MSE calculations [20, 21]. There are no available data for the complete sets of perturbing levels necessary for the calculations of halfwidths and shifts of considered spectral lines [24].

Table 3. Experimental Stark broadening parameters of Xe II spectral lines normalized to electron density $N_e = 1\times 10^{23}$ m^{-3} and electron temperature T_e = 22000 K. Results from this work [25] are marked as TW, from Nick and Helbig [26] as NH, from Di Rocco [27] as DR, from Bertuccelli et al. BE1 for Ref. [28] and BE2 for Ref. [29], and from Djurović et al.[12] as DJU and from Ćirišan et al.[30] as CI.

Transition	Term.	*J*	λ(nm)	$w_m(10^{11}s^{-1})$	$d_m(10^{11}s^{-1})$
(^{3}P$_1$)6s – (^{3}P$_1$)6p	[1]-[1]°	3/2-3/2	465.194	5.64 TW	-0.82 TW
				5.77 BE1	
				2.78 NH	
		1/2-1/2	565.938	4.20 TW	0.24 TW
				3.93 DJU	
		1/2-3/2	575.103	6.14 TW	-1.32 TW
				3.47 BE2	-2.03 DR
(^{3}P$_0$)6s – (^{3}P$_1$)6p	[0]-[1]°	1/2-1/2	426.984	3.89 TW	0.42 TW
				3.80 DJU	0.56 CI
		1/2-3/2	432.182	7.43 TW	-0.75 TW
				5.03 BE1	

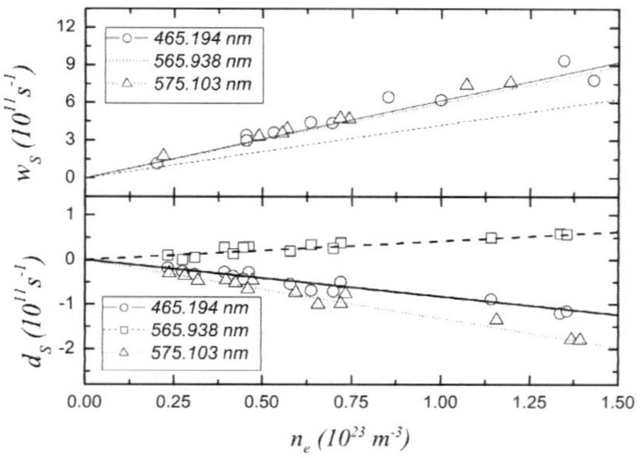

Figure 4. Measured Stark halfwidths and shifts of the lines from the multiplet ($^3P_1$)6s[1] – ($^3P_1$)6p[1]°.

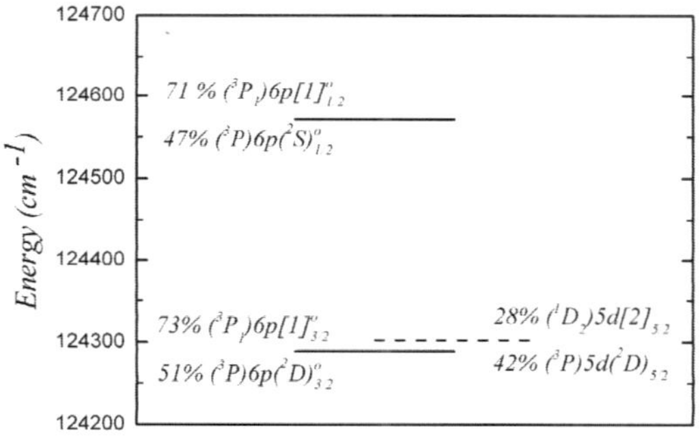

Figure 5. Energy scheme of ($^3P^1$)6p[1]° level components (solid lines) and the nearest perturbing level (dashed line). Upper designation is in *jK* and lower is in *LS* coupling. The percentage indicates the level purity.

Xe II is a multielectronic and heavy ion with a very complex energy structure, where *LS* is not an appropriate coupling scheme. The coupling conditions in the Xe II configuration are intermediate, according to the strength of the spin-orbit interaction compared to the electrostatic interaction. In this case, the *jK* coupling scheme is more appropriate and, in most cases, compositions of the energy levels configurations show more purity [31,32]. In Table 3, 5 Xe II lines from 2 multiplets are shown [14]. Experimental errors for halfwidths are estimated to be lower than ±15%, while in case of shifts these errors vary from ±15% for the highest value to ±70% for the lowest value, which is normal since the shift values are very low. It can be seen that there are large differences between the halfwidths, as well as between the shift values of the lines within the same multiplets. Measured values of the Stark broadening parameters of the lines in the multiplet ($^3P_1$)6s[1] – ($^3P_1$)6p[1]° are shown in the figure 4. Energy scheme of Xe II around 124500 cm^{-1} is shown in figure 5. The proximity of the levels ($^3P_1$)6p[1]$^o_{3/2}$ and ($^1D_2$)5d[2]$_{5/2}$ could explain the extra-broadening and extra-red shift of the lines 465.194 nm and 575.103 nm in the first multiplet and of the 432.182 nm line in the second one [14].

In case of halfwidths, results of other authors [12, 28] support the present results. There are some exceptions for 465.194 nm and 575.103 nm line. Results of Ref. [24,27] are in disagreement with our and other results. In case of shifts, comparison with the results from other works [27,30] show satisfactory agreement.

Agreement of the present measured Stark halfwidth values and the MSE calculated values is reasonably good. For the w_m values lower than $5 \cdot 10^{11} s^{-1}$, the ratio w_m/w_{MSE} is about 1.15, while for higher w_m, values the ratio is about 1.65. The ratios between the measured and calculated shifts show a relatively large variation and in some cases there are opposite directions of the measured and calculated shift. There are many difficulties in the calculations of Stark broadening parameters , particularly in shift calculations. This is even more pronounced in cases of complex energy levels structure, such as in case of Xe II, where mixing of energy levels of different multiplets occurs. For detailed explanation see Ref. [14].

4. Conclusions

In this work we analyze regular and irregular behaviour of the Stark broadening parameters of spectral lines of ionized noble gases, neon, krypton and xenon, within certain multiplets and supermultiplets.

The spectral lines were observed from the pulsed discharge plasma. Apart from taking care of experimental conditions and plasma diagnostics, attention was also paid to proper fitting and deconvolution procedure.

Clear regular behaviour of both Stark halfwidths and shifts, in agreement with Ref. [5, 6] has been observed within Ne II multiplets. In case of Kr II and Xe II, which have more complex spectra and energy levels structure, intramultiplet irregularities occur. All presented irregularities of Stark broadening parameters can be explained by the presence of perturbing energy levels close to the upper level of the observed transition. The present cases show that a careful analysis of energy levels should be always performed when trying to estimate or predict Stark broadening parameters , particularly in atoms or ions with complex energy levels structure. In the future, it would be useful to analyze these and similar irregularities both experimentaly and theoreticaly.

Acknowledgements

We thank S. González for his work on the experimental device, the Spanish Ministerio de Ciencia y Tecnología and the Consejería de Educación y Cultura de la Junta de Castilla y León for their financial support under contracts no FIS2005-03155 and VA015A05 respectively. Dr J. A. Aparicio wants to express his personal acknowledgement to the ONCE for help. S. Djurović thanks to Ministry of Science and Development of Republic of Serbia for support in Project 141024.

References

[1] Skinner, C. H.: 2009, *Phys. Scr.*, **T134**, 014022.
[2] Werner, K., Deetjen, J. L., Dreizler, et al.: 2003, in: Stellar Atmosphere Modeling, eds. I. Hubeny, D. Mihalas, K. Werner, *ASP Conference Series* **288** (San Francisco ASP),p.31.
[3] Sapar, A., Aret, A., Sapar, L., Poolamae, R.: 2009, *New Astron. Rev.*, **53**, 240.
[4] Leitherer, C.: 2010, *Proceedings of the IAU*, **262**, 73.
[5] Wiese, W. L., Konjević., N.: 1982, *J. Quant. Spectrosc. Radiat. Transfer.*, **28**, 185.
[6] Wiese, W. L., Konjević., N.: 1992, *J. Quant. Spectrosc. Radiat. Transfer.*, **47**, 185.
[7] Konjević., N., Wiese., W. L.: 1976, *J. Phys. Chem. Ref. Data*, **5**, 259.
[8] Konjević., N., Dimitrijević, M, S., Wiese., W. L.: 1984, *J. Phys. Chem. Ref. Data*, **13**, 649.
[9] Konjević., N., Wiese., W. L.: 1990, *J. Phys. Chem. Ref. Data*, **19**, 1307.
[10] Konjević., N., Lesage. A., Fuhr. J. R., Wiese., W., L.: 2002, *J. Phys. Chem. Ref. Data*, **31**, 819.
[11] Lesage, A.: 2009, *New Astron. Rev.*, **52**, 471.
[12] Djurović, S., Peláez, R. J., Ćirišan, M., Aparicio, J. A., Mar, S.: 2006, *J. Phys. B: At. Mol. Opt. Phys.*, **39**, 2901.

[13] Aparicio, J. A., Perez, C., del Val, J. A., Gigosos, M. A., de la Rosa, M. I., Mar, S.: 1998,*J. Phys. B: At. Mol. Opt. Phys.*, **31**, 4909.
[14] Peláez., R. J., Djurović, S., Ćirišan, M., Aparicio, J. A., Mar, S.: 2009, *J. Phys. B: At. Mol. Opt. Phys.*, **42**, 125002.
[15] Platiša, M., Dimitrijević, M, S., Konjević., N.: 1978, *Astron. Astrophys.*, **67**, 103.
[16] Konjević, N., Pittman, T. L. : 1986, *J. Quant. Spectrosc. Radiat. Transfer.*, **35**, 473.
[17] Purić, J., Djeniže, S., Srećković, A., Labat, J., Ćirković, Lj.: 1987, *Phys. Rev. A*, **35**, 2111.
[18] Milosavljević, V., Dimitrijević, M, S., Djeniže, S.: 2001, *Astrophys. J. Suppl.*, **135**, 115.
[19] Djeniže, S., Milosavljević, V., Dimitrijević, M, S.: 2002, *Astron. Astrophys.*, **382**, 359
[20] Dimitrijević, M, S., Konjević., N.: 1980, *J. Quant. Spectrosc. Radiat. Transfer.*, **24**, 451.
[21] Dimitrijević, M, S., Kršljanin, V.: 1986, *Astron. Astrophys.*, **165**, 269
[22] del Val, J. A., Peláez, R. J., Mar, S., Rodríguez, F., González, V. R., Gonzalo, A. B., de Castro, A., Aparicio, J. A.: 2008, *Phys. Rev. A*, **77**, 012501.
[23] Sugar, J., Musgrove, A. J.: 1991, *J. Phys. Chem. Ref. Data*, **20,** 859.
[24] NIST Atomic spectra database http://physics.nist.gov.asd.
[25] Peláez, R. J., Djurović, S., Ćirišan, M., Rodríguez, F., Aparicio, J. A., Mar, S.: 2008, *Astrophys. J.*, **687**, 1423.
[26] Nick, K. P., Helbig, V.: 1986, *Phys. Scr.*, **33,** 55.
[27] Di Rocco, H. O.: 1990, *Nuovo Cimento D*, **12**, 1485.
[28] Bertuccelli, D., Bertuccelli, G., Di Rocco, H. O.: 1991, *Rev. Sci. Instrum.*, **62**, 966.
[29] Bertuccelli, D., Bertuccelli, G., Di Rocco, H. O.: 1991, *Phys. Scr.*, **43**, 469.
[30] Ćirišan, M., Peláez, R. J., Djurović, Aparicio, J. A., Mar, S.: 2007, *J. Res. Phys.*, **31**, 11.
[31] Hansen, J. E., Persson, W. : 1987, *Phys. Scr.,* **36**, 602.
[32] Saloman, E. B.: 2004, *J. Phys. Chem. Ref. Data*, **33**, 765.

25th Summer School and International Symposium on the Physics of Ionized Gases—SPIG 2010 IOP Publishing
Journal of Physics: Conference Series **257** (2010) 012022 doi:10.1088/1742-6596/257/1/012022

Electrical breakdown in nitrogen at low pressure – physical processes and statistics

S R Gocić

Department of Physics, University of Niš, P.O. BOX 224, 18001 Niš, Serbia

gsasa@pmf.ni.ac.rs

Abstract. The results of investigation of the electrical breakdown in nitrogen, obtained in combined approach based on measuring of the current-voltage characteristic and statistical analysis of the breakdown time delay are presented in this report. Measurement of the current-voltage characteristics with additional monitoring of spatial and temporal distribution of the emission from discharge provides information concerned on development of different regime of low-pressure gas discharge and on processes of the electrical breakdown and discharge maintenance. Also, two new distributions of the statistical time delay of electrical breakdown in nitrogen, the Gaussian and Gauss-exponential ones, are presented. Distributions are theoretically founded on binomial distribution for the occurrence of initiating electrons and described by using analytical and numerical models. Moreover, the correlation coefficient between the statistical and formative time delay of electrical breakdown in nitrogen is determined. Starting from bivariate normal (Gaussian) distribution of two random variables, the analytical distribution of the electrical breakdown time delay is theoretically founded on correlation of the dependent statistical and formative time delay.

1. Introduction

The process of transition from non-conducting to conducting states of gas is known as the electrical breakdown. The research of the gas-phase and surface processes which lead to the gas breakdown has been attracting the attention of scientific community for at least one century. Starting with Paschen measurements of dependences of the static breakdown voltage U_s on pd values (pressure $\times$ interelectrode gap), it was taken further with Townsend's theory of the electrical breakdown [1]. This theory was based on the secondary electron emission from the cathode surface and charged particle multiplications in the gas phase. Recently, modifications of Townsend's theory were published in [2] for Ar. These authors developed models of electron, ion, fast atom, excited atoms and photon transport and kinetics and applied them to explain measured data of secondary electron yield in argon for various clean and dirty metal surfaces. The determination of the secondary electron yield from measured U_s *(pd)* dependencies (Paschen curves) in nitrogen for the cathode surfaces made of different metals was published in [3].

The model of oscillatory and negative differential resistance behavior of low-current H_2 discharge between parallel-plane electrodes was developed in [4]. The nonequilibrium model that includes the effects of space-charge distortion of the electric field on the secondary electron emission is used to obtain the negative differential resistance. Spatial distributions of light emission from N_2 discharge at very wide E/N range were presented in [5]. Measured distributions of characteristic bands were connected with electron excitation near the anode at lower E/N. Observed emission near the cathode at high *E/N* was postulated to be result of excitation by fast molecules. An analysis of time- and space-resolved development of different regimes of low-pressure DC discharge in argon, based on ICCD recordings of the discharge structure, was presented in [6].

Another area of investigation of the electrical breakdown of gases is represented by statistically based measurements of the electrical breakdown time delay t_d, the time that elapses from the application of voltage greater then the static breakdown voltage U_s to the breakdown. It comprises the statistical t_s and the formative time delay t_f [1]. t_s is the time which elapses from the application of sufficient voltage to the appearance of a free electron initiating breakdown. t_f is the time interval from this moment to the collapse of applied voltage and occurrence of self-sustained current [1]. The stochastic nature of breakdown time delay was experimentally proven by Zuber [7] and the exponential distribution was theoretically derived by von Laue [8]. Kiselev [9] derived mathematically the exponential distribution of t_s based on binomial distribution of electrons occurrence in interelectrode gap. In recent papers [10,11] the model of the time delay based on the convolution of independent t_s and t_f was presented. Also, the measured distribution of the statistical time delay at high electron yield is non-adequately ascribed to the formative time delay.

The results of the investigation of the electrical breakdown in nitrogen at low pressure based on measuring of the current-voltage (*I-V*) characteristic and statistical analysis of the breakdown time delay are presented in this report. The influence of the cathode surface conditions on *I-V* characteristic is examined through monitoring of axial emission profiles and 2D images of the discharge spatial structure. In statistical part, two new distributions of the statistical time delay are presented [12]. Also, the analytical distribution of the electrical breakdown time delay obtained by the correlation theory of dependent t_s and t_f, and the determination of the correlation coefficient in nitrogen are described [13].

2. Experimental details

The measurements presented in this report were carried out on a gas tube made of borosilicate glass (8245, Schott technical glass) equipped with planar copper electrode. The diameter of the electrode was $D=2.2\ cm$ and the interelectrode gaps $d=0.6$ and $0.8\ cm$. The tube was filled with research grade nitrogen at a pressure of $2.66 mbar$ (Matheson Co. with oxygen impurities of below $1 ppm$.). The static breakdown voltage was $U_s \approx 267V$ at $d=0.6cm$ and $U_s \approx 316V$ at $d=0.8cm$.

The measurement of current-voltage characteristic at $d=0.8cm$ was made by applying a voltage pulse to a DC voltage that maintains the discharge at very low current ($1\text{-}2\ \mu A$). The voltage and current signals were monitored by digital oscilloscope with appropriate probes. Development of discharge structure is traced down in time using a fast ICCD camera (Andor, iStar DH720) [6].

The time delay measurements were carried out at $d=0.6cm$ by applying rectangular (step) pulses to the discharge tube, at the glow current $I_g = 0.1\ mA$, glow time $t_g = 1s$ and at different relaxation times τ. The measurements were performed using an electronic system [12] which reached the voltage rise time and resolution limit of 0.18 μs.

3. Current-Voltage characteristics and spatial development of the dc discharge

The secondary electron emission from the cathode surface is one of the main processes governing the electrical breakdown and dc discharges maintenance. The secondary emission is induced by positive ions, photons and metastable atoms and molecules and it is usually described as an effective yield per ion γ_{eff}. The aim of this work is to explain the non-typical current-voltage characteristic (with positive slope in low-current regime) that we obtained (figure 1a). As we will show, the cause of this kind of behavior was in the inhomogenous cathode surface that was severely conditioned by the discharge that operated in constricted glow regime for a long period of time. The *I-V* characteristic deviates from the usual dependences [6,14] and a positive slope appears in the transition to normal glow instead of negative slope (dashed line). At the same time axial emission profiles (figure 1b), taken at the axes of the discharge, are in agreement with the standard behavior of low-pressure discharges [6,14], starting from low current (Townsend) mode with exponential distribution of intensity (1-3), through formation of the negative glow (4-7) with the development of the cathode fall.

Journal of Physics: Conference Series **257** (2010) 012022 doi:10.1088/1742-6596/257/1/012022

It should be noted that positive slopes in Townsend regime were predicted for secondary yields that are independent of ion energy [15].

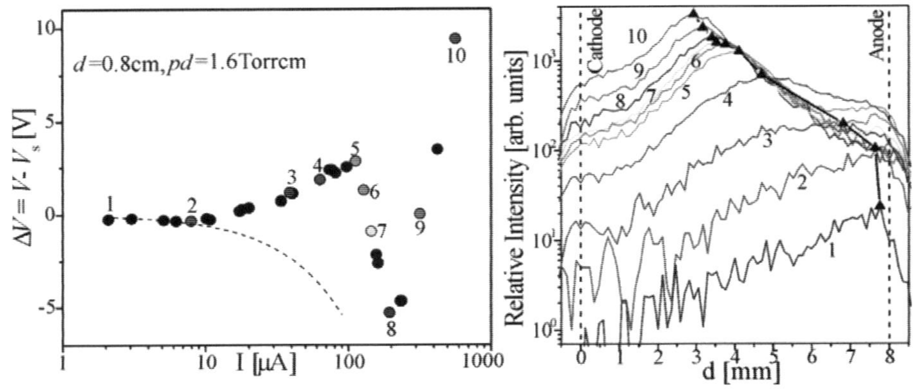

Figure 1. a) The current-voltage characteristic of N_2 discharge. b) Axial profiles of emission intensity in numbered points of *I-V* characteristic (▲- position of maximum intensity).

2D images of the discharge spatial structure (figure 2) clearly show that at lower currents (labels 1-5), discharge operates only within conditioned area of the cathode – limited circular area positioned close to the center of the cathode. From points 1 to 5, discharge goes from Townsend to glow-like discharge, while remaining within the same limited area. As the current-voltage characteristic gets negative slope, the discharge gradually spreads to the remaining cathode area.

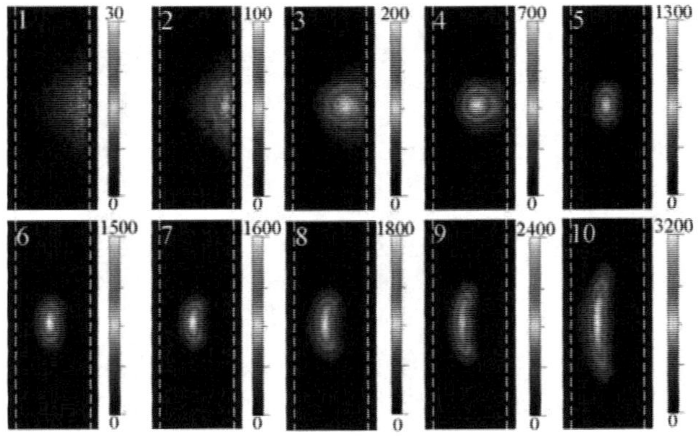

Figure 2. 2D images of discharge spatial structure. The cathode is on the left side.

Further on, the discharge resembles a typical low-pressure glow discharge behavior, except the profile is radially distorted – curved towards the cathode. Practically, due to the conditioning of the cathode, central area has different secondary electron yield than the remaining surface. This leads to the complex behavior of the discharge – it starts in the central area where, presumably, γ_{eff} is higher. The discharge will spread to the area with different γ_{eff} only when current density increases enough for the charged particles, diffused outside the conditioned area, induce sufficient space charge effects to ignite the discharge at the given voltage.

Furthermore, these results are particularly useful for studies of discharges with cathodes of different materials where the same kind of complex behavior may be expected.

4. New distributions of the statistical time delay

It this section the experimentally obtained Gaussian and Gauss-exponential distributions of the statistical time delay will be presented. They are theoretically founded on the binomial distribution for the occurrence of initiating electrons in the interelectrode space [12]. It is shown that transitions from binomial to Gaussian, Gauss-exponential and exponential distribution are caused by the value of the effective electron yield *YP* (the electron yield *Y* multiplied by the probability of breakdown *P*) [12].

The exponential distribution of the statistical time delay was mathematically strictly derived from the probability of occurrence of electrons during the time interval t_s [9]. The probability of occurrence of an effective electron in the subinterval t_s/m is YPt_s/m and for non-occurrence is $1-YPt_s/m$. According to the statistical theory, the probability of breakdown W in the time interval t_s is equal to the sum of probabilities of occurrence of j electrons in the interelectrode space, where j takes all the whole number values in the interval $[k\text{-}m]$:

$$W=\sum_{j=k}^{m} C_m^j \left(\frac{YPt_s}{m}\right)^j \left(1-\frac{YPt_s}{m}\right)^{m-j}, \tag{1}$$

where C_m^j are the binomial coefficients. Generally, the sum (1) represents Gauss-exponential distribution of waiting time t_s for the occurrence of the initial electrons. In the derivation of exponential cumulative distribution [9], it is implicitly assumed that $YPt_s/m \equiv p$ is close to zero. On the other hand, if $m \to \infty$ and if neither p nor $1\text{-}p$ is too close to zero, Gaussian distribution is obtained as a limiting case of binomial distribution for appearance of electrons or of Gauss-exponential for t_s.

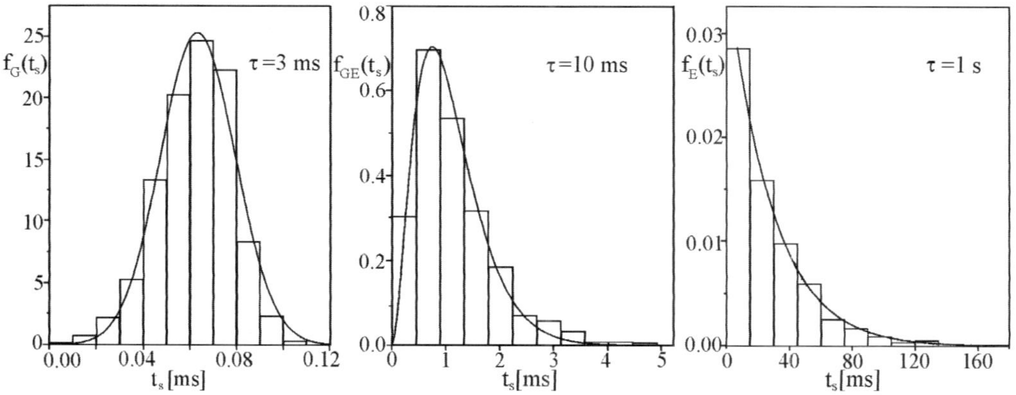

Figure 3. Experimental Gaussian $f_G(t_s)$, Gauss-exponential $f_{GE}(t_s)$ and exponential $f_E(t_s)$ density distributions fitted by numerical differentiation of equation (1) (solid line).

In figure 3, it is demonstrated how the sum (1) of binomial distributions for the electron occurrence goes to Gauss-exponential distribution $f_{GE}(t_s)$ and approaches two limiting cases of exponential $f_E(t_s)$ and Gaussian distribution $f_G(t_s)$ for the statistical time delay. The experimental distributions were obtained on basis of about *1000* measurements in series at the voltage *10%* higher then U_s. The statistic time delay values were obtained from measured t_d values, with t_f subtracted, at $\tau=10ms$, $1s$ and $3ms$ of the afterglow. The methods for determination of t_f are described in [16]. In the case of Gauss-exponential distribution all methods give approximately the same results $t_f=1.34$ *ms*, while in the case of exponential and Gaussian distribution only minimum t_d values can be used, giving for the formative time $t_f=1.35$ *ms* and $t_f=1.27$ *ms*, respectively.

The experimental density distributions $f_{GE}(t_s)$, $f_G(t_s)$ and $f_E(t_s)$ in figure 3 are fitted by numerical differentiation of relation (1). The fitting parameters for Gauss-exponential distribution $f_{GE}(t_s)$ are as follows: $k=3$, $m=82$, $p=0.008\text{–}0.16$ and $YP=2.63\times10^3 s^{-1}$. In the present case the breakdown probability is calculated to be approximately equal $P \approx 0.2$ [16], giving the value of electron yield $Y \approx 1.3 \times 10^4 s^{-1}$. When p is close to zero, the exponential distribution $f_E(t_s)$ at $\tau = 1s$ is fitted with parameters: $k=1$, $m=36$, $p=0.005\text{-}0.115$ and $YP=36s^{-1}$. Taking $P \approx 0.2$ into account the electron yield $Y=180s^{-1}$ is obtained. On the other hand, when p and $1\text{-}p$ are not too close to zero, the Gaussian distribution $f_G(t_s)$ at $\tau = 3ms$ (figure 3) is fitted with parameters: $k=747$, $m=2587$, $p=0.2535\text{-}0.32$ and $YP=1.51 \times 10^7 - 9.6 \times 10^6 s^{-1}$, which give the electron yield $Y = \overline{Y} + \sigma_Y = (6.2 \pm 1.4) \times 10^7 s^{-1}$. The presented results show that in the early afterglow, with a very high Y, the statistical time delay has a Gaussian distribution and this conclusion will be a starting point for the derivation of Gaussian distributions of the electrical brekdown and formative time delay.

5. Theory of correlation of the statistical and formative time delay

In a mathematical sense, the dependence of two continuous random variables X and Y implies that one variable, say Y, either increases or decreases as X changes. Two measures of their dependence are the covariance and the correlation coefficient $\rho(X,Y)$. If X and Y are independent continuous variables, then their covariance and correlation coefficient are equal to zero, but the opposite statement is true only in the special case of bivariate normal distribution. It can be shown that the correlation coefficient satisfies the inequality $-1 \le \rho \le 1$. Thus, a positive correlation coefficient indicates that Y increases as X increases, a negative ρ implies that Y decreases as X increases and $\rho = 0$ implies zero covariance and no correlation. The characteristic case of the joint density function of two continuous random variables is the earlier mentioned bivariate normal distribution, sometimes also called a bivariate Gaussian distribution. If we regard the statistical and formative time delay as continuous random variables $\boldsymbol{t_s}$ and $\boldsymbol{t_f}$, then their bivariate Gaussian density distribution $f(t_s,t_f)$ is given by [17]:

$$f(t_s,t_f)=\frac{1}{2\pi\sigma_{ts}\sigma_{tf}\sqrt{1-\rho^2}}\times\exp\left[-\frac{1}{2(1-\rho^2)}\left(\frac{(t_s-\overline{t_s})^2}{\sigma_{ts}^2}-2\rho\frac{(t_s-\overline{t_s})(t_f-\overline{t_f})}{\sigma_{ts}\sigma_{tf}}+\frac{(t_f-\overline{t_f})^2}{\sigma_{tf}^2}\right)\right], \tag{2}$$

with the marginal Gaussian distribution of the formative and statistical time, $f_{tf}(t_f)$ and $f_{ts}(t_s)$, which can be obtained by integration of equation (2) for all t_s and t_f, respectively. $\overline{t_s}$ and $\overline{t_f}$ denote the mean values of marginal distribution of $\boldsymbol{t_s}$ and $\boldsymbol{t_f}$, σ_{ts} and σ_{tf} are their standard deviations and ρ is the correlation coefficient of $\boldsymbol{t_s}$ and $\boldsymbol{t_f}$. Taking into account the above-mentioned dependence between $\boldsymbol{t_s}$ and $\boldsymbol{t_f}$, the density distribution of the breakdown time delay, $\boldsymbol{t_d} = \boldsymbol{t_s} + \boldsymbol{t_f}$ is calculated by [17]:

$$f(t_d)=\int_0^\infty f(t_s,t_d-t_s)dt_s\,, \tag{3}$$

which in the case of the bivariate Gaussian distribution $f(t_s,t_f)$ (equation (2)) gives [17]:

$$f(t_d)=\frac{1}{\sqrt{2\pi(\sigma_{ts}^2+2\rho\sigma_{ts}\sigma_{tf}+\sigma_{tf}^2)}}\exp\left[-\frac{1}{2}\frac{[t_d-(\overline{t_s}+\overline{t_f})]^2}{\sigma_{ts}^2+2\rho\sigma_{ts}\sigma_{tf}+\sigma_{tf}^2}\right]. \tag{4}$$

Equation (4) implies that the breakdown time delay t_d has a Gaussian density distribution with the mean value $\overline{t_d}=\overline{t_s}+\overline{t_f}$ and the standard deviation $\sigma_{td}=(\sigma_{ts}^2+2\rho\sigma_{ts}\sigma_{tf}+\sigma_{tf}^2)^{1/2}$, from which the correlation coefficient ρ can be determined.

In order to check earlier hypothesis of the independence of the statistical and formative time delay [10,11] and to demonstrate determination of the correlation coefficient, we have performed new measurements of the breakdown time delay. The dependence of the mean values of the breakdown time delay $\overline{t_d}$ on the relaxation time τ (the well known 'memory curve' [18]), obtained at the voltage 50% higher then U_s, is presented in figure 4. In the first region (I in figure 4), below $\tau=3ms$, the t_d standard deviation is nearly equal to t_f one ($\sigma_{td} \approx \sigma_{tf}$), since σ_{ts} can be neglected. In the second region (II in figure 4) the contributions of σ_{tf} and σ_{ts} to σ_{td} are of comparable value, and the correlation coefficient between t_s and t_f will be determined. In the region III of the memory curve, σ_{ts} dominates ($\sigma_{td} \approx \sigma_{ts}$) and the characteristic distributions of the statistical time delay appear (Gaussian, Gauss-exponential and exponential). The good estimation for the formative time delay in the region III is $t_{d,\min}$, which represents the minimal value of the time delay obtained in the series of 200 measurements. The $\sigma_{ts}(\tau)=1/Y(\tau)$ ($P\approx 1$ at a given voltage) dependence to about *10ms* of the region III is approximately linear in semilog scale confirming that a secondary electron yield in the interelectrode space is determined by the nitrogen ions decaying by diffusion [16]. After that, the dominant emission mechanism of electrons initiating breakdown is the surface recombination of nitrogen atoms remained from the preceding glow [18].

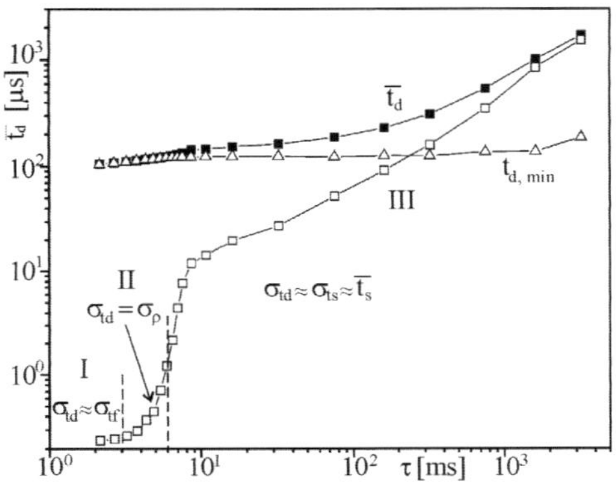

Figure 4. The memory curve, the formative and statistical time delay and their standard deviations as a function of relaxation times.

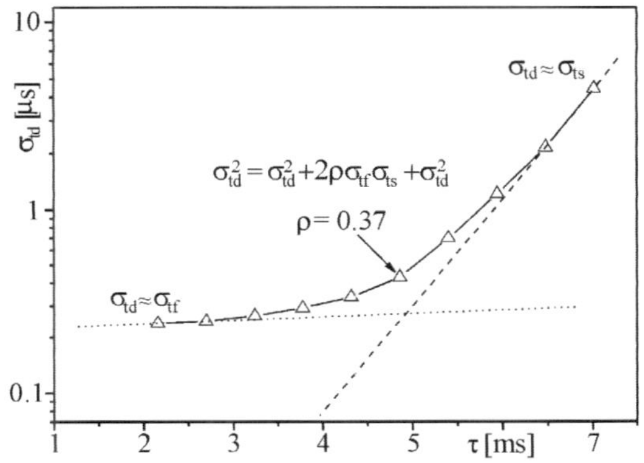

Figure 5. The $\sigma_{td}(\tau)$ dependence in the region II: (symbols) - experimental data, (dotted and dashed lines) - linear fits of σ_{tf} and σ_{ts}.

In order to determine the correlation coefficient between t_s and t_f, the $\sigma_{td}(\tau)$ dependence in the first *7 ms* of the memory curve is presented in figure 5. The experimental points below *3 ms* refer to the formative time delay and their linear extrapolation to longer τ gives an estimate of σ_{tf} in the regions II and III. Similarly, the extrapolation of the last two experimental points in figure 5 gives the estimation of σ_{ts} in the regions I and II. The experimental point at $\tau=4.9ms$ with $\sigma_{td}\approx 0.43\mu s$ was selected to determine ρ (marked with arrow in figure 5). The extrapolated values of t_s and t_f dispersions at this τ are $\sigma_{ts}=0.25\mu s$ and $\sigma_{tf}=0.27\mu s$. From the dispersion of the density distribution given by equation (4), ρ can be expressed in form:

$$\rho=\frac{\sigma_{td}^2-(\sigma_{tf}^2+\sigma_{ts}^2)}{2\sigma_{tf}\sigma_{ts}}, \tag{5}$$

which gives the value of the correlation coefficient $\rho=0.37$; that means that the statistical and formative time delay are medium correlated. The correlation coefficient, determined on the finite sample size ($n = 200$), represents the estimation of the population correlation coefficient **ρ** between t_s and t_f. In order to determine the confidence level of **ρ** we have applied the standard statistical test for the correlation coefficient on large sample which results in the probability greater then 99% that the population correlation coefficient is **ρ**>0. Thus, we can reject the hypothesis that t_s and t_f are uncorrelated, independent variables that is widely exploited in series of papers [10,11].

It is clear from the above analysis that neither the statistical and formative time delay are independent variables nor the density distribution of t_d can be calculated like their convolution [10-11]. This statement will be tested through analysis of the experimental Gaussian density distribution $f_G(t_d)$

presented in figure 6a, obtained at τ=4.9ms. It is fitted by theoretical density (equation (4)) with the parameters $\overline{t_d}$=118.3μs, ρ=0.37, σ_{ts}=0.25μs, σ_{tf}=0.27μs (the solid line in figure 6a). The agreement between experimental and theoretical distributions is obvious and confirms our statements that t_s and t_f are the dependent, correlated variables. Also, the experimental density is fitted by an analytical Gaussian distribution (the dashed line in figure 6a):

$$f_G(t_d)=\frac{1}{\sqrt{2\pi}\,\sigma_{td}}\exp\left[-\frac{(t_d-\overline{t_d})^2}{2\sigma_{td}^2}\right], \tag{6}$$

with the parameters $\overline{t_d}$=118.295μs and σ_{td}=0.431μs.

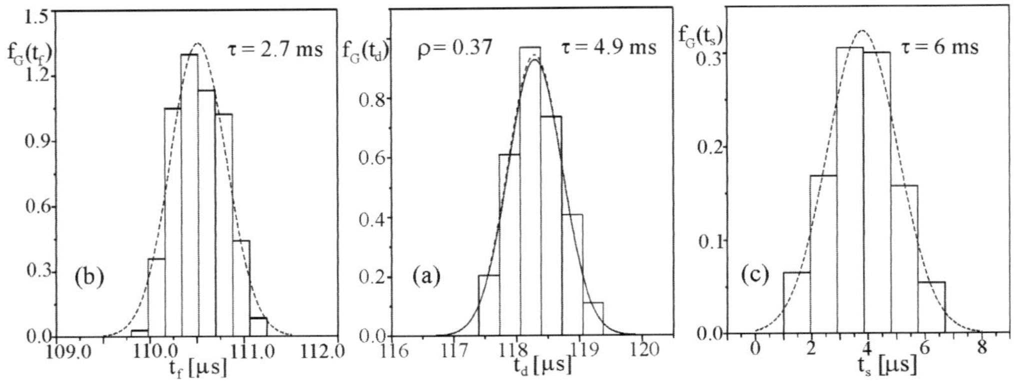

Figure 6. The Gaussian density distributions of: (a) the breakdown time t_d (solid line – fit based on equation (4), dashed line – the analytical expression (6)); (b) the formative time t_f (dashed line – the analytical expression (7)); (c) the statistical time t_s (dashed line – the analytical expression (8)).

Going to shorter relaxation times τ, the statistical time delay and its dispersion become negligible compared to the formative time dispersion due to the high level of residual ionization, and the experimental Gaussian density distribution $f_G(t_f)$ of the formative time at τ=2.7ms is obtained (figure 6b). It is fitted by an analytical Gaussian distribution (the dashed line):

$$f_G(t_f)=\frac{1}{\sqrt{2\pi}\,\sigma_{tf}}\exp\left[-\frac{(t_f-\overline{t_f})^2}{2\sigma_{tf}^2}\right], \tag{7}$$

with the parameters $\overline{t_f}$=110.5μs and σ_{tf}=0.30μs. The $f_G(t_f)$ density in figure 6b can be fitted by the theoretical distribution (equation (4)) with neglecting terms with σ_{ts} and assumptions $\overline{t_s} << \overline{t_f}$ and $t_d \approx t_f$. Moreover, the t_f Gaussian distribution (figure 6b) is obtained for $Y_{eff} \gtrsim 10^8\,s^{-1}$, similarly to the t_f Gaussian distributions in neon [19] obtained for $Y_{eff} \gtrsim 10^9\,s^{-1}$ (or $\sigma_{ts} \approx 1/Y_{eff} \sim ns$). On the other hand, for $\tau \gtrsim 6ms$, the formative time standard deviation can be neglected, i.e. $\sigma_{td} \approx \sigma_{ts}$ and the experimental distribution functions describe the statistical time delay [12]. The Gaussian density function $f_G(t_s)$ obtained at τ=6ms from the measured time delay data by subtracting $t_f \approx t_{d,\min}$=120μs is presented in figure 6c. Also, it is fitted by an analytical Gaussian distribution of the form:

$$f_G(t_s)=\frac{1}{\sqrt{2\pi}\,\sigma_{ts}}\exp\left[-\frac{(t_s-\overline{t_s})^2}{2\sigma_{ts}^2}\right], \tag{8}$$

with the fitting parameters $\overline{t_s}$=3.8μs and σ_{ts}=1.25μs (the dashed line). From relation $\overline{t_s}=\kappa\cdot\sigma_{ts}$ [18], $\kappa\approx3$ is obtained, which confirms that the Gaussian distribution in nitrogen (figure 6a) refers to the statistical time delay. Also, it was shown [12] that the statistical time delay in nitrogen has a Gaussian density distribution for $Y_{eff} \gtrsim 10^6 s^{-1}$ which is now confirmed in the case of the t_s distribution presented in figure 6c. As in case of t_f distribution, the $f_G(t_s)$ density in figure 6c can by fitted by the theoretical distribution (equation (4)) with neglecting terms with σ_{tf} and the assumption $t_f \approx \overline{t_f} \approx t_{d,\min}$.

6. Conclusion

The observed deviation of *I-U* characteristic is related to the changes of the secondary electron yield with the cathode surface conditioning. 2D discharge images clearly show significant change of the discharge structure and development of non-linear phenomena caused by space charge effects. The experimentally obtained new distributions of statistical time delay (Gauss and Gauss-exponential) based on binomial distribution for electron occurrence were presented. The experimental density distribution function of breakdown time delay t_d was modeled by the theoretical distribution of the sum of two random variables, t_s and t_f, based on the bivariate Gaussian density distribution of the dependent statistical and formative time delay. The value $\rho=0.37$ of the correlation coefficient between t_s and t_f was determined from the measured values of the standard deviations. In the limiting cases, the Gaussian density distribution functions of the formative and statistical time delay were experimentally obtained and successfully modeled.

Acknowledgments

The author expresses gratitude to professors V Lj Marković and Z Lj Petrović for their contributions to this study and to S Stamenković, D Marić and N Škoro for the help regarding the measurements. The work was partially supported by MNTR of Serbia (project 141025).

References

[1] Meek J M and Craggs J D 1978 *Electrical Breakdown of Gases* (John Wiley&Sons, Chichester)
[2] Phelps A V and Petrović Z Lj 1999 *Plasma Sources Sci. Technol.* **8** R21
[3] Marković V Lj, Gocić S R, Stamenković S N, Petrović Z Lj and Radmilović M 2001 *Eur. Phys. J. AP* **14** 171
[4] Phelps A V, Petrović Z Lj and Jelenković B M 1993 *Physical Review E* **47** 2825.
[5] Jelenković B M and Phelps A V 1987 *Physical Review A* **36** 5310
[6] Marić D, Malović G and Petrović Z Lj 2009 *Plasma Sources Sci. Technol.* **18** 034009
[7] Zuber K 1925 *Ann. Phys.* **381** 231
[8] von Laue M 1925 *Ann. Phys.* **381** 261
[9] Kiselev Ju 1965 *Proc. 7th Int. Conf. on Phenomena of Ionized Gases* (Belgrade, Yugoslavia) 838 (in Russian)
[10] Maluckov Č A, Karamarković J P, Radović M K and Pejović M M 2004 *Phys. Plasmas* **11** 5328
[11] Maluckov Č A, Karamarković J P, Radović M K and Pejović M M 2006 *IEEE Trans. Plasma Sci.* **34** 2
[12] Marković V Lj, Gocić S R and Stamenković S N 2006 *J. Phys. D: Appl. Phys.* **39** 3317 (and references theirin)
[13] Gocić S R , Marković V Lj and Stamenković S N 2009 *J. Phys. D: Appl. Phys.* **42** 212001 (and references theirin)
[14] Malović G, Strinić A, Živanov S, Marić D, Petrović Z Lj 2003 *Plasma Sources Sci. Technol.* **12** S1
[15] Nikolić M M, Dordević A R, Stefanović I, Vrhovac S and Petrović Z Lj 2003 *IEEE Trans. Plasma Sci.* **31** 717
[16] Marković V Lj, Petrović Z Lj and Pejović M M 1997 *Plasma Sources Sci. Technol.* **6** 240
[17] Bronstein I N and Semendjajev K A 1979 *Taschenbuch der Mathematik.* (Teubner Verlagsgesellschaft, Leipzig)
[18] Marković V Lj, Petrović Z Lj and Pejović M M 1994 *J. Chem. Phys.* **100** 8514
[19] Marković V Lj, Gocić S R and Stamenković S N 2009 *J. Phys. D: Appl. Phys.* **42** 015207

25th Summer School and International Symposium on the Physics of Ionized Gases—SPIG 2010 IOP Publishing
Journal of Physics: Conference Series **257** (2010) 012023 doi:10.1088/1742-6596/257/1/012023

New models and distributions of the electrical breakdown time delay in neon

S N Stamenković

Department of Physics, University of Niš, P.O. BOX 224, 18001 Niš, Serbia

ssuzana@pmf.ni.ac.rs

Abstract. The measurements of the electrical breakdown time delay t_d for a wide range of working voltages and at different preionization levels are presented. The statistical breakdown time delay t_s and the discharge formative time t_f are experimentally separated and theoretical models of their dependencies on the overvoltage and number densities of residual charges during relaxation are suggested. Several empirical and semiempirical models are used to describe the formative time delay dependence on working voltages $t_f(U)$. The empirical and theoretical models from the literature are also applied to the experimental data, without and with empirical corrections. Moreover, several new distributions are experimentally obtained: Gauss-exponential, Gaussian and double Gaussian ones for the statistical time delay, as well as Gaussian and double Gaussian distributions for the formative time. The measurements of the breakdown time delay at different preionization levels (afterglow periods) $t_d(\tau)$ obtained with a galvanic layer of gold and a sub-layer of nickel on the copper cathode are compared to the measurements with a vacuum deposited gold layer on the cathode surface. It was found that the surface charges retaining on a galvanic layer of gold influence the breakdown time delay which leads to double Gaussian distributions of the formative and statistical time delay.

1. Introduction

It is well known that the electrical breakdown in gases does not occur at the moment of voltage application to the gas tube, but with some delay. A time interval between the moment of application of voltage greater than the static breakdown voltage U_s and the electrical breakdown is the breakdown time delay t_d [1]. It has a stochastic character, which was experimentally proved by Zuber [2] and its exponential statistical distribution was theoretically derived by von Laue [3]. The breakdown time delay comprises the statistical t_s and the formative time delay t_f, i.e. $t_d = t_s + t_f$. The time that elapses from the application of working voltage U_w ($U_w > U_s$) to the appearance of free electron(s) initiating breakdown is the statistical time delay and from this moment to the collapse of the applied voltage and occurrence of a self-sustained current is the formative time delay. The best understanding of breakdown processes may be obtained by combining breakdown time delay measurements and the appropriate theoretical models.

In this paper, several empirical models for the formative time delay dependence on working voltages $t_f(U)$ are suggested: models based on power functions, exponential functions and the product of power and exponential functions. A few semiempirical models are presented such as models without and with singularity at the static breakdown voltage U_s. The theoretical models from the literature are also applied but good agreements with experimental data are found only at low overvoltages and therefore empirical corrections are introduced [4].

The measurements at different preionization level were carried out and Gaussian and double Gaussian distributions for the formative time were obtained, while for the statistical time delay three characteristic (Gaussian, Gauss-exponential and exponential) distributions were identified. The measurements of the breakdown time delay at different afterglow periods $t_d(\tau)$ obtained with a galvanic layer of gold on the copper cathode are compared to the measurements with a vacuum deposited gold layer on the cathode [5,6]. It was found that the surface charges retaining on a galvanic layer of gold influence the breakdown time delay. The surface charges mask the exponential Ne_2^+ decay in afterglow, as well as, the conversion maximum due to molecular nitrogen ions production in Ne_2^+ collisions with nitrogen impurities. The formative and statistical time delay distributions are double Gaussians due to combined effects of surface and gas-phase charges from the preceding glow. Presence of surface regions with reduced conductivity on a galvanic layer of gold is confirmed by scanning electron microscopy (SEM) images and energy dispersive X-ray (EDX) spectrum.

2. Experimental details

The breakdown time delay measurements were carried out on gas tubes made of borosilicate glass (8245, Shott technical glass) with volume of $V \approx 300cm^3$. One group of measurements was performed on the gas tube with copper cathode, gold plated by a vacuum sputter deposition ($U_s = 271V$) while the other group of measurements was performed on the gas tube whit a $0.5\mu m$ hard galvanic layer of gold with $7\mu m$ sub-layer of nickel on the copper cathode ($U_s = 197V$). The electrodes were cylindrical, with the diameter $D = 6mm$ and the interelectrode distance was $d = 6mm$. The tubes were filled with research purity neon at the pressure of $13.3mbar$ (Matheson Co. with a nitrogen impurity below $1ppm$). The time delay measurements were carried out by applying step pulses to the tubes at glow current $I_g = 0.1mA$, glow time $t_g = 1s$ and at different working voltages U_w and afterglow periods τ. The measurements were performed by using an electronic automatic system which achieved the voltage rise time and resolution limit of $0.18\mu s$. More details about the experimental procedure and measuring system can be found in [7].

3. Empirical and semiempirical models of $t_f(U)$ dependence

In this section, several empirical and semiempirical models were applied for modelling the experimental formative time delay dependence on overvoltages $t_f(U)$ obtained on the gas tube with a vacuum deposited gold layer on the cathode. The formative time delay was determined in three different ways [8]: from the Laue diagrams where linear graph of $\ln(n/N)$ intersects the time delay axis; from histograms, as the minimum value of t_d and from a difference $t_f = \overline{t_d} - \overline{t_s} \approx \overline{t_d} - \sigma(t_d)$, where standard deviation $\sigma(t_d)$ is approximately equal to the mean value of statistical time delay $\overline{t_s}$.

It is known that t_f decreases when overvoltages increase, i.e. $t_f \propto 1/\Delta U$ [9]. Applied to our experimental data it takes the form:

$$t_f = \frac{7500}{U - U_s} \; [\mu s], \tag{1}$$

where $U_s = 271V$ and fit is shown by dashed line in figure 1. The better agreement at higher overvoltages is achieved with relation (figure 1, solid line):

$$t_f = \frac{3700}{(U - U_s)^{0,7}} \; [\mu s], \tag{2}$$

as an improvement of relation (1). Also, the fitting relation [10]:

$$t_f = \frac{270}{\alpha - \alpha_s} \; [\mu s], \tag{3}$$

could be used (figure 2, dashed line), where α is the electron ionization coefficient and $\alpha_s = 6.42 cm^{-1}$ is the value of ionization coefficient at the static breakdown voltage U_s [4]. A better description of the experimental $t_f(U)$ dependence than with relation (3) is obtained with:

$$t_f = \frac{330}{(\alpha - \alpha_s)^{0,7}} \; [\mu s], \tag{4}$$

and this fit is presented by solid line in figure 2.

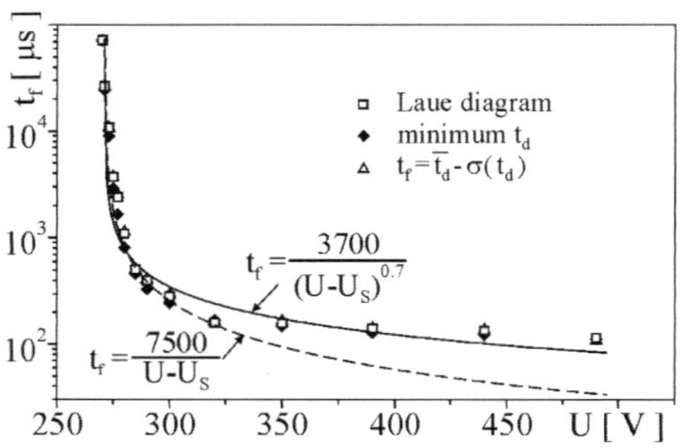

Figure 1. The formative time delay (symbols) with the fits based on relations (1) (dashed line) and (2) (solid line).

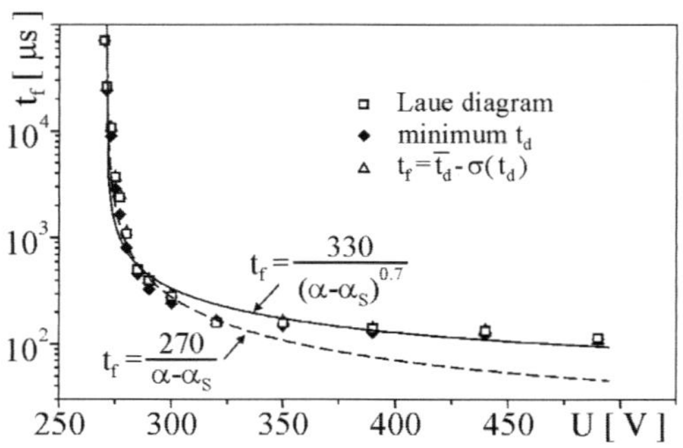

Figure 2. The formative time delay (symbols) with the fits based on relations (3) (dashed line) and (4) (solid line).

The next empirical relations can also be used to describe the experimental data:

$$t_f = 120 e^{20/(U-268)} \; [\mu s], \tag{5}$$

$$t_f = \frac{2700}{U - U_s} e^{0.0045U} \; [\mu s], \tag{6}$$

$$t_f = \frac{1.6 \times 10^5}{U - U_s} e^{-900/U} \; [\mu s]. \tag{7}$$

Appropriate fits based on relations (5), (6) and (7) are shown in figure 3 by dotted, dashed and solid lines, respectively.

The early stage of temporal growth of electron $n_e(t)$ and ion number density $n_i(t)$ when electron-ion recombination can be neglected is expressed by:

$$\frac{dn_e}{dt} = \frac{dn_i}{dt} = \alpha \, w_e \, n_e , \tag{8}$$

where w_e is the electron drift velocity. Taking into account the boundary condition at the cathode, an approximate solution for the ion number density is:

$$n_{it} \approx n_{i0} e^{\alpha \gamma w_i t} , \tag{9}$$

where n_{it} is the ion number density in Townsend's dark discharge before the collapse of the applied voltage, n_{i0} is the initial number density of ions, γ is the secondary electron yield and w_i is the ion drift velocity [4]. From relation (9), the formative time delay can be derived as:

$$t_f \approx \frac{\ln (n_{it} / n_{i0})}{\alpha \, \gamma \, w_i} . \tag{10}$$

The best fit of experimental data is shown by dashed line in figure 4, but, as could be seen from the figure, this semiempirical model is applicable at relatively high overvoltages, far from the singularity at U_s. The singularity ($t_f \to \infty$) at $U \to U_s$ could be introduced into the previous model, if we define an approximate preionization level at the static breakdown voltage n_{is}. Then, the relation (10) takes the following form:

$$t_f \approx \frac{\ln(n_{it}/n_{is})}{(\alpha-\alpha_s)\gamma w_i}, \tag{11}$$

and the best fit is presented in figure 4 by solid line.

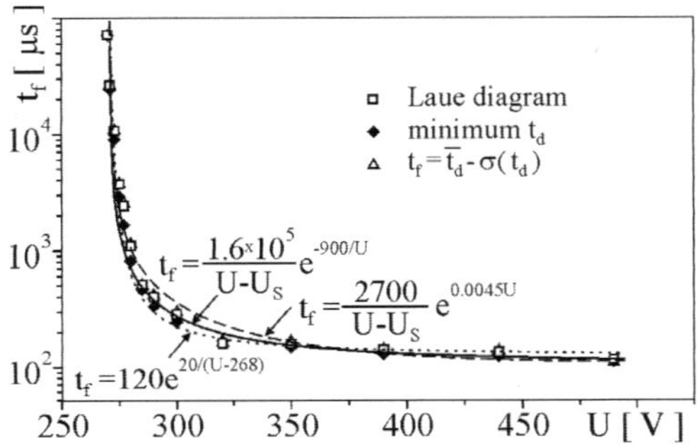

Figure 3. The formative time delay (symbols) with the fits based on relations (5) (dotted line), (6) (dashed line) and (7) (solid line).

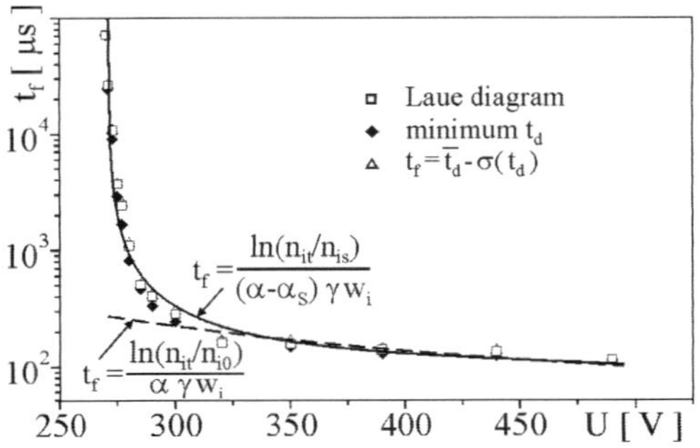

Figure 4. The formative time delay (symbols) with the fits based on relations (10) (dashed line) and (11) (solid line).

In the figure 5 the fit based on the model from Schade [11] is shown together with our empirical corrections. The Schade's model:

$$t_f = \frac{a(U)}{U-U_s}e^{b/U}, \tag{12}$$

whit parameters $a(U)=\dfrac{U_s^2}{AB\,p^2 d\,w_i}\ln\left[1+\dfrac{AB\,p^2(U-U_s)}{U_s^2}\exp\left(-\dfrac{B\,pd}{U}\right)\left(\dfrac{i}{i_0}\right)\right]$ and $b = B\,pd$ describes the formative time only at low overvoltages (dotted line in figure 5). Therefore, we made empirical corrections and the suggested fits of $t_f(U)$ with corresponding $a(U)$ dependencies are shown in the figure 5. The best agreement is obtained with empirical $a(U)$ dependence that decreases exponentially at low overvoltages and increases linearly at high overvoltages (solid lines in figure 5) [4]. Another model from the literature [12-14] is applied to experimental data and appropriate fit together with our correction are shown in the figure 6. We have used the expression for the temporal growth of ion current and derived the formative time in our conditions [4]:

$$t_f = \frac{1}{\lambda}\ln\left[\frac{\alpha-\lambda/w_i}{\alpha}\times[e^{\alpha d}-1-(1-q)(\frac{i_+}{i_0})][e^{(\alpha-\lambda/w_i)d}-1]^{-1}\right], \tag{13}$$

where i_+ and i_0 are ion and initial ion currents, while λ is the growth parameter which is determined from the characteristic equation:

$$1=\gamma\frac{\alpha}{\alpha-\lambda/w_i}\left[e^{(\alpha-\lambda/w_i)d}-1\right]. \tag{14}$$

Davidson's model with linear growth parameter is a good representation of experimental data only at low overvoltages (dashed lines in figure 6) and we made empirical correction again i.e. we introduced the exponential growth parameter (solid lines in figure 6) that describes the experimental data for all overvoltages. In general, the good agreement of models from the literature with experimental data was found at low overvoltages, while at high overvoltages the empirical corrections are necessary.

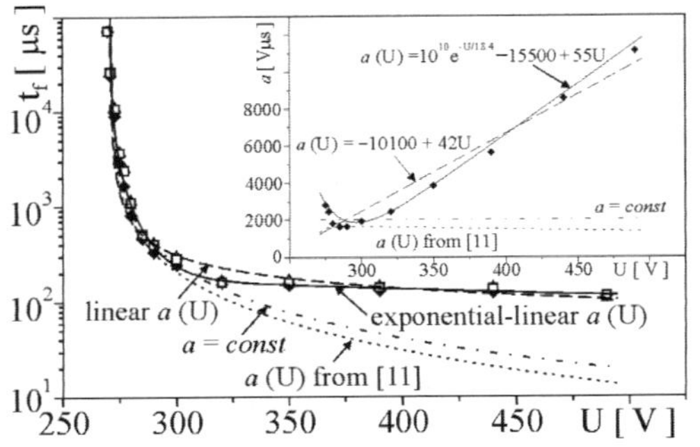

Figure 5. The formative time delay (symbols) with the fits based on model from [11]; inset a(U) dependences.

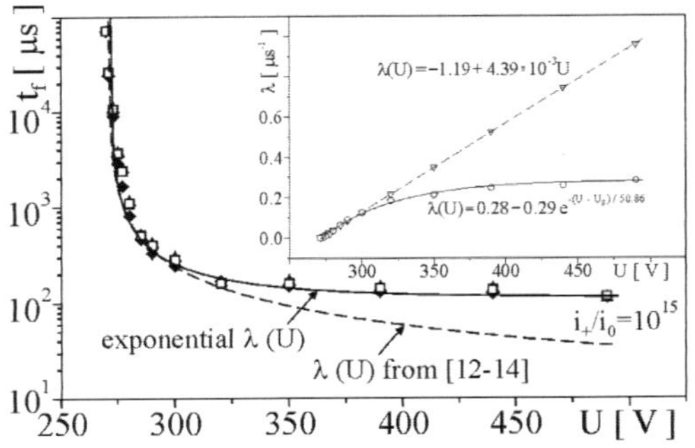

Figure 6. The formative time delay (symbols) with the fits based on model from [12-14]; inset λ(U) dependences.

4. Distributions of formative and statistical time delay

The $\overline{t_d}(\tau)$ dependences (the memory curves), as well as the standard deviations $\sigma_{td}(\tau)$ are shown in figure 7. The measurements carried out on the gas tube with a galvanic gold layer on the cathode (×,*), were compared to the measurements performed on the gas tube with a vacuum deposited gold layer (□,∇). In the case of the cathode with galvanic layer of gold and sub-layer of nickel, nickel atoms from sub-layer diffuse into the gold layer causing defects i.e. surface regions with reduced conductivity. Presence of these regions on a galvanic gold layer that retain the surface charges is confirmed by scanning electron microscopy (SEM) images and energy dispersive X-ray (EDX) spectrum (Figure 8).

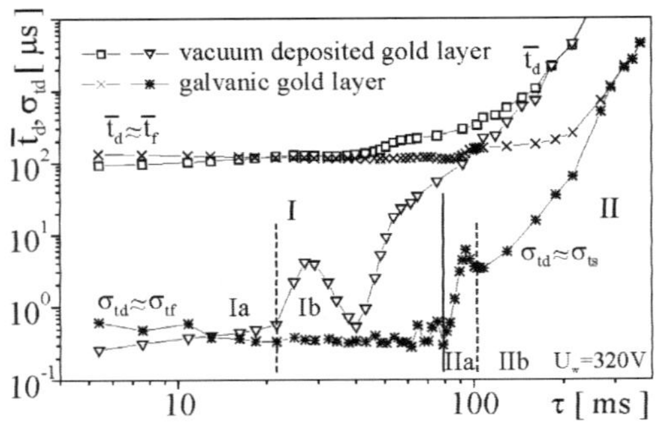

Figure 7. The breakdown time delay t_d and the standard deviations σ_{td} in neon.

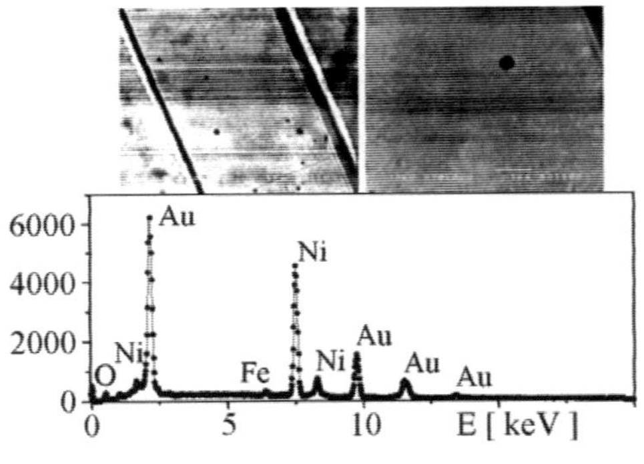

Figure 8. SEM images and energy dispersive X-ray (EDX) spectrum.

In the region Ia, experimental distributions of the formative time are obtained providing that the statistical time delay and its fluctuations σ_{ts} are reduced to a negligible values [5,6]. It was found for vacuum deposited gold layer that the formative time delay in the ionic region Ia increases linearly with the afterglow period ($\overline{t_f} \propto \tau$) (figure 7, □) consistent with an exponential Ne_2^+ decay. The standard deviation of the formative time also increases linearly with τ, but faster than t_f (figure 7, ∇) [6]. The experimental distributions of the formative time delay for a vacuum deposited gold layer in this region are Gaussian (shaded histograms, figure 9) and they are fitted by Gaussian distribution functions [6].

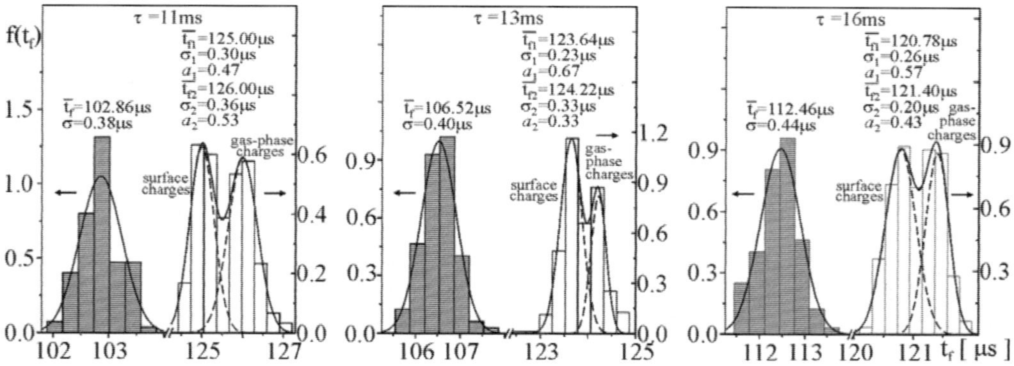

Figure 9. The double-Gaussian density distribution functions of t_f measured at the gas tube with a galvanic gold layer on the cathode surface compared to those of the gas tube with a vacuum gold layer (shaded distributions) in the region Ia at *U=320V*.

The charged particle decay is followed by the simultaneous conversion (Ib) of molecular neon ions to nitrogen ions, originating from nitrogen present as impurities $Ne_2^+ + N_2 \rightarrow N_2^+ + 2Ne$. The appearance of nitrogen ions is manifested by double-Gaussian distributions (shaded histograms, figure 10) and the experimental distributions are fitted by double-Gaussian distribution function [6]:

$$f_{GG}(t_i) = \frac{a_1}{\sqrt{2\pi}\,\sigma_1}\exp[\frac{-(t_i - \overline{t_{i1}})^2}{2\sigma_1^2}] + \frac{a_2}{\sqrt{2\pi}\,\sigma_2}\exp[\frac{-(t_i - \overline{t_{i2}})^2}{2\sigma_2^2}]. \tag{15}$$

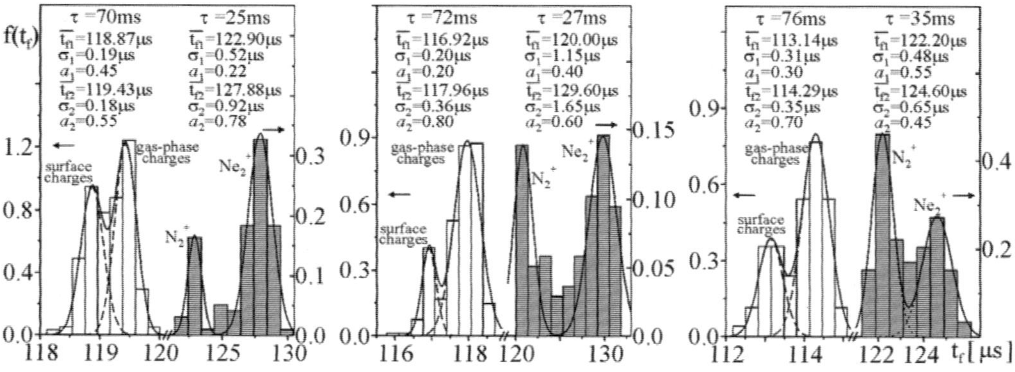

Figure 10. The double-Gaussian density distribution of t_f measured at the gas tube with a galvanic gold layer on the cathode compared to those of the gas tube with a vacuum gold layer (shaded distributions) in the region Ib at *U=320V*.

In the case of a galvanic gold layer on the copper cathode, the situation is different (figure 7, ×, ∗). The formative time delays and their standard deviations in the early afterglow region of the memory curve are almost flat and lengthened to about $60ms$. The exponential Ne_2^+ decay in afterglow (confirmed experimentally and theoretically for a vacuum deposited gold layer [6]) is masked by the surface charges on the cathode surface. Also, the conversion maximum, obtained in the region Ib for the gas tube with vacuum deposited gold layer, is significantly reduced and shifted to greater afterglow times. In the case of galvanic gold layer, the formative time distributions are double-Gaussian due to combined effect of surface and gas-phase charges (figures 9 and 10). The first Gaussians correspond to surface charges involved in discharge formation, while the second Gaussian distributions (shifted to greater formative times) correspond to gas-phase charges remained from the preceding glow. The experimental distributions of the formative time are fitted by double-Gaussian distribution functions (15) and appropriate parameters are shown in figures 9 and 10.

With regard to the statistical time delay, its experimental distributions are obtained in region II of the memory curves (figure 7), where the time delay distributions are dominated by the fluctuations of the statistical time, i.e. $\sigma_{td} \approx \sigma_{ts}$. The shapes of t_s distributions depend on electron yields Y originating from residual ionization and nitrogen atoms. The analysis shows that distributions of the statistical time delay t_s, for a vacuum deposited gold layer, change from Gaussian and Gauss-exponential to exponential distribution due to influence of residual ionization [6].

In the case of a galvanic gold layer on the cathode, surface charges affect the statistical time delay distributions. The influence of the surface charges on the experimental t_s distributions is shown in the figure 11. The appearance of the surface charge initiation is manifested by double-Gaussian distributions obtained in the first part of the region II. The experimental distributions of the statistical time delay are fitted by double-Gaussian distribution function (15) (solid lines in figure 11) and appropriate parameters are shown in the figure 11. The first Gaussians in the figure are related to the surface charge initiation, while the second ones to the gas-phase charge initiation because the ions from the gas-phase take a time drifting to the cathode after the application of voltage. Based on the distributions of the statistical time delay shown in the figure 11, it can be concluded that influence of the surface charge initiation decreasing. After the exhausting of surface charges, the breakdown initiation is dominated by the decay of gas-phase charges (region IIb), i.e. by N_2^+ ions present as impurities in neon. From the slope of the $\overline{t_s}(\tau)$ dependence in region IIb, the loss frequency of dominant species ($\nu \cong 30s^{-1}$) and the diffusion coefficient $D \approx 35cm^2s^{-1}$ could be determined [7]. The last value is in a good agreement with the value from the literature obtained for the ambipolar diffusion coefficient of N_2^+ ions in neon [15]. In this part of the memory curve, the three characteristic distributions of the statistical time delay are obtained when the afterglow period increases: Gaussian, Gauss-exponential and exponential distribution (figure 12). The analytic forms of the density distribution functions used for the fitting of the corresponding experimental distributions are also shown in the figure 12.

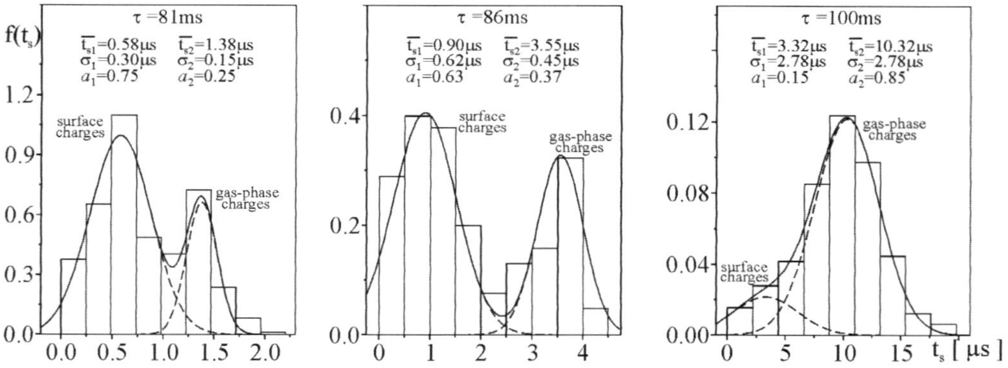

Figure 11. The double-Gaussian density distribution functions of t_s in the region IIa for the gas tube with a galvanic gold layer on the cathode.

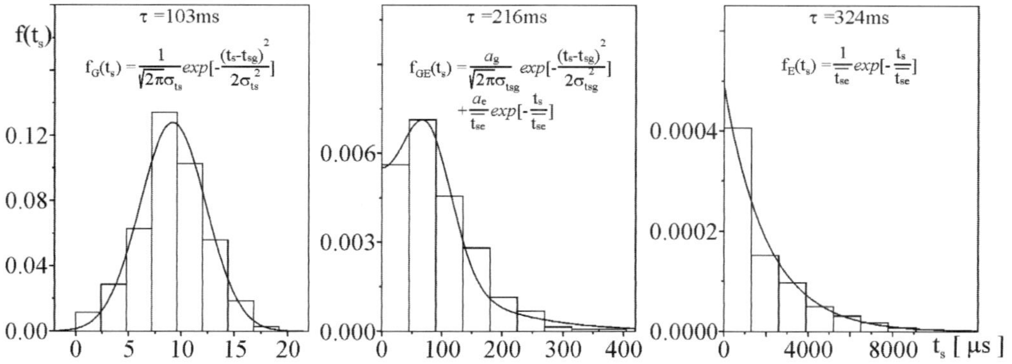

Figure 12. The three characteristic distribution functions of t_s in the region IIb of figure 7: Gaussian, Gauss-exponential and exponential distribution function.

5. Conclusion

In this study, several empirical and semiempirical models are applied for description of the experimental formative time delay dependence on working voltages in neon. Applied models from the literature agree with experimental data only at low overvoltages. Therefore, we made empirical corrections that give good agreement with the experimental data over the whole range of overvoltages. Moreover, the statistical breakdown time delay t_s and the formative time t_f are experimentally separated and several new distributions are experimentally obtained: Gauss-exponential, Gaussian and double Gaussian ones for the statistical time delay, as well as Gaussian and double Gaussian distributions for the formative time.

Acknowledgements

I would like to express my thanks to professor V Lj Marković who contributed to this study. The work was partial supported by MNZŽS of Serbia (project 141025).

References

[1] Morgan C G: 1978, in *Electrical Breakdown of Gases*, ed. J M Meek, J D Craggs (John Wiley & Sons, Chichester).
[2] Zuber K: 1925, *Ann. Phys.* **381**, 231.
[3] von Laue M: 1925, *Ann. Phys.* **381**, 261.
[4] Stamenković S N, Marković V Lj and Gocić: 2009, *Eur. Phys. J: Appl. Phys.* **45**, 11003.
[5] Marković V Lj, Stamenković S N, Gocić S R and Đurić S M: 2007, *Eur. Phys. J: Appl. Phys.* **38**, 73.
[6] Marković V Lj, Gocić S R and Stamenković S N: 2009, *J. Phys. D: Appl. Phys.*, **42**, 015207.
[7] Marković V Lj, Gocić S R, Stamenković S N and Petrović Z Lj: 2007, *Phys. Plasmas*, **14**, 103504.
[8] Marković V Lj, Petrović Z Lj and Pejović M M: 1997, *Plasma Sources Sci. Technol.* **6**, 240.
[9] Gänger B: 1953, *Der elektrische Durchschlag von Gasen* (Springer-Verlag, Berlin).
[10] Deutsch F: 1965, *Proc. VII Int. Conf. Phenomena in Ionized Gases*, ed. B. Perović, D. Tošić (Institute of Nuclear Sciences, Belgrade, Yugoslavia) 313
[11] Schade R: 1937, *Z. Phys.* **104**, 487.
[12] Dutton J, Haydon S C and Jones F L, with math. appendix by Davidson P M: 1953, *Brit. J. Appl. Phys.* **4**, 170.
[13] Davidson P M: 1955, *Phys. Rev.* **99**, 1072.
[14] Morgan C G: 1956, *Phys. Rev.* **104**, 566.
[15] Märk T D and Oskam H J: 1971, *Z. Phys. Rev.* **247**, 84

Line profile variations in selected Seyfert galaxies

W. Kollatschny, M. Zetzl, K. Ulbrich

Institut für Astrophysik, Universität Göttingen, Friedrich-Hund Platz 1, D-37077 Göttingen, Germany

E-mail: wkollat@astro.physik.uni-goettingen.de

Abstract. Continua as well as the broad emission lines in Seyfert 1 galaxies vary in different galaxies with different amplitudes on typical timescales of days to years. We present the results of two independent variability campaigns taken with the Hobby-Eberly Telescope. We studied in detail the integrated line and continuum variations in the optical spectra of the narrow-line Seyfert galaxy Mrk 110 and the very broad-line Seyfert galaxy Mrk 926. The broad-line emitting region in Mrk 110 has radii of four to 33 light-days as a function of the ionization degree of the emission lines. The line-profile variations are matched by Keplerian disk models with some accretion disk wind. The broad-line region in Mrk 926 is very small showing an extension of two to three light-days only. We could detect a structure in the rms line-profiles as well as in the response of the line profile segments of Mrk 926 indicating the BLR is structured.

1. Introduction

Seyfert galaxies belong to the class of active galaxies or active galactic nuclei (AGN). They are emitting a nuclear nonthermal continuum emission that cannot be caused by stellar sources. Seyfert galaxies are less luminous than the quasars which show nearly identical emission line spectra. However, in most cases Seyfert galaxies are close-by and therefore they can be studied in more detail. All spiral and elliptical galaxies contain a supermassive black hole of the order of 10^6 to 10^{10} solar masses in the center. The central luminosity in active galaxies is probably caused by accretion of material from the outer regions. Many details of the transformation process from kinetical energy into radiation energy are still not understood.

The line emitting black hole is surrounded by the broad emission-line region - the innermost line emitting region in AGN. The emission lines are caused by photoionization of the central nuclear region. The broad-line region (BLR) has an extension of light-days to light-months and can not be resolved spatially on direct images.

Studying the variability of the broad emission lines - caused by the central variable source - can give us information on the extension and structure of the broad-line region. This is based on the delayed response of the integrated line intensities with respect to the central ionizing source.

For a set of about 35 AGN reliable distances of the Hβ line emitting regions exist [1]. Based on these distances in combination with assumptions regarding the rotation velocities of the broad-line clouds it is possbile to derive the central black hole masses.

A study of the variations of the emission line profiles can give us information on the kinematics of the broad-line region: e.g. a faster response of the individual line wings gives us information on inflow or outflow motions. A comparison of the chronology of line-profile variations with theoretical model calculations gives us the possibility to discriminate betwen radial, turbulent

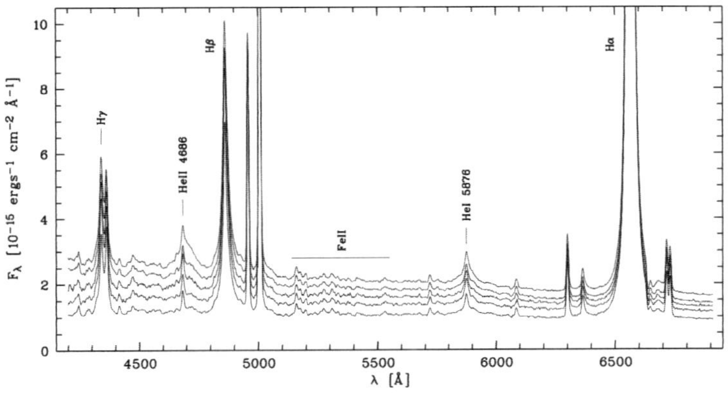

Figure 1. Five spectra of the variable narrow-line Seyfert galaxy Mrk 110. These spectra have taken with the 10m Hobby-Eberly Telescope at McDonald observatory with a typical time interval of one month (from [2])

or Keplerian motions. However, only excellent spectra with high signal/noise ratio can be used for this kind of study.

Detailed reverberation studies exist for a handful of galaxies only. Here we want to present line-profile studies of the the narrow-line Seyfert 1 galaxy Mrk 110 and of the broad-line Seyfert gakaxy Mrk 926.

2. Emission line variability in Mrk 110

Mrk 110 is a narrow-line Seyfert galaxy with a Hβ line width of 1400 km/s. We studied their broad-line variability with the 10m Hobby-Eberly Telescope (HET) at McDonald Observatory over a period of 7 months ([2]) with a typical interval of 5 days. Fig. 1 shows some optical spectra of Mrk 110 taken with the 10m Hobby-Eberly Telescope at McDonald observatory.

Light curves we derived from our variability campaign of Mrk 110 are given in Fig. 2. Shown are the light curves of the continuum flux at 5135 Å (in units of 10^{-15} erg cm^{-2} s^{-1} Å^{-1}) and of the integrated emission line fluxes of Hγ, Hβ, Hα, HeIIλ4686, HeIλ5876, HeIλ5016, and HeIλ4471 (in units of 10^{-15} erg cm^{-2} s^{-1}).

In a next step we calculated cross-correlation functions of the 5135 Å continuum light curve with the emission line light curves. The delays τ of the integrated HeII, HeI and Balmer lines with respect to continuum variations are presented in Table 1. The time lags can be interpreted as the light-travel time across the emission line region for the different broad emission lines. The size of the line emitting region corresponds to 4 to 33 light days for the individual lines (Table 1). To get some information about the velocity field of the line emitting region we derived the widths of the broad emission line profiles (FWHM) from the rms spectra of the variability campaign. The rms spectra are free of contamination from the superimposed narrow emission lines. These narrow emission lines originate at radii of 100 to 1000 pc from the center and remain constant over timescales of years. The time lags of the strongest Balmer and Helium lines in Mrk 110 are plotted as a function of their FWHM in Fig. 4.

The line widths and time lags are clearly correlated. One can interprete the light-travel time as the characteristic distance R of the line emitting region and the FWHM of the rms emission line as the characteristic velocity v of the line emitting clouds. Those emission lines that originate closer to the central black hole exhibit broader emission line widths.

It is possible to determine the central black hole mass under the assumption of Keplerian

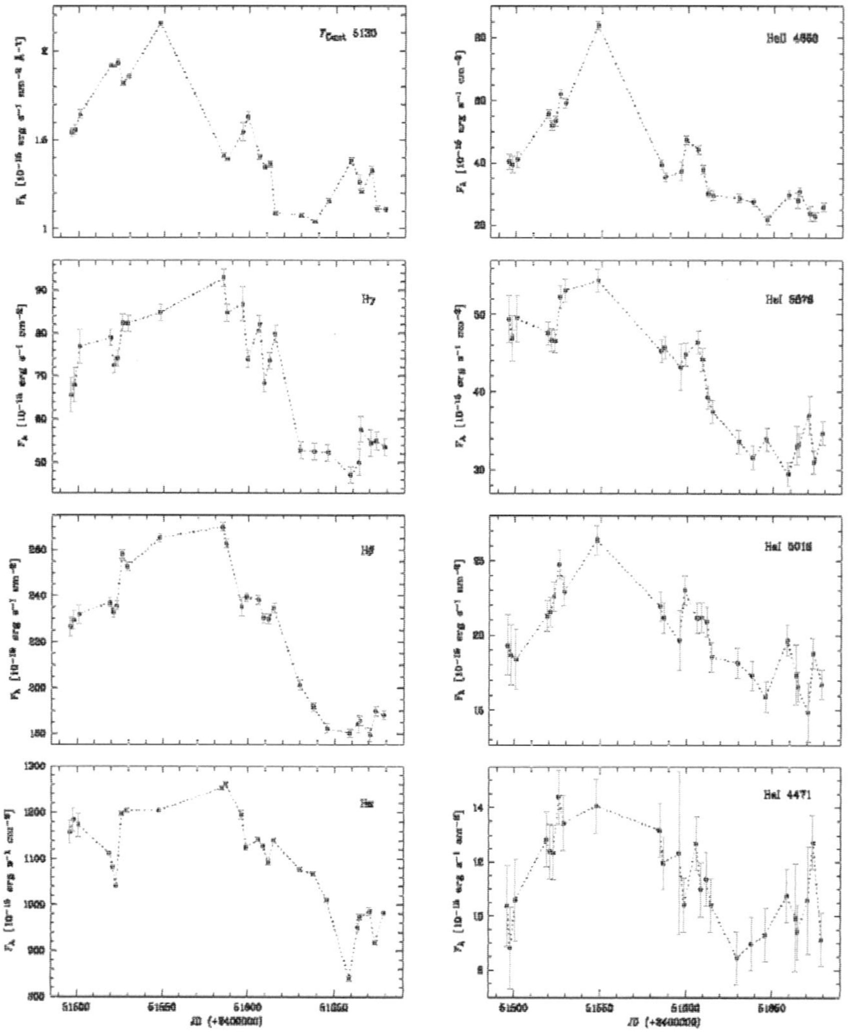

Figure 2. Light curves of Mrk 110 for the continuum flux at 5135 Å (in units of 10^{-15} erg cm^{-2} s^{-1} Å^{-1}) and for integrated emission line fluxes of Hγ, Hβ, Hα, HeIIλ4686, HeIλ5876, HeIλ5016, and HeIλ4471 (in units of 10^{-15} erg cm^{-2} s^{-1}). The points are connected by a dotted line to aid the eye (from [2]).

orbits of the line emitting clouds:

$$M_{\text{orbital}} = f v^2 G^{-1} R.$$

Here is v the characteristic emission-line velocity (rms line width), G the gravitational constant, and $R = \tau c$ the distance of the line emitting region from the center (τ = calculated cross-correlation lag of line emitting region).

From all the emission lines we calculated a mean black hole mass of $1.8 \cdot 10^7 M_\odot$ in Mrk 110 (Table 1, [2]).

Table 1. Rms line widths (FWHM) of the strongest emission lines: Hα, Hβ, HeIλ5876, and HeIIλ4686; cross-correlation lags τ and virial mass estimations M of the central black hole on Mrk 110

Line (1)	FWHM(rms) [km s^{-1}] (2)	τ [days] (3)	M [$10^7 M_\odot$] (4)
HeII	4444 ± 200	3.9 ± 2	2.25 ± 3
HeI	2404 ± 100	10.7 ± 6	1.81 ± 4
Hβ	1515 ± 100	24.2 ± 4	1.63 ± 4
Hα	1315 ± 100	32.3 ± 5	1.64 ± 5

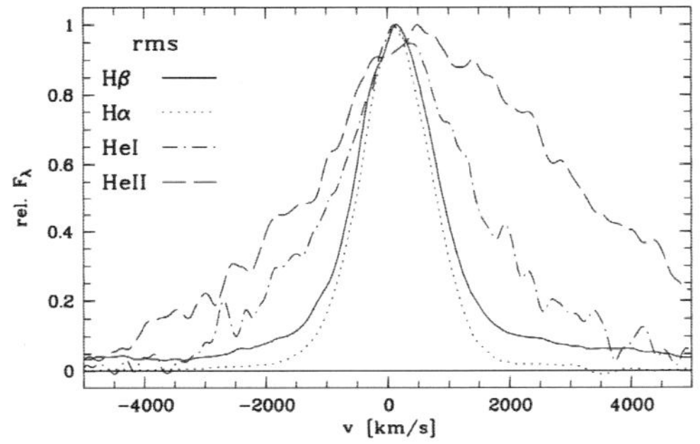

Figure 3. Normalized Balmer, HeI and HeII lines of Mrk 110 in velocity space(from [3]).

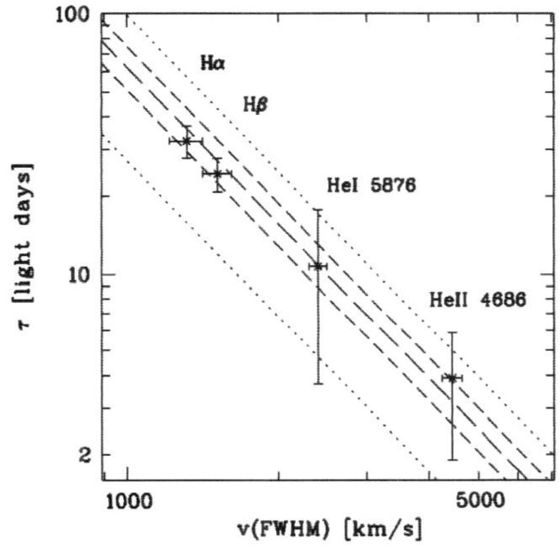

Figure 4. Distance of the Balmer and Helium emitting line regions from the central ionizing source in Mrk 110 as a function of FWHM derived from their rms line profiles. The dotted and dashed lines are model calculations for central masses of 0.8, 1.5, 1.8, 2.2, and 2.9 $\cdot 10^7 M_\odot$ (from bottom to top). (from [3]).

The relation between radius and velocity for different black hole masses is shown in Fig. 4. The dotted and dashed lines correspond to virial masses of 0.8, 1.5, 1.8, 2.2, and 2.9 $\cdot 10^7 M_\odot$ (from bottom to top).

In a next step we sliced the broad line profiles of Mrk 110 in segments of $\Delta v = 400$ km/s. This corresponds to the spectral resolution. Then we computed cross-correlation functions (CCF) of all segment light curves of the Balmer and Helium lines with respect to the 5100Å continuum light curve. The 2-D CCF give the correlations of the Hβ and HeIλ5876 line-profile segments with continuum variations as a function of radial velocity and time delay (gray scale) (Figs. 5 and 6). The contours of the correlation coefficient are overplotted.

The light curve of the central line region in Hβ shows the longest delay: about 30 to 40 light days. The outer line wings segments respond much faster to continuum variations when

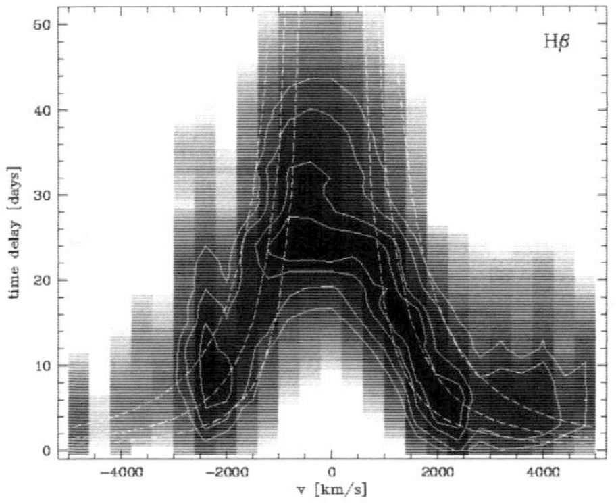

Figure 5. This 2-D CCF gives the correlation of Hβ line-profile segments with continuum variations as a function of radial velocity and time delay (gray scale). Contours of the correlation coefficient are overplotted at levels of 0.85 to 0.93 (solid lines). The dashed curves show computed escape velocities for central masses of 0.5, 1.0, and 2.0 $10^7 M_\odot$ (from bottom to top) ([3]) .

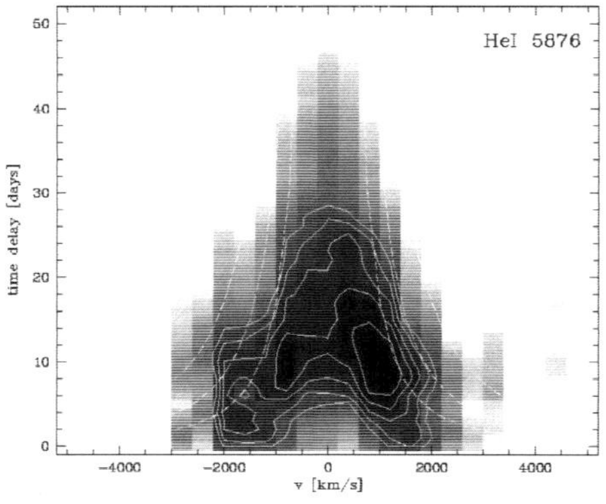

Figure 6. This 2-D CCF gives the correlation of HeIλ5876 line-profile segments with continuum variations as a function of radial velocity and time delay (gray scale). Contours of the correlation coefficient are overplotted at levels of 0.85 to 0.93 (solid lines). The dashed curves show computed escape velocities for central masses of 0.5, 1.0, and 2.0 $10^7 M_\odot$ (from bottom to top) ([3]).

compared to the inner ones with a delay of less than five days.

The same trend of the velocity-delay pattern is to be seen in the HeIλ5876 line. However the delays of the HeIλ5876 segments are only half of the Hβ delays as has been noted before for the integrated lines.

If one compares the observed velocity-delay pattern with model calculations [4] one can make some statements about the dominant velocity field in the broad-line region. One can rule out some simple kinematical models for the BLR in Mrk 110: the line wings show the shortest delay with respect to the continuum and react nearly simultaneously. Therefore we can rule out radial inflow or (biconical) outflow motions as the dominant motions. A Keplerian disk BLR model fulfills the observed velocity-delay pattern of a faster response of both line wings compared to the center. In addition one can see in Figs. 5 and 6 that the red line wing shows a slightly faster and stronger response compared to the blue wing. This is an indication for an additional accretion disk-wind in the BLR (see [5]).

3. Emission line variability in Mrk 926

Mrk 926 is a very broad-line Seyfert galaxy. The Balmer and Helium lines have width of about 10 000 km s^{-1}. We are monitoring Mrk 926 since many years ([6]). Here we present results of a variability campaign taken with the Hobby-Eberly Telescope in the years 2004 and 2005 ([7]). Fig. 7 shows 15 spectra of Mrk 926 that have been taken between August and December 2005. Fig. 8 shows the continuum light curves at 4600 and 5180 Å (in units of 10^{-15} erg cm^{-2} s^{-1} Å^{-1}) and the lightcurves of the integrated emission line fluxes of Hβ and Hα (in units of 10^{-15} erg cm^{-2} s^{-1}) for the year 2005. Fig. 9 gives the mean spectrum of Mrk 926 for both campaigns in 2004 and 2005. The rms spectrum is shown at the bottom with two different vertical scalings. It has additionally been scaled by a factor of 6 to enhance weaker line structures. The rms spectrum shows the variable part of the line profiles: In the Hα and Hβ rms spectra are an inner double-

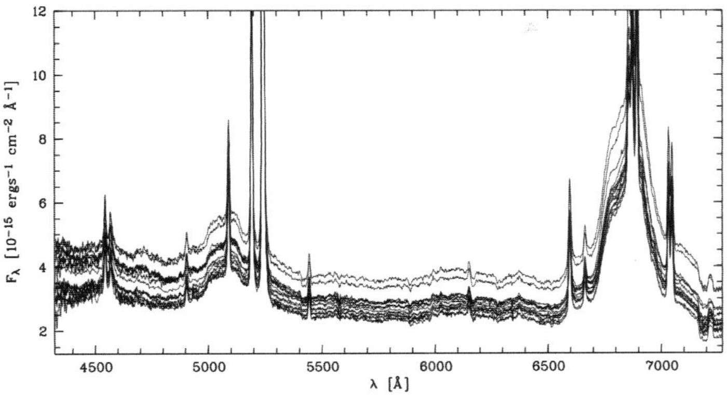

Figure 7. Spectra of Mrk 926 taken with the Hobby-Eberly Telescope between August and December 2005 ([7]).

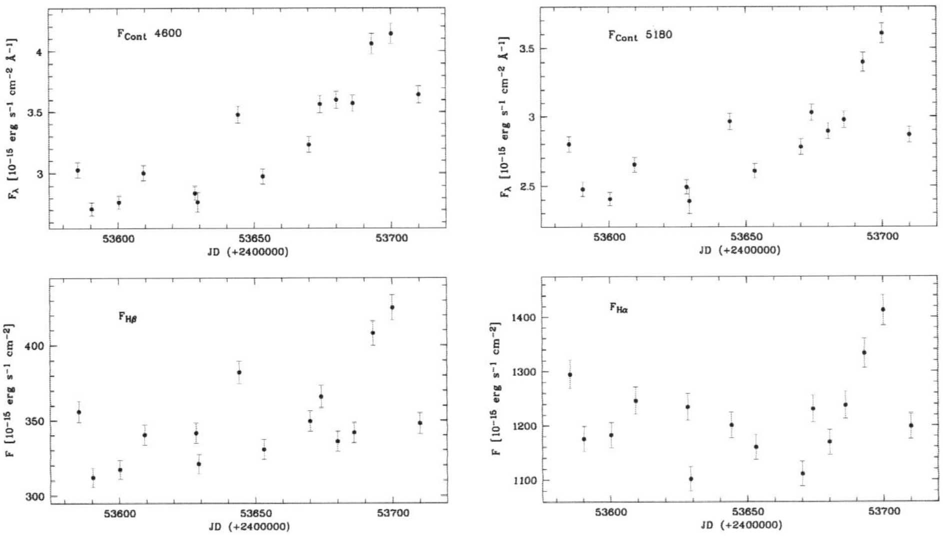

Figure 8. Light curves of the continuum fluxes at 4600 and 5180 Å (in units of $10^{-15}\,\mathrm{erg\,cm^{-2}\,s^{-1}\,Å^{-1}}$) and of the integrated emission line fluxes of Hβ and Hα (in units of $10^{-15}\,\mathrm{erg\,cm^{-2}\,s^{-1}}$) for the year 2005 ([7]).

peaked structure as well as two further outer blue and red components visible. Normalized mean and rms line profiles of Hα and Hβ in velocity space are presented in Figs. 10 to 12. The rms profiles of Hα and Hβ show detailed subcomponents while the mean profiles does not show these details. There is a narrow central component in the Balmer rms spectra ($v \leq \pm 600\,\mathrm{km\,s^{-1}}$) and two broad inner ($v \leq \pm 6\ 000\,\mathrm{km\,s^{-1}}$) as well as outer ($v \geq \pm 6\ 000\,\mathrm{km\,s^{-1}}$) line components. Comparing the mean and rms spectra with each other one can see that these broad outer and inner line segments varied with different amplitudes.

The radius of the BLR is very small with an upper limit of 2 light-days. This is the result of the cross-correlation functions CCF(τ) of the continuum light curve at 5180 Å with the Hβ and

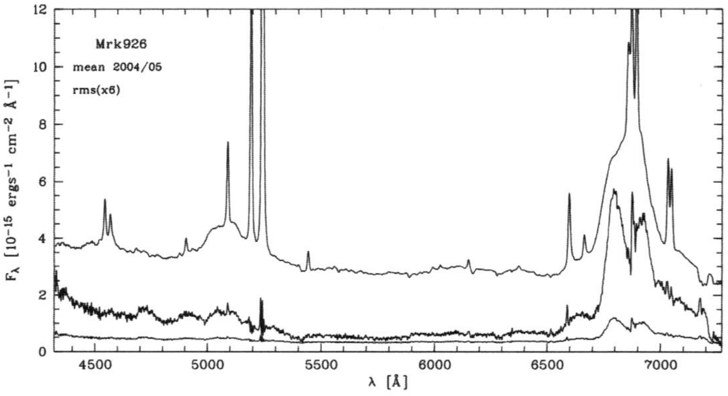

Figure 9. Mean spectrum of Mrk 926 for the years 2004 and 2005 (top). Rms spectrum at the bottom. The spectrum has been scaled by a factor of 6 to enhance weaker line structures ([7]).

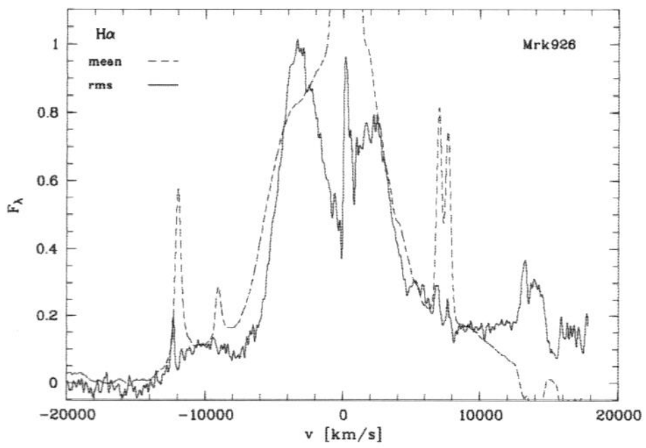

Figure 10. Normalized mean (dashed line) and rms (solid line) line profiles of Hα in velocity space ([7]).

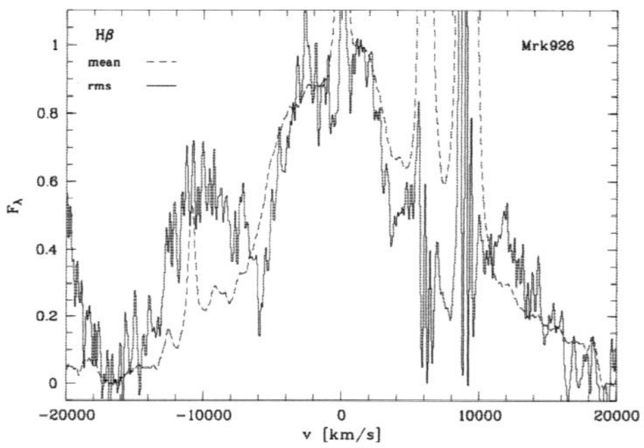

Figure 11. Normalized mean (dashed line) and rms (solid line) line profiles of Hβ in velocity space ([7]).

Hα light curves (Fig. 13). We derived an upper limit of the central black hole mass in Mrk 926 of $M = 11.2 \times 10^7 M_\odot$ based on the delays and line-widths ([7]).

Again we sliced the Hα and Hβ velocity profiles of Mrk 926 into velocity segments of widths Δv = 400 km s^{-1}. Then we measured the intensities of all subsequent velocity segments from v = −15 000 until +15 000 km s^{-1} and compiled their light curves. The light curves of all line segments including the light curves of the continuum at 5180 Å look similar. There were no clear-cut differences between the outer and inner segments or between the red and blue wings as seen in Mrk 110 ([3]) . Subsequently we computed cross-correlation functions (CCFs) of all line segment (Δv = 400 km s^{-1}) light curves with the 5180 Å continuum light curve.

The derived delays of the Hα line segments with respect to the 5180 Å continuum light curve are given in Figs 14 as a function of distance to the line center. The thin solid line in Fig. 14 delineates the contour lines of the correlation coefficient at levels of 0.85, 0.75, and 0.65 for Hα. The heavy dashed line connects the centers of all cross-correlation functions for the different line profile segments. The time delay of the line segment light curves was calculated from the uppermost 10 percent of the cross-correlation functions. One can see that all segments show a

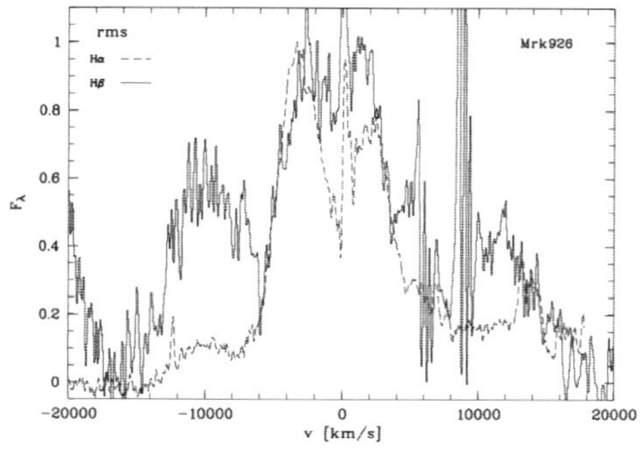

Figure 12. Normalized rms line profiles of Hα (dashed line) and Hβ (solid line) in velocity space ([7]).

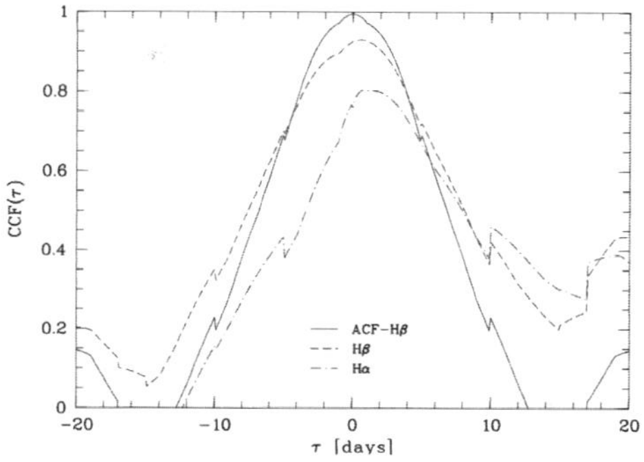

Figure 13. Cross-correlation functions CCF(τ) of the continuum light curve at 5180 Å with the Hβ and Hα light curves as well as the Hβ autocorrelation function ([7]).

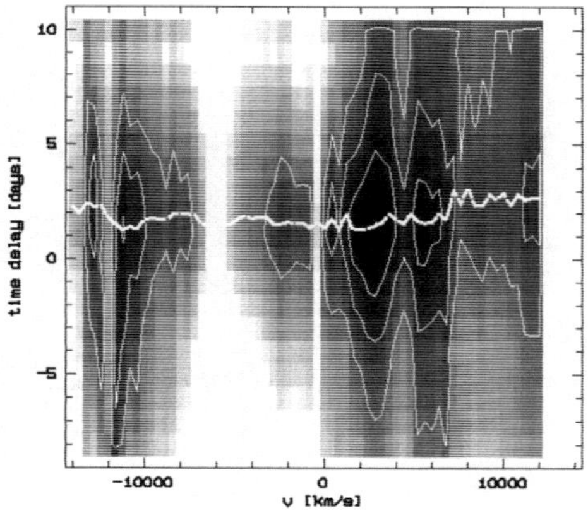

Figure 14. The 2-D CCF(τ,v) of Mrk 926 shows the correlation of the Hα line segment light curves with the continuum light curve at 5180 Å as a function of velocity and time delay (grey scale). Contours of the correlation coefficient are overplotted at levels of 0.85, 0.75, and 0.65 (solid lines). The heavy dashed line connects the centers of all individual cross-correlation functions ([7]).

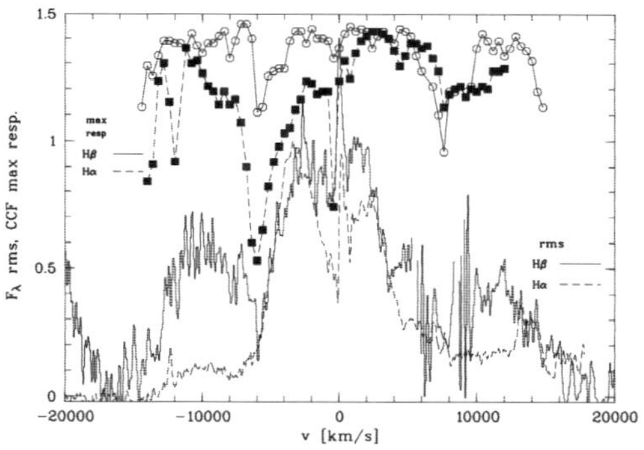

Figure 15. Maximum response of the correlation functions of the Hα (solid line, filled square) and Hβ (dashed line, open circle) line segment light curves with the continuum light curve at 5180 Å as well as their normalized rms profiles. The response curves are shifted by 0.5 to avoid strong overlap of the curves ([7]).

more or less constant delay of two to three days. Further details regarding the delay could not be resolved with our sampling rate.

Fig. 15 shows the maximum response of the Hα and Hβ line segment cross-correlation functions. It is surprising that the characteristics of the Hα and Hβ line segment responses have the same pattern. The response is independent of the intensity of the line segments. There are two pronounced outer minima at -6 000 and +7 600 km s^{-1} in the response distribution as well as two central minima at -400 and +800 km s^{-1}. It is interesting that the double structure in the Hα and Hβ response curves which contains two separate inner and outer components has been found before in the rms profiles. The same pattern was derived in a completely different way. The structured response may be considered as an independent evidence of a structured BLR in addition to the structure found in the rms profiles.

4. Acknowledgments

Based on observations obtained with the Hobby-Eberly Telescope, which is a joint project of the University of Texas at Austin, the Pennsylvania State University, Stanford University, Ludwig-Maximilians-Universität München, and Georg-August-Universität Göttingen. This work has been supported by the Niedersachsen - Israel Research Cooperation Program ZN2318.

References

[1] Peterson, B.M., Ferrarese, L., Gilbert, K.M., et al. 2004, ApJ, 613, 682
[2] Kollatschny, W., Bischoff, K., Robinson, E. L., et al. 2001, A&A, 379, 125
[3] Kollatschny, W. 2003, A&A, 407, 461
[4] Welsh, W.F., & Horne, K. 1991, ApJ, 379, 586
[5] Chiang, J., & Murray, N. 1996, ApJ, 466, 704
[6] Kollatschny, W., Zetzl, M., Dietrich, M. 2006, A&A, 454, 459
[7] Kollatschny, W., Zetzl, M. 2010, A&A, in press

High density and high temperature plasmas in Large Helical Device

Akio Komori

National Institute for Fusion Science
322-6 Oroshi, Toki, Gifu 509-5292 Japan

komori@LHD.nifs.ac.jp

Abstract. Recently a new confinement regime called Super Dense Core (SDC) mode was discovered in Large Helical Device (LHD). An extremely high density core region with more than ~ 1×10^{21} m^{-3} is obtained with the formation of an Internal Diffusion Barrier (IDB). The density gradient at the IDB is very high and the particle confinement in the core region is ~ 0.2 s. It is expected, for the future reactor, that the IDB-SDC mode has a possibility to achieve the self-ignition condition with lower temperature than expected before. Conventional approaches to increase the temperature have also been tried in LHD. For the ion heating, the perpendicular neutral beam injection effectively increased the ion temperature up to 5.6 keV with the formation of the internal transport barrier (ITB). In the electron heating experiments with 77 GHz gyrotrons, the highest electron temperature more than 15 keV was achieved, where plasmas are in the neoclassical regime.

1. Introduction

For the realization of the fusion reactor, it is necessary to confine high density and high temperature plasma for a time. For the deuterium-tritium reaction, the triple product of density, temperature and confinement time must be greater than ~ 10^{21} keV s / m^3 to realize the ignition, which is well known as the Lawson criterion [1,2]. To achieve the Lawson criterion, tremendous efforts have been made in many devices in the world for about a half-century [1]. By the end of the last century, large tokamaks, i.e. JT60-U and JET, have achieved fusion energy breakeven [1]. On the other hand, steady progress has also been made in stellarator or helical devices [3,4]. In the Large Helical Device (LHD) which is the largest superconducting heliotron device with R = 3.9 m r = 0.6 m, B_t = 3 T in National Institute for Fusion Science, vigorous experiments have been performed to improve the plasma performance [3].

Recently a promising confinement regime called Super Dense Core (SDC) mode was discovered in LHD. In this mode, the core density increases to more than ~ 1×10^{21} m^{-3} and the central plasma pressure exceeds atmospheric pressure. This high pressure core plasma is stably maintained within the internal diffusion barrier (IDB). With the IDB-SDC mode, there is a possibility for LHD type reactors to access the self-ignition condition with relatively lower temperature than expected before. The IDB-SDC mode is also favorable from the engineering point of view since one can moderate demands for heating devices and plasma facing components.

In this paper, recent results from high density studies in LHD with IDB-SDC mode, together with the conventional approaches to increase the temperature, are described. The highest ion temperature T_i

of 5.6 keV and highest electron temperature T_e of 15 keV have been obtained in LHD, although those records were achieved in respective discharges. Another interesting phenomenon called "impurity hole" was observed in high ion temperature discharges, which is characterized with the hollow impurity profile. Discussions about the formation mechanism of the impurity hole are presented in the following section.

In section 2, the high density plasma study in LHD is summarized. After describing high T_e discharges in section 3, experimental results from high T_i study are presented in section 4. Finally summary is given in section 5.

2. High density plasma

In this section a new operational regime called SuperDense Core (SDC) mode with an Internal Diffusion Barrier (IDB) [5-7] is described. In order to obtain stable high density discharges, fuelling and pumping for neutral particles should effectively be performed in a functionally-combined system. Especially for pumping, it is known that the divertor plays an important role. Thus, before showing the experimental results, explanation for two divertor configurations is expressed as an introduction to the high density study in LHD.

2.1. Two Divertor configurations in LHD

In LHD two different divertor configurations, i.e. helical divertor (HD) and local island divertor (LID),

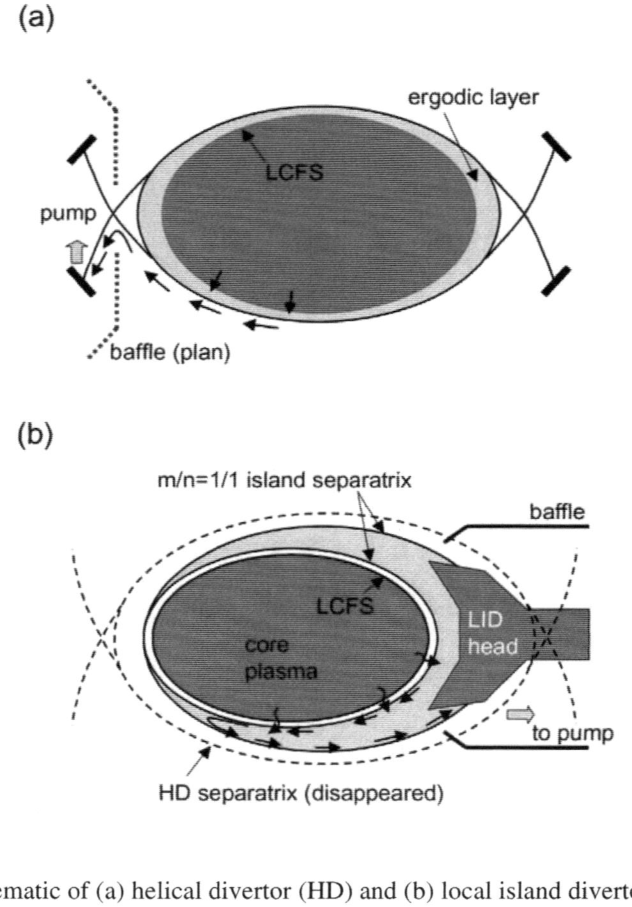

Figure 1. Schematic of (a) helical divertor (HD) and (b) local island divertor (LID).

are alternatively employed to control the heat and particle flux in the edge region. The HD is a poloidal divertor inherently built in the heliotron devices [8], which is similar to the tokamak double null configuration, as shown in figure 1 (a). On the other hand, the LID is a kind of island divertor which utilizes an externally superimposed static magnetic island with m/n=1/1, where m and n are poloidal and toroidal mode numbers, respectively [9,10]. The schematic of the LID is shown in figure 1 (b). Particles diffusing out from the confinement region cross the island separatrix, and flow along the periphery of the island to the backside of the LID head. The particles neutralized there are pumped out by a strong pumping system with a baffle which realizes a closed divertor configuration with overall pumping efficiency of larger than 30%. Unlike the conventional pump limiters, blades of the divertor head are located inside the island, thereby being protected from the high heat flow along the island separatrix. Thus there is no leading-edge problem (see figure 1 (b)). Although the LID has the efficient pumping capability with its closed configuration and strong pumping system, it has difficulties in handling the high heat load to the target plates because of the small wetted area on them. For the steady-state operation, therefore, the HD with wider wetted area is thought to be suitable. However, the HD has not yet been equipped with baffles to form the closed configuration, thus experiments in the low recycling regime are alternatively performed in the LID configuration now.

2.2. IDB-SDC mode

The IDB-SDC mode was discovered for the first time in the LID configuration [5], which was obtained when a series of pellets were injected into the neutral beam (NB) heated plasma (P_{NB} ~ 10 MW). The time evolutions of some principal parameters are presented in figure 2. The density profile

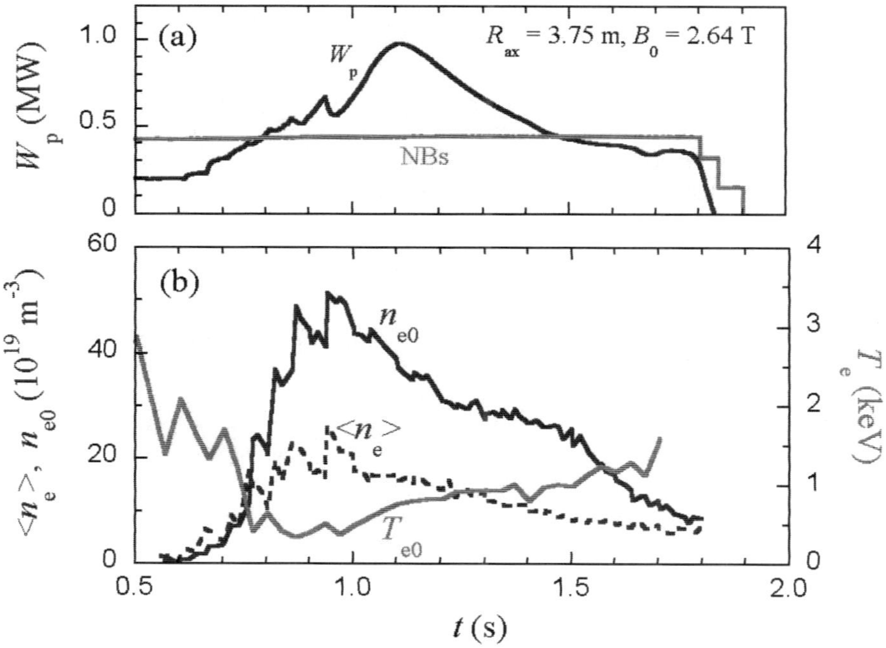

Figure 2. Time evolutions of (a) stored energy W_p together with the injection sequence of NBs, (b) central electron temperature T_{e0} and central electron density n_{e0} and averaged density $<n_e>$.

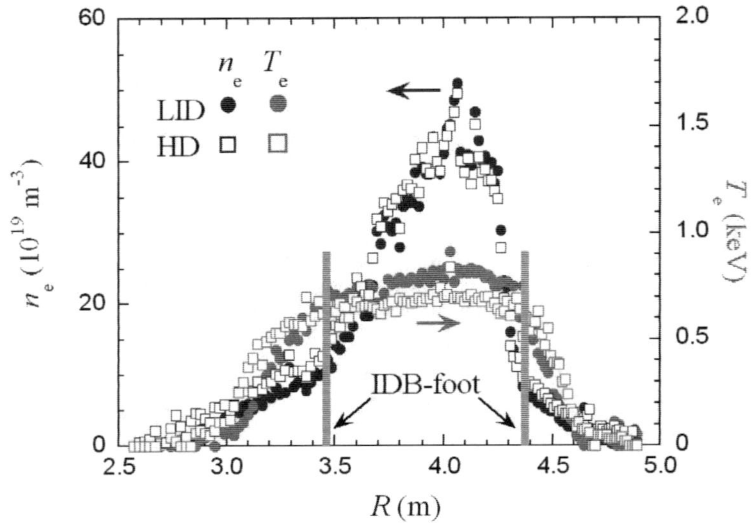

Figure 3. Electron density (blue) and temperature (red) profiles in the IDB-SDC mode. Open and closed symbols represent HD and LID configurations, respectively.

in the initial phase is flat and low. After several pellets are injected to the plasma, the density profile takes on a peaked shape, which can be seen from the abrupt rise in central density n_{e0} and its separation from the averaged density $< n_e >$ at $t \sim 0.76$ s, as shown in figure 2 (b). Subsequent pellet injection enhances the further central deposition of the fueling particles, thus n_{e0} continuously rises until the final pellet is injected at $t \sim 0.95$ s. Since NBs are still heating the plasma, central electron temperature T_{e0} starts to rise, according to the decrease of the electron density, as shown in figure 2 (b). The plasma stored energy W_p keeps on rising until $t = 1.1$ s because the relative increase in temperature is larger than the simultaneous decrease in density.

The radial electron density n_e (blue closed symbols) and temperature T_e (red closed symbols) profiles at the time when W_p reaches its maximum are depicted in Fig. 3, which are the typical profiles of the IDB-SDC mode. A core region with electron density $\sim 5 \times 10^{20}$ m^{-3} and temperature ~ 0.85 keV is maintained by the IDB. It can be seen that the extremely steep density gradient is formed on the IDB foot which is depicted with thick solid lines (gray) in figure 3. Outside the IDB, the density gradient is moderate. Due to the high central pressure, a large Shafranov shift of ~ 0.3 m is observed.

It is considered that the central fueling by pellet injection and strong edge pumping by the LID are essential for the improved mode. In the latest experiments, it has been demonstrated that such high density discharges with an IDB can be achieved even in the HD configuration if only the low recycling condition is realized. The n_e and T_e profiles depicted with open symbols in figure 3 are actually obtained in the HD configuration. It is found that both profiles between HD and LID are almost the same, except that the central temperature in the HD configuration is slightly low. From these experimental results, we have confirmed that the IDB-SDC mode is not particular in the LID configuration but a universal mode in helical devices. Similar high density mode called PEP mode has been observed in tokamaks [11].

2.3. Quasi-steady-state operation of IDB-SDC mode

The IDB-SDC mode is essentially a transient phenomenon, since it establishes in the density decay phase after the pellet injection. However, in order to approach the reactor plasma with the IDB-SDC mode, possibility of the steady-state operation is strongly required. Thus a trial to maintain the IDB-SDC mode in a steady-state was performed. A continuous pellet injector [12] with smaller pellet barrels than those for the density build-up phase was utilized to maintain the peaked pressure profile after the formation of the IDB. By injecting the pellet every 0.13 s, the quasi-steady-state IDB-SDC mode could successfully be obtained for nearly 1 s, as shown in figure 4. In the quasi-steady-state phase, perturbations by the pellets on stored energy and the density were ~ 15 %.

However another issue that has to be solved is still left. The heat load to the LID head is too high to be removed in the steady-state operation, since the wetted area of the LID head for the diverted plasma is very small. Although the LID has made a great contribution to the discovery of the IDB-SDC mode, it cannot avoid this problem as far as LID is used in the steady-state operation. For the solution of this problem, the HD configuration with wider wetted area has to be employed. In addition, an enough capability of the pumping to realize the low recycling condition is also necessary. In the following section, the effect of the edge neutral density on the formation of the IDB is discussed.

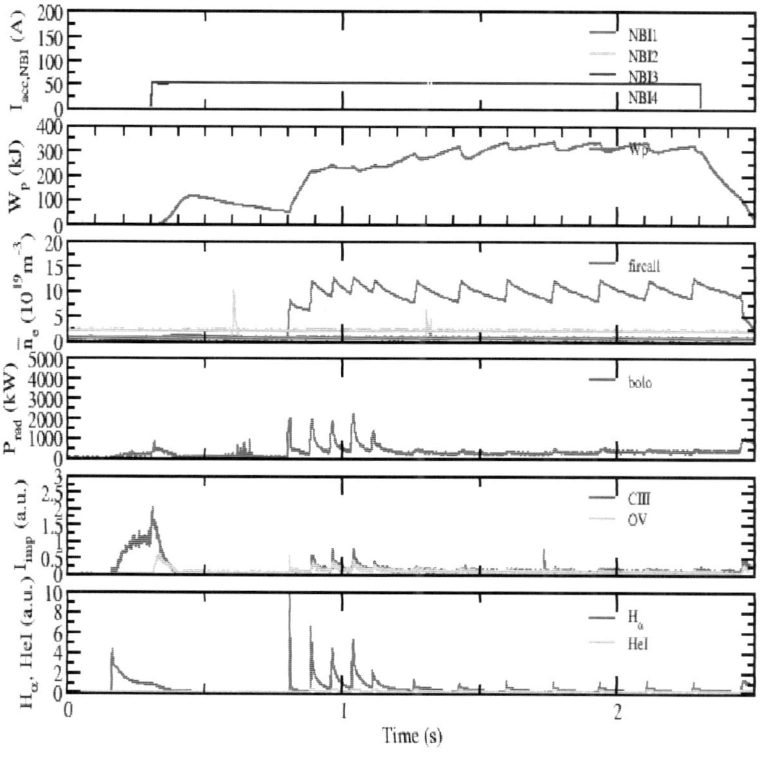

Figure 4. The time trend of the quasi-steady-state IDB-SDC discharge. From top to bottom: NBs, stored energy W_p, line averaged density n_e, radiation power P_{rad}, CIII and OV emissions, and Hα light emission.

3. High T_e discharges

High power 77 GHz gyrotrons have been installed in LHD since 2006. These gyrotrons increase T_e significantly and enable us to explore new collision regime. In the last experimental campaign in 2009, the highest central T_e more than 15 keV has been recorded with relatively low density plasma of $n_e \sim 3 \times 10^{18}$ m^{-3}.

3.1. Gyrotrons and transmission system

Figure 5 shows a schematic view of the electron cyclotron heating (ECH) system in LHD [13]. The system is composed of seven gyrotrons (one 82.7 GHz/ 0.45 MW/ 2 s, one 84 GHz/ 0.8 MW/ 3 s, one 84 GHz/ 0.2 MW/ CW, three 77 GHz/ 1 MW/ 5 s and one 168 GHz/ 0.5 MW/ 1 s). Those different types of gyrotrons are operated by two types of high voltage power supplies. One is a solid-state power supply for the 77 GHz, 84 GHz, and 168 GHz gyrotrons with depressed collectors. The other is a conventional power supply for 82.7 GHz gyrotrons, which precisely regulates the beam voltage by a regulator tube.

The transmission system for microwaves consists of matching optics units (MOU), over-sized corrugated waveguides with miter-bends, waveguide switches, polarizers, quasi-optical antennas, a DC breaks and vacuum windows. For the transmission from each gyrogtron and antenna in LHD, eight corrugated waveguides with inner diameters of 88.9 mm and 31.75 mm are utilized. The distance between each gyrotron and antenna is about 100 meters along the waveguides. Five waveguides are evacuated and the rest of eight waveguides are filled with dry air. According to the high power transmission test, total transmission efficiency is about 74 % on average, though it depends on microwave frequencies and waveguide diameters. Two antennas are installed in an upper port (U-port). The U-port antenna consists of four mirrors. The final plane mirror for a U-port antenna steers the injection angle of microwaves by a remote-controlled supersonic motor. Antennas installed in lower and outer ports consist of two mirrors, i.e. a fixed focusing mirror and a flat steering one.

3.2. Experimental results

The most remarkable result in the recent ECH experiments was the achievement of exceeding 15 keV central electron temperature with the condition of relatively low electron density of 3×10^{18} m^{-3} [14]. Figure 6 shows the radial profile of T_e under the conditions that R_{ax} = 3.53 m, B_T = 2.705 T and the power of ECH P_{ECH} = 2.7 MW. The averaged electron density at this timing is 2.3×10^{18} m^{-3}. It can be

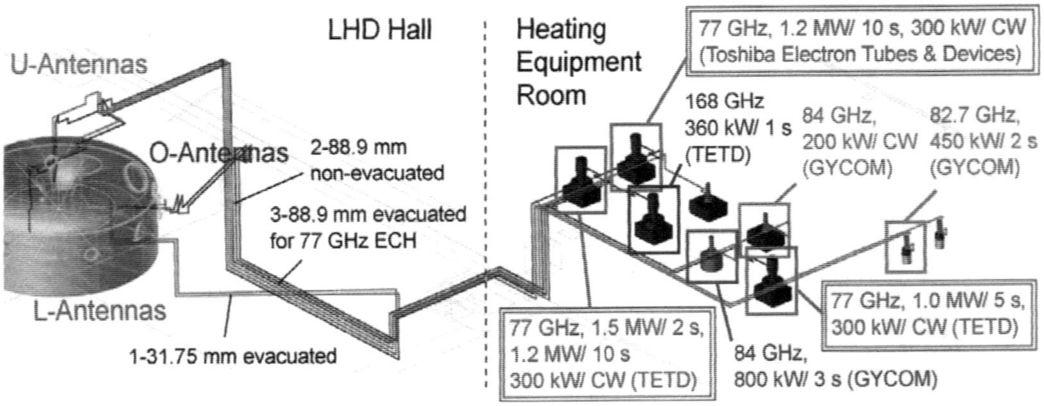

Figure 5. Schematic of the electron cyclotron heating (ECH) system in LHD

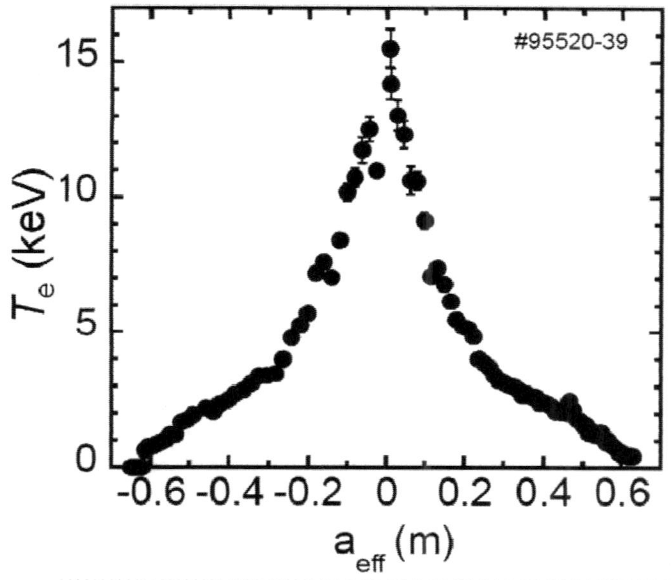

Figure 6. Electron temperature profile with the highest central electron temperature. a_{eff} is the effective minor radius.

seen that the central electron temperature is about 15 keV. Such a high T_e discharge was obtained with the simultaneous operation of three high power 77 GHz gyrotrons. Evacuation and cooling of the waveguide transmission line for 77 GHz system were also performed to keep the stable operation without unfavourable discharges in waveguides.

It was experimentally found that the highest T_e can be obtained in the configuration at R_{ax} = 3.53 m, which is consistent with the theoretical prediction of the neoclassical transport.

4. High T_i discharges

Ion temperature is one of the most important parameters to realize a continuous fusion reaction in magnetically confined fusion plasmas. In LHD, high T_i experiments have intensively performed with NB injection. The highest T_i obtained so far in LHD is 13.5 keV with argon plasma and 5.6 keV in hydrogen plasma. In this paper, experimental results with hydrogen plasma are highlighted, together with related phenomena in high T_i discharges [15].

4.1 History of high Ti experiments in LHD

In the initial phase of the LHD experiment, ion heating was performed by tangential neutral beam injection. There are three beam injectors of negative-ion-based NB with a very high beam energy of 180 keV, which are the main heating devices in LHD, of which total port-through power is more than 14MW. However, NB energy is too high (critical energy for ion heating E_c is ~ 120 keV at T_e ~ 4 keV) to heat ions, thus electrons are mainly heated with tangential NBs. In order to increase the net heating power absorbed in each ion, high Z ions were introduced to reduce ion density. With argon injection, highest T_i of 13.5 keV was obtained. In this experiment, the central ion temperature was measured from the Doppler broadening of X-ray lines obtained with a crystal spectrometer. The X-ray lines of

Ar XVII, Ti XXI, Cr XXIII, and Fe XXV were used to evaluate the ion temperature at the center of the plasma.

On the other hand, a perpendicular NB (p-NB) with low beam energy of 40 keV was installed to heat low Z ions (hydrogen) efficiently, which is the final goal of the high T_i experiment. The p-NB can also be utilized for T_i measurements with charge exchange recombination spectroscopy (CXRS), which is an important diagnostics for high T_i experiments, in addition to the crystal spectrometer.

4.2. Experimental results

A p-NB injector with relatively low beam energy of 40 keV was installed in LHD for high-power ion heating experiments in low Z plasmas. The improvement of ion confinement was observed under the low Z_{eff} conditions [16-20], and the central ion temperature of 5.6 keV was recently achieved [15]. In this section, characteristics of the improved ion confinement mode observed in high T_i experiments with low Z plasma are discussed.

Figure 7 (a) shows the typical discharge in the improved confinement regime. The plasma was generated by ECH, then sustained by negative NBs. The carbon pellet was injected to the plasma at t = 1.8 s [21]. In the density decay phase after the carbon pellet injection, additional plasma heating with negative and positive NBs was performed, hence the ion and electron temperatures increased. The power of p-NB was modulated to obtain the background signal of the CXRS for the precise T_i profile measurement. It can be seen from figure 7 (a) that even after the central electron temperature reaches

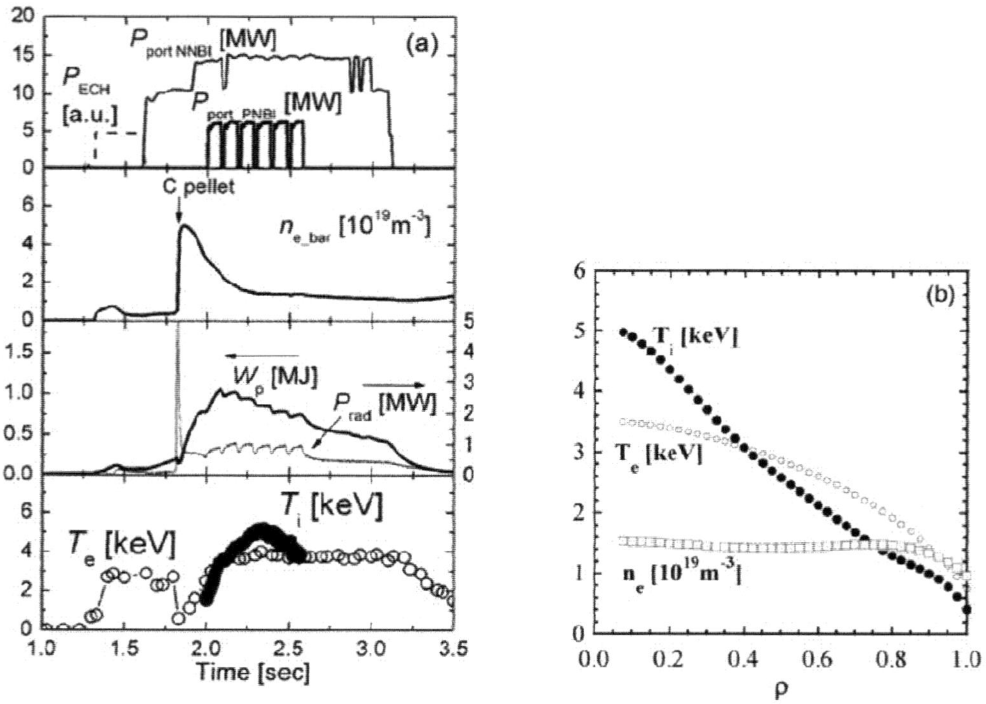

Figure 7. (a) Time evolution of typical high-Ti discharge. From top to bottom, heating power of ECH P_{ECH}, tangential NB $P_{port\ NNBI}$, perpendicular NB $P_{port\ PNBI}$, line-averaged density n_{e_bar}, stored energy W_p, radiation power P_{rad}, central electron temperature T_e and ion temperature T_i. (b) Radial profiles of T_i, T_e and n_e at t = 2.38 s. (From Ref. [20])

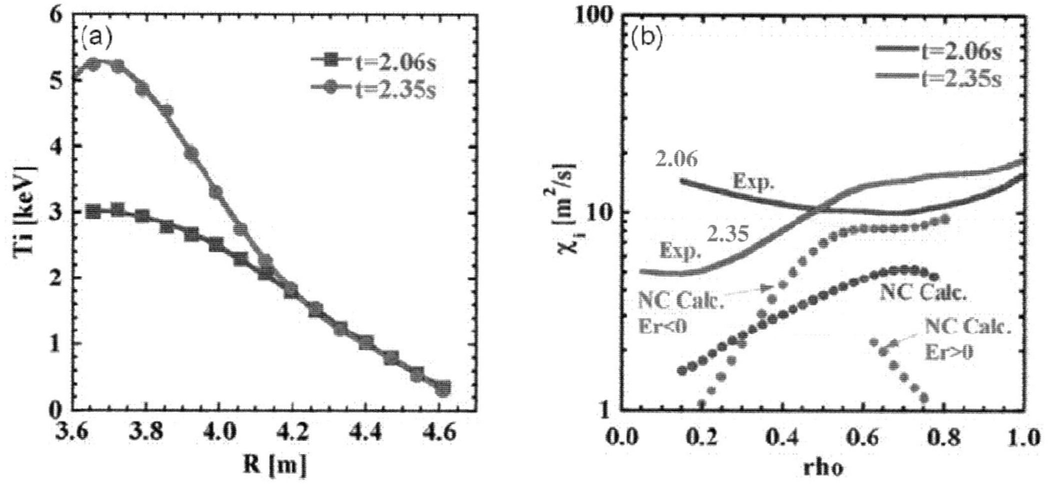

Figure 8. (a) T_i profiles at L-mode (t = 2.06 s) and ion ITB phase (t = 2.35 s). (b) Profiles of ion thermal diffusivities obtained with experiments and theoretical calculations. (From Ref. [20])

its peak at T_e ~ 4 keV, the central ion temperature T_i still increases more than 5 keV. At this moment, the ion temperature shows the peaked profile with steep gradient in the core region, which is so called the ion ITB, whereas the electron temperature T_e profile is still broad, and the density profile is also broad or flat in the core region, as shown in figure 7 (b).

The strong pump-out of carbon impurity is observed with increase in ion temperature, and the ion ITB is realized under the low-Z_{eff} condition, i.e. n_C / n_e < 0.01. Although the central ion temperature slightly decreases after reaching its maximum, the ion ITB is sustained during p-NB injection, which is longer than the energy confinement time.

As shown in figure 8, the ion thermal diffusivities χ_i are evaluated in low and high T_i phases. It is found from figure 8 (b) that χ_i in high T_i phase is significantly small. In figure 8 (b), χ_i from neoclassical transport theory was also plotted. The radial electric field E_r is also evaluated theoretically. Inside the ITB, the neoclassical ambipolarity electric field E_r is negative, although the E_r direction in the peripheral region is not determined. On the other hand, it is negative in the whole region in the low T_i phase. The improvement of the ion heat transport observed in this discharge was realized without transition between neoclassical ion root and electron root. In the L-mode, the negative E_r was experimentally observed, however no information was obtained in high T_i, i.e. the ion ITB phase. due to the strong exhaust of impurity for CXRS. The enhanced negative E_r in the ion ITB was predicted by neoclassical theory. The significant decrease of carbon impurity is observed in ion ITB plasmas, whereas it is not observed in L-mode discharges [22]. As seen in figure 9 (a) – 9 (d), the carbon density n_C at the plasma center increases during the L-mode phase. After the electron temperature is saturated, on the contrary, n_C decreases drastically. At this moment, n_C presents extremely hollow profile, which is called "the impurity hole". The n_C decreases down to ~ one-tenth of the maximum value, while central electron density n_e decreases as small as 30% with the existence of the impurity hole. These results clearly show that the carbon impurity is strongly exhausted from the core region of ion ITB plasma. Figure 10 presents the central carbon density as a function of the ion temperature gradient. It is clearly shown that the carbon density begins to decrease just after the ITB formation and lasts by the end of the ITB phase. The decrease of argon, iron and helium densities are also observed in the core region. Consequently Z_{eff} is kept low during the ion ITB phase with the impurity hole.

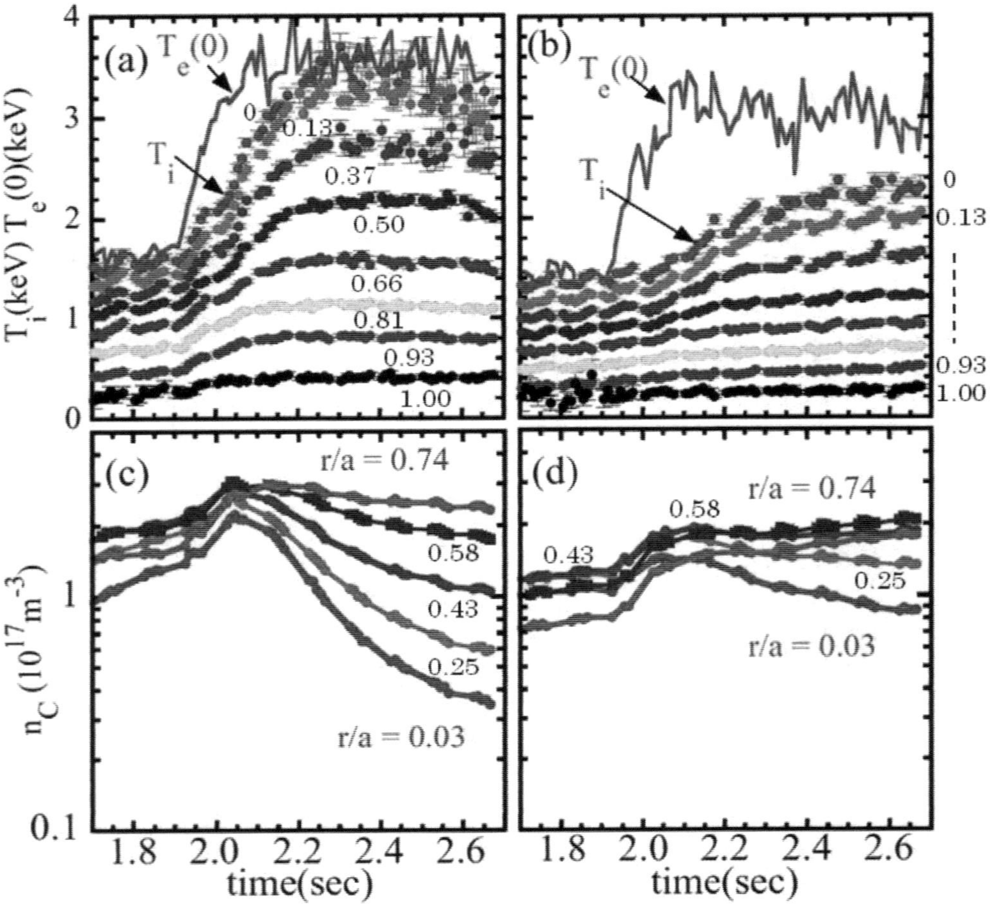

Figure 9. Time evolution of T_i and T_e in (a) ion ITB and (b) L-mode discharges at various radii of $\rho = 0, 0.13, 0.37, 0.50, 0.66, 0.81, 0.93$ and 1.0. Carbon impurity density n_C in (c) ion ITB and (d) L-mode discharges at various radii of $\rho = 0.03, 0.25, 0.43, 0.58$ and 0.74. (From Ref. [22])

Significant outward convection of impurities, which is considered to be the mechanism of the impurity hole formation, is a characteristic of the ion ITB in LHD. On the other hand, such an obvious outward convection has not been observed in tokamak ITB plasmas. According to the neoclassical transport theory, the direction of E_r plays an important role to determine the direction of impurity flux (inward for $E_r < 0$, and outward for $E_r > 0$). However it is not consistent with the experimental results observed in LHD, i.e. the impurity pump-out occurs with negative E_r. Therefore further investigation is necessary to clarify the impurity transport in the ion ITB plasma.

5. Summary

High density and high temperature plasmas have successfully been obtained in LHD. Although it is not a simultaneous achievement, the parameter range to be explored has been widely enlarged.

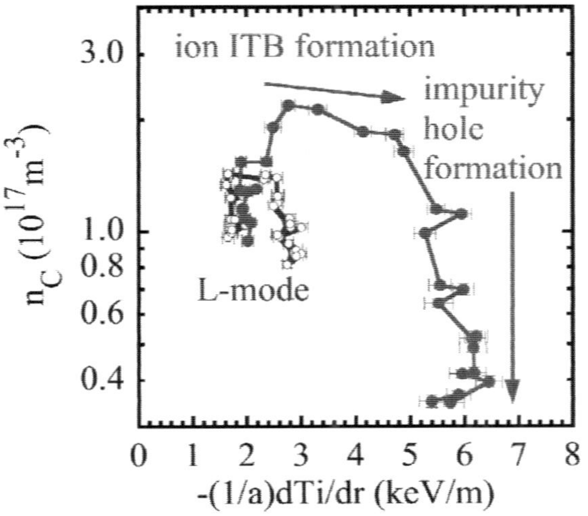

Figure 10. Relation between central carbon density and temperature gradient at r/a =0.6 for the L-mode discharge and ITB discharge. The data are plotted every 20 ms. (From Ref. [22])

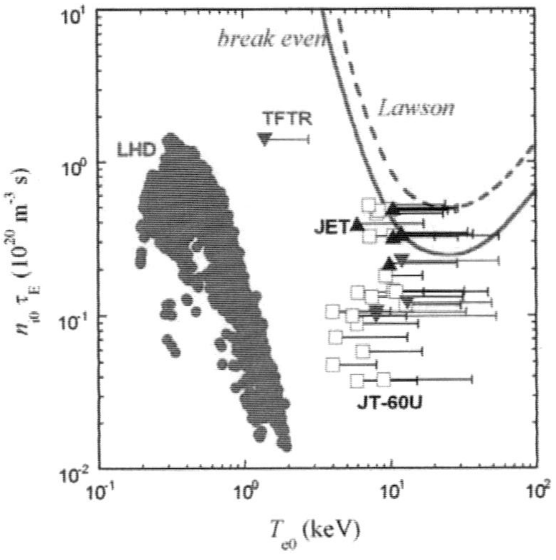

Figure 11. $n\tau$-T diagram for LHD and large tokamaks. Higher ends of error bars for tokamaks indicate ion temperature, while symbols including LHD indicate electron temperature.

In high density plasma experiments, an improved confinement regime named IDB-SDC mode was first discovered in the LID configuration. This improved mode is achieved by the effective edge pumping by LID and the central fuelling by repetitive pellet injection. It has also been confirmed that the IDB-SDC mode can be achieved even in the open HD configuration, if only the neutral pressure is kept low by the wall pumping effect. The highest density obtained so far is 1.2×10^{21} m^{-3}.

In high temperature plasma experiments, central electron temperature of 15 keV was achieved with three high power 77 GHz gyrotrons and evacuated waveguide transmission lines. The highest ion temperature of 5.6 keV was recorded in hydrogen plasma with the positive perpendicular NB injection. In such a high T_i regime, an impurity pump-out phenomenon called "impurity hole" is observed, in spite of the negative E_r.

As mentioned above, the highest density and the highest temperature cannot be obtained simultaneously so far in LHD. In the so-called $n\tau$-T diagram shown in figure 11, it is found that LHD is further from the Lawson criterion than large tokamaks at this stage. However there is a possibility to find another path to the ignition by optimizing the IDB-SDC mode [23]. Further investigation and efforts are necessary to get closer to the final goal.

References

[1] Wesson J 2004 *Tokamaks 3rd Edition* (Oxford University Press)
[2] Miyamoto K 1989 *Plasma Physics for Controlled Fusion* (The MIT Press)
[3] Yamada H et al 2010 *Fusion Sci. Technol. Special Issue on LHD* (Amecian Nuclear Society)
[4] Wakatani M 1998 *Stellarator and Heliotron Devices* (Oxford Univ. Press)
[5] Morisaki T, Masuzaki S, Kobayashi M et al 2006 *Fusion Sci. Technol.* **50** 216
[6] Ohyabu N, Morisaki T, Masuzaki S et al 2006 *Phys. Rev. Lett.* **97**, 055002
[7] Morisaki T, Ohyabu N, Masuzaki S et al 2007 *Phys. Plasmas* **14**, 056113
[8] Ohyabu N, Watanabe T, Ji H, et al 1994 *Nucl. Fusion* **34** 387
[9] Komori A, Morisaki T, Masuzaki S et al 2004 *Fusion Sci. Technol.* **46** 167
[10] Morisaki T, Masuzaki S, Komori A et al 2005 *J. Nucl. Mater.* **337-339** 154
[11] Hugon M, Milligen B, Smeulders P et al 1992 *Nucl. Fusion* **32**, 33
[12] Sakamoto R, Yamada H, Tanaka K et al 2001 *Nuclear Fusion* **41,** 381
[13] Shimozuma T et al 2009 *Proc. Topical Conf. Radio Frequency Power in Plasmas 2009*
[14] Takahashi H et al 2010 *Annual Report of NIFS 2009-2010*
[15] Nagaoka K et al 2010 *Fusion Sci. Technol.* **58** 46
[16] Yokoyama M et al 2008 *Phys. Plasmas* **15** 056111-1
[17] Nagaoka K et al 2008 *Fusion Res.* **3** S1013-1
[18] Takeiri Y et al 2008 *Plasma Fusion Res.* **3** S1001-1
[19] Ida K et al 2009 *Nucl. Fusion* **49** 015005
[20] Kaneko O et al 2009 *Plasma Fusion Res.* **4** 027
[21] Morita S et al 2003 *J. Plasma Fusion Res.* **79** 641
[22] Ida K et al 2009 *Phys. Plasmas* **16** 056111
[23] Miyazawa J et al 2010 *Fusion Sci. Technol.* **58** 29

25th Summer School and International Symposium on the Physics of Ionized Gases—SPIG 2010 IOP Publishing
Journal of Physics: Conference Series **257** (2010) 012026 doi:10.1088/1742-6596/257/1/012026

Multi-scale simulation for plasma science

S. Ishiguro[1,2], S. Usami[1], R. Horiuchi[1,2], H. Ohtani[1,2], A. Maluckov[3], M. M. Škorić[1,2]

(1) National Institute for Fusion Science, Toki, Japan
(2) Graduate University for Advanced Studies, Toki, Japan
(3) Faculty of Sciences and Mathematics, University of Nis, Serbia

E-mail: ishiguro.seiji@nifs.ac.jp

Abstract. In order to perform a computer simulation of a large time and spatial scale system, such as a fusion plasma device and solar-terrestrial plasma, macro simulation model, where micro physics is modeled analytically or empirically, is usually used. However, kinetic effects such as wave-particle interaction play important roles in most of nonlinear plasma phenomena and result in anomalous behavior. This limits the applicability of macro simulation models. In a past few years several attempts have been performed to overcome this difficulty. Two types of multi-scale simulation method for nonlinear plasma science are presented. First one is the Micro-Macro Interconnected Simulation Method (MMIS), where micro model and macro model are connected dynamically through an interface and macro time and space simulation is performed. Second one is the Equation Free Projective Integration Method (EFPI), where macro space and time scale simulation is performed by using only a micro simulator and a sophisticated numerical algorithm.

1. Introduction

Currently, development of self-consistent simulation which covers large space and time scales is being performed in variety of fields. Full torus Magneto-hydrodynamic (MHD) simulation of a large scale fusion device became realistic. Here, micro physical quantity such as transport coefficient is modeled analytically or empirically. However, micro-scale kinetic effects such as wave-particle interaction play important roles in most of nonlinear plasma phenomena limiting the applicability of macro simulation models. Therefore the macro model cannot fully explain physical mechanisms and experimentally observed results. We have to develop new methods to describe observed plasma phenomena. In this paper we present two typical methodologies which try to perform multi-scale self-consistent simulation. First one is the Micro-Macro Interconnected Simulation Method (MMIS), where micro model and macro model are connected dynamically through an interface [1]. Second one is the Equation Free Projective Integration Method (EFPI), where macro space and time scale simulations are performed by using only a micro simulator [2]. MMIS is useful for phenomena where a small scale event is concentrated in the small region. This is successfully applied to multi-scale simulations of magnetic reconnection [3], shock [4] and aurora formation [5]. MMIS needs a reliable macro simulation model. However, it is difficult to prove a reliability of macro model because of the so-called "closure problem". EFPI does not use macro equation or model. EFPI has been proposed and applied to a variety of multi-scale phenomena in different fields, in which coarse-scale (macro) behavior can be realized through short-time simulations within fine-scale models. The EFPI was first time applied to the plasma phenomenon by Shay [6] who investigated the nonlinear ion acoustic wave propagation and steepening applying a Particle-in-cell (PIC) code as a micro simulator. First, to initialize the PIC, the macro variables are "lifted" to a fine microscopic representation. The PIC code is stepped forward for a short time, and kinetic results are restricted – smoothed back to macro space. The time derivatives are estimated by numerical extrapolation, and coarse variables are projected in large steps. The process is repeated. The macro-step forward in time was performed using only the first three moments of the ion velocity probability density function (PDF). It is claimed that EFPI can reproduce PIC results, although important differences arise as a consequence of the assumptions proposed in the lifting procedure (from macro to micro scale). In particular, it was assumed that the ion PDF is Maxwellian, and that electrons are adiabatic. In this paper, we would like to explain the concept of EFPI and a primal-EFPI plasma simulation scheme to simulate the ion acoustic wave paradigm, which includes nonlinear wave steeping and kinetic effects in a plasma.

2. Micro-macro interconnected simulation method

Micro-macro interconnected simulation method (MMIS) has been proposed as a method for a variety of multi-scale problems. Here, we would like to present an application to magnetic reconnection. Magnetic reconnection is a typical multi-hierarchy phenomenon. When magnetic reconnection occurs, the field topology changes on a macroscopic scale and global plasma transport takes place. On the other hand, an electric resistivity controlled by a microscopic process is necessary as a trigger. Therefore, for complete understanding of magnetic reconnection, multi-hierarchy simulation model, which calculates both microscopic and macroscopic physics consistently and simultaneously, should be developed.

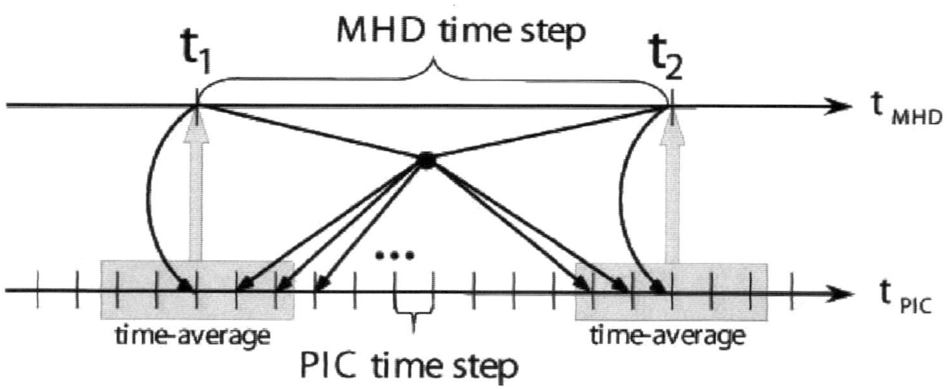

Figure 1. Time flow of Micro-Macro Interconnected Simulation Model.

In the multi-hierarchy structure of magnetic reconnection, the characteristic space-time scale changes with distance from the neutral sheet. Reconnection dynamics is controlled by the particle kinetic effects within the ion-meandering orbit scale l_{mi} while it can be expressed by an one-fluid model far from the neutral sheet.

Based on the above features, the domain decomposition method is employed. The domains differ in algorithm. Physics in the domain where microscopic kinetic effects play a crucial role is solved by particle-in-cell simulation method. This domain is called the PIC domain. On the other hand, dynamics on the periphery of the PIC domain is expressed by magneto hydrodynamics simulation method. We shall refer to this domain as the MHD domain. Between the PIC and MHD domains, an Interface domain with a finite width is inserted. The physics in the Interface domain is calculated by both the PIC and MHD algorithms. Macroscopic physical quantities in the Interface domain (for instance, magnetic field and fluid velocities) are obtained by a hand-shake scheme, $Q_{\text{inteface}} = aQ_{\text{MHD}} + (1\text{-}a)Q_{\text{PIC}}$, where Q_{MND} and Q_{PIC} indicate the values of Q calculated by the MHD and PIC algorithms, respectively. The parameter a is a function of the coordinates. Individual particle velocities in the Interface domain are newly determined so as to satisfy the (shifted) Maxwellian using the obtained macroscopic quantities at every time step. The PIC and MHD domains can be smoothly interlocked via the Interface domain.

Time flow in the multi-hierarchy model is shown in Fig. 1. We employ multi-time scale scheme, where MHD and PIC domains have different time steps. Large time steps are for MHD model, and small ones are for PIC model. For advancing time from t_1 to t_2, PIC model receives interpolation values of data at t_1 and at t_2 from MHD model at every step. On the other hand, at t_1, MHD model gets PIC data averaged over several steps around t_1.

The following procedure indicates how time advances from t_1 to t_2. (1) At $t = t_1$, physical quantities of MHD and PIC exist. (2) MHD model sends MHD information at t_1, to PIC model. (3) PIC model advances to $t = t_1 + \delta t$, where δt is time which correspond to several PIC steps. (4) PIC information averaged over the period form t_1-δt to t_1+δt is sent to MHD model. (5) MHD model goes ahead one step and reaches t_2. (6) PIC model receives MHD data at t_1 and at t_2. (7) PIC information from t_1 to $t_1 + \delta t$ which were obtained at (3) is removed. (8) PIC model advances to t_2. Figure 2 shows a schematic diagram of the multi-hierarchy model. The PIC domain covers the central region close to the neutral sheet, and the MHD domains are outside the PIC domain. The Interface domain is located between the PIC and MHD domains.

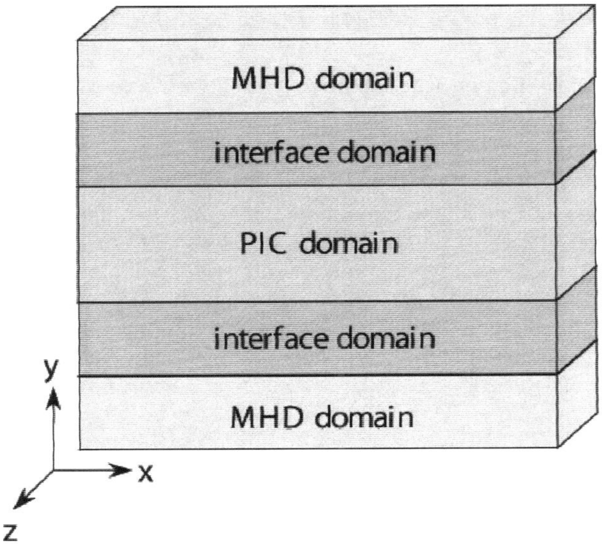

Figure 2. Schematic diagram of Micro-Macro Interconnected Simulation Model.

Before demonstrating MMIS of magnetic reconnection, some test problems are performed. The first test is the propagation of a linear Alfven wave. Smooth wave propagation was observed [7, 8]. The second test is the plasma injection from the MHD domain to the PIC domain. A smooth and continuous plasma flow propagation is observed, as shown in Fig. 3[9].

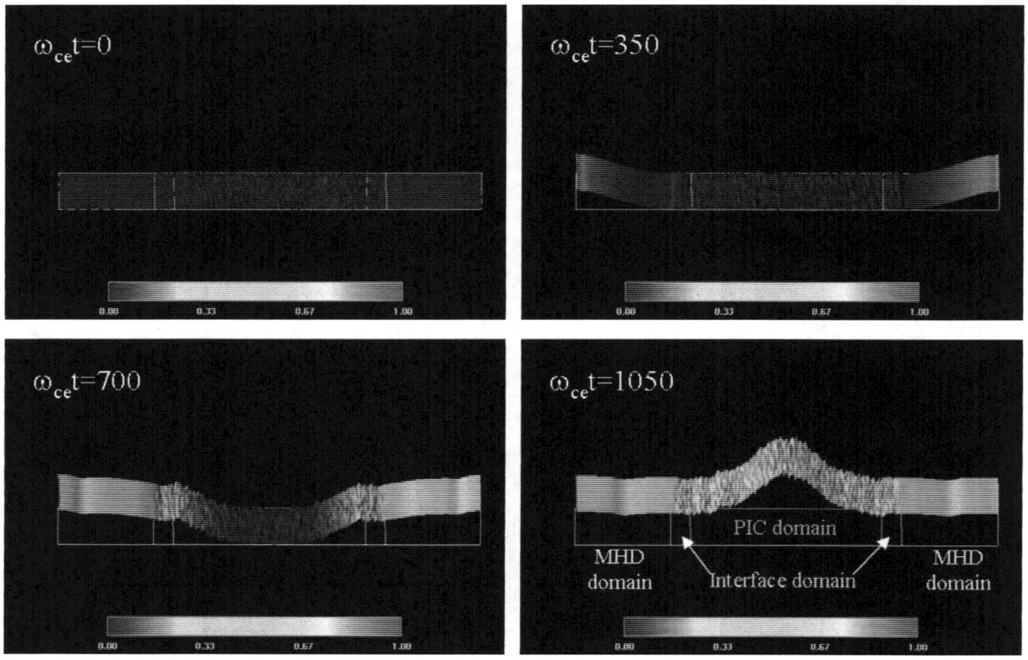

Figure 3. Time evolution of plasma density profile. Plasmas are injected from the MHD domains to the PIC domain.

Here, we present a multi-hierarchy simulation of magnetic reconnection [3]. The initial condition is given by an one-dimensional Harris-type equilibrium as $B_x(y) = B_0 \tanh(y/L)$ for the magnetic field, where B_0 is a constant and L is the scale height along the y axis. The system is periodic in the x and z directions and is free in the y direction. An external electric field $E_{zd}(t)$ is applied at the entire outside boundary of the MHD domain. Plasmas are injected inward in the y direction. The electric field $E_{zd}(t)$ in the MHD domain is programmed to evolve from zero to a constant value E_0. Figure 4 displays the magnetic lines of force (left panel) and fluid velocity vectors (right panel) in the (x,y) plane in the whole multi-hierarchy simulation domain at $\omega_{ce}t = 1323$. The lines of force and the fluid velocity are smoothly connected between the MHD and PIC domains via the Interface domain. An X point exists at the center. We can see that inflows come inward from the MHD domain and drive magnetic reconnection at the center of the PIC domain.

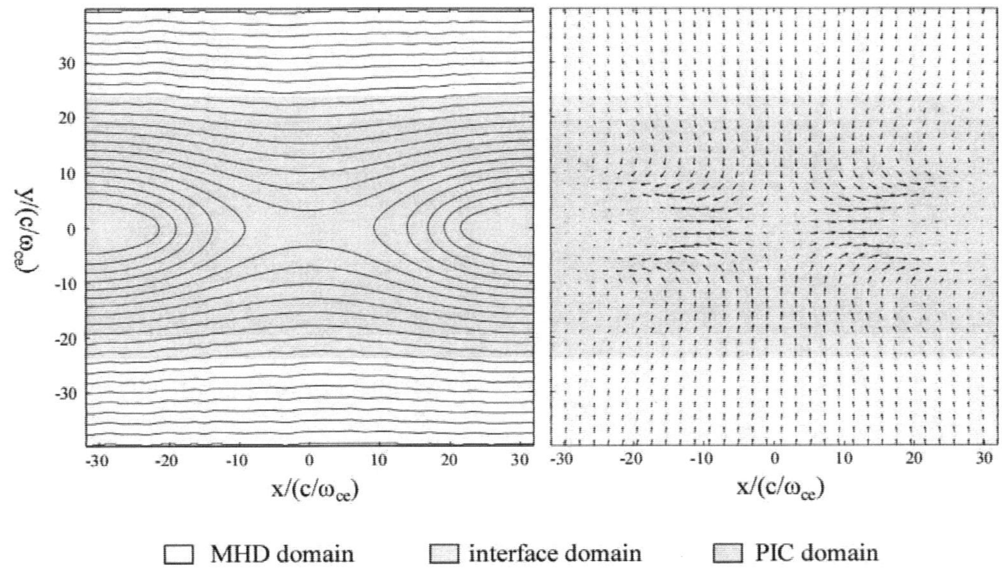

Figure 4. Magnetic lines of force (left panel) and fluid velocity vector (right panel) at $\omega_{ce}t = 1323$.

3. Equation-free projective integration method

The so-called Equation-free-projective integration (EFPI) method has been proposed and applied to a variety of multi-scale phenomena in engineering problems in which coarse-scale behaviour can be obtained through short-time simulations within the fine-scale models [2]. In EFPI, the simulation acts on two scales. The macro scale dynamics is determined by repeatedly extrapolating forward coarse scale estimates obtained from the short micro scale simulations. In the EFPI framework, the main tool that allows the performance of numerical tasks at the macroscopic level using the microscopic simulation codes is the so-called coarse time-stepper. It consists of three parts: lifting (mapping from coarse-macroscopic to fine-microscopic level), short time micro calculations around which the macroscopic calculations are wrapped and restriction (mapping from fine-micro scale to macroscopic level). The coarse time stepper is combined with time projection at macroscopic level, i.e. time projection of the coarse observables on the macroscopic scale. This structure is schematically presented in Fig. 5.

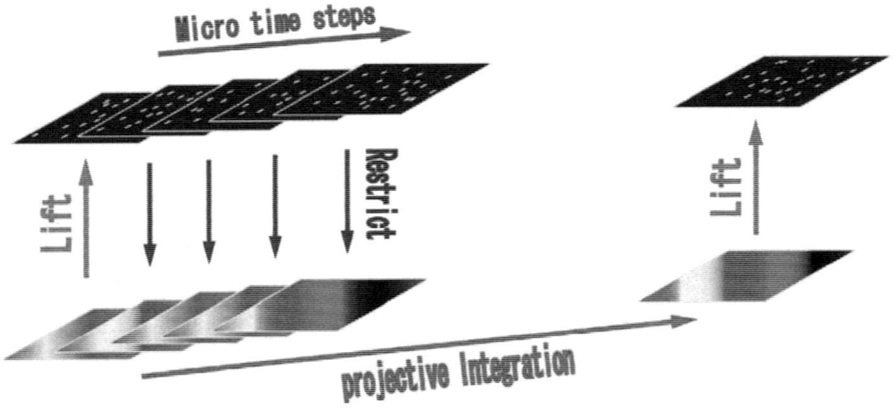

Figure 5. Structure of Equation Free Projective Integration (EFPI) method [13].

First attempt to apply EFPI to plasma physics was performed by Shay et al. , in the context of the ion acoustic wave paradigm [6]. In their EFPI scheme the electrons are assumed to be adiabatic, both the electron and ion velocity distributions are assumed to be the shifted Maxwellian and quasi-neutrality is proposed. The results of EFPI calculations are discussed with respect to the full micro-PIC simulations. At the first step the coarse observables are determined. First three moments: ion density, ion velocity and pressure are taken as the 'active' coarse observables, i. e. those macro variables which are directly computed forward in time. On the other hand, the electron density, electron velocity and electric field are taken as the 'passive' coarse variables, i.e. variables which are not calculated directly but from active macro observables. These observables are defined on the coarse mesh by the linear interpolation procedure. The micro quantities, the ion and electron positions are obtained through the lifting from corresponding densities and the ion and electron velocities are lifted from corresponding velocity distributions (approximated by the shifted Maxwellian). The PIC solver is then applied for the short time in order to ensure the system to stay near the so-called slow manifold. In other words, the implementation of the micro solver has to ensure the reconstruction of the values of the macro quantities which would be obtained under the same conditions but using only the micro solver. The coarse-macro observables are generated by the reversed operation – restriction. After the linear interpolation they are projected in time. Approximately time interval for the micro calculation is around 20 micro time steps and the macro time step is two order of magnitude larger than the micro time step. However, above procedure basically neglects kinetic effects in a plasma. The problems which appear in the reconstruction of the ion acoustic wave when the wave steepening is noticed were related to the particle trapping, non-Maxwellian features are violating quasi-neutrality. Instead of the moments, the wavelet technique for reconstructing the particle probability distribution function was indicated as a possible solution [10]. Furthermore, schemes to test an applicability of the EFPI framework, attempting to include the correlation effects in the nonlinear ion acoustic wave paradigm was undertaken by some of these authors [11]. That approach uses the marginal and conditional cumulative particle distribution functions in the ion phase space as the macro scale observables, which results in a better agreement with a fully kinetic electrostatic PIC solver.

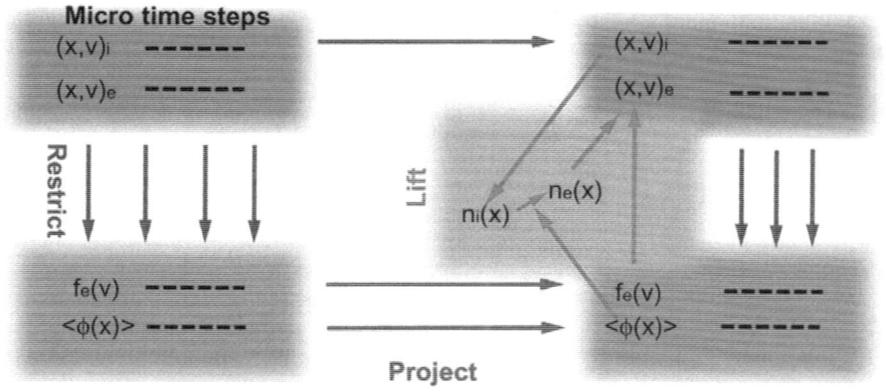

Figure 6. Structure of primal Equation Free Projective Integration Plasma Simulation [13].

Here, as an illustration aimed at exploring the feasibility of the EFPI method, we implement an original scheme, called, primal-EFPI to account for nonlinear and kinetic plasma effects [12, 13]. The basic platform for micro-simulation is the standard electrostatic PIC. Still, the working hypothesis is that the separation of scales allows us to assume that the ion dynamics is inherently coarse grained, or macroscopic, compared to electron micro-scale motion. Hence, individual ion orbits are tracked and extrapolated in time to make the projection. In contrast to the original EFPI, here, ions are not restricted via three PDF moments. Rather, ions are kept as they are, i.e., theoretically preserving nonlinear kinetic effects fully. Indeed, by inspecting ion orbits for an ensemble of test particles, we found that most projected individual orbits agree well with the original PIC prediction. We also note that a typical coarse projection time step (Δt_p) of , e.g., 100 times the micro-step (Δt), is still close to the intrinsic ion time step. Further, we find a non-uniform ion density from the projected ion orbits, while to lift ions, we actually just restart the ion motion. Next, we track the electric potential and coarse grain, i.e., the average over the electron plasma period ($2\pi/\omega_{pe}$) to smooth micro-scale fluctuations in order to extrapolate and project. Moreover, a simplifying (adiabatic) approximation for

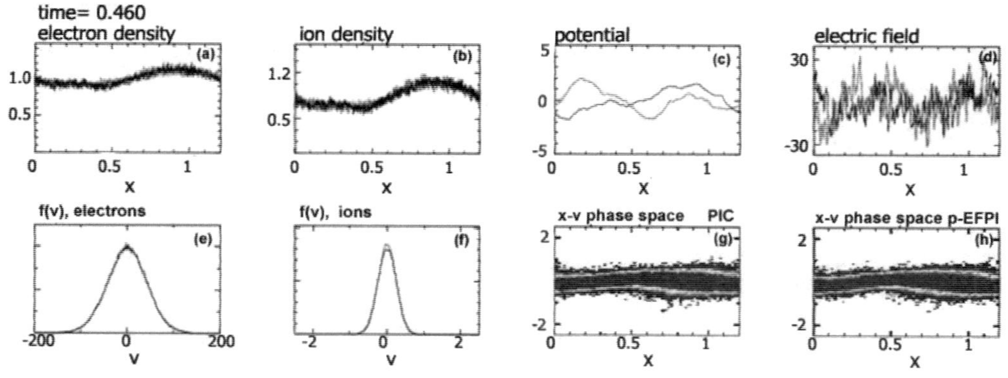

Figure 7. Snapshots of (a) electron, (b) ion density, (c) electric potential, (d) electric field and (e) electron, (f) ion PDF, for PIC (red) and p-EFPI (black) at $t = 0.46$; and ion phase space, PIC (g) versus p-EFPI[12].

electrons is not used. Instead, we self-consistently solve for non-uniform electron density from the Poisson equation, and user corresponding projected values for potential and ion density. By sampling the electron phase space, we find the standard electron velocity PDF, which appears rather smooth and easy to interpolate and project. Finally, to lift electrons, we use two projected PDFs, called marginal DFs, representing the velocity and real space distributions (number density). We performed a number of simulation runs within the same plasma variable ranges as in [6], by varying the other p-EFPI scheme parameters. Comparative snapshots at an earlier time $t = 0.46$ (in ion periods) are given in Fig. 7, to present the PIC data and the p-EFPI code results (red curve). Snapshots of the electron (1) and ion (2) density and PDFs, electric potential, and electric field at $t = 0.46$ are provided along with ion phase space plots, PIC versus p-EFPI (bottom right). As expected, a discrepancy appears in the potential and electric field (non-coarse, both macro- and micro-scale) as a phase mismatch, due to the interruptive nature of the p-EFPI simulation cycle. This stems from a difficulty in accurately reconstructing the phase relation in coherent particle dynamics, in particular, with intrinsically noisy PIC data. Our very recent optimization of the numerical scheme has improved the phase space matching. However, the smoothed variables, similar to particle density, PDFs, and even the ion phase space, compare well. Snapshots repeated at a later time, t=0.98, shown in Fig. 8, show less agreement in the nonlinear kinetic region. We also note that a difference in particle density defies simple electron adiabaticity.

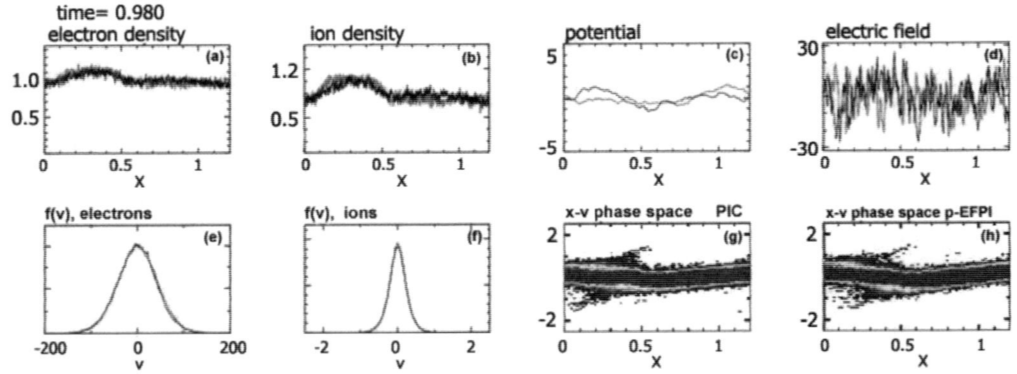

Figure 8. Snapshots of (a) electron, (b) ion density, (c) electric potential, (d) electric field and (e) electron, (f) ion PDF, for PIC (red) and p-EFPI (black) at $t = 0.98$; and ion phase space, PIC (g) versus p-EFPI [12].

Finally, to check the important energy conservation of the scheme, Figure 9 shows a plot of the comparative time evolution of the ES field energy, electron and ion kinetic and drift energies, and the total energy for PIC and p-EFPI, in total energy units. A phase-mismatch time lead in the ion wave kinetics compared to PIC is typically observed. While the p-EFPI projection step was modest (20-30 Δt), the actual agreement with full PIC is reasonable, which gives a speed-up factor of 2 in the 1D case. For EFPI feasibility as a multi-scale code, the speed-up depends on the smallest number of micro-steps (PIC) combined with the largest projection step possible. However, fundamental stability, as stated by the Courant condition, requires that $\Delta x/\Delta t > c_s$ for both micro- (PIC) and macro-projection grids, where c_s is the characteristic speed in a problem (here, the ion acoustic wave). Unexpectedly, this was found to appear as a physical effect, even if no explicit macro-scale equation was solved. Accordingly, for a large projection time step, it is necessary to change the fine (micro) to a coarse (macro) spatial grid. However, with p-EFPI, we maintained the original PIC fine resolution (512 grid points in space). Although we are in an early stage, the preliminary results seem to be promising.

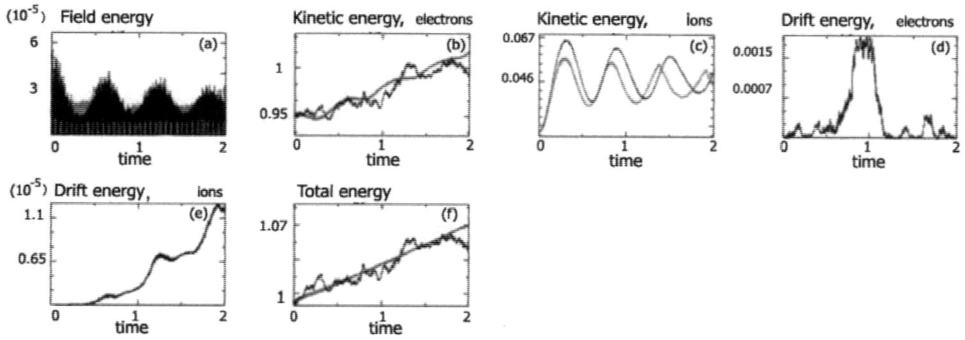

Figure 9. Time evolution of electric field, particle, and total energies for PIC (red) and p-EFPI (black) [12].

4. Summary

Development of multi-scale simulation methods is presently among hot topics in a variety of fields. In this paper, we show two typical methodologies, i.e. Micro-Macro Interlocked Simulation Method (MMIS) and Equation Free Projective Integration Method (EFPI). These method are under construction and we hope that these activity bridge to future holistic self-consistent simulation of fusion plasma device.

Acknowledgements

Stimulated discussions with T. Sato, K. Kusano, T. Sugiyama, and G. Stantchev are gratefully acknowledged. This study was performed under the auspices of and with the support of the National Institute of Natural Science (NINS) – National Institute for Fusion Science (NIFS) Collaborative Research Project (NIFS05KEIN0041, NIFS06KEIN0041, NIFS08KTAL015, NIFS09KNXN174, NIFS10KEIN1201) and NIFS Collaborative Program (NIFS04KDAT007, NIFS05KDAT007, NIFS07KDAN001). Partial support under the Ministry for Science and Technology of Serbia, Project No. 141034 is acknowledged.

References

[1] Sato T 2005 *J. Phys. Conference Series* **16**, 310
[2] Kevrekidis I G *et al* 2003 *Comm. Math. Sci.* **1**, 715
[3] Usami S, Ohtani H, Horiuchi R and Den M 2009 *Plasma and Fusion Res.* **4** 049
[4] Sugiyama T 2009 *J. Plasma and Fusion Res.* **85** 580.
[5] Sato T, Hasegawa H and Ohno N 2009 *Computational Science and Dicovery* **2** 015007
[6] Shay M A, Drake J F and Dorland B 2007 *J. Comp. Phys.* **226** 571
[7] Sugiyama T and Kusano K 2007 *J. Comp. Phys.* **227** 1340
[8] Usami S, Ohtani H, Horiuchi R and Den M 2008 *Comm. Comput. Phys.* **4**, 537
[9] Usami S, Ohtani H, Horiuchi R and Den M 2009 *J. Plasma and Fusion Res.* **85**, 585
[10] Maluckov A, Stanchev G, Shay M, Drake J F, Dorland W B, Ishiguro S and Škorić M M 2006 *33rd EPS Conf. on Plasma Physics* (Rome, Italy, 2006) http://epsppd.epfl.ch/Roma/pdf/P5_068.pdf
[11] Maluckov A, Škorić M M and Ishiguro S 2008 *Comm. Comp. Phys.* **4**, 556
[12] Škorić M M Maluckov A and Ishiguro S 2008 *Plasma and Fusion Res.* **3** 021 and 2009 *Plasma and Fusion Res.* **4**, 056
[13] Ishiguro S and Škorić M M 2009 *J. Plasma and Fusion Res.* **85** 593 (in Japanese)

25th Summer School and International Symposium on the Physics of Ionized Gases—SPIG 2010 IOP Publishing
Journal of Physics: Conference Series **257** (2010) 012027 doi:10.1088/1742-6596/257/1/012027

Influence of inelastic Rydberg atom-atom collisional process on kinetic and optical properties of low-temperature laboratory and astrophysical plasmas

A.N.Klyucharev[1], N.N.Bezuglov[1], A.A.Mihajlov[2], Lj.M.Ignjatović[2]

[1]Department of Physics, Saint-Petersburg University, Ulianovskaya 1, 198504, St-Petersburg, Petrodvorets, Russia

[2]Institute of Physics. P.O. Box 57, 11001, Belgrad, Serbia

E-mail: anklyuch@gmail.com

Abstract. Elementary processes in plasma phenomena traditionally attract physicist's attention. The channel of charged-particle formation in Rydberg atom-atom thermal and sub-thermal collisions (the low temperature plasmas conditions) leads to creation of the molecular ions – associative ionization (AI). atomic ions – Penning-like ionization (PI) and the pair of the negative and positive ions. In our universe the chemical composition of the primordial gas consists mainly of Hydrogen and Helium ($H, H^-, H^+, H_2, He, He^+$). Hydrogen-like alkali-metal Lithium (Li, Li^+, Li^-) and combinations (HeH^+, LiH^-, LiH^+). There is a wide range of plasma parameters in which the Rydberg atoms of the elements mentioned above make the dominant contribution to ionization and that process may be regarded as a prototype of the elementary process of light excitation energy transformation into electric one. The latest stochastic version of chemi-ionisation (AI+PI) on Rydberg atom-atom collisions extends the treatment of the "dipole resonant" model by taking into account redistribution of population over a range of Rydberg states prior to ionization. This redistribution is modelled as diffusion within the frame of stochastic dynamic of the Rydberg electron in the Rydberg energy spectrum. This may lead to anomalies of Rydberg atom spectra. Another result obtained in recent time is understanding that experimental results on chemi-ionization relate to the group of mixed Rydberg atom closed to the primary selected one. The Rydberg atoms ionisation theory today makes a valuable contribution in the deterministic and stochastic approaches correlation in atomic physic.

1. Introduction

Elementary processes in low-temperature plasma have traditionally been of interest for the physics of atom-atom collisions. There is a wide range of plasma parameters at which atoms and molecules play an important, and sometimes even major, part in ionization and may be considered as a prototype of elementary processes of transformation of radiant energy into electrical one. The program of Russian science research for 2008-2012 in the area "Current problems in physics" includes "... low-temperature plasma physics and the essentials of its use in production processes".

The advantages of such technologies include:
1. Relatively easy operation and control of plasma parameters.
2. Wide range of reagent energy in plasma.
3. Comprehensive processing of raw materials, environmental safety.

Journal of Physics: Conference Series 257 (2010) 012027 doi:10.1088/1742-6596/257/1/012027

4. Control of the composition of the resulting product.
5. The possibility of processing substances of various chemical composition and phase state (gaseous, liquid, solid).

Some examples of plasma technologies:
Plasma power engineering.
MHD converters of thermal energy into electrical one, thermionic energy converters, gas lasers.
Coal-based plasma power plants.

Over 50% of electrical and thermal energy in the world is produced based on coal (in Russia – 30%), while brown coal reserves are sufficient for 500-600 years.
Producing nano-coatings, such as nitrides, carbonitrides, borides, etc., fullerenes, nanotubes.
Producing superhard boron nitride, which is not found in nature.

Comprehensive utilization of toxic wastes, including fluorine containing compounds, etc.
The low temperature plasma physics includes branches devoted to elementary processes and their applications in up-to-date science-driven technologies. The interest to the work in these areas, that first showed over 100 years ago, today not only has not decreased, but quite the opposite – it is growing. First of all, it is caused by the search for new renewable energy sources, issues of energy saving, development of plasma chemistry and nanotechnologies. Even in the seemingly non-related issues of hot plasma physics, the low-temperature plasma in near-wall areas of fusion devices is the source of objectionable admixtures of heavy particles in the reactor core. One should mention also the projects on recovering the heat, emitted in nuclear reactors based on the fission principle, using various versions of plasma devices up to creating plasma generators of laser emissions, and also plasma methods of utilizing waste products of human life.

In spite of a large number of published works, the interest to this field of research is not waning. In this connection one cannot but mention the series of monographs by B. M. Smirnov on the role of excited atoms in low temperature nonequilibrium plasma, published in the period from 1968 to 1982 and introducing a new scientific field – plasma chemistry, that arose at the interface between physics of atomic collisions and chemistry of gas reactions leading to a change in the type of colliding particles or, in other words, to chemical transformations involving ions and excited atoms and molecules. Studying the physics of ionization processes in heavy particles collisions, we, in turn, are dealing with fundamental assumptions of the physics of collisions involving excited atoms, the physics of low temperature plasma, thus obtaining the opportunity to discover new phenomena and relationships. One such result has been the observation of the phenomenon of low temperature nonequilibrium plasma formation when gaseous medium is exposed to light with wavelengths corresponding to resonant optical transitions in atoms (photoresonant plasma). Such plasma at the initial stage of its development cannot be described using a traditional scheme of gaseous discharge development, which made it possible to research nonideal nonequilibrium plasma in the absence of an external electric field.

2. Ionization processes involving Rydberg atoms

In physics literature, Rydberg name began to be used for highly excited states of atoms and molecules after the publication of Rydberg's pioneer work on atom spectroscopy in 1890.

Atoms in highly excited states exist in the interstellar medium exposed to ultraviolet and X-rays from space objects, and to high energy particles with shock waves resulting from supernova explosions and interstellar clouds collisions. Thus, astronomers observe infrared radiation caused by transitions between excited states of a hydrogen atom with the values of principal quantum number n of the order of 10. Based on infrared radiation, one can judge on the population of Rydberg atom states in stellar atmospheres, atmospheres of cooling stars (dwarfs), planetary nebulas, and clouds of interstellar gas, as the lines of recombination spectrum of Rydberg states correspond to this range of wavelengths.

In the 1980s laser methods of investigating Rydberg atoms (RA) became widely accepted. This not only allowed increasing the effectiveness of classical single-photon light excitation, but made it possible to use two-photon excitation "violating" the rule of non-conservation of orbital quantum number L in the optical transition, and the methods of intra-Doppler spectroscopy free from the necessity of accounting for Doppler broadening of spectrum lines, etc. New, unconventional for classical spectroscopy, research methods emerged, for example, the method of "photon echo".

Interest to the processes of collisional ionization in highly excited atoms is caused by their fundamental significance for atomic physics and the wide possibilities of their practical use in laser atom cooling, Bose-Einstein condensation, controlling excitation processes of single atoms, creating quantum computer logical units, quantum teleportation, etc. It should be noted that the first results of quantitative research on ionization processes involving RA within the wide variation range of principal quantum number values were obtained in Refs. [1] (experiment) and [2,3] (theory). Contemporary systems of laser excitation allow obtaining RA in concentrations comparable to those of atoms in unexcited states. Even with the selective laser excitation of resonant (first excited) atomic levels, Rydberg states may be populated due to what is known as "energy pooling processes".

Based on the state of colliding particles in the entrance and exit reaction channels, the whole aggregate of ionization processes in atomic and molecular collisions can be divided into processes of collisional ionization and those of chemi-ionization. This somewhat arbitrary differentiation is caused in the first place by the ratio of the potential energy (excitation energy) of colliding particles to their kinetic energy (relative motion energy).

The range of kinetic energies of atoms, being of interest for low temperature plasma physics, is 10^{-9}–10^{3} K, which corresponds to conditions from Bose-Einstein condensate to the plasma of high-current arc. In experiments on laser atom cooling, the following division of collisions is accepted, based on the relative energy motion: thermal collisions (0.1-1 eV), subthermal collisions (10^{-3} eV) and cold collisions ($< 10^{-3}$ eV); each range of gaseous medium temperatures has its own corresponding set of primary processes leading to ionization.

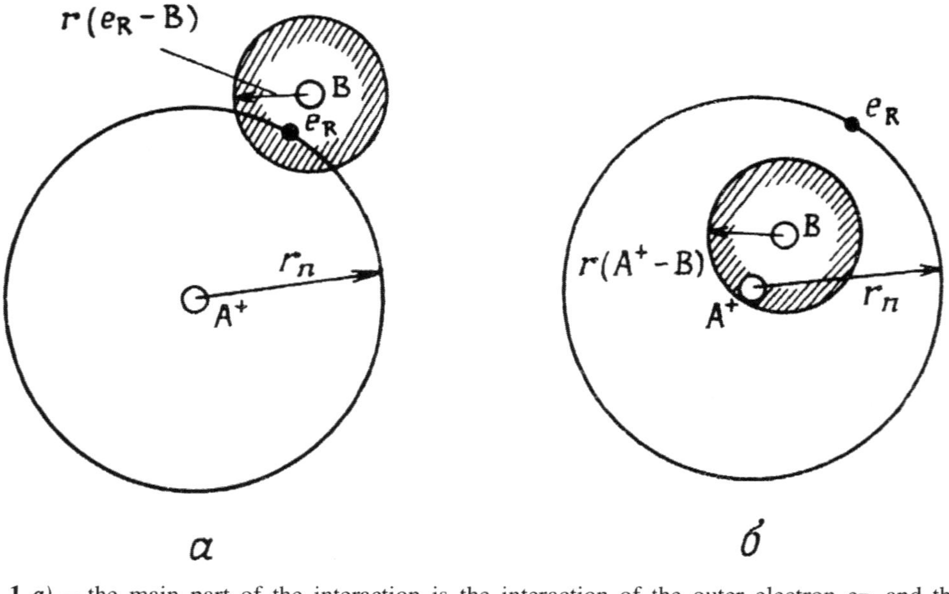

Figure 1 ***a)*** – the main part of the interaction is the interaction of the outer electron e_R and the colliding atom B; ***b)*** – e_R interacts with A^+ – B subsystem.

Authors of Ref. [4] suggested a dipole resonant mechanism leading to quantum transitions of a highly excited (Rydberg) electron in inelastic collisions involving Rydberg atoms. In the course of such a collision it is possible to identify the charge transfer mechanism of the Rydberg atom's positive nucleus on the unexcited colliding atom, at the distances lesser than the radius of the excited atom (see Figure 1). In this case the loosely bound outer electron there is in the field of a variable dipole moment depending on the internuclear distance. In the zone of thermal particle velocities such a field, induced by the charge transfer process, may be evident at significant distances. It should be noted that the principal contribution into the theoretical development of such processes was made by E. Fermi in 1934, who suggested treating a Rydberg atom as a system comprising an ion core and a quasi-free Rydberg electron (RE).

Collisions involving excited atoms and resulting in ionization are accompanied by interactions between the initial discrete state and not only ionization continuum, but also the Coulomb condensation of Rydberg atom energy terms. Thus, transitions in which upper highly excited states become populated are possible in addition to ionization. They, in turn, participate in the processes of associative and direct Penning ionization.

$$X^{**} + X \rightarrow X_2^+ + e \qquad \text{(AI)} \tag{1}$$

$$X^{**} + X \rightarrow X + X^+ + e \quad \text{(PI)} \tag{2}$$

and also in the production of a pair of a positive and negative ions

$$X^{**} + X \rightarrow X^+ + X^-. \tag{3}$$

2.1. *Chemi-ionization in RA-A collision*

Due to unfailing interest in AI and PI processes involving RA, several methods of calculating respective reaction constants have been developed in comparatively short period of time.

In Duman and Shmatov's [2] theory, quantum mechanical approach was used accounting for the character of nucleus-nucleus interaction in the course of collision. Calculation methods suggested by Mihajlov and Janev (1980) are based on quasi-classical process description. Both these theories can be applied to the range of internuclear distances of a collisional process $R \ll r_n$, where $r_n \sim n^2$ is the mean value of the RA effective radius.

In this case inelastic ionization processes can be treated as the result of the interaction between the valence electron H and the dipole moment of the subsystem $A^+ + A$, which may be interpreted as the mechanism of resonant charge transfer of the RA ion core on the colliding particle. It is possible to apply adiabatic approximation and the classical trajectory concept in these cases, provided the ratio of de Broglie wavelength λ of A atom to the characteristic dimension of RA is $\lambda/2 \cdot r_n \ll 1$. Besides, to differentiate between the degree of freedom of the atom and that of the electron within the scope of Born-Oppenheimer approximation, it is necessary for the velocity value of the relative motion of atoms $\upsilon_c \sim E^{1/2}$ to be much smaller than the motion velocity on the orbit of the Rydberg electron. Here the atomic system of units is used (a.u.). The electron states in $A^+ + A$ subsystem may be treated as the electron states of molecular ion A_2^+, correlating adiabatically with the electron states of the given subsystem with $R \rightarrow \infty$, where R is the internuclear distance.

Let us assume that for internuclear distances R at which inelastic AI and PI processes occur, the valence electron of A atom may be considered as simultaneously belonging to $A^+ + A$ subsystem. Then the main contribution to the inelastic process under consideration is made by the interaction within the range of R values, where the energy exchange between the RE and the electron component of $A^+ + A$ subsystem is of almost resonant character.

Later on, Mihajlov and coauthors [5] extended the theory of such processes, that had originally been developed for hydrogen-like alkali atoms, to hydrogen and helium atoms with regard to astrophysical tasks. At the same time, the interest has not waned in AI and PI processes involving alkali atoms. See [6,7].

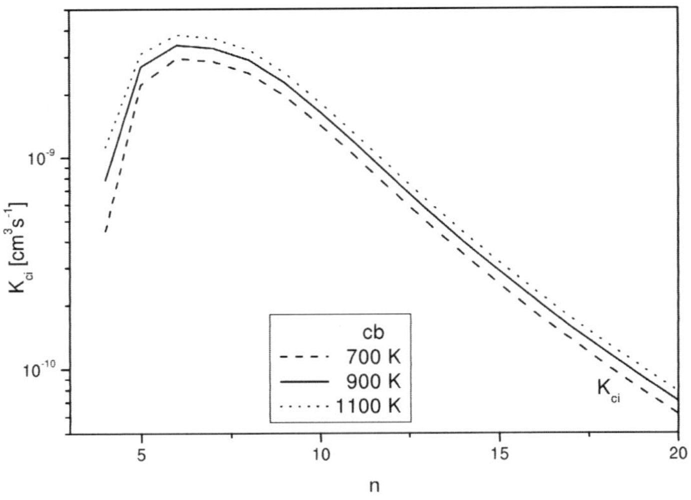

Figure 2 The total rate coefficients for chemi-ionization [19] in Li^*(n)+Na collisions in the case of crossed beam.

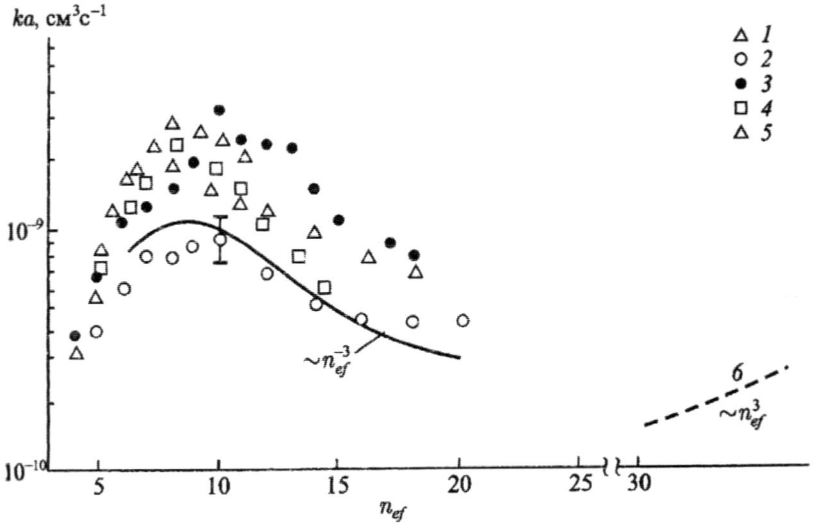

Figure 3 Summarizing picture of the hemi-ionization constants in RA-A collisions as a function of the effective quantum number of excited states [8].

It should be noted that theoretical works appearing immediately after the publication of the first experimental data to a large extent encouraged further experimental studies in this field using the opportunities provided by contemporary laser technologies. Figure 2 shows the relationship K(n) of chemi-ionization rate constants for Rydberg states of lithium atoms (Li^{**} + Na – asymmetrical case), calculated under the dipole resonant model. The Figure 3 shows relationships K(n) obtained in experiments of various kinds before 1990 by Klyucharev (1993).

2.2. Ionization in binary collisions of Rydberg atoms.

Ionization in collisions of two RAs remains a relatively little studied field of the physics of atom-atom collisions. In this case the total energy of atoms exceeds the A atom ionization potential. As a result, atomic RA-states, that are stable with respect to ionization, are converted into quasi-molecules, that can later be autoionized forming $A^+ - A^-$ states, whose excitation energy is less than the energy of $(RA)_2$ quasi-molecule with $R \to \infty$ (See Figure 4).

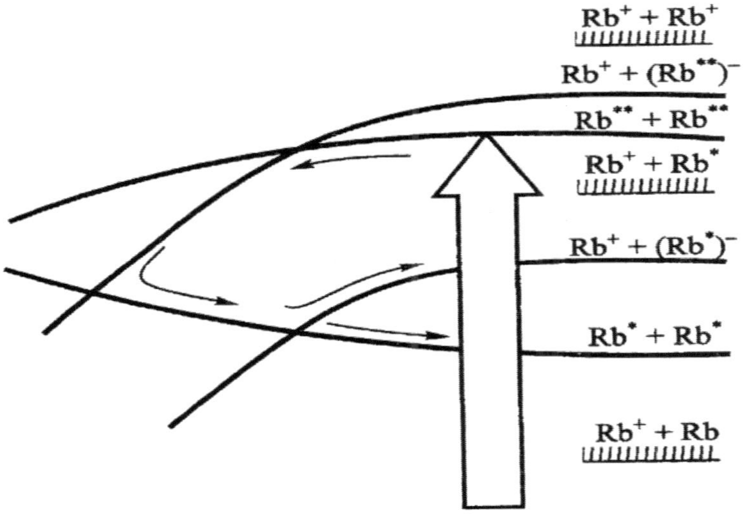

Figure 4 Schematic illustration of potential curves quasicrossing for the collisions of two highly excited alkali-metal atoms.

$$RA + RA \to A^+ + A^* + e \tag{4}$$

$$RA + RA \to (A_2^+)^* + e \tag{5}$$

$$RA + RA \to A^+ + (A^-)^* \tag{6}$$

In [6] the data are given on Penning ionization constants in collisions of alkali Rydberg atoms. Autoionization widths of intermediate quasi-molecular states were expressed through cross-sections of photo-ionization from Rydberg atom levels and the strengths of transition lines between them. Cross-sections and line strengths were calculated in quasi-classical approximation, taking into account the quantum character of RE motion and approximation of a straight-line trajectory of particle motion. The dipole-dipole approximation of atom-atom interaction works better for subthermal collisions, as in this case the time of interaction increases at large R.

2.3. Ionization-recombination processes involving RA in astrophysical objects.

In plasma conditions with the temperature of plasma electrons T_e, highly excited atomic states may be produced due to recombination processes of various kinds. One can give the example of photo-recombination corresponding to the case of low electron concentrations.

$$A^+ + e \to RA(n, L) + h\nu \tag{7}$$

Processes of ion-electron recombination in low-temperature weakly ionized plasma are the effective channel of atom formation in highly excited states. This can be well illustrated by ionization reactions involving RA in the helium enriched atmospheres of white dwarfs and in the solar photosphere, see e.g. [7]. In these conditions chemi-ionization proves to be an open, reverse to recombination, channel [9]:

$$He^{*}(n) + He \leftrightarrow He_2^{+} + e \tag{8}$$

$$He^{*}(n) + He \leftrightarrow He^{+} + He + e \tag{9}$$

The same statement is valid for the hydrogen of the solar atmosphere. As the temperature of particles increases, the role of these processes decreases. Thus, in white dwarf atmospheres, their contribution into ionization decreases relatively from 100 at T = 12000 K and 10 at T = 16000 K to 0.1 at T = 20000 K and 0.01 at T = 24000 K, when electron-atom collision processes come to the foreground

$$He^{*}(n) + e \rightarrow He^{+} + e \tag{10}$$

In theory, according to [10], parameters of reactions (8, 9) dominate in the balance of recombination and ionization processes at temperatures within the range of $12000 \leq T \leq 30000$ K. In the course of photorecombination leading to mutual neutralization of charged particles of hydrogen plasma, about 100% of the total photorecombination flux results in the population of lower excited states of the hydrogen atom: $1 \leq n \leq 7$ at $T_e \geq 3500$ K, 80% at $T_e \approx 1150$ K and 70% at $T_e \approx 350$ K. At $T_e > 24000$ K, excitation and ionization by electron impact, as well as three-body recombination, are dominating in the balance of Rydberg atom population.

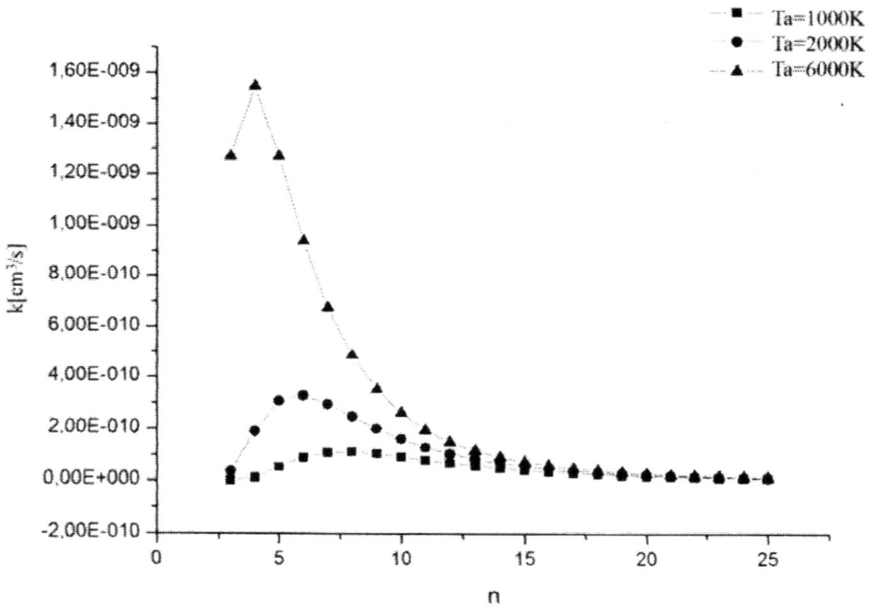

Figure 5 The chemi-ionization rate constants in hydrogen.

The role of processes (8, 9) in the RA population in the helium-enriched atmosphere of white dwarfs is noticeably bigger than it is in the hydrogen case of the solar photosphere, as helium ionization potential noticeably exceeds the same potential for the hydrogen atom. Therefore, at comparable values of pressure and temperature, the degree of helium atmosphere ionization is considerably less than in the hydrogen situation. In the solar photosphere, near its temperature minimum $T \approx 4000$ K, chemi-ionization–recombination balance plays an important role in the wide range of temperatures. In the helium-enriched atmosphere, the ratio of ionization fluxes to a dissociative recombination channel in the RA population balance decreases from 0.6 (7000 K) to 0.16 (30000 K) for n = 3 and from 0.12 (7000 K) to 0.03 (30000 K) for n = 10. Figure 5 shows results from [5] of the aggregate chemi-ionization constant calculation for hydrogen for the ranges: $n = 4 \div 25$ and $10^3 \leq T \leq 6 \cdot 10^3$ K, while analogous results for helium $n = 3 \div 10$ and $4{\cdot}10^3 \leq T \leq 3 \cdot 10^4$ K can be found in [10].

In the space plasma with low density of charged particles in the balance of recombination population of Rydberg atoms, the processes of collisions with atoms and molecules of interstellar medium can be neglected compared to the emission to lower levels. Besides, the distribution of highly excited atoms over *n* in space plasma is affected by the relict radiation with temperature 2.8 K. In these conditions, the profile of RA spectrum lines is controlled by Doppler effect and processes of Stark broadening in electric fields, as well as by collisions with protons and electrons resulting in Lorenz type of profile. In its turn, Zeeman effect leading to level splitting and polarization of RA radiation makes it possible to receive a large body of data on magnetic properties of space objects. It should be also noted that astrophysicists have at their disposal the possibility of investigating the spectrum of radio emission from other galaxies, based on the effect of RA induced emission when such lines are absorbed.

2.4. *Cold and ultracold collisions.*

Cold and ultracold collisions hold a special place in current research on atomic and molecular physics, and processes in condensed matter. Deep understanding of the physics of such processes encourages the development of new methods of controlling elementary processes, precise measuring of their parameters, and processes of weak interactions between particles. This was the beginning of new trends in atomic physics research: Bose-Einstein version of matter condensation at low gas pressures, atomic interferometry, optic lattices for atomic motion control, etc. Today all these are included under new optical physics, where cold atomic collisions play the central part. If the values of de Broglie wavelength are of the order of 10^{-3} nm, the processes of particle collisions are traditionally classified as classical physical interactions (low temperature plasma, molecular beams). While the temperature interval 1μK÷1mK of atomic collisions can be realized by the methods of intra-Doppler spectroscopy, the temperatures 1μK÷1mK obtained in magneto-optical traps have made it possible to realize the idea of Bose-Einstein condensate (BEC) in the physics of optical phenomena. Even in case of "intra-Doppler" cooling, the scale of interatomic interactions becomes comparable with the de Broglie wavelength of particles, which leads to the observation of new wave interferential interactions. At below 1μK, when critical concentration of condensed phase is reached, the effects of quantum degeneracy begin to show in the ensemble of particles. Besides, the kinetic energy imparted to the particle in the atomic collision becomes lower than the energy imparted to it by a scattered photon. This type of collisions is known as ultracold ones. The first publication on quantum effects in cold and subcold collisions is Ref. [11].

It is clear that in the conditions when the collisions involving excited (non-metastable) particles may be considered as cold (or ultracold), emission time becomes much less than colliding particles interaction time, which prohibits the processes of atomic collisions. For cold and ultracold atoms in the light field, the scales of time length and spectrum line widths are different from their classical analogs. Thus, nonhomogeneous Doppler broadening of the spectrum line may become less than its natural width, interaction time – several times longer than the time of particle spontaneous emission, de Broglie wavelength – several orders greater than atomic dimensions. Compared to thermal range, only small number of partial scattered waves will contribute into the wave function of Schrodinger equation. Extremely small values of Doppler broadening in these cases allow to obtain ultrahigh resolution in the methods of molecular spectroscopy. In case of cold atoms, multiple interaction between pseudo-bound particles and photons is possible, allowing investigation of the fine effects of the emission intensity and polarization on the interaction mechanisms in collisional processes.

Physics terminology in the field of cold and ultracold collisions, distinguishes two ionization types: photoassociative photoionization (PAPI)

$$Na(3^2S) + Na(3^2S) + h\nu_1 \rightarrow Na_2^*, \tag{11}$$

$$Na_2^* + h\nu_2 \rightarrow Na_2^+ + e, \tag{12}$$

and photoassociative autoionization (PAAI)

$$Na(3^2S) + Na(3^2S) + h\nu_1 \rightarrow Na_2^* \tag{13}$$

$$Na_2^* + h\nu_2 \rightarrow Na_2^{**} \tag{14}$$

$$Na_2^{**} \rightarrow Na_2^+ + e \tag{15}$$

Blangé an coauthors in [12] demonstrated that in the thermal range of collision energy ($T \approx 10^2$ K), AI process involving two resonant excited 3^2R sodium atoms results in the population of the same lower vibrational excited states of Na_2^+ ion, as PAAI reaction (13) does at $T = 10^2$ K. This appears to confirm that in both cases the pattern of molecular terms involved in reaction is the same.

3. Characteristic properties of stohastic ionization of Rydberg collision complex

Theoretical papers cited above did not take into account the possibility of multiple crossings of the initial energy-term of the excited quasi-molecule A_2^* of the ion molecular attraction term A_2^+, although, in [4] dipole resonant mechanism was initially suggested as a mechanism of transitions between the levels of a highly excited atom in collisions with the ground-state atoms of the same type.

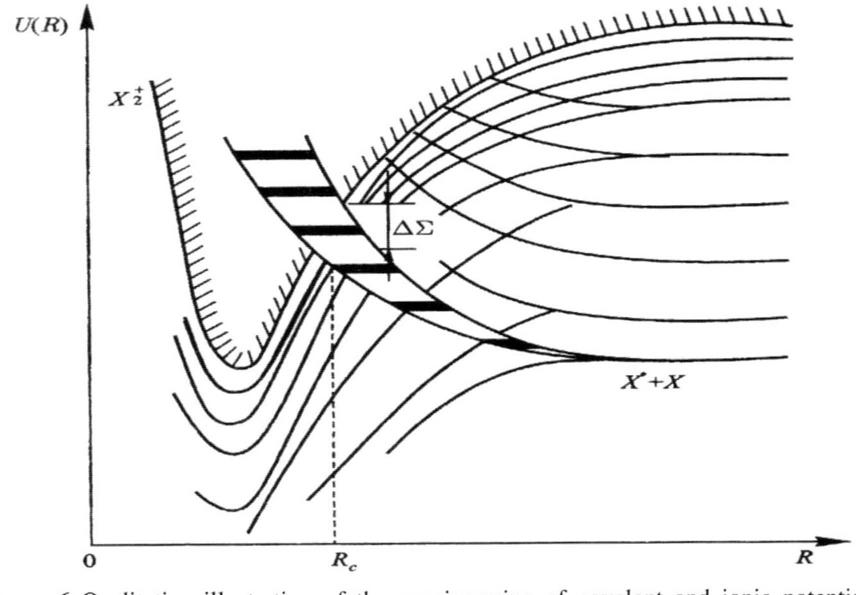

Figure 6 Qualitative illustration of the quasicrossing of covalent and ionic potential curves in "diffusion" model

The attempt to account for this possibility by traditional means of the physics of atom-atom collisions could not yield any observable result due to a large number of energy-terms crossing points. A change in RE energy at each of the multiple crossings of energy-terms $\delta E \leq 1/n^3$, so that $\delta E/E_n \sim 1/n$, where $E_n = 1/n^2$ is the valence electron binding energy. This circumstance led the authors of [13] to the idea of "diffusion" approach to the problem of RA-A collisions (see Figure 6). They meant the diffusion in energy states of an excited quasi-molecule in the course of one collision. Besides, authors of [14] had successfully applied RE diffusion in discrete RA spectrum to explain the abnormally high value of photoionization cross-section in the experiments on laser irradiation of a gas target. It was logical to suppose that quasi-monochromatic electric dipole field produced in the course

Journal of Physics: Conference Series **257** (2010) 012027 doi:10.1088/1742-6596/257/1/012027

of the resonant charge transfer, which is built into the model of a dipole resonant mechanism, may lead to the RE diffusion in discrete energy spectrum in the case of RA chemi-ionization as well. Thus, the dipole resonant model of chemi-ionization processes could be further developed, taking into account redistribution of initial excitation over the grid of energy terms of a Rydberg quasi-molecule. This redistribution was modeled as RE diffusion in the Rydberg energy spectrum of excited states, using diffusion equation of Fokker-Planck-Kolmogorov type. Results subsequently obtained in this field allowed making a conclusion on the resonant character of RE diffusion and it being influenced by the internal nonlinear dynamic resonances arising when overtones of RE motion frequency coincide with the electron transition frequency. As a result, RE motion in the energy spectrum turns into a random wandering over a grid of crossing quasi-molecule potential energy terms. This results in the time evolution of distribution f(n, t) for RE.

The results of research in the field of non-linear mechanics indicate that what is known as dynamic chaos conditions should be treated as typical rather than exceptional situation for the time dynamics of Hamiltonian systems with a small number of degrees of freedom. In the physics of atom collisions, the chaos conditions may arise as a consequence of local instability of valence electron trajectory orbits relative to indefinitely small perturbations. In general, there are always regions of the phase space and parameters of interacting particles, for which the dynamics of Hamiltonian system is stochastic (see[15,16]).

An example of an important physical system with the properties of significant trajectory instability is a quasi-classical model of an alkali atom Rydberg electron (RE) in the quasi-monochromatic electric microwave field of a dipole resonant mechanism suggested by [4]. In [17] gives the results of numerical simulation of RA diffusion ionization for alkali atoms induced by RE trajectory chaotization on a Keplerian orbit, induced by the electric microwave field generated in the quasi-molecule system in a single atom-atom collision.

Let us briefly discuss the parameters of stochastic diffusion model (the region where dynamic chaos conditions exist).

1. For stochastic instability to develop, the intensity of the internal electromagnetic field should exceed the critical value that can be estimated using what is known as resonance overlapping criterion, while the range of values n_{eff}, in which stochastic diffusion of RE is possible, is determined by Chirikov's criterion [18].

2. At the lower limit of *n* stochasticity region $n_{eff} \geq n_{min}(R)$, the diffusion is limited due to an increasing energy gap between neighboring energy-terms of excited atomic states.

3. The corresponding upper limit $N_{max}(R) \geq n_{eff}$ is related to the emergence of the effective ionization channel for Rydberg quasi-molecule at large values of *n*.

4. Effective diffusion time T_{diff} has to be less than the time of collision τ itself.

Thus, RE stochastic diffusion during the time of one collision has first of all to become evident in the range of subthermal and cold collisions.

4. Conclusion

Dipole resonant mechanism of RA deexcitation, suggested by Smirnov and Mihajlov (1971) and later extended by Mihajlov and Janev [3] to chemi-ionization processes, implies the generation of a microwave electric field in a single RA+A collision, which leads to the instability of RE motion in Keplerian orbit and eventually to the stochastic diffusion of RE over excited states and consideration of a larger number of initial quasi-molecular energy-terms than it follows from the "classical" consideration taking into account orbital moments of colliding particles. For RA with an alkali atom structure at certain conditions in a weak electric field it is possible that particular transitions become forbidden, leading to the closure of some "stochastic" ionization channels, which in principle may lead to the disappearance of infrared lines in the spectra of astrophysical objects.

5. Acknowledgements
The presented material to a large extent is a result of joint research carried out recently within the scope of the agreement on cooperation in science between the Institute of Physics of Belgrade University and the Department of Physics of St. Petersburg University.

The authors express their gratitude to Prof. Yu. N. Gnedin (Pulkovo Observatory of the Russian Academy of Science) and Prof. M. S. Dimitrijević (Belgrade Observatory) for the discussion of astrophysical aspects of the work.

References
[1] Devdariani, A. Z., Klyucharev, A. N., Lazarenko, A. V. and Sheverev, V. A.: 1978, Pis'ma Zh. Tekh. Fis., 4, 1013.
[2] Duman, E. A. and Shmatov, I. P.: 1980, Sov. Phys. JETP, 51, 1061.
[3] Mihajlov, A. A., Janev, R.: 1980. Phys. Rev. A, 12, 819.
[4] Smirnov, V.M., Mihajlov, A. A.: 1971, Opt. Spectrosc., 30, 984.
[5] Mihajlov, A. A., Ignjatović, Lj. M., Vasilievć, M. M. and Dimitrijević, M. S.: 1997, Astron. Astrophys., 324, 1206.
[6] Klyucharev, A. N., Bezuglov, N. N., Matveev, A. A., Mihajlov, A. A., Ignjatović, Lj. M., Dimitrijević, M. S.: 2007, New Astron. Rev., 51, 547.
[7] Gnedin, Yu. N., Mihajlov, A. A., Ignjatović, Lj. M., Sakan, N. M., Srećković, V. A., Zakharov, M. Yu., Bezuglov, N. N., Klyucharev, A. N.: 2009, New Astron. Reviews, 53, 259.
[8] Klyucharev, A. N.: 1993, Physics-Uspekhi (USSR), 36, 486.
[9] Mihajlov A. A, Ignjatović Lj. M., Srećković V. A. , Djurić Z. G.: 2008, JQSRT, 109, 853.
[10] Mihajlov, A. A., Ignjatović, Lj. M., Dimitrijević, M. S. and Djurić, Z.: 2003, Astrophys. J. Suppl. Ser., 147, 369.
[11] Vigué, J.: 1986, Phys. Rev., A., 34, 4476.
[12] Blangé, J. J., Uzbain, X., Rudolph, H., Dijkerman, H. A., Beijerinck, H. K. W. and Heideman, H. G.: 1996, J. Phys. B. 29, 2763.
[13] Devdariani, A. Z., Klyucharev, A. N., Penkin, N. P. and Sebyakin, Yu. N.: 1988, Opt. Spectrosc., 64, 706
[14] Delone, N., Kraynov, V., Shepelyansky, D.: 1983, Physics-Uspekhi (USSR), 26, 551.
[15] Zaslavsky, G. M., Sagdeev, R. Z., Usikov, D. A., Chernikov, A. A.: 1993, Weak chaos and quasi-regular structures. M.: Science, (in Russian).
[16] Bezuglov, N. N., Borodin, V. M., Ekers, A. and Klyucharev, A. N.: 2002, Opt. Spectrosc., 93, 661.
[17] Zakharov, M. Yu., Bezuglov, N. N., Klyucharev, A. N., Matveev, A. A., Beterov N. N., Dulye, O.: 2010, Russian Journal of Physical Chemistry, in press.
[18] Chirikov, B. V.: 1979, Phys. Rev., 52, 265.
[19] Ignjatović, Lj. M., Mihajlov, A. A. and Klyucharev, A. N.: 2008, J. Phys. B., 41, 025203.

Journal of Physics: Conference Series **257** (2010) 012028 doi:10.1088/1742-6596/257/1/012028

Case studies on recent Stark broadening calculations and STARK-B database development in the framework of the European project VAMDC (Virtual Atomic and Molecular Data Center)

S Sahal-Bréchot[1]
Paris Observatory, CNRS-UMR 8112 and University Pierre et Marie Curie, LERMA, 5 Place Jules Janssen, 92190 Meudon, France

E-mail: Sylvie.sahal-brechot@obspm.fr

Abstract. Stark broadening theories and calculations have been extensively developed for about 50 years. The theory can now be considered as mature for many applications, especially for accurate spectroscopic diagnostics and modelling. In astrophysics, with the increasing sensitivity of observations and spectral resolution, in all domains of wavelengths from far UV to infrared, it has become possible to develop realistic models of interiors and atmospheres of stars and interpret their evolution and the creation of elements through nuclear reactions. For hot stars, especially white dwarfs, Stark broadening is the dominant collisional line broadening process. This requires the knowledge of numerous profiles, especially for trace elements, which are used as useful probes for modern spectroscopic diagnostics. Hence, calculations based on a simple but enough accurate and fast method, are necessary for obtaining numerous results. Ab initio calculations are a growing domain of development. Nowadays, the access to such data via an on line database becomes crucial. This is the object of STARK-B, which is a collaborative project between the Paris Observatory and the Astronomical Observatory of Belgrade. It is a database of calculated widths and shifts of isolated lines of atoms and ions due to electron and ion collisions. It is devoted to modelling and spectroscopic diagnostics of stellar atmospheres and envelopes. In addition, it is relevant to laboratory plasmas, laser equipments and technological plasmas. It is a part of VAMDC (Virtual Atomic and Molecular Data Centre), which is an European Union funded collaboration between groups involved in the generation and use of atomic and molecular data.

1. Introduction

Atomic physics in plasmas has been a basic tool in Astrophysics for many years. In addition, a number of its developments have been stimulated not only by the advances in Astrophysics, but also by the needs in laboratory and technological plasmas (tokamaks, laser produced plasmas…). We will focus to astrophysical plasmas needs in the present paper. A major interest consists in the understanding of the evolution of the Universe, of the birth, growing, evolution and death of stars, which eject their material into the interstellar medium at the end of their life. The Big Bang explosion was responsible for the production of very light elements hydrogen and helium (and also lithium for a little part) and next the first generation of stars was born. These stars were extremely metal-poor. Then other elements (the

[1] To whom any correspondence should be addressed.

Journal of Physics: Conference Series **257** (2010) 012028 doi:10.1088/1742-6596/257/1/012028

"metals") were synthesized inside stars, and later on ejected back into the interstellar medium. The molecular regions are the nurseries at the source of birth of the new stars. Then the next generations of stars are the result of seeds from the previous ones, and so on. And the "metallicity" of the new generation increases.

Fusion reactions synthesize the elements up to iron. Heavier elements, such as platinum or gold, cannot be formed in such a manner, and are created by neutron-capture reaction: this is due to the fact that for heavy elements the high number of protons prevents nuclear reactions between charged particles.

Thus, to be more specific and to make this scenario scientifically precise with a quantitative modelling, it is necessary to accurately describe the stellar evolution, and the formation of elements, which are closely connected. Low-mass stars, such as the sun, have a long lifetime (billions of years) and end as planetary nebulae and then white dwarfs. More massive stars evolve at a faster rate and live only millions of years. They end their life in a gigantic supernova explosion, and the remnant is a neutron star (a pulsar). To make progress in these developments chemical abundances are crucial parameters to be determined. This needs an accurate interpretation of the detailed line spectra of the stellar objects. This also needs a quantitative modelling of stellar interiors and to calculate accurate radiative opacities in stellar envelopes.

In fact, thanks to the fantastic developments of the spectral resolution, of the sensitivity (high S/N) of the recent past years, and to large ground-based telescopes and space-born missions which allow to observe in all ranges of wavelengths (from γ to radio), this has become possible. Therefore an accurate analysis of very weak lines has been more and more possible. In particular the analysis of the spectra of the heavy elements which are only traces, will give the necessary information for understanding their creation through neutron capture. In another domain, interpretation of the spectra of white dwarfs, which are very faint, allows to understand the evolution of these very old stars, which are close to death.

For interpreting the spectra, identification of lines, and atomic parameters responsible for the intensity and line profiles of the lines are required. Apart oscillator strengths and excitation impact cross-sections when LTE is not fulfilled, line broadening parameters due to interactions (collisions) with the particles of the medium are required. For moderately hot (A-stars) to very hot (B and O) stars, "Stark" broadening is the main pressure broadening mechanism. It has to be known with accuracy. Nowadays collisional broadening data are needed, not only for strong lines of abundant elements (H, He, C, N, O, Ne) as in the past, but also for weak lines of abundant elements, and in addition for elements of lower abundance (the Iron-peak), and then, that is more recent, the heavy elements of very low abundance which follow on the Mendeleev table which are always very weak. Even some radioactive elements, such as thorium and uranium have been recently discovered in low-metallicity stars (stars of the first or second generation), and are used for chronometric age determination.

Section 2 will precise the physical conditions where Stark broadening plays a role in astrophysical plasmas. Section 3 will be devoted to a very brief review of the standard impact Stark broadening theory. In section 4 we will present some examples of astrophysical applications and recent results showing the new interest for weak lines.

In addition, the development of powerful computers also stimulates the development of atomic data on a large scale. The modelling of stellar atmospheres and of the stellar interiors needs extensive sets of atomic data, including collisional broadening and especially Stark broadening. For example, the PHOENIX computer code [1] developed for stellar modelling includes a database containing more than 10^7 atomic and ionic spectral lines. The access to these atomic data via on line databases becomes essential. In section 5 we will present the STARK-B database, which is a database of calculated widths and shifts of isolated lines of atoms and ions due to electron and ion collisions. This database is devoted to modelling and spectroscopic diagnostics of stellar atmospheres and envelopes. In addition, it is relevant to laboratory plasmas, laser equipments and technological plasmas. It is a collaborative project between the Paris Observatory and the Astronomical Observatory of Belgrade and has been opened online since the end of 2008, though not complete and currently under development, particularly in the framework of VAMDC (Virtual Atomic and Molecular Data Center).

2. Physical conditions for Stark broadening importance in astrophysical plasmas

Pressure broadening of spectral lines arises when an atom, ion, or molecule which emits or absorbs light in a gas or a plasma, is perturbed by its interactions with the other particles of the medium. Interpretation of this phenomena is currently used for modelling of the medium and for spectroscopic diagnostics, since the broadening of the lines depends on the temperature and density of the medium. In astrophysics, the physical conditions are very various, and collisional broadening with charged particles appears to be important in many domains. For example, in the very low dense interstellar ionized regions where the temperature is around 10^4 K and the electron density of about 10^4 cm^{-3} (the so-called H II regions), the radiorecombination lines of H and C II [1] arising from Rydberg levels (n-n±1 lines) are broadened by interactions with electrons and protons. On an opposite extreme domain of temperatures and densities (about 10^6 - 10^7 K, and electron densities of about 10^{24} cm^{-3}), the interiors of stars, which cannot be observed but are mirrored via asterosismology, can be modelled and electron collisional broadening of abundant ions plays an important role for the radiative opacities. In the very dense and hot atmospheres of neutron stars, the physical conditions are typical of those of stellar interiors, and highly-ionized atoms such as H and He-like Fe have been observed in X rays and the measured line strengths of lines indicate that the lines are significantly broadened by the Stark effect [2], and this should provide an opportunity to determine both the mass and the radius of these exotic objects.

On less extreme conditions of temperatures (10^4 to a few 10^4 K) and densities (10^{13} to 10^{15} cm^{-3}), Stark broadening is efficient for modelling and analyzing spectra of moderately hot (A), hot (B) and very hot (O) types of stars. It is dominant in comparison with the thermal Doppler effect in deep layers of the stellar atmosphere. In white dwarfs, collisional broadening and especially Stark broadening is dominant in all layers of the atmosphere (temperatures in the region of 10^4 K, electronic densities of the order of 10^{18}-10^{19} cm^{-3}) Even in stars like sun, and especially in the chromosphere which lays above the photosphere, Stark broadening is operative for lines arising from high excited levels, and especially for hydrogen lines. For interpretation of these spectra, the standard theory of Stark impact broadening of isolated lines is convenient.

3. The standard theory of Stark impact broadening of isolated lines

Following the founding work by Baranger [3] [4] [5], the theory and calculation of collisional line broadening in the impact approximation for electron and ion interactions underwent a great expansion in the sixties and the seventies, and experiments as well.

3.1. The impact approximation

The impact approximation forms the base for the theory: the interactions are separated in time. In other words, the studied radiating atom interacts with one perturber one at a time. Consequently the duration of an interaction must be much smaller than the mean time interval between two collisions, which is of the order of the inverse of the collisional line width. The effect of perturbers are independent and are additive.

The radiation field is weak, so the dressed atom theory is not necessary. It is a general case in astrophysics.

3.2. The complete collision approximation

In addition the collision is assumed to be complete. This means that the atom has no time to emit or absorb a photon during the collision process. This is valid if the duration of an interaction is much smaller than the interval between two successive emissions (or absorptions) of photons, which is of the order of the inverse of the detuning, when larger than the line width.

Consequently radiative and collision processes are decoupled and the impact broadening theory becomes an application of the theory of collisions between an atom (or ion) and interacting particles.

3.3. The case of "isolated lines"
We will be interested in "isolated lines" and exclude the case of "overlapping lines" [4], in the following. This means that the levels next to the upper or lower level of the studied transition and likely to modify the broadening by introducing optical coherences do not overlap with them. So we will consider neither hydrogen and hydrogenic ionic lines in the present paper, nor some specific helium lines and some lines arising from Rydberg levels.

3.4. The result is a Lorentz profile
Therefore the profile of the *i-f* line emitted or absorbed between the *i* and *f* levels studied is Lorentzian with a full width at half maximum W (in angular frequency units) and a shift d [5]. W can be expressed in terms of inelastic cross-sections and elastic processes as

$$W = N\int v\,f(v)\left(\sum_{i'\neq i}\sigma_{ii'}(v)+\sum_{f'\neq f}\sigma_{ff'}(v)+\sigma_{el}(v)\right), \quad (1)$$

where N is the density of the colliding perturbers, $f(v)$ the Maxwell distribution of the relative atom-perturber velocity v, $\sigma_{ii'}$ and $\sigma_{ff'}$ the inelastic cross-sections between the initial level i (resp. f final level) and the perturbing levels i' (resp f') of the *i-f* transition. $\sigma_{el}(v)$ represents the contribution of elastic collisions and include Feshbach resonances when ion-electron collisions are studied.
The shift is not given in the present paper, we refer to [5] for its quantum expression.

3.5. Orders of magnitude of the effect of electron collisions compared to the one due to ion collisions
For isolated lines, the widths and shifts due to electron collisions are generally the times higher than the widths and shifts due to ion collisions. This is due to the fact that inelastic collisions with positive ions are generally negligible, owing to the mass effect which decreases the relative velocity, and in addition, if the radiator is an ion, the Coulomb repulsion acts. This can be not the case when the perturbing levels are very close to the levels of the studied transition. This can also be not the case at very high temperatures, and especially for ion radiators when the Coulomb repulsion becomes weak. An example can be found in [6] for Cr I lines widths at very high temperatures.

3.6. Effect of fine structure and hyperfine structure on the widths and shifts
The following remark holds in LS coupling. For electronic collisions, the collision time (of the order of ρ_{typ}/v, where ρ_{typ} is a typical impact parameter) is very much smaller than the fine structure splitting. So the electronic spin has no time to rotate during the collision and can be ignored. Thus all the fine structure components have the same widths and shifts, which are equal to those of the multiplet. This is a fortiori the same for the hyperfine components. So our calculations have most often been performed for multiplets only. This remark does not apply for heavy atoms where departures from LS coupling can be important.

3.7. The calculation of the cross-sections, and then widths and shifts
3.7.1. The Semi-Classical Perturbation (SCP) approach. The calculation of the cross-sections is a major quantum mechanical problem. For most astrophysical purposes, the semi-classical-perturbation treatment is adapted and gives results with a sufficient accuracy (about 20%). The method is accurate if the perturbing levels are not too far from the levels of the studied line. The formalism has been developed and discussed in detail by [7] [8] and a very fast computer code has been created. Classical straight rectilinear paths for neutral radiators, and hyperbolic paths are introduced for ion-electron et ion-ion collisions. Then the S-matrix is obtained within the second order perturbation theory. The needed cross-sections are obtained through an integration over the impact parameter of the transition probabilities. The needed cut-offs are determined in order to maintain the unitarity of the scattering S-

matrix for low impact parameters, and Debye screening is taken into account. This formalism, as well as the computer code, have been updated and optimized several times: [9] for complex atoms, [10] for the inclusion of Feshbach resonances in the elastic ion-electron cross-sections, [11] and further papers, and [12] for transitions arising from very complex configurations.

3.7.2. The atomic structure. It is the first ingredient of the calculations.
In the quantum picture, the wavefunctions enter the calculation of the *S*-matrix.
In the semiclassical picture, oscillator strengths and needed energy levels enter the expressions of the inelastic cross-sections.
— In the past, the Coulomb approximation with quantum defect (the Bates and Damgaard approximation [13]) was used, together with measured or calculated energy levels. Now we use modern ab initio methods (cf. the review by Ben Nessib (2009) [14]. The computer codes or the data can be downloaded on line. Thus the calculations of widths and shifts can be made from the beginning to the end without any additional external input or experimental adjustment. The chosen atomic structure package enters our computer semi-classical code and that allows, when these methods are applicable, to obtain widths and shifts for one-two hundred of lines.
— TOPbase, the Opacity Project atomic database, contains accurately calculated energy levels, f-values and photoionisation cross sections for astrophysically abundant neutral atoms and ions. They have been computed in the close-coupling scattering theory by means of the R-matrix method with innovative asymptotic techniques [15] [16]. L-S coupling is assumed. So the use of TOPbase is especially adapted to light and low and moderately ionized atoms and ions.
— The Cowan code, originally written by Cowan [17] is an atomic structure package consisting of a set of computer programs for calculation of energy levels, radiative transition wavelengths and probabilities, electron impact excitation and photoionization cross sections etc. The Hartree-Fock-Slater multi-configuration expansion method with statistical exchange is the normal option since it is most computationally efficient. The relativistic corrections are treated by perturbations. So the method is especially suited to moderately heavy atoms which are little and moderately ionized.
— SUPERSTRUCTURE (SST) [18] is well suited for computation of large quantities of atomic data for highly charged ions. The wave functions are determined by diagonalization of the nonrelativistic Hamiltonian using orbitals calculated in a scaled Thomas-Fermi-Dirac-Amaldi potential. Relativistic corrections are introduced according to the Breit-Pauli approach. Atomic data are obtained in intermediate coupling.

3.7.3. The quantum code in intermediate coupling for Stark broadening calculations. The combination of SST with the *S*-matrix calculated with the Distorted Wave approximation (DW) code of Eissner [19] and the code of Saraph JAJOM [20] have permitted to create a quantum code for electron-impact widths of ions [21] based on the quantum formalism of impact Stark broadening in intermediate coupling [22]. This quantum method has been compared to experimental results of some Be-like and Li-like ions with success [21] [23] but has not still been applied to astrophysics needs. It should be especially useful for UV and XUV resonance lines of highly ionized atoms when the semiclassical method cannot give accurate results, owing to the large distance of the levels of the studied transition with the perturbing levels. On the example of stellar interiors, the electron density is high enough that electron impact is the most important source of line broadening and contributes significantly to the radiative accelerations needed for evolutionary models and opacity calculations [24], whereas many data are currently missing. However, the quantum method, though in principle more accurate than the semi-classical one, is cumbersome to undertake. It can be usefully applied to particular lines, provided that the levels be not too excited: for excited levels the number of necessary configurations become excessive and the computer code cannot converge in the current state of the art. So numerous astrophysical applications are currently achieved within the semi-classical perturbation (SCP) theory.

3.7.4. The Modified Semi-Empirical Method (MSE). A number of applications were also achieved with the Modified Semi-Empirical (MSE) method [25] [26] [27] and other papers cited in [28]. It is less accurate, though more simpler to use due to the considerably smaller set of atomic data needed in comparison with the SCP theory [28]. It can effectively replace the SCP method when this one cannot be used due to a lack of atomic data.

3.8. Systematic trends and regularities

Systematic trends for energy levels and oscillator strengths in the Coulomb approximation are well known. Asymptotic behaviours of the cross sections at very small or very high incident energy are also well known. Systematic trends and regularities as a function of temperature, density, quantum numbers, have been found for Stark broadening widths and shifts. This has been an active subject of research since many years, cf. [29] and references therein. This topic has a growing interest with the calculations of widths and shifts at a large scale for many temperatures and densities, especially for the use of databases. Cf. the recent example [30], that concerns the behaviour of the widths and shifts of Ne I, 1837.8 nm as a function of temperature. Many other examples can be found in papers concerned with Stark broadening, it is not possible to enumerate them here.

4. Astrophysical applications within the SCP and MSE theories

During the recent past years, the work of the French-Serbian-Tunisian team was focused toward calculations of numerous Stark widths and shifts of astrophysical interest. On the atomic physics side, a special effort was turned towards ab initio calculations. On the astrophysical side, recent calculations have been performed for trace elements, for line stratification and for lines of white dwarfs. A very few selected examples chosen among a great number of published results are given below.

4.1. Trace elements and abundance determinations

About 50% of the element abundances beyond iron are produced via slow neutron capture nucleosynthesis (s-process) [31], which starts at iron-peak seed, synthesizing the elements between iron and bismuth.

The three isotopes of tellurium Te-122, 123 and 124 provide a unique opportunity to investigate s-process. The SCP method was used to calculate the Stark broadening parameters for 4 Te I multiplets [32] within the ultraviolet, visible and infrared wavelength range, in the range of temperatures and densities for A and B type stars and DA, DB and cool DO white dwarfs where Stark broadening is of interest. Such temperatures may be of interest also for the modelling of subphotospheric layers even in cooler stars. If Doppler broadening dominates the line centre, Stark broadening may influence line wings in the upper layers of the atmosphere. Yet the Stark broadening mechanism is absolutely dominant in comparison with the thermal Doppler mechanism in deeper layers of the stellar atmosphere. The importance of taking into account accurate Stark broadening data for modelling and analyzing of neutral tellurium spectra confirms previous findings for a number of trace elements in chemically peculiar stellar spectra. Other examples are given and discussed in [28]: Cr II [33], Ge I [35], Ga I [34], Cd I [36], Cr I [33], and others. In particular, neglecting the Stark broadening mechanism introduces an error of between 10% and 45% in the equivalent widths and corresponding errors in the abundances for the Nd II lines in A type stars.

4.2. Chemical stratification

Accurate Stark broadening data are also crucial for studying chemical stratification by fitting the calculated and observed spectra. The following concerns new Stark broadening calculations made with the SCP method, most with the Bates and Damgaard approximation [13] for the oscillator strengths.

For Si I lines, it is interesting to note that the electron impact is dominant, but collisions with proton and He II impacts are also important for broadening of the lines [37]. From the comparison of our calculations with the observations it was found that Stark broadening and inclusion of stratification can explain asymmetry of the Si I 6142.48 Å and 6155.13 Å lines in the atmosphere of a rapidly oscillating roAp star, where these lines are asymmetrical and shifted. The obtained results are the first calculated data for the considered lines.

In fact, the Stark broadening effect can explain the asymmetry as well as the shift of these lines in many stars including the Sun. In hotter Ap stars, besides Stark broadening, the stratification plays a very important role in producing line asymmetry. The sensitivity of the line asymmetry to changes in the abundance variation through the stellar atmosphere can be used for such spectroscopic diagnostics. Chromium is one of the most anomalous elements in Ap stars. Cr I lines from 4p–4d transitions are known to have fairly large Stark damping. When interpreting observations of a Cr-rich Ap star, it was shown [33] that Stark broadening is important to synthesize the profiles. In addition, the SCP calculations showed that the contribution of proton and He II collisions to the line width and shift is significant and comparable, and is sometimes even larger than electron-impact contribution depending of the electron temperature. Moreover, not only the Stark line width, but also the Stark shift may contribute to the blue as well as to the red asymmetry of the same line depending on the electron-, proton-, and HeII density in stellar atmosphere. In addition, although they belong to the same multiplet, the widths and shifts of the different lines can be quite different.

Recent abundance and stratification analysis [38] based on Cr II lines in the spectrum of the Ap star HD 133792 was revisited in [39], by using new SCP Stark broadening data for synthetic spectrum calculations. These new theoretical results agree with the scarce experimental results and provide an accurate fit to the line profiles observed in Ap stars. In conclusion, accurate Stark broadening are essential to take into account for the study of abundance stratification, not only for the Cr I abundance, but also for all trace elements, especially in hot stars.

Other examples concerning Cu III, Se III, Au II, Co III, Nd II, etc. can be found in [32].

4.3. White dwarfs

White dwarfs are denser and sometimes hotter than stars of the principal sequence. The electronic density lies between 10^{17} and a few 10^{19} cm^{-3}, the temperature is about 10^4 to a few 10^4 K, so Stark broadening is the dominant mechanism, and often more important than the Doppler profile in the line centre.

4.3.1. Stark width compared to the Doppler width. For example, the influence of Stark broadening on 6 Cu III, 6 Zn III and 3 Se III spectral lines in DB white dwarf atmospheres was investigated [40]. MSE Stark broadening calculations were achieved because the SCP approach could not be applied due to the lack of reliable atomic data. As an illustration of the relative order of magnitude of the different causes of broadening, it was shown that the Stark width of the Se III 3815.5 Å line is larger than the Doppler one by up to two orders of magnitude within the range of optical depths considered for the plasma conditions in the DB white dwarf atmospheres.

A detailed investigation of Stark broadening of 84 UV spectral lines of Cd III was performed by the MSE approach. It was applied to F0–B0 type stars and DA and DB white dwarfs [41]. As for A type stars, this work shows that the neglect of Stark broadening introduces an error of between 10% and 45% in the equivalent widths and corresponding errors in the abundances of white dwarfs.

4.3.2. Ionized manganese lines and effect of the hyperfine splitting. Ionized manganese lines are of interest for the analysis and modelling of stellar spectra as for example for HgMn stars and DB white dwarfs. SCP calculations and oscillator strengths with the Bates and Damgaard [13] approximation, due to the lack of reliable atomic data, were performed [42] and compared to recent experiments [43]. The importance of taking into account the different hyperfine structure components by summing their respective profile intensities was emphasized. They indeed have the same width and shift equal to that of the fine structure line. This falls within the same reasoning [44] [45] where the importance of adding the different fine structure lines of the Hα line of hydrogen and Lyman hydrogenic ionic lines broadened by electrons and ions collisions was shown. In spite of that, a large disagreement (up to a factor 2) remains between the Mn II SCP calculations and the experiments.

Considering that the experiments are reliable, and that the Bates and Damgaard approximation may be insufficient for heavy atoms, this may show the importance of having a reliable atomic structure for Stark broadening calculations of heavy atoms.

The obtained results were used [46] for the investigation of the influence of Stark broadening on Mn II spectral line profiles in Ap stars and also in DB white dwarfs. It was again demonstrated that Stark broadening is an essential mechanism to take into account for the analysis of DB white dwarf spectra.

4.3.3. C II lines in white dwarfs with carbon atmosphere. A new type of white dwarfs has recently been discovered [46] [47]. The surface composition of these stars is mostly composed of carbon. There is hardly neither hydrogen nor helium in the atmosphere. In order to understand the origin and evolution of this new type of stars, the determination of gravity is essential, and it is necessary to develop a new generation of accurate models. Thus it is crucial to take into account accurate Stark broadening for modelling the atmosphere. In fact there is a lack of Stark broadening data for C II. Using SCP approach and the online TOPbase R-matrix atomic structure, ab initio Stark broadening parameters for 148 C II multiplets have been obtained. More details on the calculations are discussed in [49].

4.4. Ab initio calculations for stellar spectra modelling

The Cowan code coupled to the MSE approach was applied [50] to new Stark broadening calculations of several multiplets of S II, S III and S IV and demonstrated the interest of ab initio calculations. Then method grew up and was coupled to the SCP approach and permitted to obtain data for several hundreds of lines with an accurate atomic structure in a reasonable time of computation. Besides R-matrix and TOPBase currently applied to C II lines, the SST package was used in several papers in view of stellar modelling applications: Si VI [51] especially for DO white dwarfs atmospheres, Ne V [52], Si V [53].

5. The STARK-B database

Nowadays, the access to numerous data via an on line database becomes crucial. Numerical and bibliographic Databases in Atomic and Molecular Physics are essential for both the modelling of various astrophysical media and the interpretation of astrophysical spectra provided by ground or space-based telescopes.

In this context, we are currently developing the STARK-B database [54] of calculated widths and shifts of isolated lines of atoms and ions due to electron and ion collisions in the impact approximation. It is planned to include the results of calculations made by Dimitrijević, Sahal-Bréchot, and co-workers, and contained in more than 120 publications. It is a collaborative project between the Astronomical Observatory of Belgrade and the Paris Observatory. The database is currently developed in Paris, and a mirror is planned in Belgrade. It is devoted to modelling and spectroscopic diagnostics of stellar atmospheres and envelopes. In addition, it is relevant to laboratory plasmas, laser equipments and technological plasmas. Hence, the domain of temperatures and densities covered by the tables is wide and depends on the ionization degree of the considered ion. The data are gradually implemented and the database is already on line though not yet complete.

A graphical interface is provided. First, the user clicks on the element in the Mendeleev periodic table and then on the ionization degree of interest. Next, with a few clicks, the user chooses the colliding perturber(s), the perturber density, the transition(s) by quantum numbers and the plasma temperature(s). Then a table displaying the widths and shifts is generated. Bibliographic references are given and linked to the publications via the SAO/NASA ADS Physics Abstract Service [55] and/or within DOI. They can be freely downloaded if the access is not restricted. The widths and shifts data can be downloaded in ASCII. A request by domain of wavelengths instead by transitions is planned. The implementation of remaining data is in progress. The further developments especially concern the compatibility of the output with the VO (Virtual Observatory) standards, which will be defined by VAMDC.

In fact, STARK-B is a part of VAMDC (Virtual Atomic and Molecular Data Centre) [56]. VAMDC is an European Union funded collaboration between groups involved in the generation and use of atomic and molecular data which was created in summer 2009 for three years. In fact, the free exchange of atomic and molecular data requires the definition both of standards which model the data structure and of tools that implement these standards and that help to carry out science using these data. VAMDC aims to build a secure, documented, flexible and interoperable e-science environment-based interface to existing atomic and molecular data. In particular, there is a collaborative project of standardisation of lists of lines within the IVOA (International Virtual Observatory Alliance) [57] consortium as well as a wider collaborative project of standardisation concerning many more processes of atomic and molecular data. Paris Observatory is a major partner in development of both standards and tools. All that is commented and described in more details in [58] [59].

6. Conclusion

The development of space born spectroscopy, building of giant telescopes of the new generation and increase of accuracy of computer codes for modelling of stellar atmospheres has opened up a new era for developing theoretical and experimental Stark broadening data with application to stellar spectra and stellar interiors modelling. Due to lack of space, the numerous applications to laboratory and technological plasmas have not been mentioned in the present paper. All these applications also shows the need to perform new accurate calculations for heavy atoms and ions. Ab initio methods, with sophisticated atomic structure packages coupled to the semi-classical perturbation method for Stark broadening calculations should be the best at present. The continuation of the development and service of a powerful and constantly updated online database is also indispensable.

7. Acknowledgments

A part of this work has been supported by VAMDC. VAMDC is funded under the "Combination of Collaborative Projects and Coordination and Support Actions" Funding Scheme of The Seventh Framework Program. Call topic: INFRA-2008-1.2.2 Scientific Data Infrastructure. Grant Agreement number: 239108

8. References

[1] Bildsten S, Chang P and Paerels P 2003 *Astrophys. J.* **591** L29
[2] Rovenskaya N I 2004 *Astrophys. Space Sci.* **291** 113
[3] Baranger M 1958a *Phys. Rev.* **111** 481
[4] Baranger M 1958b *Phys. Rev.* **111** 494
[5] Baranger M 1958c *Phys. Rev.* **112** 855
[6] Dimitrijević M S, Ryabchikova T, Popović L Č, Shulyak D and Khan S 2005 *A&A* **435** 1191
[7] Sahal-Bréchot S 1969a *A&A* **1** 91
[8] Sahal-Bréchot S 1969b *A&A* **2** 322
[9] Sahal-Bréchot S 1974 *A&A* **35** 321
[10] Fleurier C, Sahal-Bréchot S and Chapelle J 1977 *JQSRT* **17** 595
[11] Dimitrijević M S and Sahal-Bréchot S 1984 *JQSRT* **31** 301
[12] Mahmoudi W F, Ben Nessib N and Sahal-Bréchot S 2008 *EPJD* **47** 7
[13] Bates D R Damgaard A 1949 *Trans. Roy. Soc. Lond. Ser. A* **242** 101
[14] Ben Nessib N 2009 *New Astron. Rev.* **53**, 255
[15] Cunto W, Mendoza C, Ochsenbein F and Zeippen CJ 1993 *A&A* **275** L5
[16] http://cdsweb.u-strasbg.fr/topbase/topbase.html
[17] Cowan RD 1981 *The Theory of Atomic Structure and Spectra* (Berkeley, CA: University of California Press)
[18] Eissner W Jones M and Nussbaumer H 1974 *Comput. Phys. Commun.* **8** 270
[19] Eissner W 1998 *Comput. Phys. Commun.* **114** 295
[20] Saraph H E 1978 *Comput. Phys. Commun.* **15** 247

[21] Elabidi H, Ben Nessib N, Cornille M, Dubau J and Sahal-Bréchot S 2008 *J. Phys.B* **41** 025702
[22] Elabidi H, Ben Nessib N and Sahal-Bréchot S 2004 *J.Phys.B* **37**, 63
[23] Elabidi H, Sahal-Bréchot S and Ben Nessib N 2009 *EPJD* **54** 51
[24] Alecian G, Michaud G and Tully J 1993 *Ap. J.* **411** 882
[25] Dimitrijević M S and Konjević J 1980 *JQSRT* **24** 451
[26] Dimitrijević M.S 1982 *A&A* **112** 251
[27] Dimitrijević M S and Kršljanin V 1986 *A&A* **165** 269
[28] Dimitrijević M S 2003 *A&A Trans.* **22** 389
[29] Purić J, Dojčinović I P, Nikolić M, Šćepanović M, Obradović B M and Kuraica M M 2008 *ApJ* **680** 803
[30] Christova M, Dimitrijević M S, Simić Z and Sahal-Bréchot S 2010 *J.Phys.Conf.* **207** 012025
[31] Reifarth R 2010 *J. Phys. Conf. Ser.* **202** 012022
[32] Simić Z, Dimitrijević M S and Kovačević A 2009 *New. Astr. Rev.* **53** 246
[33] Dimitrijević M S, Ryabchikova T, Simić Z, Popović L Č and Dačić S M 2007 *A&A* **469** 681
[34] Dimitrijević M S, Dačić M, Cvetković Z and Simić Z 2004 *A&A* **425** 1147
[35] Dimitrijević M S, Jovanović P and Simić Z 2003 *A&A* **410** 735
[36] Simić Z, Dimitrijević M S, Milovanović N and Sahal-Bréchot S 2005 *A&A* **441** 391
[37] Dimitrijević M S, Ryabchikova T, Popović L Č, Shulyak D and Tsymbal V 2003 *A&A* **404** 1099
[38] Kochukhov O, Tsymbal V, Ryabchikova T, Makaganyk V, and Bagnulo S 2006 *A&A* **460** 831
[39] Dimitrijević M S, Ryabchikova T, Simić Z, Popović L Č and Dačić M 2007 *A&A* **469** 681
[40] Simić Z, Dimitrijević M S, Popović L Č and Dačić M 2006 *New Astron.* **12** 187
[41] Milovanović N, Dimitrijević M S, Popović L Č, and Simić Z 2004 *A&A* **417** 375
[42] Popović L Č, Dimitrijević M S, Simić Z, Dačić M, Sahal-Bréchot S and Kovačević A 2008 *New Astron.* **13** 85
[43] Djeniže S, Bukvić S, Srećković A and Nikolić Z 2006 *NewAstron.* **11** 256
[44] Stehlé C and Feautrier N 1985 *J. Phys. B: At. Mol. Phys.* **18** 1297
[45] Stehlé C 1985 *J. Phys. B: At. Mol. Phys.* **18** L43
[46] Simić Z, Dimitrijević M S, Popović L Č, Dačić M, Kovačević A and Sahal-Bréchot S 2008 Contrib. Astr. Obs. Ska. **38** 451
[47] Dufour P, Liebert J, Fontaine G and Behara N 2007 *Nature* **450** 522
[48] Dufour P, Fontaine G, Liebert J, Schmidt G D and Behara N 2008 *Astrophys. J.* **683**, 978
[49] Larbi-Terzi N, Sahal-Bréchot S, Ben Nessib N and Dimitrijević M S 2010 Stark broadening parameters for white dwarf atmospheres research (*JPCS preprint SPIG 25th*)
[50] Milovanović N and Dimitrijević M S 2007 Cowan Code and Stark Broadening of Spectral Lines of S II, S III and S IV *Spectral Line Shapes in Astrophysics: VI Serbian Conference on Spectral Line Shapes in Astrophysics (Sremski Karlovci, Serbia, 11-15 June 2007) AIP Conf. Proc.* vol 938, pp. 258-261
[51] Hamdi R, Ben Nessib N, Milovanović N, Popović L Č, Dimitrijević M S and Sahal-Bréchot S 2008, *MNRAS* **387** 871
[52] Hamdi R, Ben Nessib N, Dimitrijević M S and Sahal-Bréchot S 2007 *ApJS* **170** 243
[53] Ben Nessib N, Dimitrijević M S and Sahal-Bréchot S 2004 *A&A* **423** 397
[54] http://stark-b.obspm.fr
[55] http://www.adsabs.harvard.edu/
[56] http://www.vamdc.eu
[57] http://www.ivoa.net/
[58] Dimitrijević M S, Sahal-Bréchot S, Kovačević A, Jevremović D, Popović L Č 2010 European Virtual Atomic and Molecular Data Center – VAMDC (*JPCS preprint SPIG 25th*)
[59] Dubernet ML, Boudon V et al. 2010 *JQSRT* **111** 2151

25th Summer School and International Symposium on the Physics of Ionized Gases—SPIG 2010 IOP Publishing
Journal of Physics: Conference Series **257** (2010) 012029 doi:10.1088/1742-6596/257/1/012029

The disk emission in the Broad Line Region of Active Galactic Nuclei

Edi Bon[1], Luka Č. Popović[1], Nataša Gavrilović[1] and Giovanni La Mura[2]

[1]Astronomical Observatory, Volgina 7, 11060, Belgrade
[2]Department of Astronomy, University of Padova, Vicolo dell'Osservatorio, I-35122 Padova, Italy

E-mail: ebon@aob.rs

Abstract. We studied the disk emission component hidden in the single-peaked broad emission lines (BELs) of active galactic nuclei (AGN) using a two-component model. We assumed that the broad lines are formed in an accretion disk plus a surrounding non-disk region, with isotropic cloud velocities. To compare simulated line profiles with observed ones we measured the full widths (at 10 per cent, 20 per cent and 30 per cent of the maximum intensity). We found that the hidden disk emission may be present in BELs even if the characteristic of two-peaked-line profiles is absent. We found that in the case of the hidden disk emission in single-peaked broad-line profiles, the disk inclination tends to be small and that the contribution of the disk emission to the total flux should be smaller than the contribution of the surrounding region.

1. Introduction

The detailed structure of the innermost part of Active Galactic Nuclei (AGN) is still an open question. Broad Emission Lines (BELs) are often used for sonding the inermost part of AGN. They show high diversity in their shapes and widths what indicates different kinematical properties of emmiting gas. It is widely accepted that the central engine consists of a supermassive black hole fueled by an accretion disk. Using BEL shapes, one can constrain the geometry of the material in the Broad Line Region (BLR) [1]. In some rare cases, where the accretion disk clearly contributes to the BELs, we can investigate the disk properties using broad, double-peaked, low-ionization lines [2, 3, 4, 5, 6]. The rotation of the material in the disk results in one blueshifted and one redshifted peak, while the gravitational redshift produces a displacement of the center of the line and a distortion of the line profile. In some cases, BELs show shoulder-like profiles in the wings, which could indicate that some part of the flux in BELs originate in the accretion disk emission [7, 5, 8].

To explain the complex morphology of the observed BEL shapes, different geometrical models have been discussed [9]. In some cases the BEL profiles can be explained only if two or more kinematically different emission regions are considered [10, 11, 12, 13, 7, 14, 20, 15, 16, 17, 18, 19]. In particular, the existence of a Very Broad Line Region (VLBR) with random velocities at 5000-6000 km/s within an Intermediate Line Region (ILR) has also been considered to explain the observed BEL profiles [9, 17, 18, 19, 21]. Even though, the majority of AGN with BELs have only single peaked lines, it does not necessarily indicate that the contribution of the disk emission to the BELs profiles is negligible.

Here we give a short overview of our investigation about possibility of a presence of a disk emission flux in AGN emission lines.

2. Data sets and reductions

First sample of 14 AGN spectra was observed with the 2.5 m Isaac Newton Telescope (INT) on La Palma at 2002 [14]. Objects were chosen due to indication of the disk emission in the X-ray part of the spectrum. For some lines with low S/N ratio we used Hubble Space Telescope (HST) observations obtained with the Space Telescope Imaging Spectrograph (STIS) on January 2000 (NGC 3516). Details about observation and data reduction can be found in [20, 14]. The narrow and satellite lines were cleaned [13, 7]. The broad profiles from this sample were fitted with two component model [20].

Second sample contains 90 broad-line-emitting AGN [22], which have been collected from the third data release of Sloan Digital Sky Survey (SDSS)[1]. The spectra were already corrected for sky-emission, telluric absorption, the Galactic extinction and redshift [22]. Since the interest was to investigate the broad emission line profiles, we subtracted the narrow components of Hα and the satellite [NII] lines. The spectral reduction (including subtraction of stellar component) and the way to obtain the broad line profiles are explained in more details in [22].

Previously cleaned broad Hα profile was normalized, converted from wavelength to velocity scale, and smoothed using DIPSO software package[2].

The sample of 90 AGN contained the spectra with different FWHM of broad Hα from about 1000 km/s to 7000 km/s, with at least several representatives in every 500 km/s [22].

3. The Model

We assumed that the BELs can be described with two kinematically different components, one contributing to the wings (that we assumed to be emmitted from the disk) and other to the core (emitting clouds that surround the disk with isotropic distribution of the velocities).

The local broadening (σ) and shift (z_λ) of each disk element have been taken into account as in [2], i.e. the δ function has been replaced by a Gaussian function:

$$\delta \rightarrow \exp\frac{(\lambda-\lambda_0-z_\lambda)^2}{2\sigma^2}. \tag{1}$$

We express the disk dimension in gravitational radii ($R_g = GM/c^2$, G being the gravitational constant, M the mass of the central black hole, and c the velocity of light).

We assumed that the additional emission region can be described by a surrounding region with an isotropic velocity distribution, i.e. the emission line profile generated by this region can be described by a Gaussian function with broadening w_s and shift z_s. Finally, line profile can be described by the relation:

$$I(\lambda) = I_d(\lambda) + I_s(\lambda), \tag{2}$$

where $I_d(\lambda)$, and $I_s(\lambda)$ are the emissions of the relativistic accretion disk and the non-disk region, respectively.

The synthetic profiles generated with this model can fit the line profiles of BELs in AGN, but it is hard to determine the exact physical properties with so many free parameters in the model [14]. First of all, the disk model includes many parameters (the size of the emitting region, the

[1] http://www.sdss.org/dr3
[2] http://www.starlink.rl.ac.uk

emissivity and inclination of the disk, the velocity dispersion of the emitters in the disk, etc.). In order to perform numerical tests, one needs to introduce some constraints and approximations:

i) The disk inclination affects the emission obtained from the disk. The observed flux from the disk (F_d) is proportional to the disk surface (S_d), as

$$F_d \sim S_{\mathrm{eff}} \sim S_d \cdot \cos(i), \tag{3}$$

where i is the inclination, and S_{eff} is the effective disk emitting surface, therefore, one cannot expect a high contribution of the disk emission to the total line profile for a near edge-on projected disk.

ii) One should not expect emission of the low ionized lines in the part of the disk too close to the centeral black hole. Consequently, the model given by [2] can be properly used, i.e. it is not necessary to include a full relativistic calculation [24].

iii) The emissivity of the disk as a function of radius, r, is given by $\epsilon = \epsilon_0 r^{-p}$, but we fixed the value of emissivity index to $p = 3$, since while varying this parameter there were no significant difference between normalized profiles.

iv) Previous estimations of the double-peaked AGN emission lines [2, 3, 4, 5, 25] show that the typical dimensions of an accretion disk that emits low ionization lines are of the order of several thousands R_g. For that reason we did not consider dimensions on the disk larger than 100000 R_g. This was an important approximation to limit the computing time.

v) We consider a systemic velocity shift of the non-disk region (not greater than $\pm$ 3000 km/s) to test the possibility of the outflow/inflow.

We considered the following parameters to study the simulated and observed BELs profiles:

i) The flux ratio between the non-disk region (F_s) and the disk (F_d):

$$Q = \frac{F_s}{F_d}, \tag{4}$$

where

$$F_{tot} = F_s + F_d = (1 + Q)F_d.$$

Using this parameter, the total line profile (normalized to the disk flux)[3] can be written as:

$$\frac{I_{tot}(\lambda)}{F_d} = \frac{I_d(\lambda)}{F_d} + Q\frac{I_s(\lambda)}{F_s}, \tag{5}$$

where $I(\lambda)$ is the wavelength dependent intensity. The composite profile is normalized according to:

$$\Im(\lambda) = \frac{I_{tot}(\lambda)}{I_{tot}^{max}}, \tag{6}$$

where I_{tot}^{max} is the maximum intensity of the composite line profile.

ii) For the composite line profile $\Im(\lambda)$ we measured full widths at 10%, 20%, 30% and 50% of the maximum intensity, i.e. $w_{10\%}$, $w_{20\%}$, $w_{30\%}$ and $w_{50\%}$. Then we define coefficients k_i (i = 10, 20, 30) as $k_{10} = w_{10\%}/w_{50\%}$, $k_{20} = w_{20\%}/w_{50\%}$ and $k_{30} = w_{30\%}/w_{50\%}$ where $w_{50\%}$ is Full Width at Half Maximum (FWHM). It is obvious that the coefficients k_i are functions of the radius R and other parameters of the disk. Using these normalized widths we can compare results for AGN with different random velocities.

iii) We also measured the asymmetry (A_i) at i =10%, 20%, 30% of maximum intensity of the modeled and observed lines as:

[3] This is taken from technical reasons to simulate different contributions of the disk and the non-disk component. First we normalized both line profiles to their fluxes, and after that we rescaled the non-disk component by multiplying it with Q, so the whole profile is given in units of the disk flux.

$$A_i = \frac{W_i^R - W_i^B}{FWHM}, \tag{7}$$

where W_i^R and W_i^B are red and blue half widths at i = 10%, 20% and 30% of the maximum intensity, respectively.

We simulated only the disk profiles, taking into account different values of the disk parameters. In the first instance, the relative importance of the disk contribution to the core or to the wings depends on the disk inclination. The contribution of the disk to the center of the line or to the wings is not so much sensitive to the outer radius, but significantly depends on the disk inclination. A face-on disk contributes more to the core of the line, while a moderately inclined disk ($40° > i > 20°$) contributes significantly to the line wings. For $i > 40°$ the disk emission will strongly affect the far wings of the composite profile. Another very important parameter is the flux ratio between components, Q.

We found that the presence of the disk emission is difficult to detect in the line profile when the contribution of the disk is smaller than 30% of the total line emission ($Q > 2$): in the case of a low inclination both the disk and non-disk region contributes to the line core and it is very hard to separate the disk and non-disk region. In the case of a highly inclined disk, the disk emission spreads in the far wings, and could not be resolved from the continuum, especially if the observed spectrum is noisy. For the case of dominant disk emission ($Q < 0.3$), if the inclination is low, the line will be shifted to the red, and if the inclination is high, two peaks or at least shoulders should appear in the composite line profile (see Fig. 1). Consequently, further in the paper we will consider only cases where 0.3< Q <2 [18].

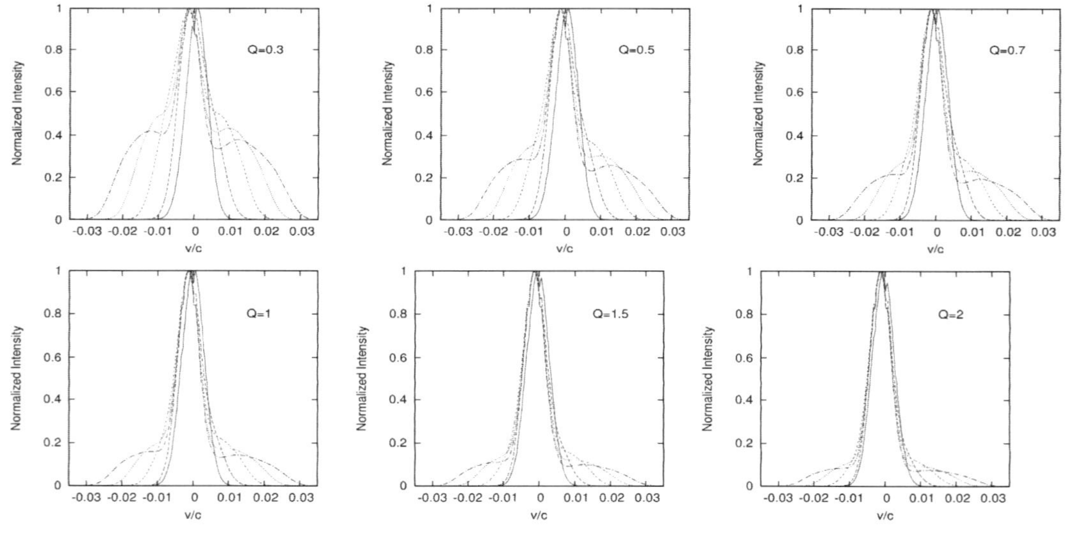

Figure 1. Simulated line profiles emitted by the two-component model for five different inclinations (i=1, 10, 20, 40 and 60 degrees, from the narrowest to the broadest line, respectively) for different contributions of the disk to the composite line profiles (denoted by Q in figures). The inner radius of the disk is taken to be 400 Rg, while the outer radius is 1200 Rg.

We set some constrains in the parameter space, and we constructed a grid of simulated line profiles. As a rough approximation, we fixed the width of the central Gaussian component to the 1000 km/s, as mean value obtained from the two component model fit [20]. As it was shown in [14], the random velocities in the disk and in the non-disk region were approximately the same, so they were also fixed to the value of 1000 km/s.

For the first sample, the inner radius of the disk was chosen to be $R_{inn} = 400\ R_g$, as the mean value of this parameter in the previous fittings [23]. For the outer radius, the chosen value was $R_{out} = 3000\ R_g$ [20], since the variation of this parameter, after values of several thousand gravitational radii, did not show significant influence to the shape of the line (especially if the line was noisy).

4. Results

In order to detect possible accretion disk emission in the flux of the single peaked broad emission lines of AGN, we applied several methods (fitting with the model, comparing the measured coefficients with the grid of simulated spectra by this model, and with the model constructed of two isotropic regions that could be represented only with Gaussians: very broad line region (VBLR) and intermediate broad line region (ILR), Gaussian fitting with three components...).

From the Gaussian analysis we could conclude that two distinct kinematic regions might be present in the BELs. The central Gaussian component indicated the existence of the emission from the region with intermediate velocities (ILR), while in the wings two shifted broad Gaussians indicated possible presence of disk-like emission (VBLR). The results for the widths of the emission lines of Fe II template showed that these lines probably originate in the ILR.

From fitting with the twocomponent model we found [19]:

- two-component model can fit well the profiles of broad emission lines from our samples, but it is very hard to determine the disk parameters, because of the large number of the free parameters,
- random isotropic velocities in the surrounding region are similar to random velocities in the disk, what implies that these two regions could be connected through some process (for example through the wind from the accretion disk),
- the values of the parameters indicate that the inclinations are smaller than 50°, and the dimensions of the disk could be from several hundred R_g for the inner radius, to several hundred thousands R_g for the outer radius.

After comparing the measurements of k_i for the sample, with the measurements of the simulated profiles, we could derive the estimates for i and Q. As the result, we found that the most of the measured points were between $0.5 < Q < 1.5$, and $10^\circ < i < 25^\circ$ (see Figs. 2 and 3). The results showed a consistence between estimations of i and Q for both Hα and Hβ, with very small discrepancies [23].

After this analysis, we tested this method to the larger sample of 90 AGN [18]. The measurements were performed in the same way as for 14 AGN, with a difference that we compared them to simulated spectra for the disk dimensions of $R_{inn} = 600\mathrm{R_g}$ and $R_{out} = 4000\ \mathrm{R_g}$ (averaged disk size obtained from the fitting of the double-peaked lines in the paper of [5]). For this sample we measured widths for Hα lines. After measuring we noticed that for higher inclinations $i > 30^\circ$ error bars were rapidly incrising (see Fig. 4), and we excluded those few measurements from further analysis. This could indicate that the model may not be suitable for higher inclinations, and that this may indicate significant torus influence. As it could be seen in Fig. 3, most of the measurements are located within $1 < Q < 2$, and $10^\circ < i < 25^\circ$ [18].

5. Conclusions

This investigation indicates the disk emission presence in the flux of single peaked BELs.

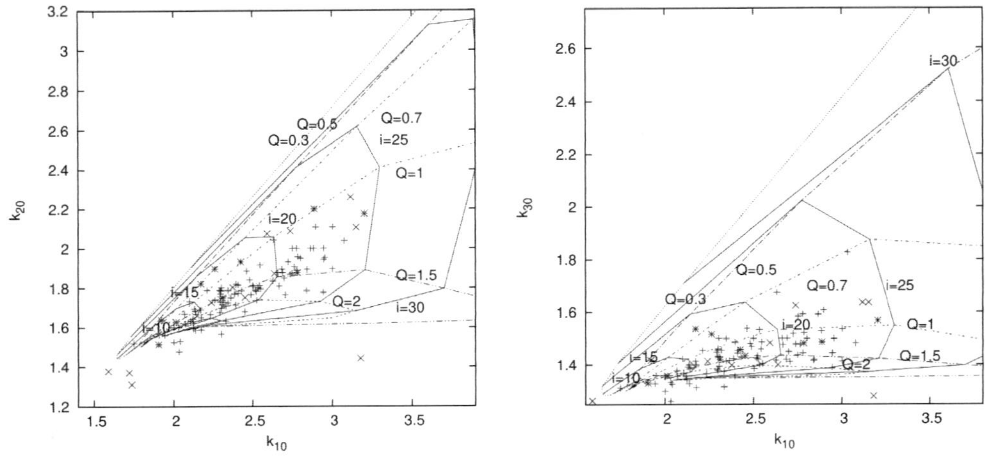

Figure 2. The measured width ratios (crosses for the Hα of the larger sample, + and asterisk signs for the Hα and Hβ of the smaller sample, respectively) and simulated values with two component model (solid isolines represent inclinations i = 10, 15, 20, 25 and 30 degrees, respectively and dashed isolines different flux ratios Q=0.3, 0.5, 0.7, 1, 1.5 and 2) for the inner disk radius of 600 Rg and outer 4000 Rg.

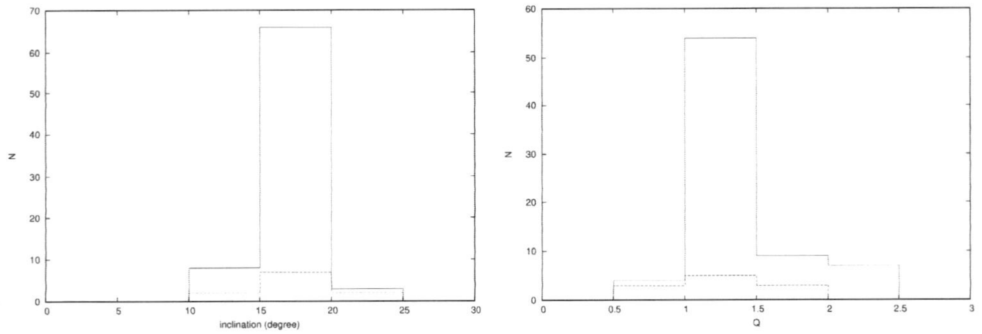

Figure 3. Histograms of the inclination (left) and $Q = F_s/F_{disk}$ (right) for the sample in the case with (dashed line) and without (solid line) systematic blue shift of the non-disk component. The cases where discrepences of the inclination estimation were larger than one degree, were not taken into account.

We carried out the following analyzes: i) fitting of single peaked BELs with Gaussians and with two-component model, and ii) comparing line parameters of a sample with simulated profiles.

From our investigations we can conclude that there could be a high probability that the disk emission flux might be present in the single peaked emission line profiles, but with the disk inclinations mainly smaller then $i < 25°$. Such small inclinations and large discrepancies from simulated profiles for higher inclinations should be discussed in the context of unified model, i.e. possible disk orientation to the torus or partial obscuration by the torus (see Fig. 5).

Both of the values ($i > 25°$ and $Q > 1$) may indicate that we have a low inclined disk which contribution to the total line flux is smaller than the non-disk emitting region.

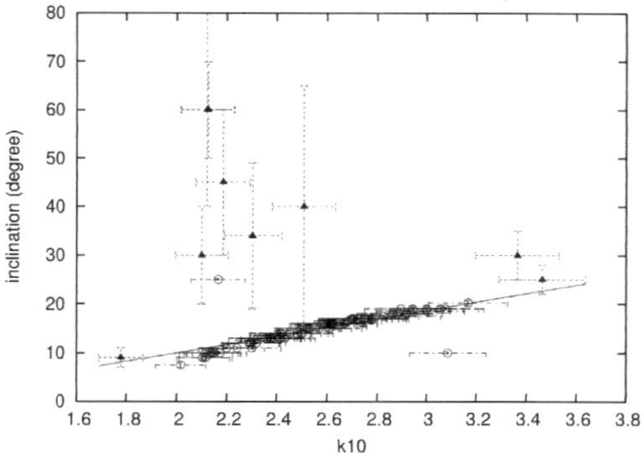

Figure 4. Disk inclination vs. full width at 10 % of line intensity normalized to FWHM (k_{10}). The points where discrepancies were more than one degree (mainly above the other points) are denoted with full triangles.

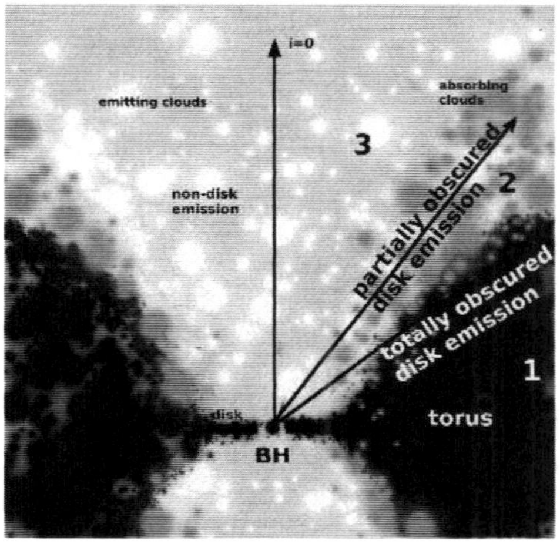

Figure 5. Obscuration of the disk emission: 1) torus, 2) absorbing material around the torus and 3) the region without absorption.

5.1. Acknowledgments

The work was supported by the Ministry of Science and Technology of Serbia through the project 146002: "Astrophysical spectroscopy of extragalactic objects".

References

[1] Popović L Č 2006 *Serb. Astron. J.* **344** 115.
[2] Chen K & Halpern J P 1989 *ApJ* **344** 115.
[3] Chen K, Halpern J P, Filippenko A V 1989 *ApJ* **339** 742.
[4] Eracleous M & Halpern J P 1994 *ApJS* **90** 1.
[5] Eracleous M & Halpern J P 2003 *ApJ* **599** 886.
[6] Eracleous M, Lewis K T, Flohic H M L G 2009 *New Ast. Rev.* **53** 133.
[7] Popović L Č, Mediavlilla E G, Bon E, Stanić N, Kubičela A, 2003 *ApJ* **599** 185.
[8] Strateva I V, Strauss M A, Hao L, Schlegel D J, Hall P B, Gunn J E, Li L-H, Ivezić Ž, Richards G T, Zakamska N L, Voges W, Anderson S F, Lupton R H, Schneider D P, Brinkmann J, Nichol R C 2003 *ApJ* **126** 1720.
[9] Sulentic J W, Marziani P & Dultzin-Hacyan D 2000 *ARAA* **38** 521.

[10] Marziani P, Sulentic J W, Calvani M, Perez E, Moles M, Penston M V 1993 *ApJ* **410** 56.
[11] Romano P, Zwitter T, Calvani M, Sulentic J 1996 *MNRAS* **279** 165.
[12] Popović L Č, Stanić N, Kubičela A, Bon E 2001 *A&A* **367** 780.
[13] Popović L Č, Mediavlilla E G, Kubičela A, Jovanović P 2002 *A&A* **390** 473.
[14] Popović L Č, Mediavlilla E G, Bon E, Ilić D 2004 *A&A* **423** 909.
[15] Ilić D, Popovć LČ, Bon E, Mediavilla E G, Chavushyan V H 2006, *MNRAS* **371** 1610.
[16] Collin S, Kawaguchi T, Peterson B M, Vestergaard M 2006 *A&AS* **456** 75.
[17] Hu C, Wang J M, Chen Y M, Bian W H, Xue S J 2008 *ApJ* **683** 115.
[18] Bon E, Popović L Č, Gavrilović N, La Mura G, Mediavlilla E G 2009 *MNRAS* **400** 924.
[19] Bon E, Popović L Č, Gavrilović N, Giovanni L M 2009 *New Ast. Rev.* **50** 716.
[20] Bon E, Popović L Č, Ilić D, Mediavlilla E G 2006 *New Ast. Rev.* **50** 716.
[21] Corbin M R & Boroson T A 1996 *ApJS* **107** 69.
[22] La Mura G, Popović L Č, Ciroi S, Rafanelli P, Ilić D 2007 *ApJ* **671** 104.
[23] Bon E 2008 *Serb. Astron. J.* **177** 9.
[24] Jovanović P & Popović LČ 2008 *Fortschr. Phys.* **56** 456.
[25] Perez E, Mediavilla E, Penston M V, Tadhunter C, Moles M 1988 *MNRAS* **230** 353.

25th Summer School and International Symposium on the Physics of Ionized Gases—SPIG 2010 IOP Publishing
Journal of Physics: Conference Series **257** (2010) 012030 doi:10.1088/1742-6596/257/1/012030

Spectral indices of radio loops

V. Borka Jovanović[1] and D. Urošević[2]

[1]Laboratory of Physics (010), Vinča Institute of Nuclear Sciences, University of Belgrade, P. O. Box 522, 11001 Belgrade, Serbia

[2]Department of Astronomy, Faculty of Mathematics, University of Belgrade, Studentski trg 16, 11000 Belgrade, Serbia

E-mail: vborka@vinca.rs

Abstract. Observations of the continuum radio emission at 1420, 820 and 408 MHz enabled estimations of the brightness of the radio loops. We calculated the mean brightness temperatures and surface brightnesses of the six main Galactic radio-continuum loops I-VI at the three frequencies. We have demonstrated the reality of Loops V and VI and presented diagrams of their spectra for the first time. We derived the radio spectral indices of Galactic radio loops from radio surveys at three frequencies. The method we have developed for large radio loops, was also used for smaller ones. In this paper we also estimated the temperatures and brightnesses of the Monoceros radio loop at 1420, 820 and 408 MHz and of Cygnus loop. The spectra (mean temperature versus frequency) between the three frequencies, as well as the $T-T$ graphs, are estimated and the spectral indices are also obtained. Using the supernova remnant (SNR) hypothesis for the origin of radio loops, distances are calculated from the surface brightnesses and the angular diameters. The obtained results confirm non-thermal origin of all radio loops and we show that our method is applicable to almost all SNRs.

1. Introduction

It is very well known that some radio spurs can be joined into small circles. The sets of spurs which form small circles are called loops. During the early seventies, four major loops were recognized. Their discoveries and studies took place in this order: Loop I [1, 2, 3, 4], Loop II [1, 5, 4], Loop III [5, 4] and Loop IV [3, 4, 6]. In 1970. Salter gave the most precise determination of these circles' parameters, which were later published in [7]. A detailed review of the subject was published in [8]. In Refs. [9] and [10] it was made an observation that some other spurs could be connected into loops V and VI. Later, we have demonstrated [11] the probable existence of Loops V and VI. In general, they display shell structure and have non-thermal spectra, similar to the main loops I-IV which are usually assumed to be SNRs.

A star in the constellation of Cygnus exploded and its remnant is Cygnus loop. It is classified as a middle-aged SNR located below (but near the plane of) the Galactic equator less than 1 kpc away from us. It is listed in Green's catalogue of SNRs as G74.0-8.5 [12, 13, 14]. The Monoceros loop was considered as an object similar to major loops [15]. Monoceros Nebula can be found in a catalog of Galactic SNRs listed as G205.5+0.5 [12, 13, 14].

Basic characteristics of Monoceros and Cygnus loop, as well as many other Galactic loops, can be found in a catalog of Galactic SNRs [12, 13, 14]. In Ref. [16], the SNRs that lie less than $\sim$ 2 kpc away from the Earth are discussed and classified by distance to the Earth. The data on Galactic supernova remnants, in all the spectral bands, also may be found in [17, 18].

In order to study the structure of Galactic radio loop emission, it is necessary to determine their spectral indices. Here we present a short overview of our recent investigations in this field.

2. Analysis and method

The aim of our work was to investigate the spectral indices of radio loops. First we derive the average brightness temperatures and surface brightnesses at 1420, 820 and 408 MHz of the six main Galactic radio-continuum loops and of the Monoceros and Cygnus loop, as well as their radio spectral indices. The temperatures, surface brightnesses and the radio spectral indices of the radio loops are computed using data taken from radio-continuum surveys at 1420, 820 and 408 MHz. Diameters and distances of Loops I-VI were also calculated, as well as diameter and distance of Monoceros loop. We also demonstrate the reality of Galactic radio Loops V and VI. The spectral indices of Loops V and VI, Monoceros and Cygnus loop are also obtained from $T - T$ plots.

As it is said in Ref. [19] contour lines, which correspond to the minimum and maximum brightness temperatures for each spur, are taken to define their borders. The contour line T_{min} is the the lower temperature limit between the background and the spur, and T_{max} is the upper temperature limit between the spur and unrelated confusing sources (superposed on the spur and hence requiring elimination from the calculation). In this manner, background radiation was considered as radiation that would exist if there were no spurs. We used averages over the data within these two curves: the contour for T_{min} and the contour for T_{max}. The loop area was divided into different sections which correspond to spurs. The background radiation was subtracted in the following way: the temperature of the loop with the background added was determined at first. After that the temperature of the background near the loop was determined. Finally, the difference of these two values was calculated.

The method for determination of the brightness temperature given in article [19] was developed for large radio loops (I-VI), but later on we applied it to much smaller loops (Monoceros and Cygnus) and showed that it is rather efficient in the case of smaller radio loops.

3. Results and discussion

We used radio-continuum surveys at three frequencies: 1420 MHz [20], 820 MHz [21], 408 MHz [22] and calculated the corresponding mean temperatures and brightnesses [23, 24, 19, 11, 25, 26, 27]. Knowing three values of brightnesses, the spectral indices could be calculated as coefficients of the linear fits of logarithmic temperature versus frequency plots.

Figs. 1 - 4 show main Galactic radio Loops I-IV at 1420, 820 and 408 MHz, with two contours representing the temperatures T_{min} and T_{max}, e.g. the borders of the loops. These borders are determined using given method. The right panel of Fig. 4 shows temperature scales in K, for 1420, 820 and 408 MHz.

The initial assumption in our investigation was the supernova remnant (SNR) hypothesis for the origin of radio loops, we are supposing all radio loops to be SNRs [28, 29, 30, 8]. We also study how results for brightnesses and distances of radio loops agree with current theories of SNR evolution. For this purpose, the ambient density and initial explosion energy of the loops are discussed [11]. We also discuss applications of different $\Sigma - D$ relations [19, 11, 26].

After deriving the brightness temperature T_{b}, we have converted these values into surface brightness Σ_ν by:

$$\Sigma_\nu = (2k\nu^2/c^2)T_{\mathrm{b}}, \tag{1}$$

where k is Boltzmann constant and c is the speed of light.

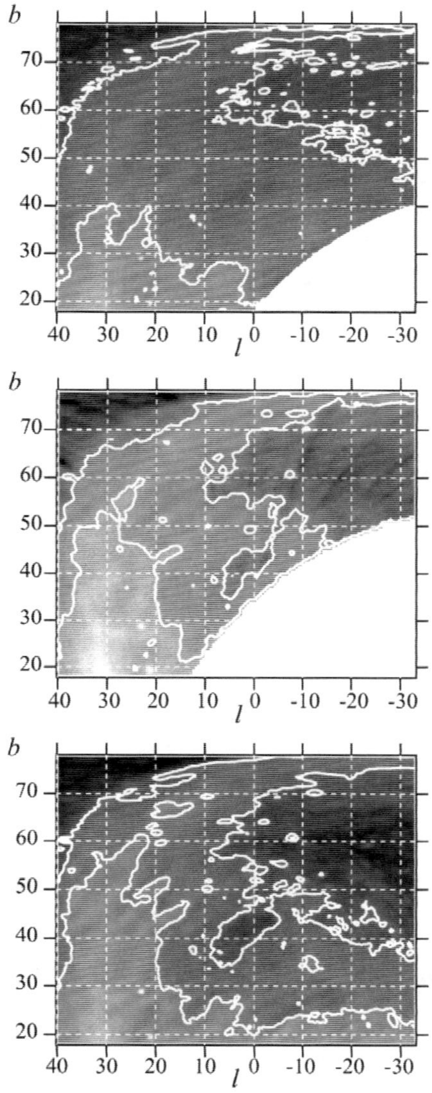

Figure 1. The area of Loop I at 1420 (top), 820 (middle) and 408 MHz (bottom), with contours of brightness temperature. Two contours are plotted, which represent the temperatures $T_{\rm min}$ and $T_{\rm max}$, as given in Table 1 in Ref. [19]. White areas in the top and bottom pictures signify that no data exist there at 1420 and 820 MHz. This contains the part of the North Polar Spur normal to the Galactic plane: $l = [40°, 0°]$; $b = [18°, 78°]$ and its part parallel to the Galactic plane: $l = [360°, 327°]$; $b = [67°, 78°]$.

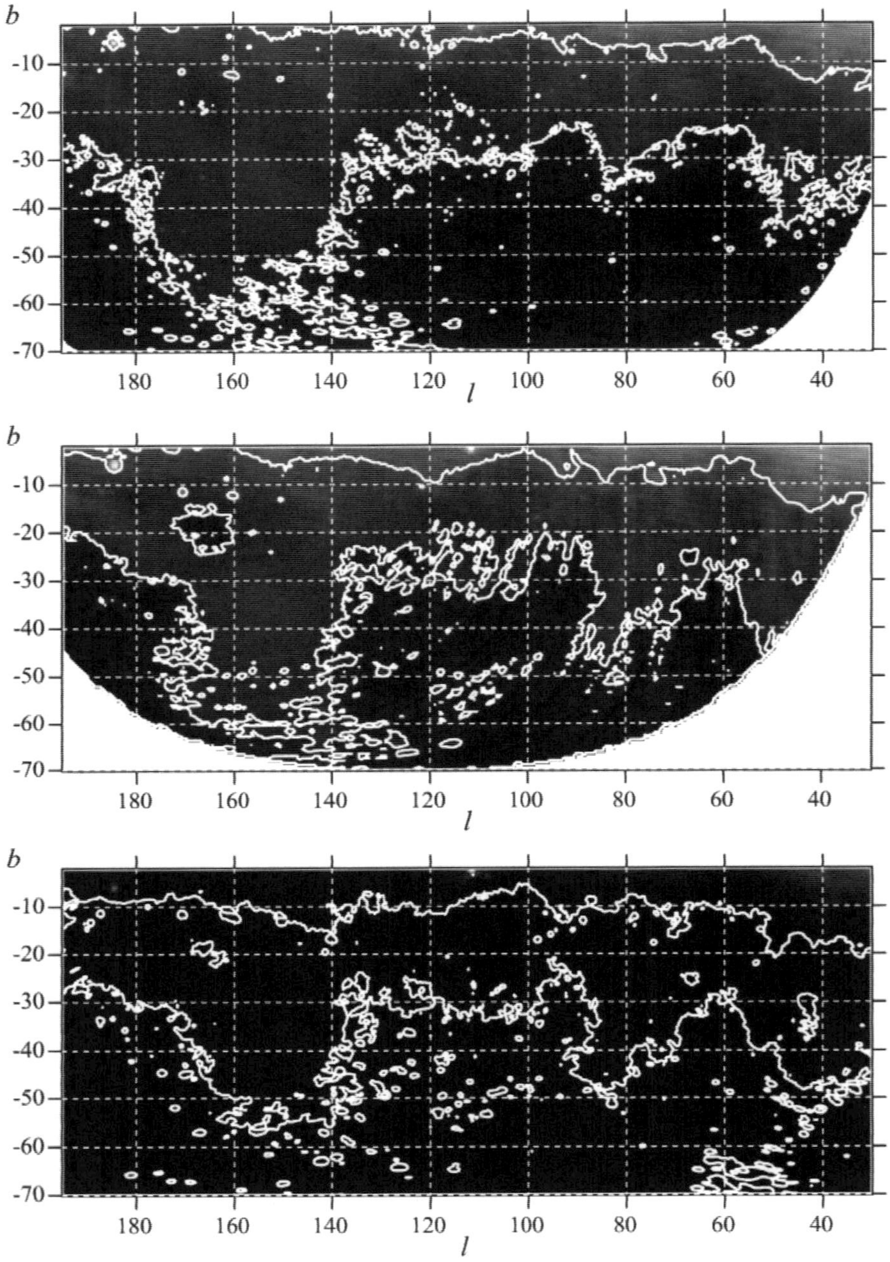

Figure 2. The same as Fig. 1, but for Loop II. White areas in the top and bottom pictures signify that no data exist there at 1420 and 820 MHz. Spurs belonging to this radio loop have positions: l = [57°, 30°]; b = [-50°, -10°] for spur in Aquarius and l = [195°, 130°]; b = [-70°, -2°] for spur in Aries.

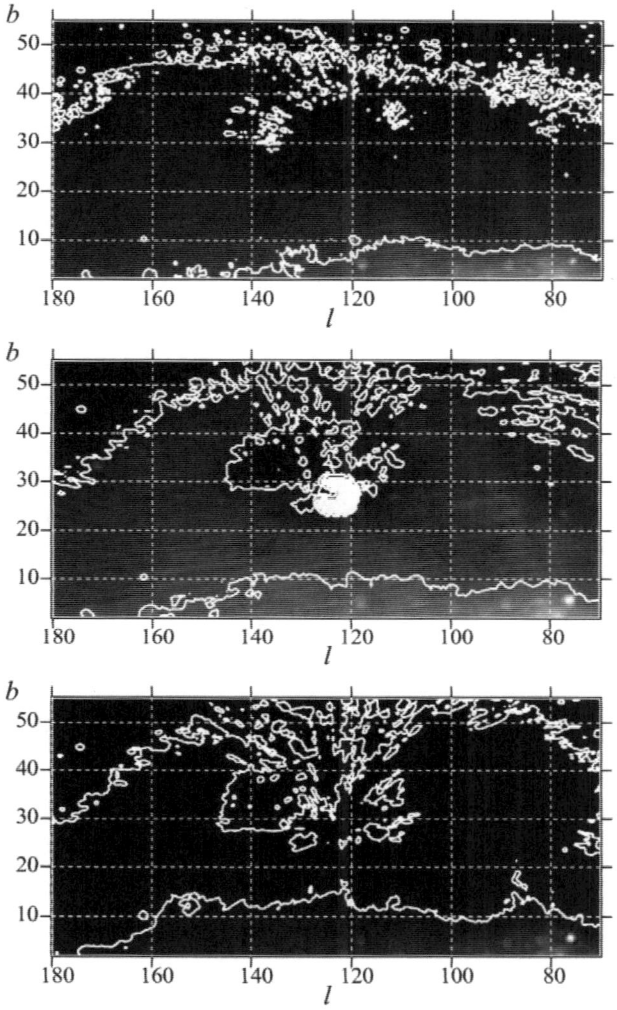

Figure 3. The same as Fig. 1, but for Loop III. The white area in the center of the picture for 820 MHz signifies that no data exist there at that frequency. Spurs belonging to this radio loop have positions: $l = [180°, 135°]$; $b = [2°, 50°]$ and $l = [135°, 110°]$; $b = [40°, 55°]$ for first spur and $l = [110°, 70°]$; $b = [6°, 50°]$ for second one.

Assuming the spectra to have a power-law form:

$$T_{\mathrm{b}} = K\nu^{-\beta}, \tag{2}$$

we get:

$$\log T_{\mathrm{b}} = \log K - \beta \log \nu, \tag{3}$$

where β is the spectral index and K is a constant.

Knowing the values of brightnesses (at 1420, 820 and 408 MHz) we were able to derive spectral indices from fitting equation (3) to the data. The spectra were generated using mean temperatures at three different frequencies. This best-fit straight line spectrum enables

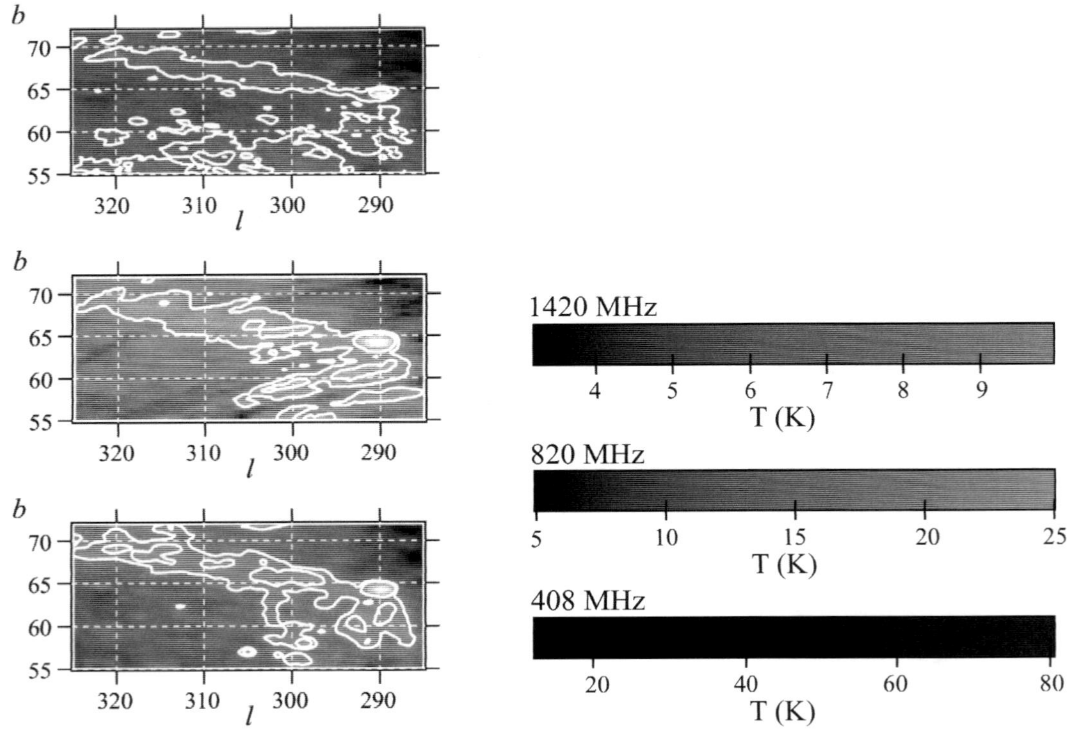

Figure 4. *Left panel*: The same as Fig. 1, but for Loop IV. This radio loop has position: l = [330°, 290°]; b = [48°, 70°]. *Right panel*: Temperature scales for 1420, 820 and 408 MHz. All are given in K and they are used for all previous pictures of the radio loops.

calculation of spectral index as negative value of the line's direction coefficient. The obtained values of radio spectral indices for large Galactic Loops I-VI are 2.74 ± 0.08, 2.88 ± 0.03, 2.68 ± 0.06, 2.90 ± 0.28, 3.03 ± 0.15 and 2.90 ± 0.09, respectively [19]. For smaller SNRs Monoceros and Cygnus, the values are: 2.70 ± 0.14 [26] and 2.76 ± 0.03 [31], respectively. The values obtained (all greater than 2.2) confirm a non-thermal origin for the emission from the radio loops.

The radio spectral index β can be also determined from $T - T$ plot, between the pair of frequencies ν_1 and ν_2, by the relation:

$$T_{\nu 1}/T_{\nu 2} = (\nu_1/\nu_2)^{-\beta}, \tag{4}$$

from which we have

$$\beta = \log(a_{12})/\log(\nu_1/\nu_2), \tag{5}$$

where a_{12} is direction coefficient of line $T_{\nu 2}(T_{\nu 1})$. In that way, for Loops V and VI obtained spectral indices from $T - T$ plots are 2.90 ± 0.11 and 2.77 ± 0.13, respectively [11]. For Monoceros loop we obtained 2.63 ± 0.30 [26] and for Cygnus 2.78 ± 0.41 [31].

By use of the spectral indices, the brightnesses have been extrapolated to 1 GHz according to relation:

$$\Sigma_{1\mathrm{GHz}}/\Sigma_{\nu\mathrm{GHz}} = (\nu/1)^{(\beta-2)}. \tag{6}$$

Journal of Physics: Conference Series **257** (2010) 012030 doi:10.1088/1742-6596/257/1/012030

Applying relation (6), the brightness at any frequency become reduced to brightness at 1 GHz, so the results become comparable.

The obtained results confirm a non-thermal origin and nearby locations for the Galactic radio loops. Therefore, we have indications that they are very old SNRs that evolve in low ambient densities, with high initial explosion energies [11].

4. Conclusions

The main results of our investigation may be summarized as follows:

- Our results are consistent with the SNR hypothesis and suggest that the main Galactic radio loops may have a SNR origin [23, 24, 19].
- We note that our values for spectral indices are in between the corresponding values for Loops I and III given in [32] and [33] ([19]). In case of the main Galactic radio loops we used $\Sigma - D$ relations for supernova remnants by [34]. The estimated distances of the main radio loops derived using the 1420 MHz data are in good agreement with the earlier results (e.g. [35]).
- In our papers, we present the first radio-continuum spectra for the main radio loops, plus Loops V and VI, made by using average brightness temperatures at three different frequencies. We find that good linear fits can be made to each of these, supplying accurate spectral indices. The radio spectra of the loops are fitted rather well by power-low spectra which is consistent with a SNR origin for these features [19].
- We have estimated the distances of the main Galactic radio loops and Loops V and VI from the northern-sky radio continuum survey at 1420 MHz [20] using two different $\Sigma - D$ relations. The estimated distances of the main radio loops derived using the 1420 MHz data are in good agreement with the earlier results (e.g. [35]) [19, 11]. From the $\Sigma - D$ diagram at 1 GHz taken from [36] with the derived values for the six loops superposed, it can be concluded that the surrounding density of the main radio loops, plus Loops V and VI, is low and that the explosion energy is high. Further, the locations of the Loops on this $\Sigma - D$ diagram agree with the general distribution and this strengthens the idea that they are old SNRs [11].
- From the spectral index analysis we can confirm that the emission from the radio loops is non-thermal in origin. We have demonstrated the probable existence of Loops V and VI. The distances and spectral indices of Loop V and Loop VI are estimated. In general, they display shell structure and have non-thermal spectra, similar to the main loops I-IV which are usually assumed to be SNRs.
- We calculated the brightness temperatures and surface brightnesses of the Monoceros radio loop at 1420, 820, and 408 MHz. The linear spectrum is estimated for mean temperatures versus frequency between 1420, 820, and 408 MHz. It is the first time that the brightness temperatures of the Monoceros loop are calculated at 820 and 408 MHz frequencies from the observational data. We sampled much more points (more than 1000) at 1420 MHz than in previous papers (95 points) [37, 38]. Also, the brightness temperature is now derived using a different method. The temperature of this radio loop at 1420 MHz is in good agreement with the result obtained in [37] ([25, 26]).
- In case of the Monoceros radio loop we used empirical $\Sigma - D$ relations for supernova remnants by Urošević [39] and Arbutina et al. [40]. The estimated distance of the Monoceros radio loop is in good agreement with the earlier results [13]. The spectral index analysis confirms the non-thermal origin of the Monoceros radio loop [26].
- We present the radio continuum spectrum of the Cygnus loop using average brightness temperatures at three different frequencies. Linear fit provides reliable spectral index.

Obtained results are consistent with the SNR hypothesis and suggest that the Cygnus radio loop may have a SNR origin. Also, we derived the $T - T$ plots which enables also calculation of spectral index. [27].

According to the results summarized in this paper, one can conclude that radio loop emission of the six main Galactic radio-continuum loops, Monoceros and Cygnus loop is in good agreement with current theories of SNR evolution.

We showed that method for defining a loop border and for determining the values of temperature and brightness, which we developed for main Galactic loops I-VI, could be applicable to all SNRs.

Acknowledgments

This research is part of the project "Gaseous and stellar component of galaxies: interaction and evolution" (No. 146012) supported by the Ministry of Science of Republic of Serbia.

References

[1] Large M I, Quigley M J S and Haslam C G T 1962 *Mon. Not. R. Astron. Soc.* **124** 405
[2] Haslam C G T, Large M I and Quigley M J S 1964 *Mon. Not. R. Astron. Soc.* **127** 273
[3] Large M I, Quigley M J S and Haslam C G T 1966 *Mon. Not. R. Astron. Soc.* **131** 335
[4] Salter C J 1970 Ph. D. Thesis, Univ. of Manchester
[5] Quigley M J S and Haslam C G T 1965 *Nat* **208** 741
[6] Reich W and Steffen P 1981 *Astron. Astrophys.* **93** 27
[7] Berkhuijsen E M, Haslam C G T and Salter C J 1971 *Astron. Astrophys.* **14** 252
[8] Salter C J 1983 *Bull. Astr. Soc. India* **11** 1
[9] Milogradov-Turin J 1972 M. Sc. Thesis, Univ. of Manchester
[10] Milogradov-Turin J and Urošević D 1997 *BA Belgrade 155* 41
[11] Borka V, Milogradov-Turin J and Urošević D 2008 *Astron. Nachr.* **329** 397
[12] Green D A 2004 *Bull. Astr. Soc. India* 32, 335
[13] Green D A 2006 A Catalogue of Galactic Supernova Remnants (2006 April version), Cavendish Laboratory, Cambridge, UK.
[14] Green D A 2009 *Bull. Astr. Soc. India* **37** 45
[15] Spoelstra T A Th 1973, *Astron. Astrophys.* **24** 149
[16] Delahaye T, Lavalle J, Lineros R, Donato F and Fornengo N 2010 arXiv:1002.1910v1 [astro-ph.HE]
[17] Guseinov O H, Ankay A, Tagieva S O 2003, *Serb. Astron. J.* **167** 93
[18] Guseinov O H, Ankay A, Tagieva S O 2004, *Serb. Astron. J.* **168** 55
[19] Borka V 2007, *Mon. Not. R. Astron. Soc.* **376** 634
[20] Reich P and Reich W 1986 *Astron. Astrophys. Suppl. Ser.* **63** 205
[21] Berkhuijsen E M 1972 *Astron. Astrophys. Suppl. Ser.* **5** 263
[22] Haslam C G T, Salter C J, Stoffel H and Wilson W E 1982 *Astron. Astrophys. Suppl. Ser.* **47** 1
[23] Borka V 2006, Book of contributed papers of the 23rd SPIG 599
[24] Borka V, Milogradov-Turin J and Urošević D 2006 *Publ. Astron. Obs. Belgrade* **80** 61
[25] Borka Jovanović V and Urošević D 2008 *Publ. Astron. Obs. Belgrade* **84** 459
[26] Borka Jovanović V and Urošević D 2009a, *Astron. Nachr.* **330** 741
[27] Borka Jovanović V and Urošević D 2009b *Publ. Astron. Obs. Belgrade* **86** 101
[28] Berkhuijsen E M, Haslam C G T and Salter C J 1970 *Nature* **225** 364
[29] Berkhuijsen E M 1971 *Astron. Astrophys.* **14** 359
[30] Shklovskii I S and Sheffer E K 1971 *Nature* **231** 173
[31] Borka Jovanović V and Urošević D, in preparation
[32] Reich P and Reich W 1988b *Astron. Astrophys.* **196** 211
[33] Berkhuijsen E M 1973 *Astron. Astrophys.* **24** 143
[34] Case G L and Bhattacharya D 1998 *ApJ* **504** 761
[35] Berkhuijsen E M 1986 *Astron. Astrophys.* **166** 257
[36] Berezhko E G and Völk H J 2004 *Astron. Astrophys.* **427** 525
[37] Urošević D Milogradov-Turin J. 1998 *Serb. Astron. J.* **157** 35
[38] Milogradov-Turin J and Urošević D 1996 *Publ. Astron. Obs. Belgrade* **54** 47
[39] Urošević D 2002 *Serb. Astron. J.* **165** 27
[40] Arbutina B, Urošević D, Stanković M and Tešić Lj 2004 *Mon. Not. R. Astron. Soc.* **350** 346

Journal of Physics: Conference Series **257** (2010) 012031 doi:10.1088/1742-6596/257/1/012031

Parametric dependence of two-plasmon decay in homogeneous plasma

Dejan R. Dimitrijević

Department of physics, Faculty of sciences and mathematics, University of Niš, Višegradska 33, 18001 Niš, Serbia

dimke@junis.ni.ac.rs

Abstract. A hydrodynamic model of two-plasmon decay in a homogeneous plasma slab near the quarter-critical density is constructed in order to improve our understanding of the spatio-temporal evolution of the daughter electron plasma waves in plasma in the course of the instability. The scaling of the amplitudes of the participating waves with laser and plasma parameters is investigated. The secondary coupling of two daughter electron plasma waves with an ion-acoustic wave is assumed to be the principal mechanism of saturation of the instability. The impact of the inherently nonresonant nature of this secondary coupling on the development of two-plasmon decay is researched and it is shown to significantly influence the electron plasma wave dynamics. Its inclusion leads to nonuniformity of the spatial profile of the instability and causes the burst-like pattern of the instability development, which should result in the burst-like hot-electron production in homogeneous plasma.

1. Introduction

Two-plasmon decay (TPD) is a nonlinear parametric instability during which a large amplitude electromagnetic pump wave decays into two electron plasma waves near the quarter-critical plasma density [1,2], $n_{cr}/4$. TPD has primarily been known for its detrimental aspect, as a source of suprathermal (hot) electrons capable of preheating the fusion targets in the inertial confinement fusion experiments [3-5]. The electrostatic nature of the daughter electron plasma waves renders the experimental verification of various aspects of the theory of TPD rather difficult, and the comparison of theoretical and numerical predictions with the experimental evidence is mainly qualitative.

Here we report the results of our investigation of the spatio-temporal evolution of the amplitudes of the waves participating in the instability, especially the daughter electron plasma waves, and its dependence on the pump laser intensity and plasma parameters. Since the secondary coupling of the daughter electron plasma waves with ion-acoustic waves is widely recognized as the principal mechanism leading to the saturation of the instability, we are particularly interested in investigating the influence of the nonresonant nature of this secondary coupling on the wave dynamics and the spatio-temporal development of TPD. A transparent hydrodynamic model of TPD in spatially homogeneous plasma slab near the quarter-critical density is constructed, taking into account numerous nonlinear features of the instability. Similar model has been utilized earlier [6, 7] in order to search for the time-dependent local solutions for the slowly varying wave amplitudes, by neglecting all the convective terms and, consequently, group velocities of the coupled waves in the corresponding coupled-wave equations, which was justified by the fact that the nature of the instability is strictly resonant. This approach has been generalized [8, 9] by simulating the complete set of coupled-wave equations numerically, in order to gain a more complete insight into the spatio-temporal behavior of the wave amplitudes inside the homogeneous plasma slab.

In the next section, we will introduce the theoretical model that describes the evolution of the slowly varying amplitudes of the waves taking part in the development and the saturation of the instability in the homogeneous plasma slab. The model hydrodynamic equations are simulated numerically and the corresponding results are presented and discussed in the final section.

2. Theoretical model

We consider the propagation of a linearly polarized high intensity laser radiation (pump wave), with frequency ω_0 and wavelength λ_0, in a homogeneous underdense plasma slab whose width is $L = 100\lambda_0$. The pump wave enters the slab perpendicularly through its left boundary. If plasma density is in the vicinity of $n_{cr}/4$, the pump electromagnetic wave decays nonlinearly into two daughter electron plasma waves, whose frequencies and wave vectors are $\omega_{1,2}$ and $\vec{k}_{1,2}$, respectively. Resonant nature of TPD implies that the well-known matching conditions for the frequencies and wave vectors, $\omega_0 = \omega_1 + \omega_2$ and $\vec{k}_0 = \vec{k}_1 + \vec{k}_2$, must be satisfied strictly. Bearing in mind that the electron plasma wave frequencies are temperature-dependant through the Bohm-Gross dispersion relations $\omega_{1,2}^2 = \omega_{pe}^2 + 3k_{1,2}^2 v_e^2$, where ω_{pe} is the electron plasma frequency and v_e the thermal velocity, it is clear that the plasma slab density must be somewhat lower than $n_{cr}/4$ in order to provide conditions necessary for the onset of TPD.

Numerous theoretical, experimental and simulation studies [8-14] have indicated that an efficient coupling of the electron plasma wave electrostatic energy to shorter-wavelength ion density fluctuations, or ion-acoustic waves, is the main factor responsible for the saturation of the instability in homogeneous plasma. Our simulations performed in the limit of time-only dependence [6, 7] confirmed that the saturation occurs simultaneously with the ion fluctuation amplitude reaching its maximum value, whose order was only a fraction of a percent of the bulk plasma density. The matching conditions for the secondary coupling are $\omega_s = \omega_1 - \omega_2$ and $\vec{k}_s = \vec{k}_1 - \vec{k}_2$, where ω_s and $\vec{k}_s$ are the frequency and the wave vector of the ion-acoustic wave. It is clear though that the nature of the secondary coupling is nonresonant, so the corresponding frequency mismatch $\Delta\omega = \omega_2 + \omega_s - \omega_1$ must be taken into account when writing down the equations of the system evolution. Our earlier work [6, 7] confirms that it leads to shifting and broadening of the frequency lines in the plasmon spectra, as well as an increase of their complexity and the appearance of new spectral lines corresponding to lower-frequency oscillations. Thus the influence of the nonresonance of the secondary coupling on the spatio-temporal evolution of TPD proved to be significant and has been further investigated [8, 9].

We start with the basic set of electron and ion fluid equations combined with the Maxwell equations, to yield the following system of normalized partial differential equations for the slowly varying amplitudes of the coupled waves, namely electromagnetic wave (a), electron plasma waves (N_1, N_2), and the ion-acoustic wave (N_s) participating in the secondary coupling:

$$\left(\frac{\partial}{\partial t} + V_0 \frac{\partial}{\partial x}\right) a = -B_0 N_1 N_2 , \quad (1)$$

$$\left(\frac{\partial}{\partial t} + V_1 \frac{\partial}{\partial x}\right) N_1 = B_1 a N_2^* + C_1 N_2 N_s \exp(-i\Omega t) - \Gamma_1 N_1 , \quad (2)$$

$$\left(\frac{\partial}{\partial t} + V_2 \frac{\partial}{\partial x}\right) N_2 = B_2 a N_1^* - C_2 N_1 N_s^* \exp(i\Omega t) - \Gamma_2 N_2 , \quad (3)$$

$$\left(\frac{\partial}{\partial t} + V_s \frac{\partial}{\partial x}\right) N_s = C_3 N_1 N_2^* \exp(i\Omega t) . \quad (4)$$

Here, only the wave amplitude dependences in the direction of the pump propagation are accounted for, and the group velocities V_0, V_1, V_2 and V_s stand for the longitudinal components of the corresponding vectors. This can be done without loss in generality in the regime of prime interest for laser fusion applications, $(v_0/v_e) \leq 0.1$, where the transverse components of plasma waves are quite small, which is the case when visible and UV lasers are utilized. From the geometry of the instability follows that in such cases one plasma wave is directed forward, almost along the x-axis, while the other is directed backward, and longitudinal components of their wave numbers relate as $k_1 \sim k_0 \gg k_2$. Our simulation results refer to this parametric range, although even in the regime $(v_0/v_e)^2 > 10$, typical for CO_2 laser driven plasmas, some qualitatively accurate predictions can be made.

In the system of equations (1)-(4) time and space variables are normalized to $1/\omega_0$ and c/ω_0 respectively, the group velocities of all the waves to the speed of light c, and the frequency mismatch is normalized as $\Omega = \Delta\omega/\omega_0$. The amplitude of the electromagnetic wave is represented by the dimensionless parameter $a = v_0/c$, where $v_0 = eE_0/m\omega_0$ is the electron quiver velocity in the laser electric field whose amplitude is E_0. The amplitudes of the electrostatic waves are normalized to the equilibrium (bulk) plasma density n_0, and the damping rates for the electron plasma waves, $\Gamma_{1,2}$, which include both collisional and noncollisional (Landau) damping, to ω_0. The coupling coefficients are given as:

$$B_0 = \alpha^3 \kappa_\perp \left(\kappa_1^2 - \kappa_2^2\right)/4\kappa_1^2\kappa_2^2,$$

$$B_{1,2} = \kappa_\perp \left(\kappa_1^2 - \kappa_2^2\right)/8\alpha\kappa_{2,1}^2,$$

$$C_{1,2} = \alpha^3 \left|\vec{\kappa}_1 \cdot \vec{\kappa}_2\right|/4\beta_e^2\kappa_{2,1}^2\kappa_s,$$

$$C_3 = \sqrt{Zm/M_i}\,\beta_e\kappa_s^3 \left|\vec{\kappa}_1 \cdot \vec{\kappa}_2\right|/4\kappa_1^2\kappa_2^2,$$

and the dimensionless plasma parameters are $\alpha = \omega_p/\omega_0 = \sqrt{n_0/n_{cr}}$ and $\beta_e = v_e/c = \sqrt{T_e(\mathrm{keV})/511}$, where v_e stands for the electron thermal velocity which corresponds to the electron temperature T_e. The value for the electron-ion mass ratio is taken to be $Zm/M_i = 1/3600$ in our simulations, and all the wave vectors and their components are normalized to ω_0/c.

We will limit our discussion to the optimal case of maximum instability increment, since the contribution of the nonresonant coupling is relatively small due to the fact that TPD is instability of a highly resonant nature. The relation between longitudinal and transversal components of the wave vectors of daughter electron plasma waves in the case of maximum instability increment is the well-known equation of hyperbola $k_\perp^2 = k_\parallel(k_\parallel - k_0)$. By combining this equation with TPD matching conditions and the dispersion relations for the electron plasma waves, and by subsequently normalizing all the variables as stated above, we easily obtain the expressions for intensities of the normalized wave vectors of both electron plasma waves and their components in the form:

$$\kappa_{1,2} = \sqrt{(1-2\alpha)/6\beta_e^2 \pm \sqrt{(1-\alpha^2)(1-2\alpha)/12\beta_e^2}}, \qquad (5)$$

$$\kappa_{1,2\parallel} = \sqrt{(1-\alpha^2)/4} \pm \sqrt{(1-2\alpha)/12\beta_e^2}, \qquad (6)$$

$$\kappa_\perp = \sqrt{(1-2\alpha)/12\beta_e^2 - (1-\alpha^2)/4}, \qquad (7)$$

while normalized intensity of the wave number of the ion-acoustic wave is

$$\kappa_s = \left|\vec{\kappa}_1 - \vec{\kappa}_2\right|. \qquad (8)$$

Finally, longitudinal components of the group velocities, calculated from the corresponding dispersion relations and normalized to the speed of light, can be written in our notation as:

$$V_0 = \sqrt{1-\alpha^2},$$
$$V_{1,2} = \left(3\beta_e^2/2\alpha\right)\left(\sqrt{1-2\alpha/3\beta_e^2} \pm \sqrt{1-\alpha^2}\right), \tag{9}$$
$$V_s = \beta_e\left(\kappa_{1\parallel} + \kappa_{2\parallel}\right)/60\kappa_s .$$

Clearly, the wave vectors and the scattering angles for the optimal case of maximum instability increment at given plasma density, as well as the group velocities of the participating waves, depend solely on plasma parameters, α and β_e.

3. Results and discussion

We simulate the system (1)-(4) in order to investigate the behavior of the shorter-wavelength, forward propagating electron plasma wave, whose potential detrimental impact in the inertial confinement fusion experiments is most prominent. We will assume that the pump wavelength is $\lambda_0 = 1\mu\text{m}$ and the time scale, expressed in picoseconds, will be dependent on that assumption.

The electron temperature dependence of the forward-propagating electron plasma wave amplitude temporal evolution is displayed in Figure 1(a). The saturation value of the amplitude N_1 clearly scales with the square root of the electron temperature, and our simulations indicate that the same can be said about the ion-acoustic wave amplitude. This is caused by the fact that the group velocities of the electrostatic waves in plasma, calculated through (9), increase linearly with β_e. It can also be inferred that the electrostatic wave amplitudes saturate much earlier when the electron temperature is higher, as the higher electron thermal energy leads to a more efficient damping of the amplitude oscillations. On the other hand, deeper into the plasma the amplitudes saturate with considerable delay: for the parameters in Figure 1(a) and the electron temperature of 600 eV, the saturation occurs after about 10 ps for $x = 25\lambda_0$, and less than 3 ps for $x = 75\lambda_0$.

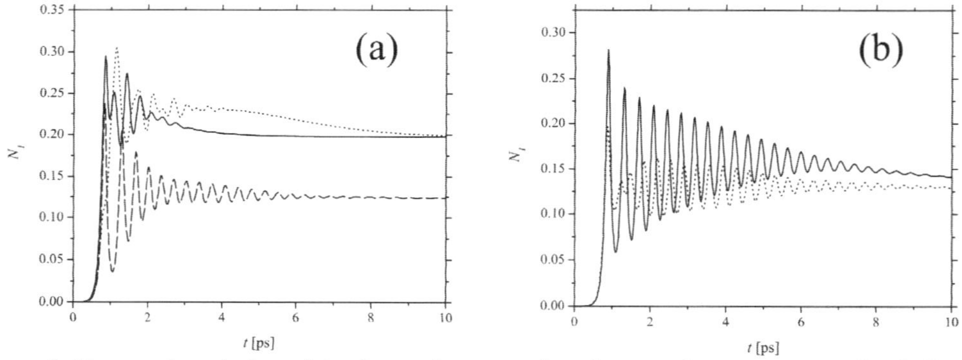

Figure 1: Temporal evolution of the forward-propagating electron plasma wave amplitude for: (a) $a = 0.03$, $n = 0.248n_{cr}$ and various electron temperatures and distances from the left plasma slab boundary: $T = 0.2\,\text{keV}$, $x = 25\lambda_0$ (dashed line), $T = 0.6\,\text{keV}$, $x = 25\lambda_0$ (solid line), $T = 0.6\,\text{keV}$, $x = 75\lambda_0$ (dotted line); (b) $a = 0.03$, $T = 0.2\,\text{keV}$, $x = 75\lambda_0$, and for plasma densities: $n = 0.240n_{cr}$ (solid line) and $n = 0.248n_{cr}$ (dotted line).

Since the electron plasma wave group velocity decreases with increasing plasma density, as inferred from (9), its saturation amplitude exerts the same dependence, which is illustrated in Figure 1(b). Even though the initial amplitude oscillations are far more rapid for lower plasma densities than for the higher ones, they reach the saturation at approximately the same time, which is of the order of 10 ps. While the saturation amplitude is about 14% for $n = 0.240$, its value is only 1% lower for

$n = 0.248$. Such a weak density dependence, however, does not apply to the ion-acoustic wave amplitude, which plunges from 0.4% to 0.06% in the same interval of densities. Our model thus confirms that even very low values of the ion fluctuations turn out to be sufficient to procure saturation of the instability, especially when the plasma density is close to $n_{cr}/4$.

Spatio-temporal evolution of the electron plasma wave amplitude is displayed in Figure 2 in three-dimensional (a) and contour plot (b) representation, for a typical set of laser and plasma parameters, in the case of resonant secondary coupling. The character of the wave dynamics is simply periodic throughout the plasma slab with natural exception of the narrow bands along the boundaries. The saturation level remains mainly constant, approximately 0.14 for the chosen parameters.

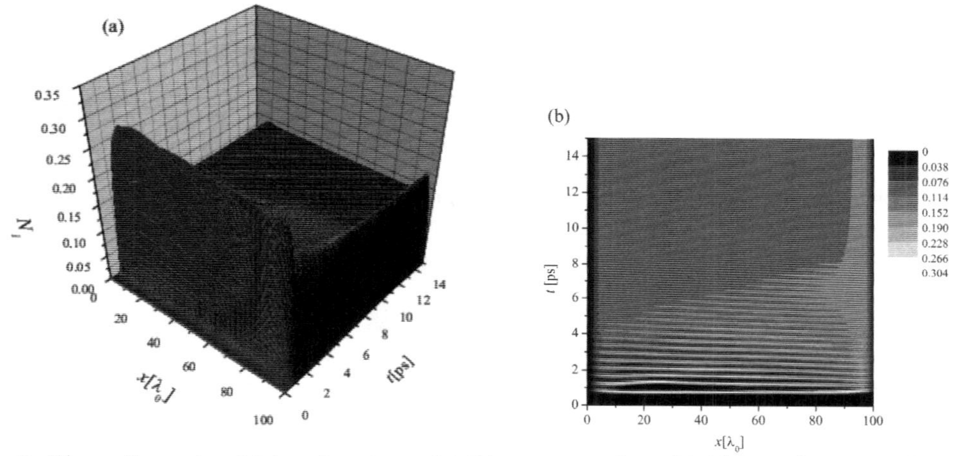

Figure 2: Three-dimensional (a) and contour plot (b) representation of initial spatio-temporal evolution of the forward-propagating electron plasma wave amplitude for $a = 0,03$, $T = 0,2\text{keV}$, $n = 0,240 n_{cr}$, when the secondary coupling is resonant.

It is rather illustrative to compare these results with those represented in Figure 3, obtained for the same set of parameters when nonresonance of the secondary coupling is taken into account. As before, the results are represented in three-dimensional (a) and contour plot diagram (b) form. The distinct nonuniformity of the electron plasma wave amplitude throughout the plasma slab is revealed for $\Omega \neq 0$. Apparently, TPD wave intensity oscillates at the frequency corresponding to that of the ion-acoustic waves and saturates after about 6 ps in the vicinity of the left plasma slab boundary, and up to 10 ps towards the right boundary. As anticipated, the saturation generally occurs later than in the case of resonant secondary coupling, and the saturation amplitude is considerably higher, because the ion-coupling mechanism is less efficient when $\Omega \neq 0$. These features of the wave dynamics clearly become more pronounced towards the right plasma slab boundary, as the impact of finite Ω causes a cumulative effect on larger space scales. This is in good agreement with benchmark results of Meyer's group [12] which were obtained by using CO_2 laser radiation with $10.6\mu m$ wavelength, where the Thomson scattering streak record of the spatial development of TPD wave intensity confirms significant spatial nonuniformity of the electron plasma wave intensity, with closely spaced local maxima of resolution limited size. Apparently, the impact of Ω is significant even though its value is of the order 10^{-3} for most parametric regimes of interest.

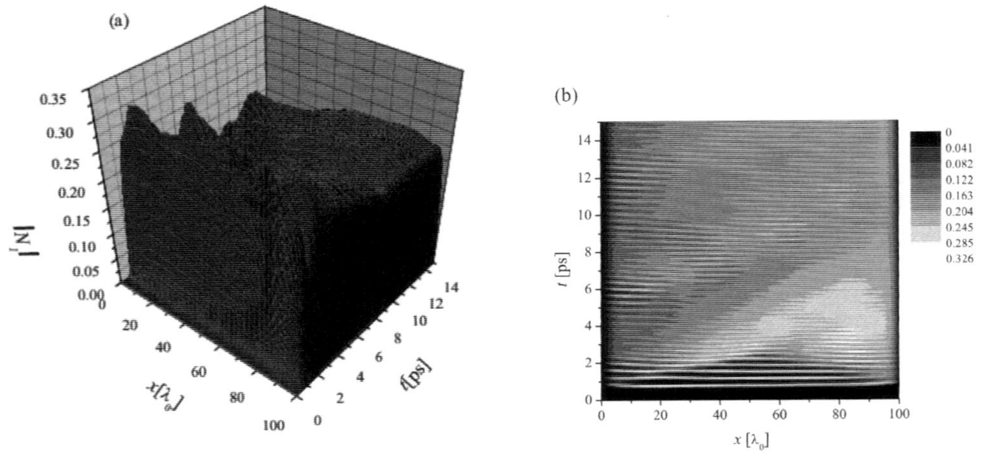

Figure 3: Three-dimensional (a) and contour plot (b) representation of initial spatio-temporal evolution of the forward-propagating electron plasma wave amplitude for $a = 0,03$, $T = 0,2\mathrm{keV}$, $n = 0,240 n_{cr}$, when nonresonance of the secondary coupling is taken into account.

Some interesting conclusions can be drawn by observing the long-term spatio-temporal evolution of the absolute value of the forward-propagating electron plasma wave amplitude, displayed in Figure 4. After the initial period of rapid oscillations, which unfolds according to the previously described scenario, the electron plasma wave amplitude saturates temporarily to a value of 0.14 to 0.17, depending on x, for the chosen laser and plasma parameters, and then enters a period of rapid oscillations. The regime of these oscillations is clearly quasi-periodic, with at least two characteristic frequencies, apparently generated by beating of the ion-acoustic frequency with the frequency mismatch of the secondary coupling. Clearly, introduction of the finite frequency mismatch Ω contributes to greater richness and complexity of the wave dynamics and causes distinct nonuniformity of the spatial profile of the instability.

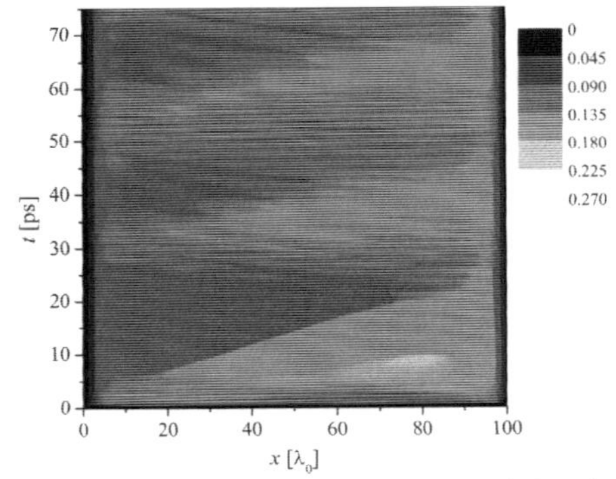

Figure 4: Long-term spatio-temporal evolution in $x = 25\lambda_0$ of the absolute value of the forward-propagating electron plasma wave amplitude for $a = 0.03$, $T = 0.2\,\mathrm{keV}$ and $n = 0.248 n_{cr}$, when nonresonance of the secondary coupling is taken into account.

These frequencies are better distinguished in Figure 5, which provides us with an insight into the long-term temporal evolution of the absolute value of the plasmon amplitude for the same laser and plasma parameters as in Figure 4, in $x = 25\lambda_0$ (a), $x = 50\lambda_0$ (b), $x = 75\lambda_0$ (c).

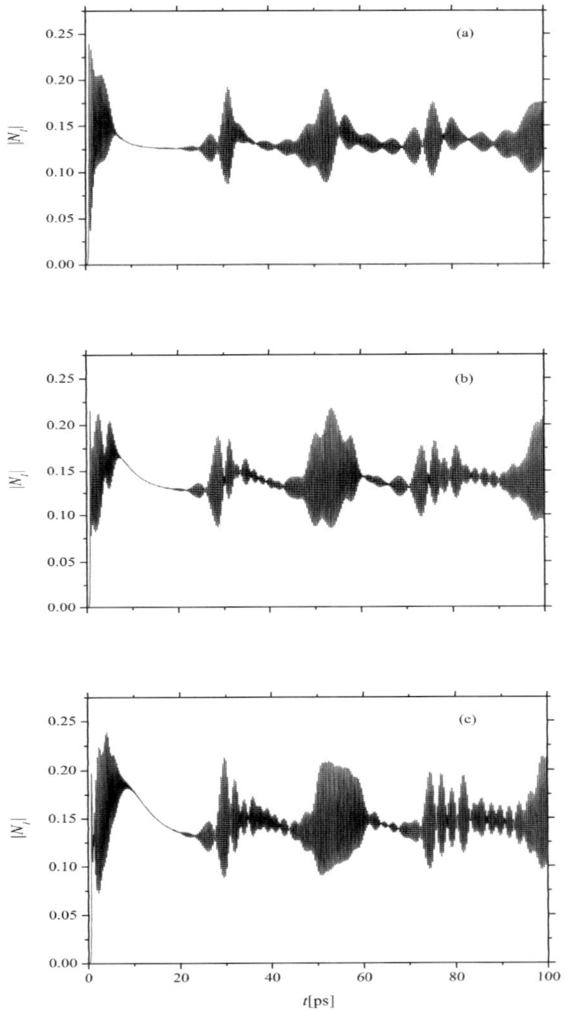

Figure 5: Long-term temporal evolution of the absolute value of the forward-propagating electron plasma wave amplitude for $a = 0.03$, $T = 0.2\,\text{keV}$, $n = 0.248n_{cr}$, in $x = 25\lambda_0$ (a), $x = 50\lambda_0$ (b), $x = 75\lambda_0$ (c), when nonresonance of the secondary coupling is taken into account.

The maximum amplitude of these rapid oscillations is significantly above the saturation amplitude and can even amount to 25% closer to the right plasma slab boundary. The complexity of the wave dynamics increases simultaneously, barring the narrow band in the vicinity of the right plasma boundary in which the boundary effect prevails, although the oscillations always preserve their quasi-periodic nature. Every maximum is followed by an interval when the amplitude slowly decreases, gradually reaching the saturation value, after which another period of rapid oscillations occurs and so on. This is somewhat similar to the burst-like behavior of the instability in inhomogeneous plasmas, when the profile modification near the quarter-critical density is taken into account, as the density profile steepens and relaxes under the influence of the ponderomotive force of the pump wave [1, 2]. The frequency of these pulses is better defined in the vicinity of the left plasma boundary, while their amplitude grows toward the right boundary. We conclude that TPD instability should occur in bursts even in homogeneous plasmas, due to the inherently nonresonant nature of the ion saturation mechanism, although the amplitudes of these bursts should be considerably smaller than in the above mentioned case of ponderomotive profile steepening. An increased, burst-like production of hot electrons is also to be expected during periods of instability enhancement.

Acknowledgments
This work has been supported by the Ministry of Science of Serbia, under Project No. 141034.

References

[1] Kruer W L 1988 The Physics of Laser Plasma Interactions, (Redwood City, CA: Addison-Wesley)
[2] Baldis H A, Campbell E M, and Kruer W L 1991 Handbook of Plasma Physics: Physics of Laser Plasma, ed A M Rubenchik and S Witkowski (Amsterdam: Elsevier) p. 361
[3] Young F C, Herbst M J, Manka C K, Obenschain S P, and Gardner J H 1985 Phys. Rev. Lett. **54** 2509
[4] Yaakobi B, Stoeckl C, Boehly T, Meyerhofer D D, and Seka W 2000 Phys. Plasmas **7** 3714
[5] Stoeckl C et al 2003 Phys. Rev. Lett. **90** 235002
[6] Dimitrijević D R, and Jovanović M S 2002 Phys. Rev. E **66** 056408
[7] Dimitrijević D R, and Jovanović M S 2002 Contrib. Papers 21st Int. Symp. On the Phys. Of Ionized Gases p. 550
[8] Dimitrijević D R, and Maluckov A A 2010 Journal of Plasma Phys. accepted for publishing
[9] Dimitrijević D R and Maluckov A A 2009 Proceedings 7th general Conference of the BPU p. 1456
[10] Karttunen S J 1981 Phys. Rev. A **23** 2006
[11] Baldis H A, and Walsh C J 1983 Phys. Fluids **26** 1364
[12] Meyer J 1992 Phys. Fluids B **4** 2934
[13] DuBois D F, Russell D A, and Rose H A 1995 Phys. Rev. Lett. **74** 3983
[14] Labaune C, Baldis H A, Bauer B S, Tikhonchuk V T, and Laval G Phys. Plasmas **5** 234

Journal of Physics: Conference Series **257** (2010) 012032 doi:10.1088/1742-6596/257/1/012032

European Virtual Atomic Data Centre - VAMDC

Milan S Dimitrijević[1], Sylvie Sahal-Bréchot[2], Andjelka Kovačević[3], Darko Jevremović[1], Luka Č. Popović[1]
[1]Astronomical Observatory, Volgina 7, 11060 Belgrade, Serbia

[2] Paris Observatory, CNRS-UMR8112 and University Pierre et Marie Curie, LERMA, 5 Place Jules Janssen, 92190 Meudon, France

[3]Department of Astronomy, Faculty of Mathematics, Studentski Trg 15, 11000 Belgrade, Serbia

mdimitrijevic@aob.bg.ac.rs, sylvie.sahal-brechot@obspm.fr, andjelka@matf.bg.ac.rs

Abstract.
The Virtual Atomic and Molecular Data Centre (http://www.vamdc.eu, VAMDC) is an European Union funded FP7 project aiming to build a secure, documented, flexible and interoperable e-science environment-based interface to existing atomic and molecular data. It will also provide a forum for training potential users and dissemination of expertise worldwide. This review describes the VAMDC project and its objectives.

1. Introduction

Reliable atomic and molecular data are of critical importance for different applications in astrophysics, atmospheric physics, fusion, environmental sciences, combustion chemistry, and in industrial applications from plasmas and lasers to lighting. Resources of such data are highly fragmented, presented in different, non-standardized ways, available through a variety of highly specialized and often poorly documented interfaces, so that their full exploitation is limited, which make difficulties in many research fields like for example the understanding the chemistry of Solar system and of the wider universe, the study of the terrestrial atmosphere, the development of the fusion rersearch, lasers etc.

The development of powerful computers stimulates the development of atomic data on a large scale. The modelling of stellar atmospheres and of the stellar interiors needs extensive sets of atomic data, including collisional broadening. For example, the PHOENIX computer code [1] developed for stellar modelling includes a database containing more than 10^7 atomic, ionic and molecular spectral lines.

The development of satellite astronomy, providing a huge amount of high quality astronomical spectra produced an information avalanche and leaded to the creation of huge data collections as e. g. IUE and HST archive. For example Sloan Digital Sky Survey SDSS, contains spectra of ~ 230 million objects. The problem is how to analyse such amount of data?

The idea of Virtual Observatory was formulated at the end of 2000, and from 2001 the FP5 project Astrophysical Virtual Observatory – AVO was the basis for creation of European Virtual Observatory - EURO-VO (http://www.euro-vo.org).

Virtual observatories today combine research in different areas of astrophysics (some of which are fairly new): multi-wavelength astrophysics, archival research, survey astronomy, temporal astronomy, theory and simulations (comparisons with observations) and information technology, digital detectors, massive data storage, the Internet, data representation standards.

In order to facilitate the international coordination and collaboration necessary for the development and deployment of the tools, systems and organizational structures necessary to enable the international utilization of astronomical archives as an integrated and interoperating virtual observatory, International Virtual Observatory Alliance (IVOA, http://www.ivoa.net) was formed in June of 2002. So the work of the IVOA mainly focuses on the development of standards.

SerVO - Serbian virtual observatory (http://www.servo.aob.rs/~darko) is a project whose funding was approved through a grant TR13022 from Ministry of Science and Technological Development of Republic of Serbia [2]. The project objectives are: a) Establishing SerVO and join the EuroVO and IVOA; b) Establishing SerVO data Center for digitizing, archiving and publishing in VO format photo-plates [3] and other astronomical data produced at Belgrade Astronomical Observatory; c) Development of tools for visualization of data; d) Publishing, together with Observatoire de Paris, STARK-B - Stark broadening data base containing as the first step Stark broadening parameters obtained within the semiclassical perturbation approach by two of us (MSD-SSB) in VO compatible format; e) Make a mirror site for DSED (Darthmouth Stellar Evolution Database [4,5] in the context of VO.

2. VAMDC – Virtual Atomic and Molecular Data Centre

In order to enable an efficacious and convenient search for available atomic and molecular data and their adequate use, we should try to solve existing problems in A&M data community, preventing productive search and data mining, which leaded to the VAMDC Idea. Main of such problems are: a) Lack of standards and common guidelines; b) Interoperability problem. c) Data exchange problem. Namely data exchange is often informal, via e-mails, ASCII files... without a standardization; d) Overlapping of efforts; e) Need of hiring computer engineers since the majority of developers are Astronomers, Physicists, Chemists; f) Data identification problem - XML schemata keys not only for data exchange but also for data identification; g) Need for a critical evaluation of data.

This means that at present, every time the same A&M database is used for a new application, the output has to be adapted. For example if we want to develop automatic tools for the visualisation of simulations of planetary, stellar or the interstellar medium spectra, this will require automatic access to different A&M databases, cross-matching the retrieved data as well as checking the quality of data. For such a purpose there is no actually a coherent and sufficiently general infrastructure to perform such tasks.

Additionally, if we want to enable an efficacious search for A&M data, we have a need for the:

-Creation of search engines that must look "everywhere" in order to map A&M Universe;

-An accessible and interoperable e-infrastructure for A&M data.

Very useful will be also the creation of a forum of data producers, data users and databases developers, as well as the training of potential users in European Research Area and wider.

All above mentioned problems will address Virtual Atomic and Molecular Data Centre (VAMDC – [6]), a FP7 founded project which started on July 1 2009 with budget of 2.9 MEuros over 42 months. Its aim is to build accessible and interoperable e-infrastructure for atomic and molecular data upgrading and integrating European (and wider) A&M database services and catering for the needs of variety of data users in science, research and development, and industry.

The VAMDC will build a secure, documented, flexible, easily accessible and interoperable e-infrastructure for A&M data that on the one hand can directly extract data from the existing depositories, while on the other hand be sufficiently flexible to be tuned to the needs of a wide variety of users from academic, governmental, industrial communities or by the general public.

The starting points are infrastructure and capabilities developed by EURO-VO (Virtual Observatory) EGEE (**E**nabling **G**rids for **E**-scienc**E**) and Astrogrid and a series of network and service activities will be organized in order to establish self-sustainable computational and data mining services. It has the specific aim of creating an infrastructure

The VAMDC can be understood as a publisher infrastructure. The VAMDC will deploy yellow pages (registries) in order to find resources, design user applications in order to meet the user needs, build data access layers above databases to provide unified outputs from these databases, care about asynchronous queries with workflows and the storage of large quantity of data with VO space, and connect its infrastructure to the grid.

VAMDC

Virtual Atomic and Molecular Data Centre

Fig. 1 VAMDC logo.

In fulfilling these aims, the VAMDC project will organise a series of Networking Activities specifically aimed at engaging data providers, coordinating activities among existing database providers, ascertaining and responding to the needs of different user communities and providing training and awareness of the VAMDC across the international A&M community and communities of other users.

Project leader is Marie-Lise Dubernet from Observatoire de Paris and core consortium is made of 15 institutions with 24 scientific groups from France, Serbia, Russia, England, Austria, Italia, Germany, Sweden and Venezuela.

Partners in the Consortium of the Project are: 1) The coordinator, Centre National de Recherche Scientifique - CNRS (Université Pierre et Marie Curie, Paris; Observatoire de Paris; Université de Reims; Université Joseph Fourier de Grenoble, Université de Bordeaux 1: Université de Bourgogne, Dijon; Université Toulouse 3); 2) The Chancellor, Masters and Scholars of the University of Cambridge – CMSUC; 3) University College London – UCL; 4) Open University – OU (Milton Keynes, England); 5) Universitaet Wien - UNIVIE; 6) Uppsala Universitet – UU; 7) Universitaet zu Koeln – KOLN; 8) Istituto Nazionale di Astrofisica – INAF (Catania, Cagliari); 9) Queen's University Belfast – QUB; 10) Astronomska Opservatorija - AOB (Belgrade, Serbia); 11) Institute of Spectroscopy RAS – ISRAN (Troitsk, Russia); 12) Russian Federal Nuclear Center - All-Russian Institute of Technical Physics - RFNC-VNIITF (Snezhinsk, Chelyabinsk Region, Russia); 13) Institute of Atmospheric Optics - IAO (Tomsk, Russia); 14) Corporacion Parque tecnologico de Merida – IVIC (Merida, Venezuela); 15) Institute for Astronomy RAS - INASAN (Moscow, Russia).

External VAMDC partner is also NIST – National Institute for Standards and Technology in Washington.

The main users of VAMDC facilities will be Astronomy, Plasma science, Atmospheric Science Radiation science and Fusion community as well as Industries using technological plasmas and Lightning industry

3. Databases

The core of the VAMDC e-infrastructure is the databases upon which it is based. The actual (the number will change) databases are shortly described below.

VALD database [7] of atomic data for analysis of radiation from astrophysical objects, developed and maintained by researchers at 7 European institutes, is created in 1995 in Vienna. The main nodes are in Vienna, Uppsala and Moscow, where are the three mirror sites . VALD contains a vast collection of spectral line parameters (central wavelengths, energy levels, statistical weights, transition probabilities, line broadening parameters) for all chemical elements of astronomical importance (http://vald.astro.univie.ac.at/).

CHIANTI [8] is an atomic database for spectroscopic diagnostics of astrophysical plasmas. Created in 1997, it contains critically evaluated set of up-to-date atomic data for the analysis of optically thin collisionally ionised astrophysical plasmas and is the preferred reference database in solar physics. It lists experimental and calculated wavelengths, radiative data and rates for electron and proton collisions (http://sohowww.nascom.nasa.gov/solarsoft, http:// www.damtp.cam.ac.uk/ user/ astro/chianti/)

EMol Database [9], at the Open University in Milton Keynes, contains a comprehensive listing of critically evaluated and regularly updated measured and calculated cross sections for electron interactions with molecular systems. It is of interest for the plasma industry, and for the disciplines of discharge physics, fusion, aeronomy and radiation chemistry. It also offers a suite of semi-empirical theoretical methods so that cross sections may be evaluated for targets for which there are currently no experimental data.

CDMS - Cologne Database for Molecular Spectroscopy (http://www.ph1.uni-koeln. de/ vorhersagen/) provides recommendations for spectroscopic transition frequencies and intensities for atoms and molecules of astronomical interest and for studying the Earth atmosphere in the frequency range 0-10 THz, i.e. 0-340 cm^{-1}. The CDMS is cross correlated with its US counterpart, the JPL Jet Propulsion Laboratory Submillimeter Catalogue (http://spec.jpl.nasa.gov/) [10].

BASECOL database [11] (http://basecol.obspm.fr) contains excitation rate coefficients for ro-vibrational excitation of molecules by electrons, He and H_2 and it is mainly used for the study of interstellar, circumstellar and cometary atmospheres.

GhoSST (Grenoble astrophysics and planetology Solid Spectroscopy and Thermodynamics, http://ghosst.obs.ujf-grenoble.fr) database service, offers spectroscopic laboratory data on molecular and atomic solids and liquids from the near UV to the far-infrared.

UMIST - University of Manchester Institute of Science and Technology (UMIST) database for astrochemistry [12] (http://www.udfa.net/), created [13] in 1991, provides reaction rate data and related software for chemical kinetic modelling of astronomical regions.

KIDA - KInetic Database for Astrochemistry will contain data on chemical reactions used in the modelling of the chemistry in the interstellar medium and in planetary atmospheres. A preliminary version was released in June 2009 (http://kida.obs.u-bordeaux1.fr).

PAHs (Polycyclic Aromatic Hydrocarbon) and carbon clusters spectral database (http://astrochemisty.ca.astro.it/database/) in Cagliari, is developed by the CESR (Centre d'Etude Spatiale des Rayonnements/CNRS), and provides a number of properties for a sample of presently about 60 species in four charge states: anion, neutral, cation and dication [14]. The properties include general energetic such as electron affinity and ionisation energies, static polarizability, permanent dipole moment, van der Waals coefficients, symmetry, multiplicity, and optimised geometry of the ground electronic state; harmonic vibrational analyses, i. e. normal modes, their frequencies and IR activities; and vertical electronic photoabsorption cross-sections and complex frequency-dependent electronic polarisabilities in the linear regime.

LASP (Laboratorio di Astrofisica Sperimentale) Database (http://web.ct.astro.it/weblab/ dbindex.html#dbindex) at the INAF (Istituto Nazionale di Astrofisica) - Catania Astrophysical Observatory, contains (i) infrared (IR) spectra of molecules in the solid phase for both pure species and their mixtures before and after processing with energetic ions and UV photons [15, 16, 17] (ii) IR optical constants of molecules in the solid phase and after processing with energetic ions [18, 19]; (iii) band strengths of the IR absorption bands [20, 21]; and (iv) density values of frozen samples [21,22].

Spectr-W[3] [23] atomic database (http://spectr-w3.snz.ru), created in 2002 in collaboration between the Russian Federal Nuclear Centre All-Russian Institute of Technical Physics (RFNC VNIITF - Snezhinsk, Chelyabinsk Region, Russia) and the Institute for High Energy Densities of the Joint Institute for High Temperatures of the Russian Academy of Sciences (IHED JIHT RAS - Moscow). It lists experimental, calculated, and compiled data on ionization potentials, energy levels, wavelengths, radiation transition probabilities and oscillator strengths, and also parameters for analytic approximations for electron-collision cross-sections and rates for atoms and ions.

The V.E. Zuev Institute of Atmospheric Optics (IAO) in Tomsk (http://www.iao.ru/) hosts the following databases:

CDSD - The Carbon Dioxide Spectroscopic Databank [24] (http://cdsd.iao.ru and ftp://ftp.iao.ru/pub/CDSD-2008), containing calculated spectral line parameters for seven isotopologues of carbon dioxide

S&MPO - Spectroscopy & Molecular Properties of Ozone) relational database [25] (http://ozone. iao.ru and http://ozone.univ-reims.fr/), developed in collaboration with the University of Reims, contains spectral line parameters for the ozone molecule, experimental UV cross-sections, information on ozone's molecular properties, updated reference lists classified by type as well as programs and extended facilities for user applications.

"Spectroscopy of Atmospheric Gases" (http://spectra.iao.ru), containing the well-known databases such as HITRAN [25], GEISA [26] and HITEMP [27]. Both "Spectroscopy of Atmospheric Gases" and S&MPO have the programs for simulation of synthetic spectra from microwave to visible wavelengths.

W@DIS - Water Internet @ccessible Distributed Information System (http://wadis.saga.iao.ru) lists experimental water-vapour spectroscopy data from the literature and calculated line lists. W@DIS contains energy levels, transition positions and line intensities, and line profile characteristics.

Databases under the management of Corporacion Parque tecnologico de Merida – IVIC (Instituto Venezoelano de Investigaciones Scientificas) and CeCalCULA (Centro Nacional de Cálculo Científico de la Universidad de Los Andes).

TIPTOPbase [28] located at the Centre de Données astronomiques de Strasbourg, France (http://cdsweb.u-strasbg.fr/topbase/home.html), contains:

TOPbase (http://cdsweb.u-strasbg.fr/topbase/topbase.html), listing atomic data computed in the Opacity Projec, namely LS-coupling energy levels, gf-values and photoionization cross sections for light elements ($Z \leq 26$) of astrophysical interest.

TIPbase (http://cdsweb.u-strasbg.fr/tipbase/home.html). Intermediate-coupling energy levels, A-values and electron impact excitation cross sections and rates for astrophysical applications ($Z \leq 28$), computed by the IRON Project.

OPserver [29], located at the Ohio Supercomputer Center, USA, (http://opacities.osc.edu/), a remote, interactive server for the computation of mean opacities for stellar modelling using the monochromatic opacities computed by the Opacity Project.

Within VAMDC e-infrastructure are also:

XSTAR database [30], used by the XSTAR code (http://heasarc.gsfc.nasa.gov/ docs/software/ xstar/xstar.html) for modelling photoionised plasmas.

HITRAN - HIgh-resolution TRANsmission molecular absorption database [25] (http://www.cfa. harvard.edu/hitran/), used extensively by the atmospheric and planetology research communities. It lists individual line parameters for molecules in the gas phase (microwave through to the UV), photoabsorption cross-sections for many molecules, and refractive indices of several atmospheric aerosols.

GEISA - Gestion et Etude des Informations Spectroscopiques Atmosphériques (Management and Study of Atmospheric Spectroscopic Information) database [26] (http://ara.lmd.polytechnique. fr/index.php?page=geisa-2 or http://ether.ipsl.jussieu.fr/etherTypo/?id=950) is a computer accessible database system, designed to facilitate accurate and fast forward, calculations of atmospheric radiative transfer.

HITEMP, a high temperature extension to HITRAN [27] (To access the HITEMP data: ftp to cfa-ftp.harvard.edu; user = anonymous; password = e-mail address). So far HITEMP contains data for water, CO_2, CO, NO and OH.

The external partner of VAMDC is NIST - National Institute of Standards and Technology, a major centre for atomic and molecular data compilation, which has developed a number of numerical and bibliographic atomic and molecular databases. Among others it hosts:

Atomic Spectra Database (http://physics.nist.gov/asd3) listing critically evaluated data on about 77,000 energy levels and 144,000 spectral lines from atoms and ions of 99 elements.

The Handbook of Atomic Spectroscopic Data [31] (http://physics.nist.gov/Handbook) listing energy levels and prominent spectral lines for neutral and singly ionised atoms.

The Spectral Data for the Chandra X-Ray Observatory database [32] (http://physics.nist.gov/chandra) presenting critically compiled wavelengths (20 A to 170 A), energy levels, line classifications, and transition probabilities for several astrophysically important elements.

SAHA, The NLTE plasma population kinetic modeling database containing theoretical data [33, 34]. (http://nlte.nist.gov/SAHA).

Three bibliographic databases providing references on atomic energy levels and spectra (http://physics.nist.gov/cgi-bin/ASBib1/ELevBib.cgi), transition probabilities (http://physics.nist.gov/ cgi-bin/ASBib1/TransProbBib.cgi) and spectral line shapes and line broadening (http://physics.nist.gov/cgi-bin/ASBib1/LineBroadBib.cgi.).

4.STARK-B database

Our contribution to the VAMDC e-infrastructure is the STARK-B database (http://stark-b.obspm.fr) [35], a collaborative project between Laboratoire d'Etude du Rayonnement et de la matière en Astrophysique of the Observatoire de Paris-Meudon and the Astronomical Observatory of Belgrade. This is a database of the theoretical widths and shifts of isolated lines of atoms and ions due to collisions with charged perurbers, obtained within the impact approximation. For the moment STARK-B contains results obtained using the semiclassical perturbation approach [36,37]. The corresponding computer code has been optimized and updated in Refs. [38, 39, 40] and following papers. For the review and all updates see e.g. Ref. [41].

This database is devoted to modelling and spectroscopic diagnostics of stellar atmospheres and envelopes. In addition, it is also relevant to laboratory plasmas, laser equipment and technological plasmas. The database is currently developed in Paris, and a mirror is planned in Belgrade. It is already on line though not yet complete. It is described in detail in Ref. [35].

The precursor of STARK-B as well as of SerVO was BELDATA and its main content was database on Stark broadening parameters. A history of BELDATA can be traced in Refs [42, 43, 44, 45, 46, 47].

The participants of AOB (Astronomical Observatory – Belgrade) VAMDC Node are: Milan S. Dimitrijević, Luka Č. Popović, Andjelka Kovačević, Darko Jevremović, Zoran Simić, Edi Bon and Nenad Milovanović.

We also have a close collaboration with Sylvie Sahal-Bréchot from Paris Observatory, Nebil Ben Nessib, Walid Mahmoudi, Rafik Hamdi, Haykel Elabidi, Besma Zmerli and Neila Larbi-Terzi from Tunisia, Magdalena Christova from Technical University of Sofia and Tanya Ryabchikova from Institute of Theoretical Astronomy in Moscow.

Initiatives like VAMDC are important for both producers of atomic and molecular data and stellar atmosphere modellers as one of prime users. VAMDC is an example of the global collaborations and innovations in e-science. It is expected to become one of major European cyber-infrastructures with a world wide impact.

Acknowledgments

A part of this work has been supported by VAMDC, funded under the "Combination of Collaborative Projects and Coordination and Support Actions" Funding Scheme of The Seventh Framework Program. Call topic: INFRA-2008-1.2.2 Scientific Data Infrastructure. Grant Agreement number: 239108. The authors are also grateful for the support provided by Ministry of Science and Development of Republic of Serbia through projects TR13022 "Serbian Virtual Observatory", 146001 "Influence of collisional processes on astrophysical plasma spectra" and 146002 "Astrophysical Spectroscopy of Extragalactic Objects".

References

[1] Bildsten S, Chang P and Paerels P 2003 *Astrophys. J.* **591** L2936.
[2] Jevremović D, Dimitrijević MS, Popović LČ, et al. 2009, *New Astron. Rev.* **53** 222.
[3] Tsvetkova K, Tsvetkov M, Protić-Benišek V and Dimitrijević MS 2009 *Publ. Astron. Soc. "Rudjer Bošković"* **9** 255.
[4] Dotter A, Chaboyer B, Ferguson JW, Lee H-C, Worthey G, Jevremović D and Baron E 2007 *Astrophys. J.* **666** 403.
[5] Dotter A, Chaboyer B, Jevremović D, Kostov V, Baron E and Ferguson JW 2008 *Astrophys. J. Suppl.* **178** 89.
[6] Dubernet ML, Boudon V, Culhane JL, et al. 2010 *JQSRT* **111** 2151
[7] Kupka F, Piskunov N, Ryabchikova TA, Stempels HC and Weiss WW 1999 VALD-2: Progress of the Vienna Atomic Line Data Base *Astron. Astrophys. Sup. Ser.* **138** 119.
[8] Dere KP, Landi E, Young PR, Del Zanna G, Landini M and Mason HE 2009 CHIANTI - an atomic database for emission lines IX. Ionization rates, recombination rates, ionization equilibria for the elements hydrogen through zinc and updated atomic data *Astron Astrophys.* **498** 915.
[9] Mason NJ 2007 Electron Induced Processing; Applications and Data Needs, Proceedings of ICAMDATA06, AIP Conference proceedings **901** 74.
[10] Müller HSP, Schlöder F, Stutzki J and Winnewisser G 2005, The Cologne Database for Molecular Spectroscopy, CDMS: a useful tool for astronomers and spectroscopists *J. Mol. Struct.* **742** 215.
[11] Dubernet ML, Grosjean A, Daniel F, Flower D, Roueff E, Daniel F, Moreau N and Debray B 2006 Ro-vibrational Collisional Excitation Database BASECOL, Proceedings Joint Meeting ITC14 and ICAMDATA 2004, Toki, Japan *Journal of Plasma and Fusion Research* **7** 356.
[12] Woodall J, Agundez M, Markwick-Kemper AJ and Millar TJ 2007 The UMIST database for astrochemistry *Astron. Astrophys.* **466** 1197.
[13] Millar TJ, Rawlings JMC, Bennett A, Brown PD and Charnley SB 1991 Gas-phase reactions and rate coefficients for use in astrochemistry – the UMIST ratefile *Astron. Astrophys. Suppl. Ser.* **87** 585.
[14] Malloci G, Joblin C and Mulas G 2007 On-line database of the spectral properties of polycyclic aromatic hydrocarbons *Chem. Phys*. **332** 353.
[15] Strazzulla G, Leto G and Palumbo ME 1993 *Adv. Space Res.* **13** 189.
[16] Palumbo ME, Baratta GA, Fulvio D, Garozzo M, Gomis O, Leto G, Spinella F and Strazzulla G 2008 *J. Phys. Conf. Series* **101** 012002.
[17] Leto G and Baratta GA 2003 *Astron. Astrophys.* **397** 7.
[18] Palumbo ME, Baratta GA, Collings MP and McCoustra MRS 2006 *Physical Chemistry Chemical Physics* **8** 279.
[19] Brunetto R and Roush TL 2008 *Astron. Astrophys*. **481** 879.
[20] Mulas G, Baratta GA, Palumbo ME and Strazzulla G 1998 *Astron. Astrophys*. **333** 1025.
[21] Brunetto R, Caniglia G, Baratta GA and Palumbo ME 2008 *Astrophys. J.* **686** 1480.
[22] Fulvio D, Sivaraman B, Baratta GA, Palumbo ME and Mason NJ 2009 *Spectrochimica Acta Part A* **72** 1007.
[23] Faenov AY, Magunov AI, Pikuz TA, Skobelev IY, Loboda PA. Bakshayev NN, Gagarin SV,

Komosko VV, Kuznetsov KS, Markelenkov SA. Petunin SA and Popova VV 2002 Spectr-W-3 online database on atomic properties of atoms and ions, in "Atomic and Molecular Data and Their Applications", Schultz DR, Krstic PS, Ownby F (Eds.) *AIP Conf. Proc.* **636** 253.

[24] Perevalov VI and Tashkun SA. 2008 CDSD-296 (Carbon Dioxide Spectroscopic Databank): Updated and Enlarged Version for Atmospheric Applications, 10th HITRAN Database Conference, Cambridge, MA, USA.

[25] Rothman LS, Gordon IE, Barbe A, Benner DC, Bernath PF, Birk M, et al. 2009 The *HITRAN* 2008 Molecular Spectroscopic Database *JQSRT* **110** 533.

[26] Jacquinet-Husson N, Scott NA, Chedin A, et al. 2008 The GEISA spectroscopic database: Current and future archive for Earth and planetary atmosphere studies *JQSRT* **109** 1043.

[27] Rothman LS, Gordon IE, Barber RJ, Dothe H, Gamache RR, Goldman A, Perevalov VI, Tashkun SA and Tennyson J 2010 HITEMP, the High-Temperature Molecular Spectroscopic Database *JQSRT* 111 2139.

[28] Cunto W, Mendoza C, Ochsenbein F and Zeippen C 1993 TOPbase at the CDS *Astron. Astrophys.* **275** L5.

[29] Mendoza C, Seaton MJ, Buerger P, Bellorin A, Melendez M, Gonzalez J, Rodriguez LS, Delahaye F, Palacios E, Pradhan AK and Zeippen CJ 2007 OPserver: interactive online computations of opacities and radiative accelerations *Mon. Not. R. Astr. Soc.* **378** 1031.

[30] Bautista MA and Kallman TR 2001 The XSTAR atomic database *Astrophys. J. Suppl.* **134** 139.

[31] Sansonetti JE, Martin WC and Young SL 2005 *Handbook of Basic Atomic Spectroscopic Data* National Institute of Standards and Technology, Gaithersburg, MD.

[32] Podobedova LI, Musgrove A, Kelleher DE, Reader J, Wiese WL, Coursey JS and Olsen K 2003 *Spectral Data for the Chandra X-Ray Observatory*, National Institute of Standards and Technology, Gaithersburg, MD.

[33] Ralchenko Yu 2006 *NIST SAHA Plasma Kinetics Database,* National Institute of Standards and Technology, Gaithersburg, MD.

[34] Rubiano JG, Florido R, Bowen C, Lee RW and Ralchenko Yu 2007, Review of the 4th NLTE Code Comparison Workshop *High Energy Density Physics* 3 225.

[35] Sahal-Bréchot S 2010 Case studies on recent Stark broadening calculations and STARK-B database development in the framework of the European project VAMDC (Virtual Atomic and Molecular Data Centre) *J. Phys. Conf. Series* This issue.

[36] Sahal-Bréchot S 1969 *Astron. Astrophys.* **1** 91.

[37] Sahal-Bréchot S 1969 *Astron. Astrophys.* **2** 322.

[38] Sahal-Bréchot S 1974 *Astron. Astrophys.* **35** 321.

[39] Fleurier C, Sahal-Bréchot S and Chapelle J 1977 *JQSRT* **17** 595.

[40] Dimitrijević MS and Sahal-Bréchot S 1984 *JQSRT* **31** 301.

[41] Dimitrijević MS 1996 *J. Aplied Spectrosc.* **63** 684.

[42] Popović LČ, Dimitrijević MS, Milovanović N and Trajković N 1999 *Publ. Astron. Obs. Belgrade* **65** 225.

[43] Popović LČ, Dimitrijević MS, Milovanović N and Trajković N 1999 *J. Res. Phys.* **28** 307.

[44] Milovanović N, Popović LČ and Dimitrijević MS 2000 *Publ. Astron. Obs. Belgrade* **68** 117.

[45] Milovanović N, Popović LČ and Dimitrijević MS 2000 *Baltic Astron.* **9** 595.

[46] Dimitrijević MS, Popović LČ, Bon E, Bajčeta V, Jovanović P and Milovanović N 2003 *Publ. Astron. Obs. Belgrade* **75** 129.

[47] Dimitrijević MS and Popović LČ 2006, in Virtual Observatory; Plate Content Digitization, Archive Mining, Image Sequence Processing, eds. M. Tsvetkov, V. Golev, F. Murtagh, R. Molina, Heron Press Science Series, Sofia, 115.

Plasma flow interaction with ITER divertor related surfaces

Ivan P. Dojčinović
University of Belgrade, Faculty of Physics, POB 44, 11000 Belgrade, Serbia

ivbi@ff.bg.ac.rs

Abstract. It has been found that the plasma flow generated by quasistationary plasma accelerators can be used for simulation of high energy plasma interaction with different materials of interest for fusion experiments. It is especially important for the studies of the processes such as ELMs (edge localized modes), plasma disruptions and VDEs (vertical displacement events), during which a significant part of the confined hot plasma is lost from the core to the SOL (scrape off layer) enveloping the core region. Experiments using plasma guns have been used to assess erosion from disruptions and ELMs. Namely, in this experiment modification of different targets, like tungsten, molybdenum, CFC and silicon single crystal surface by the action of hydrogen and nitrogen quasistationary compression plasma flow (CPF) generated by magnetoplasma compressor (MPC) has been studied. MPC plasma flow with standard parameters (1 MJ/m^2 in 0.1 ms) can be used for simulation of transient peak thermal loads during Type I ELMs and disruptions. Analysis of the targets erosion, brittle destruction, melting processes, and dust formation has been performed. These surface phenomena are results of specific conditions during CPF interaction with target surface. The investigations are related to the fundamental aspects of high energy plasma flow interaction with different material of interest for fusion. One of the purposes is a study of competition between melting and cleavage of treated solid surface. The other is investigation of plasma interaction with first wall and divertor component materials related to the ITER experiment.

1. Introduction

Intense thermal loads in fusion devices which occur during ELMs (edge localized modes), plasma disruptions and VDEs (vertical displacement events), will result in macroscopic erosion associated with the formation of cracks, melting, droplets, evaporation or sublimation. The main goals of plasma surface interaction investigation are simulation the expected ELM and disruption loads in ITER and other fusion devices [1].

Edge localized modes (ELMs) represents sudden bursts of particles and energy out of the plasma due to turbulent transport processes localized in the tokamak edge region. Peak heat loads on the divertor plates and first wall during ELMs. The ELMs occurring during ELMy H-mode called Type I ELMs lead to the largest particle and power fluxes onto the divertor plates and expel from the core plasma about 30% of the input power [2].

Power losses are such that the target surface temperature exceeds the sublimation point (for carbon) or melting point (for tungsten). ITER ELMs loads are estimated as (1-2) MJ/m^2 during (0.1-1) ms and the repetition frequency of an order of 1 Hz (~400 ELMs during each ITER pulse) [1,2]. The divertor peak power flux during ELMs is in the range of 1-10 GW/m^2. For the disruptions, the heat loads to ITER divertor components are anticipated to be of the order (10-100) MJ/m^2 with load duration (1-10) ms.

Simulation experiments are being made at pulsed plasma gun facilities such as MK-200, VIKA, QSPA TRINITI [3], QSPA Kh50, PLADIS, as well as at electron beam facilities like JUDITH and JEBIS. The basic idea is an investigation of erosion effects under realistic plasma parameters and conditions. In plasma gun experiments energy loads characteristics of Type I ELMs and disruptions in ITER are obtained and studied. The QSPA facility provides realistic heat loads (i.e., adequate pulse duration and energy density) to simulate the expected Type I ELM loads in ITER [2].

Plasma facing materials for ITER is limited to three basic elements: carbon, beryllium and tungsten. During transient events brittle destruction (carbon based materials) as well as melt layer splashing (metals) can occurs [4]. Maximum allowable ELM energy density is 0.22-0.64 MJ/m^2 for carbon and 0.54-1.6 MJ/m^2 for tungsten.

The aim of this paper is to present some results from the study of compression plasma flow (CPF) interaction with a targets, in order to demonstrate the capability of CPF from the magnetoplasma compressor to study the materials surface characteristics of interest for fusion devices. During surface treatment by high-power energy beams different phenomena can occur such as evaporation, sputtering, ablation [5], eching [6], exfoliation etc. The material may be removed from surface in the form of charged particles, vapor, liquid droplets, solid flakes etc.

These investigations are related to fundamental aspects of high energy plasma flow interaction with target surface. One of the purposes is observation of competition between melting and cleavage of treated surface. The other is investigation of plasma interaction with first wall component materials related to the ITER experiment.

2. Experimental setup and procedure

A magnetoplasma compressor (MPC) is a source of supersonic CPF and represents a magnetoplasma analog of a Laval nozzle. The electrode system of MPC consists of the conically shaped copper central electrode (cathode) and cylindrical outer electrode made of 8 copper rods, symmetrically positioned along a circle of 5 cm in diameter. In the MPC inter-electrode region the plasma is accelerated due to the Ampere force as the result of the radial component of plasma current I_r interaction with azimuth magnetic field B_φ (Fig.1). The MPC can operate with different working gasses and their mixtures [7,8]. The MPC is a quasistationary system, since the time duration of the CPF is much longer (~50 μs) than the characteristic flight time of plasma along the accelerating channel (~1 μs). With hydrogen as working gas and voltage 3-4 kV, typical plasma current is 80-100 kA, discharge duration 100-150 μs, average plasma density 5×10^{16} cm^{-3}, plasma flow velocity up to 120 km/s.

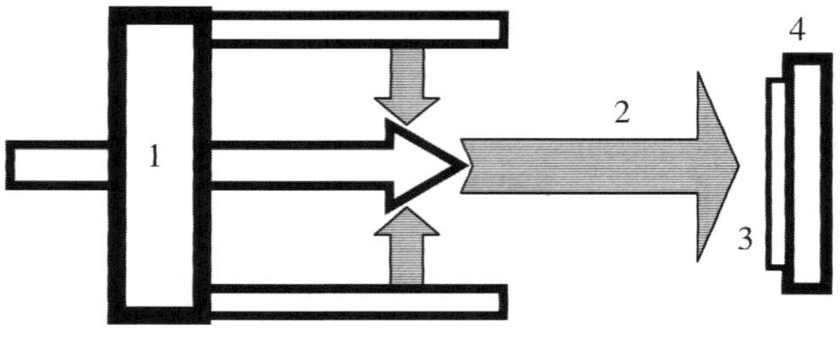

Figure 1. Scheme of experimental setup: (1) Magnetoplasma compressor (MPC), source of quasistationary compression plasma flow, (2) compression plasma flow, (3) sample, (4) sample brass holder.

MPC plasma flow parameters, i.e. plasma energy density ~1 MJ/m^2 in 0.1 ms, simulated transient peak thermal loads during Type I ELMs.

For the studies of CPF interaction with surfaces, silicon wafers (100 and 111 orientation) 1 mm thick and 10 mm in diameter, tungsten, molybdenum and CFC were used. The sample is mounted on the cylindrical brass holder of the same diameter, and placed in front of the MPC cathode at the distance of 5 cm. The samples are exposed to a single plasma pulse. To investigate the morphology of treated surface, optical microscopy (OM), scanning electron microscopy (SEM JEOL 840A) and atomic force microscopy (AFM, AutoProbe CP-Research SPM made by TM Microscope-Veeco) were used.

3. Experimental results and discussion

OM micrographs of central part of the treated Si (111) and Si (100) surfaces are given in Fig. 2a and Fig. 2b, respectively. Rhombic and triangular regular fracture features are obtained in the case of Si (111) (Fig. 2a), as expected for threefold symmetry. On Si (100) surface treated by CPF, two sets of fracture lines intersecting at 90° form a grid that divides the surface into rectangular blocks (Fig. 2b). Some of the blocks are ejected from the surface, and large holes at the surface can be seen to emerge. In this case, development of subsurface fracture, parallel to the surface, occurs. The thickness of this block is about 10 μm.

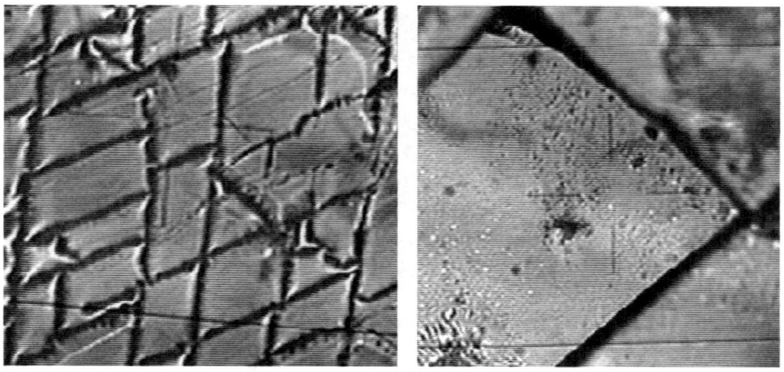

Figure 2. OM micrograph of Si (111) (left) and Si (100) (right) surface after CPF treatment, central part of the sample. Image size: 50×50 μm^2.

Surface fracture and highly-oriented periodic cylindrical structures [9-11] obtained by single plasma pulse treatment of the Si (100) sample surface are shown in Fig. 3 and the carbon fibre components (CFCs) after CPF action are shown in Fig. 4. These silicon structures are obtained in the peripheral part of the target surface. The obtained periodic cylindrical structures are frozen capillary surface waves quenched at a particular moment during a process of the melt resolidification [11]. Typical wavelength (hill-to-hill distance) of these ripple structures formed on the treated silicon surfaces are about 3-4 μm with amplitude (half-distance of hill to valley) of 0.2 μm and lengths up to 200 μm. AFM was used for periodic structures surface investigation. Structures are smooth and sinusoidally shaped with an amplitude (half hill-to-valley distance) of about 0.3 μm. The regular silicon surface structures can be obtained on the area region up to several square millimetres. The estimated average surface temperature was about 2300 K [12].

Formation of the observed surface features may be explained by energetic action of CPF on the surface. Time evolution of the compression plasma flow interaction with treated silicon sample is studied using IMACON 790 high speed camera [9]. Analyzing these results, one can conclude that the interaction of CPF with silicon sample surface causes an evaporation of a thin surface layer and formation of a shock-compressed plasma layer (plasma plume) [9]. The plasma produced within the layer consists predominantly of evaporated silicon from the sample surface. Formation of this cloud of dense target plasma results in the shielding of the processed surface from a direct action of the CPF and the surface protection from further excessive evaporation. A typical thickness of the shock-compressed plasma layer is about 1 cm. Using the high speed camera, the time of interaction was estimated to be ~40 μs [9].

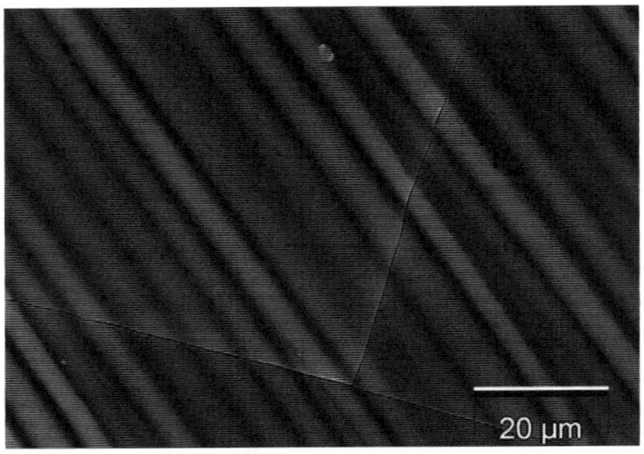

Figure 3. SEM micrograph of Si (100) surface after CPF treatment.

Energetic action of CPF (absorbed energy 10-15 J per pulse, flow power density $\sim 1 \cdot 10^5$ W/cm^2 [13]) on the surface causes the fast heating and melting of the surface layer and the presence of high dynamic pressure of CPF of the order of several atmospheres [13]. Namely, CPF kinetic energy thermalization causes the heating of shock-compressed plasma layer. The target surface is heated by convective and radiative heating and a high temperature gradient occurs. In addition, deceleration of the CPF results in the formation of current loops (vortices), due to freezing of the magnetic field into the plasma, and a magnetic field of 1-10 mT is induced at the surface [13].

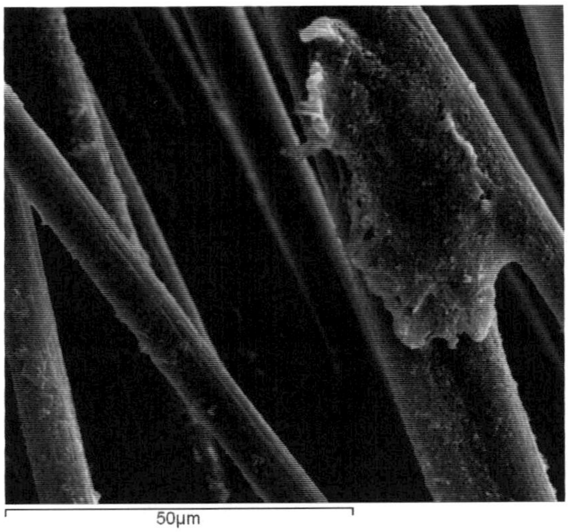

Figure 4. The carbon fibre components (CFCs) after CPF action.

Melt layer erosion at high heat loads is melt splashing due to the formation, growth, and bursting of bubbles inside the liquid layer (Figs. 5-7). This results from the continuous heating and overheating of the liquid layer during energy deposition.

Journal of Physics: Conference Series **257** (2010) 012033 doi:10.1088/1742-6596/257/1/012033

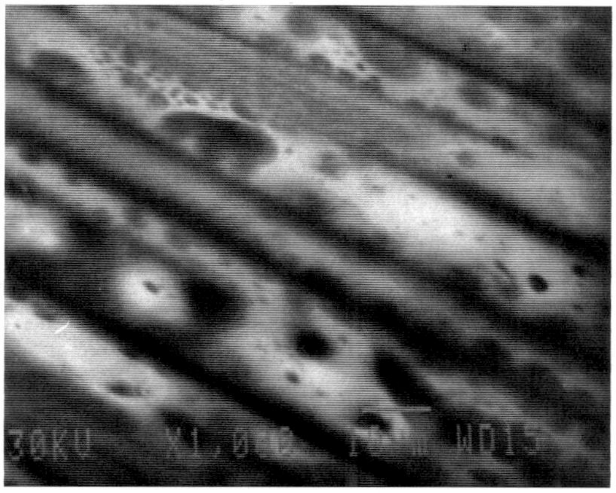

Figure 5. The large holes formation on silicon surface after CPF action.

Melting of target surfaces during high-power pulsed energy streams treatment is the normally expected effect, but only in some cases are cracks and fractures formed (Figs 2-4). In this experiment, during plasma flow treatment of silicon single crystal surfaces, melting occurs predominantly in the peripheral part of the target. In this region capillary waves are induced on the liquid silicon surface in contrast to the central part of the treated silicon sample where regular fracture features are observed. It is worth emphasizing that regular fracture features formation is a process in competition with surface melting. However, in the surface transition region, both frozen capillary waves and regular fractures are observed.

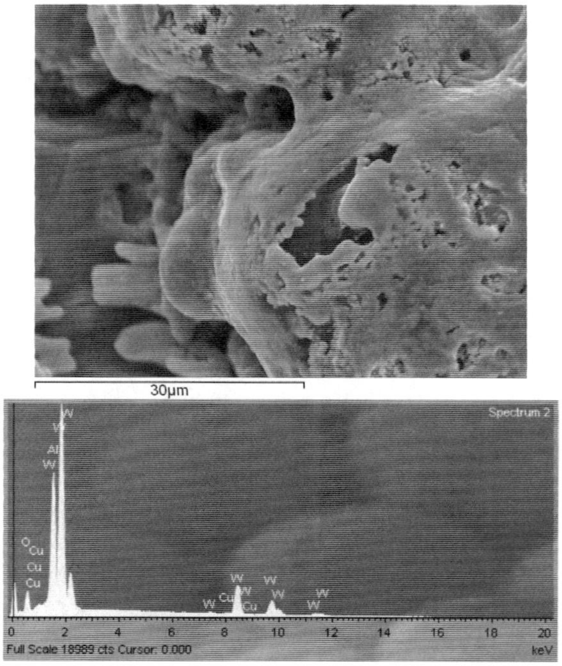

Figure 6. The tungsten surface after CPF action (up) and EDAX spectrum (down).

Formation of regular fracture features (Figs. 2 and 3) can be explained by assuming that a considerable fraction of the absorbed plasma flow energy is trapped in fractures rather than converted to heat energy. Single crystal silicon is well known as a typical anisotropic material exhibiting brittle behaviour at room temperature which can be observed as the local response of an atomically sharp crack for a critical loading condition. Covalent bonding in silicon gives rise to a high barrier to dislocation motion and silicon is, therefore, inherently brittle. Silicon single crystal becomes ductile above about 600ºC (the brittle-ductile transition), depending on strain rate [14].

The anisotropy of crack propagation direction from an energetic point of view is explained with the Griffith criterion [15]: in equilibrium, the mechanical energy released upon crack advance must be in balance with the energy required to create the two new surfaces. This is a necessary condition for fracture and leads to conclusion that crystal lattice planes with low surface energies are energetically favored as cleavage planes. The lowest energy cleavage planes in silicon are the {111} planes [16]. Silicon has two principal cleavage planes: {111} planes, usually the easy cleavage planes, and {110} planes.

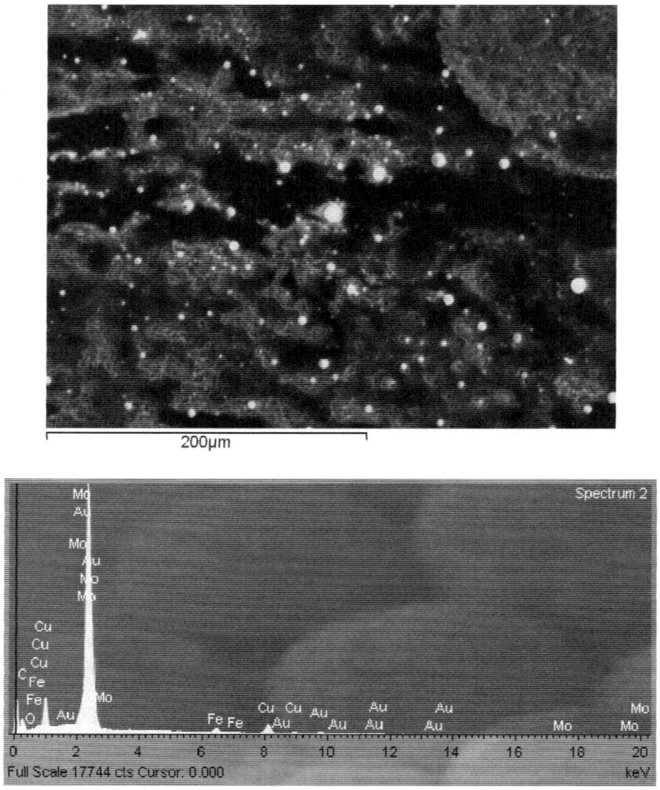

Figure 7. The molybdenum surface after CPF action (up) and EDAX spectrum (down).

Low adhesion between blocks and silicon bulk, and eventual ejection of blocks from the CPF treated surface (Fig. 2), can be explained by the development of subsurface fracture, parallel to the surface. Cracking between the block and the bulk is growing due to local energy absorption. This exfoliation phenomenon is very important for investigation of the plasma interaction with fusion reactor first wall materials, in order to find the conditions for avoiding of high target mass losses.

Detected silicon periodic cylindrical structures obtained in this experiment by interaction of CPF with the silicon surface (Fig. 3) may be presumed as capillary waves [9-11]. Capillary waves on the molten silicon surface are induced during the interaction of the plasma flow with the target. Once the plasma flow interaction with the silicon sample has ended, fast recrystallisation and quenching causes the freezing of wavy structures on the silicon surface. Rapid solidification can produce serious cracking problems due to differential stresses. During cooling and unloading of the treated surface, the residual stresses created in the subsurface region can produce surface fracturing. From Fig. 3 one can conclude that cracking occurred after solidification of periodic cylindrical structures and may be explained with the emerging of residual stresses. Brittle destruction might increase with the number of plasma discharges because of crack propagation into the depth of the sample thus resulting in material pre-damaging. The holes arise because of crack formation resulting in macroscopic destruction.

Melting and thus melt motion, as well as melt layer splashing are expected to be the main mechanisms of metallic target damage. Melt splashing occurs from boiling and gas bubbles, splashing due to absorption of plasma momentum, erosion from hydrodynamic instabilities developed in the liquid layer due to the tangential and perpendicular forces acting on the free surface of the liquid, erosion due to run-off of melt layers over the structure, and erosion caused by mechanical vibration of the machine during the disruption event. These effects are detected at the silicon surface. The plasma stream causes melt layer boiling. Pressure of the impacting plasma stream is 10 atm in the beginning, but with decreasing pressure, violent boiling and bubble formation in the overheated liquid occurs. Bubbles caused large particle emission. The majority of droplets remain at the sample surface.

4. Conclusions

From the surface structures studied here (regular fracture features, frozen capillary waves, bubble formations), it is possible to conclude that they are similar to those originating from macroscopic erosion processes observed for the first wall and divertor surface in fusion devices (predominantly, brittle destruction, melt formation with melt layer splashing, bubbling with large particle emission) [1-4]. Material in the form of droplets and grains was observed as being ejected from surfaces treated by compression plasma flow. This is the evidence that evaporation process plays only a minor role in the resulting surface erosion. The results obtained here verify the capability of compression plasma flow in study of plasma wall interaction in fusion devices with use of different materials in wall construction.

Acknowledgements

The authors are grateful to Dr M. Mitrović and Dr A. Žekić from University of Belgrade, Faculty of Physics, for the SEM analyses and Mr D. Randjelović from Institute HTM-CMTM in Belgrade for the AFM analyses. This work is supported by the Ministry of Science and Technology Development of the Republic of Serbia within the Project 141043.

References

[1] Federici G, Andrew P, Barabaschi P, Brooks J, Doerner R, Geier A et al. 2003 *J. Nucl. Mater.* **313-316** 11

[2] Federici G, Brooks J, Coster DP, Janeschitz G, Kukushkin A, Loarte A et al. 2001 *J. Nucl. Mater.* **290-293** 260

[3] Federici G, Zhitlukhin A, Arkhipov N, Klimov N, Giniyatulin R, Landman I et al. 2005 *J. Nucl. Mater.* **337-339** 684

[4] Arkhipov NI, Bakhtin VP, Kurkin SM, Safronov VM, Toporkov DA, Vasenin SG et al. 2000 *Fusion Eng. Desing* **49-50** 151

[5] Keidar M, Boyd ID, Beilis II 2004 *J. Appl. Phys.* **96** 5420

[6] Ostrikov KN, Kumar S, Sugai H. 2001 *Phys. Plasmas* **8** 3490

[7] Purić J, Dojčinović IP, Astashynski VM, Kuraica MM, Obradović BM. 2004 *Plasma Sources Sci. Technol.* **13** 74

[8] Dojčinović IP, Kuraica MM, Obradović BM, Cvetanović N, Purić J. 2007 *Plasma Sources Sci. Technol.* **16** 72
[9] Purić J, Astashynski VM, Dojčinović IP, Kuraica MM. 2004 *Vacuum* **73** 561
[10] Dojčinović IP, Kuraica MM, Purić J. 2006 *Vacuum* **80** 1381
[11] Dojčinović IP, Kuraica MM, Obradović BM, Purić J. 2006 *Appl. Phys. Lett.* **89** 071501
[12] Astashynski VV, Ananin SI, Emelyanenko AS, Kostyukevich EA, Kuz'mitski AM, Zhvavy SP et al. 2006 *Appl. Surf. Sci.* **253** 1866
[13] Uglov VV, Anishchik VM, Astashynski VV, Astashynski VM, Ananin SI, Askerko VV et al. 2002 *Surf. Coat. Technol.* **158-159** 273
[14] Spence JCH 1999 *Acta Mater.* **47** 4153
[15] Griffith AA 1921 *Philos. Trans. R.. Soc. London A* **221** 163
[16] Hauch JA, Holland D, Marder MP, Swinney HL. 1999 *Phys. Rev. Lett.* **82** 3823

Physical properties of the broad line region in active galactic nuclei

Dragana Ilić[1], Luka Č. Popović[2], Stefano Ciroi[3], Giovanni La Mura[3] and Piero Rafanelli[3]

[1] Department of Astronomy, Faculty of Mathematics, University of Belgrade, Studentski trg 16, 11000 Belgrade, Serbia
[2] Astronomical Observatory Belgrade, Volgina 7, 11000 Belgrade, Serbia
[3] Department of Astronomy, University of Padua, Vicolo dell'Osservatorio, I-35122 Padua, Italy

E-mail: dilic@matf.bg.ac.rs

Abstract. We present here the study of the plasma in the broad line region (BLR) of active galactic nuclei (AGN). In order to probe the physical properties of the emitting plasma in the BLR we analyze the fluxes of the following broad emission lines (BELs): the hydrogen Balmer lines (Hα to Hε) and the helium lines from two subsequent ionization levels (He II λ4686 and He I λ5876). The BELs are obtained from the spectral synthesis photoionization code CLOUDY. We investigate these BELs in order to find conditions in the BLR where so-called Boltzmann-plot (BP) can be applied, and we found that in a number of modeled spectra it is working. We used these spectra to explore the dependence between plasma parameters (e.g. the averaged temperature, hydrogen density, etc.) and the ratio of He II λ4686 and He I λ5876 lines.

In this progress report we present our investigation of the emitting plasma in the BLR using the most intensive broad spectral lines in AGN spectra.

1. Introduction

Active galactic nuclei (AGN) are the most intriguing objects in the Universe, since they are the brightest and the most distant objects. The most accepted scenario of the structure of AGN is the one in which AGN are powered by the accretion of matter from the host galaxy onto super-massive black hole. One of the ways to study the inner emitting region of an AGN, closest to the black hole, is by analyzing its broad emission lines (BELs). These lines originate in the broad line region (BLR), which physics is still not fully understood (see e.g. [1, 2], for a review). It is obvious from the BEL fluxes and profiles that the physics and kinematics in the BLR is more complicated than in the narrow line region (NLR) or in gaseous nebulae, where we can use the forbidden line ratios for plasma diagnostics [3, 4].

Most of the BLR is photoionized. The definitive proof comes from the reverberation studies showing a direct response of the BEL fluxes to the continuum flux changes (see e.g. [5]). So far numerous papers have been dedicated to describe the BLR physics and explain its emitting spectrum using the photoionization models (see e.g. [5, 6, 7, 8], etc.). Even though there are roughly half a dozen large codes that are now highly improved, these numerical simulations still have some important unresolved problems related to the BLR [9]. One of the most interesting problems is the observed small ratio between Lyα and Hβ lines (usually in the range 5 - 15), while models predict much higher values (30 - 50) (see [10] and all references therein).

Moreover, the BLR is a kinematically and physically complex region and can be separated into at least two distinct regions according to the origin of the "high-ionization lines" (HILs) and "low-ionization lines" (LILs) (see e.g. [11, 1]), but also according to the observed complex line profiles of a single line that cannot be explained with a simple model (see e.g. [2, 12]). It is known that different BELs are produced over a wide range of distances from the central continuum source, and under very different physical conditions (see e.g. [2]). Therefore, collisional excitation, self-absorption, dust obscuration and complicated coupled line and continuum radiative-transfer problems should be taken into account when calculating the emission line spectrum, as well as the evident stratification of the BLR and its complex geometry.

2. Methods for plasma diagnostics

Spectroscopy, in general, offers different methods for diagnostics of emitting plasma (see e.g. [13]), but these methods could not be properly used to probe the physical conditions in the BLR of AGN. Particularly, it is difficult to find a direct method which would only use the observed BEL to determine the temperature and density in the BLR. There are some indications that the temperature should be less than 35,000 K due to the observed Fe II emission, as for the higher temperatures Fe II would be almost completely collisionally ionized to Fe III [13]. Note here that new investigations show that the Fe II emission is probably connected with an intermediate line region (see e.g. [14, 15], and references therein), therefore the problem of the temperature limit mentioned above should be taken with caution. Furthermore, there are some limits to the density obtained from the observed emission lines, i.e. due to the fact that the forbidden lines (e.g. [OIII] lines) are suppressed in the BLR, the lower limit for the electron density is roughly $n_{\rm e} \approx 10^8 {\rm cm}^{-3}$, while for the upper limit a critical density for collisional deexcitation of the broad lines of around $n_c \approx 10^{14} {\rm cm}^{-3}$ can be used.

So far, there are only two proposed methods in the literature that include observations of the BELs to determine the BLR physical properties [16, 17]. Marziani et al. [16] found, using the CLOUDY photoionization computation, that the ratio between UV emission lines Si III] λ1892 and C III] λ1909 is a good density diagnostic in the density range $9.5 < \log n_{\rm e} < 12$. And since that ratio is correlated with the width of the broad Hβ line, they gave a relationship for the estimates of the electron density in the BLR using either the ratio between these semi-forbidden lines or the width of the broad component of the Hβ line [16]. Also, Laor [17] proposed a method that considers electron scattering influence on the line shapes and determines the physical parameters of the BLR by fitting the emission line exponential wings, in the case of the low luminosity AGN. The method assumes that the exponential wings are produced by the optically thin, isotropic, thermal electron scattering. In that case, by fitting the line wings with an electron-scattering model one can estimate the electron density and optical depth of the region [17]. However, the problem of this direct method for BLR diagnostics is that it can be applied only for the cases of the BEL profiles with exponential wings (e.g. the case of galaxy NGC 4395). Besides, another problem of this method is that kinematics of the emission gas in the BLR can strongly affect the BEL profiles.

On the other hand, Popović [18, 19] suggested that the Boltzmann-plot (BP) method, that is widely used in the laboratory plasma diagnostics [20], could be used in the case of the BLR of some AGN (see also [21, 22]). The BP method basically uses the flux ratios from one line series (as e.g. hydrogen Balmer line series) and the atomic parameters of the corresponding line series, to estimate the excitation temperature if the plasma is at least in the Partial Local Thermodynamical Equilibrium - PLTE [20]. The plasma is in PLTE if the populations of sufficiently highly excited levels are related to the next ion's ground-state population by Saha-Boltzmann relations [23] or to the total population in all fine-structure levels of the ground-

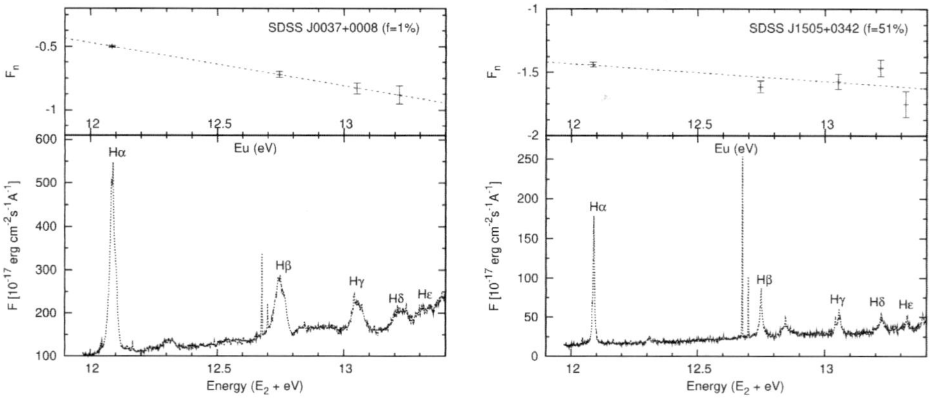

Figure 1. The examples of AGN where BP method works (left panel) and where it does not give valid results (right panel), taken from [21]. Each panel contains the spectrum and the Boltzmann-plot applied on the line fluxes of the Balmer series. For the analysis, only the broad component of the Balmer lines was used. The wavelength is converted to the energy units (eV), but shifted for the energy of the lower level in the Balmer series E_2.

state configuration [20]. Moreover, the plasma in which PLTE exists for one series can be non-stationary and/or two-temperature plasma (for more details see [23]).

The Boltzmann-plot method is a well-known method for estimating the laboratory plasma excitation temperature [20, 24]. Starting from the assumptions that the emission line region is optically thin and that for one line series (as e.g. Balmer line series) the population of the upper energy states ($n \geq 3$) can be described with the Saha-Boltzmann distribution[1], i.e. plasma is at least in the PLTE, we have that the flux F_{ul} of the transition $u \rightarrow l$ can be written as [20, 25]

$$\log_{10}(F_{\rm n}) = B - AE_u \tag{1}$$

where $F_{\rm n}$ is the "normalized" line-flux with respect to the atomic constants which characterize the transition $F_{\rm n} = \frac{F_{ul}\cdot\lambda}{g_u A_{ul}}$ (where λ is the transition wavelength, g_u is the statistical weight of the upper level, A_{ul} is the transition probability), and $A = \log_{10}(e)/kT_{\rm exc} \approx 5040/T_{\rm exc}$ is the temperature indicator. Applying Eq. (1) to one line series and obtaining the value of the parameter A, we can estimate the excitation temperature of the region where these lines are originating (for more details on the BP method see [20] and for discussions on its application to the astrophysical plasma see [25, 22, 14]).

For some group of AGN (e.g. NGC 5548 or Mrk 817, see [22, 26]), it has been shown that the BP method could be applicable to the Balmer line series and might be used, at least roughly, to estimate the BLR physical properties [18, 22]. Fig. 1 gives an example of AGN where BP method works (left panel) and where it does not give valid results (right panel), taken from [21]. Each panel contains the AGN spectrum and the BP applied on the fluxes of the broad Balmer lines.

The question that comes naturally here is why BP works for some type 1 AGN and for some not? Therefore, the main motivation of this work is to investigate under which circumstances the BP method might be used to explore the physical conditions of the BLR, since photoionization is

[1] We note here that since the emission deexcitation goes as $u \rightarrow l$ it is not necessary that level l has the Saha-Boltzmann distribution.

Journal of Physics: Conference Series **257** (2010) 012034 doi:10.1088/1742-6596/257/1/012034

expected to control the plasma physical properties. Studying the BEL ratios, especially including the hydrogen Balmer lines (from Hα up to Hε), the HeII λ4686, and the HeI λ5876, we investigate the influence of BLR temperature and density, in order to assess whether particular combinations of these parameters can meet the fundamental assumptions of the BP method. To clarify this, we calculate several grids of numerical models, using the CLOUDY spectral synthesis code [27], to estimate the properties of the resulting spectra.

3. The numerical simulations

In order to investigate different BEL ratios, we generate grids of photoionization models of the BLR using the CLOUDY code (version C07.02.01: [27, 28]). Input parameters for the simulations are chosen to match the standard conditions in the BLR [28, 29, 8]. Assuming solar chemical abundances and constant hydrogen density, and using the code's AGN template for the incident continuum shape (a continuum similar to the typical radio quiet active galaxies is chosen), we compute an emission-line spectrum for the coordinate pair of hydrogen gas density $n_{\rm H}[{\rm cm}^{-3}]$ and hydrogen-ionizing photon flux $\Phi_{\rm H}[{\rm cm}^{-2}{\rm s}^{-1}]$. The grid dimensions spanned over 4 orders of magnitude in each direction, from origin at $\log n_{\rm H} = 8$, $\log \Phi_{\rm H} = 17$ and with 0.2 dex increments, resulting with 441 simulations per grid. A column density $N_{\rm H}[{\rm cm}^{-2}]$ was kept constant while producing the grid of simulations. Many authors claim that the most probable value of the column density in the BLR is $N_{\rm H} = 10^{23}{\rm cm}^{-2}$ (see e.g. [30, 29, 8]). Since the column density defines the optical thickness of the region, i.e. for higher column density the optical thickness in the emission lines increases, here we produce 5 different grids of models changing the column density in the range $N_{\rm H} = 10^{21} - 10^{25}{\rm cm}^{-2}$. Therefore, the total number of simulations was 2205, i.e. 441 per grid.

Our further work on the results of the CLOUDY grids of models consists of an analysis of the emission line fluxes generated by the code. The code gives fluxes[2] of all lines formed in the emitting region with given chemical abundance, hydrogen density and hydrogen-ionizing photon flux. We consider in our analysis the hydrogen Balmer lines (Hα to Hε) and the flux ratio R between the helium lines He II λ4686 and He I λ5876, defined as $R = F({\rm HeII}\lambda4686)/F({\rm HeI}\lambda5876)$, where F is the line flux. We particularly consider these helium lines, since they are the lines of the same element but in two different ionization states, thus their ratio He II λ4686/He I λ5876 is sensitive to the change in the temperature and density [20]. Besides, these lines are in the same spectral range as Balmer lines. Additionally, from the grids of models we consider in our analysis an averaged temperature, which is the electron temperature averaged over the BLR radius ($T_{\rm av}$ further in the text).

3.1. The analysis of the simulated BELs

For every grid of models, we analyzed the emission line fluxes generated by the code. The first step in analyzing the BEL ratios is to apply the BP method on the Balmer line ratios obtained by CLOUDY. Particularly, applying Eq. (1) we estimate the parameter A, from which we then calculate the BP temperature, i.e. the excitation temperature (hereinafter denoted as $T_{\rm BP}$) of the region where Balmer lines are formed. Also, from the best-fitting of the normalized line ratios, we obtain the error of the BP fit (hereinafter denoted as f). In many cases a satisfactory fit of the Eq. (1) is not obtained, i.e. f has pretty large values. This is more noticeable in the case of higher values of the hydrogen density and ionizing-photon flux, hence we plot the error of the BP fit f in the hydrogen density vs. ionizing flux plane for all 5 grids of models (Fig. 2). In Fig. 2 we give only the contours inside which f is less than 10%, 20% and 30%.

[2] The CLOUDY code gives all line fluxes normalized to the Hβ flux. Since it was not important for our analysis, we have used these values.

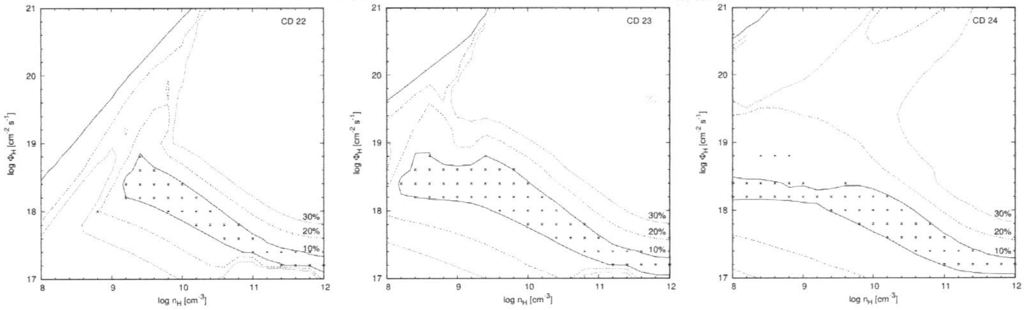

Figure 2. The error of the BP fit f in the hydrogen-density vs. ionizing flux plane for three different column densities ($N_{\mathrm{H}} \in [10^{22} - 10^{24}]$ cm^{-2}) with contours inside which the BP fit error is smaller than 10%, 20% and 30%. The asterisks represent the simulations which satisfied three defined constraints (see text for details).

It can be seen in Fig. 2 that there is a certain regularity in the values of the input parameters Φ_{H} and n_{H} for which the f is smaller than 10%. Therefore, the BP method could be considered valid if f is less than 10% (eventually 20% in the measured spectra) and that the parameter space where this is valid is pretty constrained for different column densities N_{H} (Fig. 2). Therefore, we introduced the constraint for our CLOUDY simulations that the error of the BP fitting of modeled spectra f is less than 10%.

Since we have photoionization as the main energy input mechanism and the assumption of the solar chemical abundances, we take into account only models where T_{av} and T_{BP} were not higher than 20000 K. Moreover, we introduce the additional constraint of $R < 2$ since this is the expected observational limit [31, 32] and an expected theoretical value for the ratio between HeII λ4686 and HeI λ5876 for the temperatures and densities we are using. The number of simulations that met our requirements within different uncertainty ranges was reduced to 22, 34, 54, 48, 73 for column densities $N_{\mathrm{H}} = 10^{21} - 10^{25}\mathrm{cm}^{-2}$, respectively. From these we formed five sets of simulated spectra of single column density, labeled as e.g. CD21 for column density $N_{\mathrm{H}} = 10^{21}\mathrm{cm}^{-2}$, etc. The models that satisfy these 3 constraints are represented with asterisks in Figure 2.

4. The relations between the BEL ratios and the BLR physical properties

We investigate in more details the five sets of simulated spectra defined above. First, we plot the average temperature T_{av} of the emitting region, one of the outputs of the model, with respect to the ratio R of the helium lines for all 5 sets (Fig. 3, left), and also the hydrogen density n_{H} versus the ratio R (Fig. 3, right). If the energy levels approximate a Saha-Boltzmann distribution, we expect the data sets to show a specific dependence on T_{av} and n_{H}. We fit our data sets with the functions: $T_{\mathrm{av}} = A + B \cdot R$ and $\log n_{\mathrm{H}} = \frac{D}{(C+R)}$, where n_H is given in units of $10^7\mathrm{cm}^{-3}$. The best fitting results are summarized in Table 1. In the case of the low column density ($N_{\mathrm{H}} = 10^{21}$ and $10^{22}\mathrm{cm}^{-2}$), the samples were divided in two according to the limit $R = 0.5$, as around this value there is an obvious break. The best-fitting parameters for $R > 0.5$ are given in brackets. It can be seen from Fig. 3 that, even when the column density changes, the dependence of hydrogen density on helium line ratio R does not. Indeed, the ratio between the helium lines represents the ratio of the two ionization states that, according to the Saha-Boltzmann equation, should decrease with the electron density (in this case as hydrogen is nearly totally ionized $n_{\mathrm{e}} \approx n_{\mathrm{H}}$) as $R \sim 1/n_{\mathrm{e}}$ and increase with the temperature.

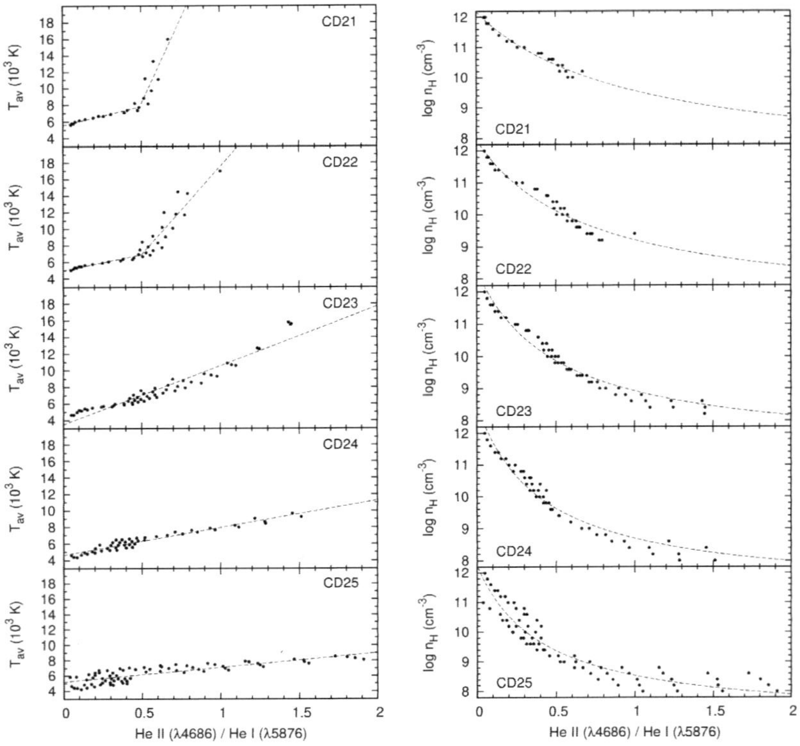

Figure 3. The average temperature $T_{\rm av}$ of the emitting medium (left) and the hydrogen density (right) as the function of the ratio R between the He II $\lambda4686$ and He I $\lambda5876$ lines, for five selected sets of simulated spectra of different column density $N_{\rm H} = 10^{21} - 10^{25}{\rm cm}^{-2}$. The dashed line represents the best-fitting with the function $T_{\rm av} = A_i + B_i \cdot R$ for the average temperature and $\log n_{\rm H} = D_i/(C_i + R)$ for the hydrogen density. The name of the set is specified on each plot. In the case of the low column density ($N_{\rm H} = 10^{21}$ and $10^{22}{\rm cm}^{-2}$), the sample was divided and fitted separatly for the average temperature.

Table 1. The best-fitting parameters A_i and B_i of the function $T_{\rm av} = A_i + B_i \cdot R$, and C_i and D_i of the function $\log n_{\rm H} = D_i/(C_i + R)$ for different column densities $N_{\rm H}$. The best-fitting parameters for $R > 0.5$ are given in brackets.

log$N_{\rm H}$ [cm^{-2}]	A_i [K]	B_i [K]	D_i [cm^{-2}]	C_i
21	5556±102	4478±346	5.04±0.22	0.97±0.05
	(-12533±7747)	(41351±13460)		
22	4998±98	3801±300	3.76±0.17	0.70±0.04
	(-3721±1939)	(21256±2872)		
23	3486±200	7116±288	2.91±0.10	0.51±0.03
24	4634±87	3326±134	2.37±0.10	0.40±0.03
25	5208±119	1899±152	2.16±0.12	0.41±0.04

5. Discussion

In this progress report we give some results of our recent investigations of different broad line parameters (the flux ratios, EW, FWHM, etc.) in order to probe the physics of the BLR, i.e. the hydrogen Balmer lines (Hα to Hε) and the helium lines from two subsequent ionization levels (He II λ4686 and He I λ5876).

From the analysis of the hydrogen Balmer lines, we can say that the BP method gives valid results if the error of the BP fit f is less than 10% (eventually 20% in the measured spectra). The parameter space of CLOUDY simulations where $f < 10\%$ is well constrained and in a similar range for different column densities $N_{\rm H}$ (Fig. 2). This area of parameters contains lower ionizing fluxes and higher hydrogen densities, but depending on the column density the area increases keeping the same trend between the $n_{\rm H}$ and $\Phi_{\rm H}$. Therefore, for those parameters the BP method could be applied for the estimation of the excitation temperature. This indicates that for some physical conditions, even if we have the photoionization as the main heating process in the BLR, the hydrogen Balmer lines are produced in such way that they obey the Saha-Boltzmann equation, i.e. the BLR plasma is at least in the Partial Local Thermodynamical Equilibrium.

The physical parameters of the simulations for which the BP method could be applied ($f < 10\%$) follow some relations even when the column density changes. There is a linear relation between the average temperature $T_{\rm av}$ of the emitting region and helium lines ratio R: $T_{\rm av} = A + B \cdot R$ (Fig. 3, Table 1). Also, the hydrogen density depends on the helium line ratio R as: $\log n_H = D/(C + R)$, where n_H is given in the units of $10^7{\rm cm}^{-3}$ (Figure 3, Table 1). Having in mind problems such as different emitting regions of helium and Balmer lines or the multicomponent origin of BELs, the relations given in Table 1 could be used as a rough estimate of the BLR physical parameters from direct measurements of the BELs.

5.1. Acknowledgments

D.I. would like to thank to the Department of Astronomy of the University of Padova in Italy for their hospitality. This work was supported by the Ministry of Science of Serbia through the project Astrophysical Spectroscopy of Extragalactic Objects (#146002).

References

[1] Sulentic J W, Marziani P and Dultzin-Hacyan D 2000 *ARA&A* **38** 521
[2] Gaskell C M 2009 *New Astronomy Review* **53** 140
[3] Osterbrock D E 1989 *Astrophysics of Gaseous Nebulae and Active Galactic Nuclei* (Sausalito: University Science Books)
[4] Krolik J H 1999 *Active Galactic Nuclei: From the Central Black Hole to the Galactic Environment* (Princeton: Princeton University Press)
[5] Peterson B M et al. 2004 *ApJ* **613** 682
[6] Davidson K and Netzer H 1979 *RvMP* **51** 715
[7] Kwan J and Krolik J 1981 *ApJ* **250** 478
[8] Korista K T and Goad M R 2004 *ApJ* **606** 749
[9] Ferland G J 1999 in *Quasars and Cosmology* (ASP Conf. Ser.) vol 162 eds G Ferland and J Baldwin p 147
[10] Netzer H 1975 *MNRAS* **171** 395
[11] Baldwin J A 1997 in *IAU Colloq. 159, Emission Lines in Active Galaxies: New Methods and Techniques* (ASP Conf. Ser.) vol 113 eds B M Peterson, F Cheng and A S Wilson p 80
[12] Popović L Č, Mediavilla E G, Bon E and Ilić D 2004 *A&A* **423** 909
[13] Osterbrock D E and Ferland G J 2006 *Astrophysics of Gaseous Nebulae and Active Galactic Nuclei (2nd ed.)* (Sausalito, California: University Science Books)
[14] Ilić D, Kovačević J, Popović L Č 2009 *New Astronomy Review* **53** 149
[15] Kovačević J, Popović L Č, Dimitrijević M S 2010 *ApJS* **189** 15
[16] Marziani P, Sulentic J W, Zwitter T, Dultzin-Hacyan D and Calvani M 2001 *ApJ* **558** 553
[17] Laor A 2006 *ApJ* **643** 112
[18] Popović L Č 2003 *ApJ* **599** 140
[19] Popović L Č 2006 *ApJ* **650** 1217 (an Erratum)

[20] Griem H R 1997 *Principles of plasma spectroscopy* (Cambridge: Cambridge University Press)
[21] La Mura G, Popović L Č, Ciroi S, Rafanelli P and Ilić D 2007 *ApJ* **671** 104
[22] Popović L Č, Shapovalova A I, Chavushyan V H, Ilić D, Burenkov A N, Mercado A, Ciroi S and Bochkarev N G 2008 *PASJ* **61** 1
[23] van der Mullen J A M, Benoy D A, Fey F H A G, van der Sijde B and Vlcek J. 1994 *PhRvE* **50** 3925
[24] Konjević N 1999 *PhR* **316** 339
[25] Popović L Č 2006 *SerAJ* **173** 1
[26] Ilić D, Popović L Č, Bon E, Mediavilla E, Chavushyan V H 2006 *MNRAS* **371** 1610
[27] Ferland G J, Korista K T, Verner D A, Ferguson J W, Kingdon J B and Verner E M 1998 *PASP* **110** 761
[28] Ferland G J 2006 *Hazy, A Brief Introduction to Cloudy 06.02* (University of Kentucky Internal Report)
[29] Korista K T and Goad M R 2000 *ApJ* **536** 284
[30] Dumont A-M., Collin-Souffrin S and Nazarova L 1998 *A&A* **331** 11
[31] Kollatschny W, Bischoff K, Robinson E L, Welsh W F and Hill G J 2001 *A&A* **379** 125
[32] Vanden Berk D E, Richards G T, Bauer A, Strauss M A, Schneider P et al. 2001 *AJ* **122** 549

25th Summer School and International Symposium on the Physics of Ionized Gases—SPIG 2010 IOP Publishing
Journal of Physics: Conference Series **257** (2010) 012035 doi:10.1088/1742-6596/257/1/012035

Studying the complex BAL profiles in the BALQSOs spectra

E. Lyratzi[1,2], E. Danezis[1], L. Č. Popović[3], A. Antoniou[1], M. S. Dimitrijević[3], and D. Stathopoulos[1]

1 University of Athens, Faculty of Physics, Department of Astrophysics, Astronomy and Mechanics, Panepistimioupoli, Zographou 157 84, Athens, Greece,

2 Eugenides Foundation, Syngrou 387, 175 64 P. Faliro, Greece

3 Astronomical Observatory, Volgina 7, 11060 Belgrade, Serbia

elyratzi@phys.uoa.gr

Abstract. Most of Broad Absorption Lines (BALs) in quasars (QSOs) present very complex profiles. This means that we cannot fit them with a known physical distribution. An idea to explain these profiles is that the dynamical systems of Broad Line Regions (BLRs) are not homogeneous but consist of a number of density regions or ion populations with different physical parameters. Each one of these density regions gives us an independent classical absorption line. If the regions that give rise to such lines rotate with large velocities and move radially with small velocities, the produced lines have large widths and small shifts. As a result they are blended among themselves as well as with the main spectral line and thus they are not discrete. Based on this idea we study the BALs of UV C IV resonance lines in the spectra of a group of Hi ionization Broad Absorption Line Quasars (Hi BALQSOs) using the Gauss-Rotation model (GR model).

1. Introduction

Approximately about 10% of all quasars present broad, blue shifted absorption lines. The outflow velocity can reach up to 0.1 0.2 c. Usually, in their spectra we observe the lines of high ionization species, as C IV λ 1549 Å, Si IV λ 1397 Å, N V λ 1240 Å and Lya. Rarely, some low ionization lines, such as Mg II λ 2798 Å and Al III λ 1857 Å, also exhibit broad absorption lines [see e.g. 1, 2]. Broad Absorption Lines (BALs) can have different shapes. Also different types of these objects may have differences in their continua [3].

One proposed explanation of the Broad Absorption Line (BAL) phenomenon is that BALQSOs (Broad Absorption Line Quasars) and non-BALQSOs are distinct populations of objects [4]. Similarly, some have argued that only low-ionization BALQSOs (LoBALs) are a different class of quasars [5].

Others suggest that BALQSOs and non-BALQSOs are the same type of quasar but viewed from different orientations [6-8] or at different stages in their life cycles [9].

The spectrum of a BALQSO is usually interpreted as a combination of (i) a broadband continuum arising from the central engine, (ii) the broad emission lines coming from the Broad Emission Line Regions (BELR), emerging near the center of the QSO and (iii) the broad absorption lines that are superposed, originating in a separate outlying region, so called Broad Absorption Line Region

(BALR) (see also [10]). However, it is also possible, that emission and absorption occur in the same line-forming region [11].

An important question is: Which are the physical connections between the BLR (Broad Line Region) and BALR? This is also important, since at least a part of the BLR seems to originate from wind of accretion disk [see 12, 13].

Another question is: Where is the BALR placed with respect to the center of a BALQSO and the BLR? To answer this question, one should investigate the kinematical properties of the emission and absorption lines.

Disk wind models [12, 14, 15] explain many properties of BAL quasars, but it is unclear if they can explain the full range of BAL profiles and column densities. BALs are caused by outflowing gas intrinsic to the quasar and are not produced by galaxies along the line of sight (as is the case for most narrow-absorption systems).

Determining whether a quasar is a BALQSO is a complicated task. The standard method is to calculate the "balnicity" index (BI), defined by Weymann et al. in [6]. A BI of zero indicates that broad absorption is absent, while a positive BI indicates not only the presence of one or more broad absorption lines but also the amount of absorption. The BI is essentially a modified equivalent width of the broad absorption line, expressed in velocity units and is defined as follows: (i) absorption should appear between 3000 and 25000 km/s blueward of C IV emission redshift and (ii) at this place and for at least 2000 km/s the absorption must fall at least 10% below the continuum. BIs can range from 0 to 20000 km/s.

We can classify the QSOs using the BI criterion in the following categories [16]:

- HiBALs: BALQSOs that present broad absorption trough just blueward of C IV emission.
- LoBALs: BALQSOs that present broad absorption troughs just blueward of both the C IV and Mg II emission lines.
- Non-BALs: BALQSOs that present no broad absorption troughs just blueward of the C IV and Mg II emission lines.
- FeLoBALs: BALQSOs with excited iron absorption features.

The 25000 km/s limit for the BI is chosen to avoid emission and absorption from Si IV. Absorption lines within 3000 km/s with width smaller than 2000 km/s are excluded to avoid contamination from absorption that might not be due to an outflow. These lines are called "associated absorption lines" [17]. Some of these associated systems are known to be intrinsic outflows, but others may simply be the result of absorption in the host galaxy or a nearby galaxy.

Here, we present some ideas in order to explain the complex structure of BALs in QSOs and especially we study the BALs of UV C IV resonance lines in the spectra of a group of Hi BALQSOs using the Gauss-Rotation model (GR model).

2. The multi-structure of BALs in QSOs

Most of BALs in QSOs present very complex profiles. This means that we cannot fit them with a known physical distribution. An idea is that the dynamical systems of BLRs are not homogeneous but consist of a number of density regions or ion populations with different physical parameters.

Each one of these density regions gives us an independent classical absorption line. If the regions that give rise to such lines rotate with large velocities and move radially with small velocities, the produced lines have large widths and small shifts. As a result they are blended among themselves as well as with the main spectral line and thus they are not discrete. A similar phenomenon can explain the very complex profiles of a great number of very broad lines in the spectra of hot emission stars [10, 18, 19].

Based on this idea we study the BALs of UV C IV resonance lines in the spectra of a group of Hi BALQSOs using the GR model [10, 18, 19].

The proposed model is relatively simple, aiming to describe the regions where the spectral lines are created. We assume that the BALR and BELR are composed of a number of successive independent absorbing/emitting density layers of matter (that originate in a disk wind). The absorbing regions have three apparent velocities (projected on the line-of-sight of an observer): radial velocity (V_{rad}) of the BALR, random velocity of the ions (V_{rand}) in the BALR and the rotational velocity (V_{rot}) of the BALR.

3. Data

In order to study the C IV resonance lines (λλ 1548.187, 1550.772 Å) we apply the GR model to the spectra of 15 Broad Absorption Line Quasars (BALQSOs) taken from the Sloan Digital Sky Survey's Data Release 7. The SDSS imaging survey uses a wide-field multi-CCD camera [20]. The spectra cover the optical range 3800–9200 Å at a resolution of 1800–2100.

In table 1, column 1 lists the name of the QSOs, using the SDSS format of J2000.0 right ascension (hhmmss.ss) and declination (±ddmmss.s), column 2 lists the modified Julian date-plate-fiber, column 3 lists the redshift and column 4 lists the dates of observations.

Table 1. Studied BALQSOs

Object Name (SDSS)	MJD-Plate-Fiber	Redshift	Date
J015024.44+004432.99	51793-0402-485	2.00596	9/6/2000, 10:06
J015048.83+004126.29	51793-0402-505	3.70225	9/6/2000, 10:06
J021327.25-001446.92	51816-0405-197	2.39948	9/29/2000, 9:57
J023252.80-001351.17	51820-0407-158	2.03289	10/3/2000, 9:41
J023908.99-002121.42	51821-0408-179	3.74	10/4/2000, 9:38
J025331.93+001624.79	51816-0410-391	1.8214	9/24/2000, 11:26
J025747.75-000502.91	51816-0410-117	2.19139	9/24/2000, 11:26
J031828.91-001523.17	51929-0413-170	1.98447	1/20/2001, 4:23
J102517.58+003422.17	51941-0272-501	1.88842	2/1/2001, 9:30
J104109.86+001051.76	51913-0274-482	2.25924	1/4/2001, 11:00
J104152.62-001102.18	51913-0274-159	1.70876	1/4/2001, 11:00
J104841.03+000042.81	51909-0276-310	2.03044	12/31/2000, 11:08
J110041.20+003631.98	51908-0277-437	2.01143	12/30/2000, 11:19
J110736.68+000329.60	51900-0278-271	1.74162	12/22/2000, 12:12
J112602.81+003418.23	51614-0281-432	1.7819	3/11/2000, 6:52

4. Method

The relevant broadening mechanism, in the case of BALs, is the random motion of the absorbing gas, but also a part comes from the rotation caused by the massive black hole. In order to find the limits for the rotational and random velocities, we fitted the observed lines using two approaches:

1. Gauss-Rotation approach (GR approach): We assume that random motion is dominant, i.e. the random velocity is maximal and the rotational component is minimal,
2. Rotation-Gauss approach (RG approach): We assume that the rotational component is dominant.

After that we used F-test to conclude on which approach of the model is more appropriate to explain the complex absorption line profiles.

In figure 1 we give some examples of the fitted spectra. The black line corresponds to the observed spectra and the blue line corresponds to the theoretical line profiles given by the GR model. Below each fitting, the green line represents the residual, which gives the differences between the observed spectrum and the theoretical profile.

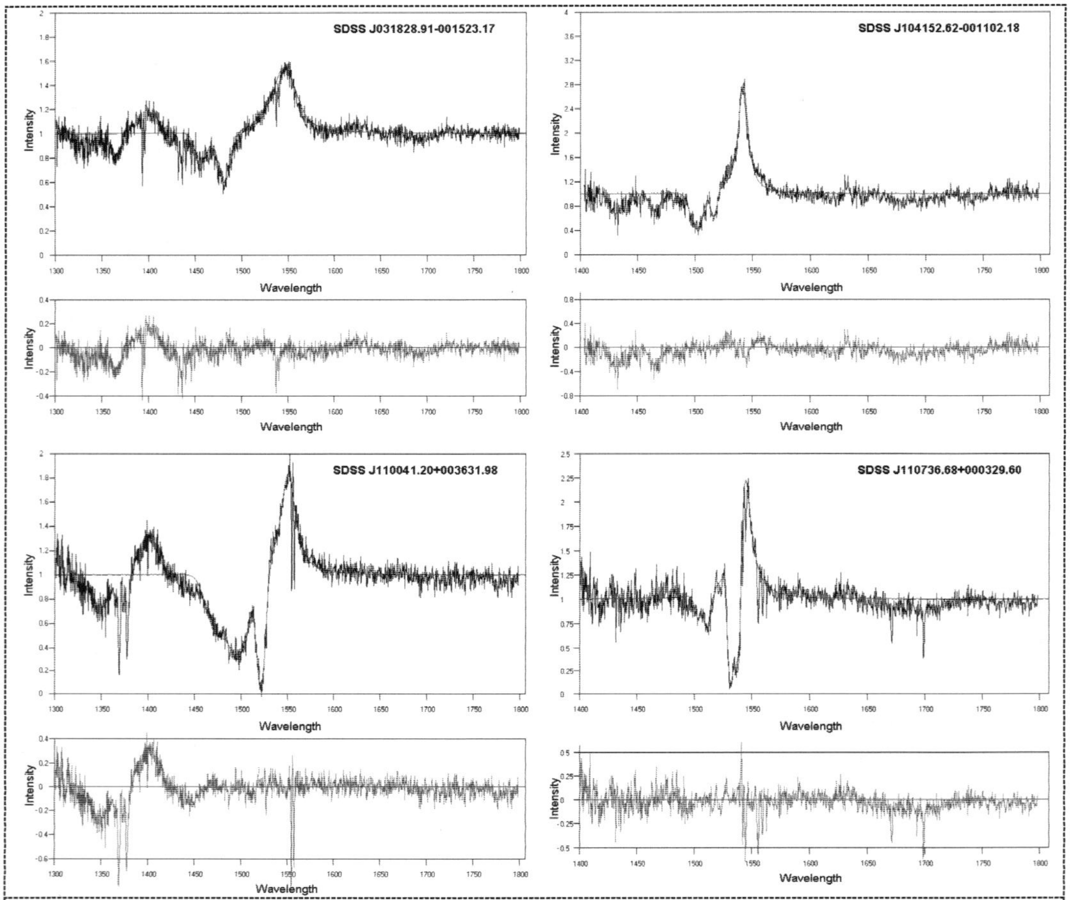

Figure 1. Examples of the fitted spectra. The black line corresponds to the observed spectra and the blue line corresponds to the theoretical line profiles given by the GR model. Below each fitting, the green line represents the residual, which gives the differences between the observed spectrum and the theoretical profile.

5. Results

Using the GR model, we were able to fit the studied C IV spectral lines with 2 up to 4 absorption components and we calculated the values of the kinematical parameters (rotational, radial and random velocities), as well as the column density and the absorbed energy. The calculated values are given in tables 2-5. As one can see in tables 2-5, the radial velocity of the studied BALRs takes values between -2225 km/s and -17994 km/s, the values of the rotational velocity of the BALRs are between 30 km/s and 2400 km/s and the random velocity of the ions take values between 228 km/s and 2280 km/s. In most cases, the best fit of the observed spectral features is given by the GR approach. However, there were 4 cases (indicated with italics in tables 2-5), where the best fit is accomplished with the RG approach.

Table 2. Radial Velocities (km/s)

Object Name (SDSS)	$Vrad_1$	$Vrad_2$	$Vrad_3$	$Vrad_4$
J015024.44+004432.99	-3483	-7546		
J015048.83+004126.29	-3473	-5611		
J021327.25-001446.92	-3096	-5321		
J023252.80-001351.17	-5224	-8649		
J023908.99-002121.42	-8513			
J025331.93+001624.79	-4257	-5611		
J025747.75-000502.91	-7352			
J031828.91-001523.17	-13157	-17994		
J102517.58+003422.17	*-6772*	*-8126*	*-10545*	
J104109.86+001051.76	-2225	-5417	-6675	-9867
J104152.62-001102.18	-6385	-9287		
J104841.03+000042.81	*-5998*	*-8320*	*-11609*	
J110041.20+003631.98	*-5417*	*-10061*	*-14704*	
J110736.68+000329.60	*-2360*	*-3618*	*-7352*	*-8900*
J112602.81+003418.23	-6288	-7642		

Table 3. Rotational and Random Velocities (km/s)

	Rotational Velocities				Random Velocities			
Object Name (SDSS)	$Vrot_1$	$Vrot_2$	$Vrot_3$	$Vrot_4$	$Vrand_1$	$Vrand_2$	$Vrand_3$	$Vrand_4$
J015024.44+004432.99	2200	1500			1140	1368		
J015048.83+004126.29	500	200			456	228		
J021327.25-001446.92	100	100			935	912		
J023252.80-001351.17	100	1800			388	388		
J023908.99-002121.42	30				2280			
J025331.93+001624.79	250	200			570	410		
J025747.75-000502.91	1000				2052			
J031828.91-001523.17	350	150			1482	1368		
J102517.58+003422.17	*800*	*750*	*100*		*228*	*228*	*342*	
J104109.86+001051.76	200	50	50	100	342	342	570	798
J104152.62-001102.18	50	50			456	1140		
J104841.03+000042.81	*1000*	*500*	*500*		*388*	*410*	*456*	
J110041.20+003631.98	*1200*	*2400*	*1500*		*456*	*1482*	*2052*	
J110736.68+000329.60	*900*	*300*	*500*	*600*	*228*	*251*	*342*	*360*
J112602.81+003418.23	50	50			502	502		

Table 4. Column Density ($\times 10^{10}$ cm^{-2})

	λ 1548.187 Å				λ 1550.772 Å			
Object Name (SDSS)	CD_1	CD_2	CD_3	CD_4	CD_1	CD_2	CD_3	CD_4
J015024.44+004432.99	-8.81	-3.51			-8.15	-3.20		
J015048.83+004126.29	-2.29	-0.78			-2.11	-0.71		
J021327.25-001446.92	-4.12	-2.03			-3.80	-1.85		
J023252.80-001351.17	-3.11	-3.67			-2.95	-3.38		
J023908.99-002121.42	-7.67				-7.04			
J025331.93+001624.79	-1.31	-0.87			-1.19	-0.80		
J025747.75-000502.91	-6.43				-5.90			
J031828.91-001523.17	-2.45	-1.39			-2.23	-1.26		
J102517.58+003422.17	*-1.82*	*-2.04*	*-0.51*		*-1.68*	*-1.89*	*-0.46*	
J104109.86+001051.76	-2.12	-1.58	-2.00	-2.81	-1.98	-1.46	-1.84	-2.58
J104152.62-001102.18	-0.94	-2.72			-0.85	-2.48		
J104841.03+000042.81	*-2.56*	*-1.73*	*-0.72*		*-2.36*	*-1.59*	*-0.66*	
J110041.20+003631.98	*-3.80*	*-5.65*	*-4.04*		*-3.51*	*-5.18*	*-3.68*	
J110736.68+000329.60	*-3.46*	*-1.87*	*-0.78*	*-0.53*	*-3.23*	*-1.74*	*-0.71*	*-0.48*
J112602.81+003418.23	-1.52	-2.21			-1.39	-2.04		

Table 5. Absorbed Energy (eV)

	λ 1548.187 Å				λ 1550.772 Å			
Object Name (SDSS)	Ea_1	Ea_2	Ea_3	Ea_4	Ea_1	Ea_2	Ea_3	Ea_4
J015024.44+004432.99	-13.55	-5.40			-12.53	-4.91		
J015048.83+004126.29	-3.52	-1.19			-3.25	-1.09		
J021327.25-001446.92	-6.34	-3.12			-5.85	-2.84		
J023252.80-001351.17	-4.78	-5.65			-4.53	-5.19		
J023908.99-002121.42	-11.81				-10.82			
J025331.93+001624.79	-2.01	-1.34			-1.83	-1.22		
J025747.75-000502.91	-9.90				-9.06			
J031828.91-001523.17	-3.77	-2.14			-3.42	-1.94		
J102517.58+003422.17	*-2.80*	*-3.14*	*-0.78*		*-2.57*	*-2.90*	*-0.70*	
J104109.86+001051.76	-3.27	-2.43	-3.08	-4.32	-3.04	-2.24	-2.83	-3.96
J104152.62-001102.18	-1.44	-4.19			-1.31	-3.81		
J104841.03+000042.81	*-3.94*	*-2.67*	*-1.11*		*-3.62*	*-2.45*	*-1.01*	
J110041.20+003631.98	*-5.84*	*-8.70*	*-6.21*		*-5.40*	*-7.95*	*-5.65*	
J110736.68+000329.60	*-5.32*	*-2.87*	*-1.20*	*-0.81*	*-4.97*	*-2.68*	*-1.09*	*-0.73*
J112602.81+003418.23	-2.33	-3.40			-2.13	-3.14		

6. Final remarks

The main purpose of this study is to answer to the question: Is a multi-structure of BLR regions able to explain the complex profile of Hi BALQSOs spectra?

Additionally, if the answer of the above question is positive, our next step is to test whether the GR model is able to reproduce the complex profiles of Hi BALQSOs and to calculate a group of physical parameters of the corresponding BLRs.

Concluding the results of our study we could remark that:

1. It is possible that the dynamical systems of BLRs are not homogeneous but consist of a number of density regions or ion populations with different physical parameters. Each one of these density regions gives us an independent classical absorption line. If the regions that give rise to such lines rotate with large velocities and move radially with small velocities, the produced lines have large widths and small shifts. As a result they are blended among themselves as well as with the main spectral line and thus they are not discrete.
2. As we can see in tables 2-5, the GR model is able to reproduce accurately the complex profiles of Hi BALQSOs and to calculate a group of physical parameters of the corresponding BALRs.

Acknowledgements

This research project is progressing at the University of Athens, Department of Astrophysics, Astronomy and Mechanics, under the financial support of the Special Account for Research Grants, which we thank very much. This work also was supported by Ministry of Science Technological Development of Serbia and, through the projects "Influence of collisional processes on astrophysical plasma line shapes" and "Astrophysical spectroscopy of extragalactic objects".

Funding for the SDSS and SDSS-II has been provided by the Alfred P. Sloan Foundation, the Participating Institutions, the National Science Foundation, the U.S. Department of Energy, the National Aeronautics and Space Administration, the Japanese Monbukagakusho, the Max Planck Society, and the Higher Education Funding Council for England. The SDSS Web Site is http://www.sdss.org/.

The SDSS is managed by the Astrophysical Research Consortium for the Participating Institutions. The Participating Institutions are the American Museum of Natural History, Astrophysical Institute Potsdam, University of Basel, University of Cambridge, Case Western Reserve University, University of Chicago, Drexel University, Fermilab, the Institute for Advanced Study, the Japan Participation Group, Johns Hopkins University, the Joint Institute for Nuclear Astrophysics, the Kavli Institute for Particle Astrophysics and Cosmology, the Korean Scientist Group, the Chinese Academy of Sciences (LAMOST), Los Alamos National Laboratory, the Max-Planck-Institute for Astronomy (MPIA), the Max-Planck-Institute for Astrophysics (MPA), New Mexico State University, Ohio State University, University of Pittsburgh, University of Portsmouth, Princeton University, the United States Naval Observatory, and the University of Washington.

References

[1] Hamann F, Korista K T and Morris S L 1993 *ApJ* **415** 541
[2] Crenshaw M D, Kraemer S B and George I M 2003 *ARAA* **41** 117
[3] Reichard T A, Richards G T, Hall P B, Schneider D P, Vanden Berk D E, Fan X, York Donald G, Knapp G R and Brinkmann J 2003 *AJ* **126** 2594
[4] Surdej J and Hutsemekers D 1987 *A&A* **177** 42
[5] Boroson T A and Meyers K A 1992 *ApJ* **397** 442
[6] Weymann R J, Morris S L, Foltz C B and Hewett P C 1991 *ApJ* **373** 23
[7] Ogle P M, Cohen M H, Miller J S, Tran H D, Goodrich R W and Martel A R 1999 *ApJS* **125** 1
[8] Schmidt G D and Hines D C 1999 *ApJ* **512** 125
[9] Becker R H, White R L, Gregg M D, Brotherton M S, Laurent-Muehleisen S A and Arav N 2000 *ApJ* **538** 72
[10] Lyratzi E, Popović L·Č, Danezis E, Dimitrijević M S, Antoniou A 2009 *NewAR* **53** 179

[11] Branch D, Leighly K M, Thomas R C and Baron E 2002 *ApJL* **578** L37
[12] Murray N and Chiang J 1998 *ApJ* **494** 125
[13] Popović L·Č, Mediavilla E, Bon E, Ilić D 2004 *A&A* **423** 909
[14] Proga D, Stone J M and Kallman T R 2000 *ApJ* **543** 686
[15] Williams R J R, Baker A C, Perry J J 1999 *MNRAS* **310** 913
[16] Richards G T et al. 2001 *AJ* **121** 2308
[17] Foltz C B, Weymann R J, Peterson B M, Sun L, Malkan M A and Chaffee F H 1986 *ApJ* **307** 504
[18] Danezis E, Popović L·Č, Lyratzi E and Dimitrijević M S 2006 *AIP Conference Proceedings* **876** 373
[19] Danezis E, Nikolaidis D, Lyratzi E, Popović L·Č, Dimitrijević M S, Antoniou A and Theodosiou E 2007 *PASJ* **59** 827
[20] Gunn J E et al. 1998 *AJ* **116** 3040

The Calculation of the Photo Absorption Processes in Dense Hydrogen Plasma with the Help of Cut-Off Coulomb Potential Model

Nenad M. Sakan

Institute of Physics, Pregrevica 118, Zemun, Belgrade, Serbia

E-mail: nsakan@ipb.ac.rs

Abstract. Extensive work was done in the application of a cut-off Coulomb model on the description of the optical processes of the photo ionization and inverse bremsstrahlung. Presented work deals with a usage of a cut-off Coulomb model pseudo potential for the calculation of the optical absorption process in dense hydrogen plasma as a entirely quantum mechanical process. Although the mentioned processes are strongly influenced by the collective process in dense plasma, the used pseudo potential enables to model the described interaction with the plasma system as a binary process. There are several advantages of such approach; the existence of the exact analytical solutions for the wave functions in the described potential enables to eliminate one of the several sources of numerical error. Also, more complex processes of the interaction inside plasma could be considered, and they have been added in presented work. The work on description of such processes has been started. The collective phenomena of the plasma are here described as an additional shifting and broadening of a bond states levels. Furthermore, with the adding of mentioned broadening and additional shifting of the bond states as free external parameters the good agreement between the analyzed experimental data and our model solutions occurs. The method of determination of the cut-off radius was developed and applied in our considerations. The presented model is a good approach for the description of dense hydrogen plasma of moderate and high non-ideality. It presents an easily extendable model, in which is easy to introduce additional processes and effects.

1. Introduction

In this paper is studied a new model method of the describing of the continuous absorption of electromagnetic (EM) radiation in dense strongly ionized hydrogen plasma, caused by the atomic photo-ionization processes

$$E_{h\nu} + H^*(nl) \rightarrow H^+ + e_{\vec{q}'}, \tag{1}$$

and electron-ion inverse "bremsstrahlung" processes

$$E_{h\nu} + e_{\vec{q}} + H^+ \rightarrow e_{\vec{q}'} + H^+, \tag{2}$$

where $E_{h\nu}$ is the energy of the photon with the wavelength λ, n and l - principal and orbital quantum numbers of hydrogen excited states, $\vec{q}$ and $\vec{q}'$ - the momentum of the free electron before and after scattering on the considered ion H^+.

While in weakly and moderately non-ideal plasma, this absorption is caused by the neutral atoms and electron-ion collision complex which interaction with the neighborhood can be neglected, as for example in Solar photosphere [5, 6], or described within the framework of a perturbation theory [18, 12, 13, 15, 16] in the dense strongly non-ideal plasma the situation is in principle different.

By now a lot of effort was aimed to the development of the quantum-statistical methods for the description of the thermodynamical and transport properties of dense strongly non-ideal plasma [9, 11, 8, 7, 10, 14] while the absorption processes was treated only for plasma with electron densities $N_e < 10^{18} cm^{-3}$, where the approximation of electron-atom and electron-ion binary collisions is still applicable. The area of really dense plasma with $N_e > 10^{19} cm^{-3}$ was not systematically studied from the aspect of the bound-bound, bound-free and free-free absorption processes, excluding some efforts of semi-empirical describing of such processes [25, 26]. Because of that the development of a model method which describes the mentioned absorption process in dense strongly non-ideal plasma on a simple and physically acceptable way is the one of the actual tasks. Within this work as a landmark is taken the hydrogen plasma with the electron density $N_e = 1.5 \cdot 10^{19} cm^{-3}$ and the temperature $T = 23000K$, which was experimentally studied in [26]. The direct result of this work is a new model method for the determination of absorption coefficients $\kappa_{bf}(\lambda)$ and $\kappa_{ff}(\lambda)$, characterizing the bound-free and free-free absorption processes (1) and (2) in the strongly non-ideal hydrogen plasma, which is based on a cut-off Coulomb pseudo-potentials, similar to the one used for the determination of the non-ideal plasma conductivity. The presented method is tested in the optical range of photon wavelengths $350nm \leq \lambda_{h\nu} \leq 550nm$.

2. Theory

2.1. The cut-off Coulomb potentials

The obvious way of simplification of principally many body processes of photo absorption transitions inside plasma was transformation to the corresponding transitions of the electron in an adequately chosen pseudo-potential, which replaces the considered ion and the rest of the system. In [22], in order to obtain the method of the describing of such process which would be practically applicable, generally non-local pseudo-potential in usual way was soughed in the form of the corresponding local one-particle potential. As such potential was chosen one of model screening Coulomb potential, namely cut-off potential (4).

On the occasion of the choosing of the model potential it was taken into account the argumentation from the [22], which shows that often used model Debye-Hückel (DH) potential is not adequate for strongly non-ideal plasma. Let us draw attention that we here do not have in mind some undesirable properties of the DH potential [28, 27], but the way of the obtaining of that potential itself. Namely, in accordance with [19] the DH potential is the average electrostatic potential which is generated by the observed ion and all charged particles from its neighborhood, which are often treated as the screening cloud. Consequently, the electron, that is involved in scattering on the considered ion, also is the part of that cloud. In spite of this fact the DH potential, as it is known, is used often in weakly non-ideal plasma when the number $n_D \gg 1$, where n_D is the number of the electrons inside the sphere with the Debye radius r_D.

However, in the case of strongly non-ideal plasmas, when $n_D \cong 1$, as it is in the considered plasma, practically, the complete cloud is consisted of the free electron that is involved in scattering, and the DH potential could not be used any more. Contrary to that, in the case $n_D \cong 1$ the application of the cut-off Coulomb potential, as it was noted in [22], is physically completely justified, since it automatically provides: just Coulomb behavior of the potential in the close vicinity of the considered ion; the lowering of the atom ionization potential caused by the influence of the neighborhood, which is equal to the average potential energy of a free electron in plasma; non Coulomb asymptotic of the wave function of a free electron.

All mentioned have caused that one of the considered here model cut-off Coulomb potentials has the form, which is shown, in the Fig. (1a), where e is the absolute value of electron charge, r - the distance from the origin of the chosen reference frame, r_c - corresponding screening radius, and the value $U_p = -e^2/r_c$ has to be interpreted as the above mentioned the average potential energy of a free electron in plasma. Other model cut-off Coulomb potential is considered here because the fact that in the case of the first model the average potential energy of the electron in the region $0 < r < r_c$, for the difference of the region $r_c < r < \infty$, is not equal to the energy U_p, which is illustrated by Fig. (1 a). However, in the plasma the moving of the electron from the region occupied by the one ion to the region occupied by the nearest neighbor ion is realized in the potential with the maximal value (between the position of the mentioned ions), which is greater than average values of potential. Because of that the average potential energies of the electron in the region occupied by the one ion and in the rest of the plasma have to be equal to the average energy of the free electron in the whole system denoted here by U_p. One can see that this condition can be satisfied in the case of other cut-off Coulomb potential, which is shown in Fig. (1b), when the parameter $k = 1/2$. Namely, it can be shown that

$$\int_0^{(k+1)r_c} U(r)4\pi r^2 dr = U_p V = -\frac{e^2}{r_c} \cdot \frac{4\pi}{3} \left[(k+1)r_c\right]^3, \tag{3}$$

is only valid for $k = 1/2$, where V is the volume of sphere with radius r_c, which is determined on the basis of the result from [28].

In further consideration we will take the value $-e^2/r_c$ as the zero of the energy. After that, the potentials shown in the Figs. (1a) and (1b) are transformed to the forms $U_0(r;r_c)$ and $U_k(r;r_c)$, respectively, where

$$U_0(r;r_c) = \begin{cases} -\frac{e^2}{r} + \frac{e^2}{r_c} & : \quad 0 < r \le r_c, \\ 0 & : \quad r_c < r, \end{cases} \tag{4}$$

$$U_k(r;r_c) = \begin{cases} -\frac{e^2}{r} + \frac{e^2}{r_c} & : \quad 0 < r \le (k+1)r_c \\ 0 & : \quad (k+1)r_c < r \end{cases}, \tag{5}$$

where $U_0(r;r_c)$ is the same potential as in [22]. Because of the above mentioned, in the case of the potential $U_k(r;r_c)$ we will consider that $k = 1/2$.

Let us denote that the form of the potential (5) is not caused by the presence of some new mechanism that increases the barrier in the region $r > r_c$ for the electron in the complex $(H^+ + e)_{nl}$ or $(H^+ + e)_{\vec{q}}$, but exclusively by the requirement for the satisfying of the condition (3).

2.2. The photo-ionization and inverse "bremsstrahlung" cross-sections

Since under the condition from [26] the considered wavelength $\lambda \gg r_s$, where $r_s = (3/4\pi N_e)^{1/3}$ is the corresponding Wigner-Seitz radius, the dipole approximation in the case of considered processes is valid. According to that, the cross section for these bound-free and free-free absorption processes are given by the expressions from [24], namely

$$\sigma(nl;E') = \frac{4\pi^2 e^2 k}{3(2l+1)} \sum_{l'=l\pm1} l_{max} \left(\int P_{nl} r P_{E'l'} dr \right)^2, \tag{6}$$

$$\sigma(E;E') = \frac{8\pi^4}{3} \frac{\hbar e^2 k}{q^2} \sum_{l'=l\pm1} l_{max} \left(\int P_{El} r P_{E'l'} dr \right)^2, \tag{7}$$

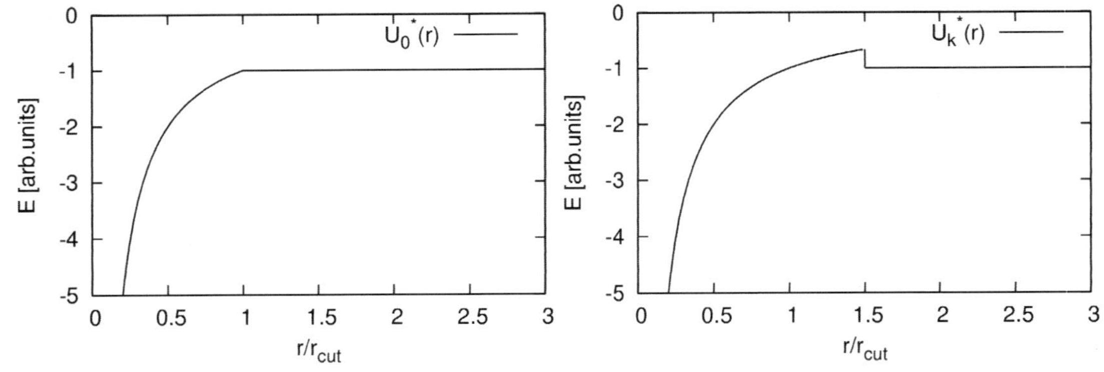

a) b)

Figure 1. The behavior of the used potentials: **a)** - from (4), **b)** - from (5).

where $k = \epsilon_\lambda/\hbar c$ is the momentum of the absorbed photon with the given λ, $E = \hbar^2 q^2/2m$ and $E' = \hbar^2 q'^2/2m$ - the energies of the free electron, l_{max} - maximal value of l and l', m - the electron mass, and c -the light velocity. Here the radial wave function of the electron in the model potentials (4) and (5) with $k = 0.5$ is denoted with P_{nl}/r, for the bound states with given n and l, and with P_{El}/r and $P_{E'l'}/r$ for the free states with the given E and l or E' and l'. The functions P_{nl} and P_{El} are obtained in strict analytical form by the means of the expressions for the Whittaker, Coulomb, spherical Bessel, and modified Bessel functions.

In further calculations for the determination of the photo-ionization cross section $\sigma(nl; E')$ is used Eq. 6, while in the case of inverse "bremsstrahlung" cross section $\sigma(E; E')$ is used the expression which is obtained by means of the known relations [24], which connect the matrix elements of the j-th components ($j = 1, 2, 3$) of the radius-vector $\vec{r}$, electron momentum $\vec{p}$, and gradient of the potential $\vec{\nabla} U(\vec{r})$, namely

$$< in|\vec{\nabla}_j U(\vec{r})|fin >= \frac{i}{\hbar}(E_{in} - E_{fin}) < in|\vec{p}_j|fin >, \tag{8}$$

$$< in|\vec{p}_j|fin >= \frac{i\ m}{\hbar}(E_{in} - E_{fin}) < in|\vec{r}_j|fin >, \tag{9}$$

where $U(\vec{r})$ in the considered case is equal to $U_0(r)$ or $U(r;k)$. Namely, from Eqs. (7), (8) and (9) it follows the expression

$$\sigma(E; E') = \frac{4\pi^4}{3} \frac{\hbar^6 e^2}{m^3 c E E_{h\nu}^3} \sum_{l'=l\pm1} l_{max} \left(\int_0^{(k+1)r_c} P_{El} \nabla_r U(r) P_{E'l'} dr \right)^2, \tag{10}$$

where $E_{h\nu} = E' - E$, and with $U(r) = U(r; k = 0) \equiv U_0(r)$ or $U(r; k = 1/2)$, which enables to use the shape of the potentials (4) and (5) and to avoid all difficulties connected with the calculation of the dipole matrix element in Eq. (7) in the whole region of space $0 < r \leq \infty$. Just Eq. (10) is used here for the calculation of the inverse "bremsstrahlung" cross-section $\sigma(E; E')$.

2.3. The partial and total absorption coefficients

The expressions (6) and (10) for the photo-ionization and inverse "bremsstrahlung" cross-sections enable the direct determination of the partial absorption coefficients, characterizing the bound-free and free-free absorption processes (1) and (2), given by the relations

$$\kappa_{bf}^{(0)}(\lambda; N_e, T) = \sum_{n=1}^{n_{max}} \sum_{l=0}^{n-1} N_{nl} \cdot \sigma(nl; E'), \tag{11}$$

$$\kappa_{ff}^{(0)}(\lambda; N_e, T) = N_i N_e \cdot \int_0^{\infty} \sigma(E; E') v f(v) dv, \tag{12}$$

where N_{nl} is the density of the atoms H^*, e.g electron-ion pairs in the bound states with the given quantum numbers n and l, T - the plasma temperature, and n_{max} - the principal quantum number of the last realizing bond state for the given N_e and T. However, while the expression (12) for the free-free absorption coefficient $\kappa_{ff}^{(0)}(\lambda; N_e, T)$ should generate the purely acceptable results, the situation in connection with Eq. (11) is different. Namely, the results obtained by means of Eq. (11) should be similar to the ones for the diluted plasma (see for example [5]), since, contrary to the existing experimental results [26], the unique serious difference would ensue from the lowering of the photo-ionization limits for the realizing bound states for the value close to e^2/r_c.

The plasma-ion interaction at the considered densities is mainly of Stark type, and also it was made a transition from many particle model towards the two particle model. Because of that there should be included and additionally considered a shift and the broadening of a bond state levels, as a result of a many particle interactions.The mentioned shifts and broadenings are treated as the semi-empirical quantities, which appear as the external parameter of the theory. Here, the shift of (nl)-level is denoted by Δ_{nl}^{sh}, and broadening by Δ_{nl}^{br}. As it is usual we assume that the electron in atom H_{nl}^* in the plasma could be in the state with the energies which are dominantly grouped around the energy $\varepsilon_{nl}^{max} = \varepsilon(nl) + \Delta_{nl}^{sh}$, inside the interval $(\varepsilon_{nl}^{max} - \Delta_{nl}^{br}/2, \varepsilon_{nl}^{max} + \Delta_{nl}^{br}/2)$. Let $P_{nl}(\varepsilon)$ is the probability density which characterizes the distribution of the energies of the mentioned state within the interval $(\varepsilon_{nl}^{max} - \Delta_{nl}^{br}/2, \varepsilon_{nl}^{max} + \Delta_{nl}^{br}/2)$, which satisfies the conditions

$$max\{P_{nl}(\varepsilon)\} = P(\varepsilon = \varepsilon_{nl}^{max}), \qquad \int_{\varepsilon_{nl}^{max} - \Delta_{nl}^{br}/2}^{\varepsilon_{nl}^{max} + \Delta_{nl}^{br}/2} P_{nl}(\varepsilon) d\varepsilon = 1. \tag{13}$$

In accordance with above consideration, here we will characterize the bound-free and free-free processes by the photo-ionization and inverse "bremsstrahlung" partial absorption coefficients

$$\kappa_{bf}(\lambda; N_e, T) = \int_{\varepsilon_{nl}^{max} - \Delta_{nl}^{br}/2}^{\varepsilon_{nl}^{max} + \Delta_{nl}^{br}/2} P_{nl}(\varepsilon) \cdot \tilde{\kappa}_{bf}^{(0)}(\lambda; N_e, T; \varepsilon) d\varepsilon, \tag{14}$$

where $\tilde{\kappa}_{bf}^{(0)}(\lambda; N_e, T; \varepsilon)$ is obtained from (6) and (11) by replacing free electron energy E' with $\tilde{E}' = E' + (\varepsilon - \varepsilon_{nl})$,

$$\kappa_{ff}(\lambda; N_e, T) = \kappa_{ff}^{(0)}(\lambda; N_e, T), \tag{15}$$

where $\kappa_{ff}^{(0)}(\lambda; N_e, T)$ is given by Eq. (12), as well as the corresponding total absorption coefficient

$$\kappa_{tot}(\lambda; N_e, T) = (\kappa_{ff}(\lambda; N_e, T) + \kappa_{bf}(\lambda; N_e, T)) \cdot \left[1 - \exp\left(-\frac{\epsilon_\lambda}{kT}\right)\right], \tag{16}$$

where it is taken into account the influence of the stimulated emission.

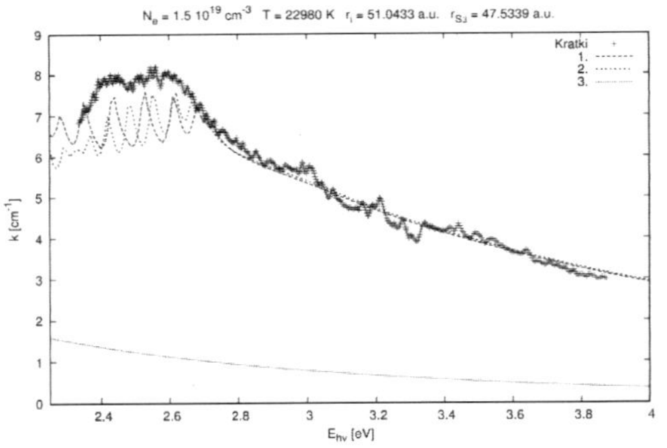

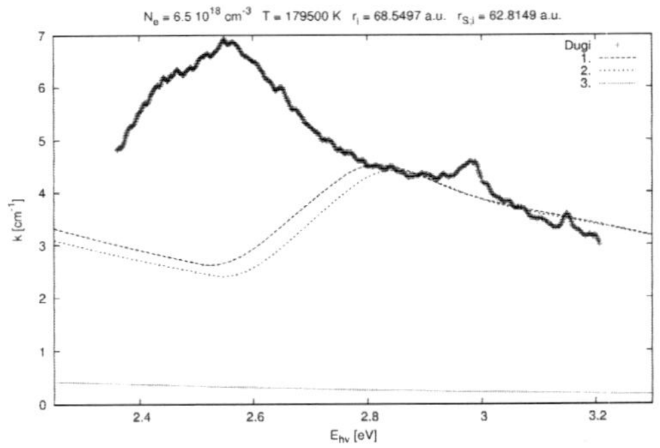

Figure 2. The calculated data for the potential (4), left figure The short pulse, comparison with the results in the case of the potential U_0 with $r_c = 44.964\ a.u.$ Curve **1.** - model with changeable shift and broadening, $\Delta E = 0.6\ eV$ and $\delta E = 1\ eV$ for $n = 2$. Curve **2.** - model with constant shift and broadening, case $\Delta E = 0,5\ eV$ i $\delta E = 0,75\ eV$. **3.** - κ_{ff}. The right figure, long pulse, comparison with the results in case of the potential U_0 with $r_c = 55.0523\ a.u.$. Curve **1.** - model with changeable shift and broadening $\Delta E = 0.275\ eV$ and $\delta E = 0.25\ eV$ for $n = 2$. Curve **2.** - model with constant shift and broadening, case $\Delta E = 0.25\ eV$ i $\delta E = 0.25\ eV$, **3.** - κ_{ff}.

3. Results and discussion

In this paper the calculations of the total absorption coefficient $\kappa_{tot}(\lambda; N_e, T)$ with the cut-off Coulomb potential (4) were made for the strongly non-ideal hydrogen plasma $N_e = 1.5 \cdot 10^{19} cm^{-3}$ and $T = 23000K$, as well as $N_e = 6.5 \cdot 10^{18} cm^{-3}$ and $T = 18000K$ taken from [26].

After process of selection of adequate shift and broadening parameters and comparison with the experimental data, good agreement was found. The good agreement with the experimental data in area where only continuous absorption is present, e.g. at the energies $E_{h\nu} \geq 2.8\ eV$, and the form of the total continuous absorption coefficient gives a space for bond-bond transition absorption.

Without further research on bond-bond transition within the frame of this model, there is not much to be said and analyzed for the model of broadening and shifting of bond state levels. Allthow, at this moment, it is just a parameter without further involvement into the processes behind it, it should be emphasized again that good agreement with experimental data exists.

4. Conclusion

Besides the fact that the presented model is still in process of development, a good agreement with the experimental data was shown.

There is a need to develop a model of bond-bond absorptions, which would enable the investigation of form of broadening and shifting of bond state levels. It would enable the studies of the broadening and shifting effects more in detail and develop a more concise model.

Also there is still a need for developing of both faster numerical procedures and code parallelism to improve speed and accuracy.

Acknowledgments

The work presented in this progress report was done under the MNTRS project 141033.

References

[1] K. Suchy. *Beitz. Plasmaphysik*, 4:71, 1964.

[2] C. Deutsch, M. M. Gombert, and H. Minoo. *Phys. Letters*, 66A:381, 1978.
[3] K. Gunther and R. Radtke. *Electric Properties of Weakly Nonideal Plasmas.* Akademie, Berlin, 1984.
[4] W. Kraeft, D. Kremp, W. Ebeling, and G. Ropke. *Quantum Statistics of Charged Particle System.* Academie-Verlag, Berlin, 1986.
[5] D. Mihalas. *Stellar Atmospheres* W. H. Freeman, San Francisco 1978
[6] A. A. Mihajlov and Lj. M. Ignjatović and N. M. Sakan and M. S. Dimitrijević *Astronomy and Astrophysics,* 437:1023-1025, 2007.
[7] G. Rinker, Phys.Rev.A. 37 (1988) 1284, Tehnical Report No. LA-10608-MS, Los Alamos National Laboratory (1986), unpublished.
[8] S. Ichimaru H. Iyetomi and S. Tanaka, Phys. Rep. 149 (1987) 91.
[9] W. Ebeling, W. D. Kraeft, and D. Kremp, *Theory of bound states and Ionization Equilibrium in Plasmas and Solids,* Academia-Verlag, Berlin (1976).
[10] V. E. Fortov and I. T. Iakubov, *Physics of Nonideal Plasma* Hemisphere, New York (1989).
[11] W. D. Kraeft, D. Kremp, W. Ebeling, and G. Röpke *Quantum Statistics of Charged Particle Systems* Akademie-Verlag, Berlin, (1986).
[12] I. M. Tkachenko and P. Fernández de Córdoba *Phys.Rev.E.* **57**, No.2, (1997) 2222-2229
[13] H. Reinhollz, Yu. Zaporoghetz, V. Mintsev, V. Fortov, I. Morozov, and G. Röpke, *Phys.Rev.E.* **68**, (2003) 036403-1-036403-10
[14] V. M. Adamyan, Z. Djuric, A. A. Mihajlov, N. M. Sakan and I. M. Tkachenko *J.Phys.D.* **37** (2004) 1896-1903.
[15] V. B. Mintsev and V. E. Fortov *J.Phys.A* **39** (2006) 4319-4327
[16] Gabor J. Kalman, J. Martin Rommel amd Krastan Blagoev *Strongly Coupled Coulomb Systems* Kluwer academic publishers
[17] A. A. Mihajlov, D. Djordjevic, M. M. Popovic, T. Meyer, M. Luft, and W.D. Kraeft. *Contrib. Plasma Phys.*, 29(4/5):441, 1989.
[18] G.A. Kobzev, I.T. Jakubov, and M.M. Popovich, editors. *Transport and Optical Properties of Non-Ideal Plasmas.* Plenum Press, New York, London, 1995.
[19] P. Debye and E. Hückel; *Physikalishe Z.,* 1923, vol. 24, p.185
[20] A. A. Mihajlov, D. Djordjević, M. M. Popović, T. Meyer, M. Luft, and W. D. Kraeft. Determination of electrical conductivity of a plasma on the basis of the coulomb cut-off potential model. *Contrib. Plasma Phys.*, 29(4/5):441–446, 1989.
[21] Mihajlov, A. A., Dimitrijević, D., Vučić, S,, Djordjević, D., Luft, M., Kraeft, D.. *Contrib Plasma Phys.* **26** (1986) 19
[22] Mihajlov, A. A., Dimitrijević, D., Djordjević, D., Luft, M., Kraeft, D., *Contrib Plasma Phys.* **27** (1987) 1
[23] W. D. Kraeft, M. Luft, and A. A. Mihajlov. Scattering properties and electrical conductivity for the coulomb cut-off potential. *Physica A*, 120:263–278, 1983.
[24] I. I. Sobel'man *Atomic Spectra and Radiative Transitions.* Springer Verlag, Berlin, 1979.
[25] Gavrilova, T. V.; Aver'yanov, V. P.; et.al. Optics and Spectroscopy **98**, Issue 5, pp/667-674
[26] Y. Vitel, T. V. Gavrilova, L. G. D'yachkov and Yu.K. Kurilenkov. Spectra of dense pure hydrogen plasma in Balmer area *JQSRT*, 83 (2004): 387..405
[27] A. A. Mihajlov, Y. Vitel and Lj. M. Ignjatović, The new screening characteristics of strongly non-ideal and dusty plasmas. Part 1: Single-Component Systems, *High Temperature,* 46, No. 6, 737-745, (2008)
[28] A. A. Mihajlov, Y. Vitel and Lj. M. Ignjatović, The new screening characteristics of strongly non-ideal and dusty plasmas. Part 2: Two-Component Systems, *High Temperature,* 47, No.1, 5-16, (2009)
[29] V. M. Adamtyan, *private communication* 2008.

Stark broadening of heavy metal spectral lines in atmospheres of chemically peculiar stars

Zoran Simić
Astronomical Observatory, Volgina 7, 11060 Belgrade, Serbia
E-mail: zsimic@aob.bg.ac.rs

Abstract. Data on the Stark broadening of heavy metal spectral lines are of interest not only for laboratory but also for astrophysical plasma research as e.g. for stellar spectra analysis and synthesis. Here, we investigated theoretically the influence of collisions with charged particles on heavy metal spectral line profiles for Te I, Cr II, Mn II, Au II, Cu III, Zn III, Se III, In III and Sn III in spectra of A stars and white dwarfs. We applied semiclassical theory of Sahal-Bréchot since the most of published results in literature until now are determined using this method. When it can not be applied in an adequate way, due to the lack of reliable atomic data, we used modified semiempirical theory of Dimitrijević & Konjević, Dimitrijević & Kršljanin.

Stark broadening parameters, widths and shifts, were obtained for spectral lines of neutral emitter Te I, singly charged emitters Cr II, Mn II and Au II and doubly charged emitters Cu III, Zn III, Se III, In III and Sn III. We considered as well the contributions of different collision processes to the total Stark width in comparison with Doppler one. In this case we obtained contributions for elastic, strong and inelastic collisions for upper and lower levels.

For example, chromium lines are interesting due to their presence in stellar atmospheres, so that they give possibility to determine chromium abundance and investigate chromium stratification in stelar atmospheres and to be used for the diagnostics of stellar plasma and for more refined synthesis of stellar spectra. We consider the effect of Stark broadening on the shapes of Cr II spectral lines observed in the spectra of stars in the middle part of the main sequence. Stark broadening parameters were calculated by the semiclassical perturbation approach. For stellar spectra synthesis, the improved version SYNTH3 of the code SYNTH for synthetic spectrum calculations was used. Stark broadening parameters for Cr II spectral lines of seven multiplets belonging to 4s-4p transitions were calculated. New calculated Stark broadening parameters were applied to the analysis of Cr II line profiles observed in the spectrum of Cr-rich star HD 133792. We found that Stark broadening mechanism is very important and should be taken into account, especially in the study of Cr abundance stratification.

1. Introduction

Data on the Stark broadening of neutral atom and ion lines are of interest not only for laboratory but also for astrophysical plasma research, particularly for stellar spectra synthesis and analysis.

Numerous analyses of stellar spectra with new space techniques point to the importance of trace elements, especially in spectra of chemically peculiar stars. Therefore, determination of heavy metal spectral lines in atmospheres of these stars is relevant for laboratory and astrophysical plasma investigations. High-resolution spectra allow us to study different broadening effects using well-resolved line profiles. Stark broadening is the most important pressure broadening mechanism for A-type stars and especially for white dwarfs. Neglecting this mechanism may therefore introduce significant errors into abundance determinations and spectra modelisation.

Tellurium lines are of astrophysical interest due to their presence in stellar atmospheres. For example, in Ref [1], is reported, that in Procyon photosphere spectrum one line of tellurium is identified, and used to determine the abundance of this element.

Data on the Stark broadening of single ionized chromium spectral lines are of interest not only for laboratory but also for astrophysical plasma research as e.g. for chromium abundances determination and opacity calculations. They have been identified in A-type star spectra, as e.g. 7 Sex [2], ϕ Aqu [3]. As an example, in the spectrum of ϕ Aqu, authors [3] identified 28 Cr II spectral lines and noted overabundace with value log Cr/H = -5.85±0.27. Particularly interesting are resonance lines investigated here, since they are often present in stellar spectra.

Ionized manganese lines are of interest for the analysis and modelling of stellar spectra as for example for HgMn stars. For plasma conditions in such star atmospheres, hydrogen is mainly ionized and Stark broadening is the main pressure broadening mechanism, influencing spectral line shapes. It is shown as well [4, 5] that in atmospheres of A stars exist conditions where Stark widths are comparable and even larger than the corresponding thermal Doppler widths.

Many examples show presence in hot stellar atmospheres of ionized gold, copper, zinc, selenium, indium, tin, rare earths and other trace elements which before the epoch of space born stellar spectroscopy were astrophysically insignificant, but now the need for their atomic data is increasing.

Here, we investigated theoretically the influence of collisions with charged particles on heavy metal spectral line profiles for Te I, Cr II, Mn II, Au II, Cu III, Zn III, Se III, In III and Sn III in spectra of A stars and white dwarfs. We applied semiclassical theory [6, 7] and when it can not be applied in an adequate way, due to the lack of reliable atomic data, modified semiempirical theory [8, 9] was used.

We obtained Stark broadening parameters, widths and shifts, for spectral lines of neutral emitter Te I, singly charged emitters Cr II, Mn II and Au II and doubly charged emitters Cu III, Zn III, Se III, In III and Sn III. In the case with the available experimental and other theoretical data for the considered spectral lines we analyzed an agreement or a disagreement with our theoretical results. Also, here we considered the contributions of different collision processes to the total Stark width in comparison with Doppler one.

We consider the effect of Stark broadening on the shapes of CrII spectral lines observed in stellar atmospheres of the middle part of the main sequence. Stark broadening parameters were calculated by the semiclassical perturbation approach. For stellar spectra synthesis, the improved version SYNTH3 of the code SYNTH for synthetic spectrum calculations was used. New calculated Stark parameters were applied to the analysis of Cr II line profiles observed in the spectrum of Cr-rich star HD 133792. We found that Stark broadening mechanism is very important and should be taken into account, especially in the study of Cr abundance stratification.

2. Theory

Calculations have been performed within the semiclassical perturbation formalism, developed and discussed in detail in Refs. [6, 7]. This formalism, as well as the corresponding computer code, have been optimized and updated several times [10, 11, 12, 13].

Within this formalism, the full width of an isolated spectral line of a neutral emitter broadened by electron impacts (W) can be expressed in terms of cross sections for elastic and inelastic processes as

$$W = \frac{\lambda^2}{\pi c} N \int v f(v) dv (\sum_{i' \neq i} \sigma_{ii'}(v) + \sum_{f' \neq f} \sigma_{ff'}(v) + \sigma_{el} + W_R), \qquad (1)$$

and the corresponding line shift d as

$$d = \frac{\lambda^2}{2\pi c} N \int v f(v) dv \int_{R_3}^{R_D} 2\pi\rho d\rho \sin 2\phi_p. \tag{2}$$

Here, λ is the wavelength of the line originating from the transition with initial atomic energy level i and final level f, c is the velocity of light, N is the electron density, $f(v)$ is the Maxwellian velocity distribution function for electrons, ρ denotes the impact parameter of the incoming electron, and ϕ_p is the phase shift due to the polarization potential. The inelastic cross sections $\sigma_{jj'}(v)$ (where $j = i$ or f) and elastic cross section σ_{el} are determined according to Chapter 3 in [7]. The cut-offs (needed for the calculation of inelastic and elastic cross sections and the shift), included in order to maintain for the unitarity of the S-matrix, and to take into account Debye screening are described in Section 1 of Chapter 3 in [7]. W_R gives the contribution of the Feshbach resonances [14] and this term is zero if the emitters are neutral atoms. Other differences between neutral and ionized emitters is that for calculations of the cross sections rectilinear perturber paths are taken for neutral ones and hyperbolic paths for ionized species.

The formulae for the ion-impact broadening parameters are analogous to the formulae for electron-impact broadening. We note that the fact that the colliding ions could be treated using impact approximation in the far wings should be checked, even for stellar atmosphere densities.

When the semiclassical perturbation formalism can not be applied in an adequate way, due to the lack of reliable atomic data, a modified semiempirical formalism has been used.

According to the modified semiempirical (MSE) approach [for review see [15]][8, 9] the electron impact full width at half maximum (FHWM) of an isolated ion line is given by

$$W = N\frac{4\pi}{3c}\frac{\hbar^2}{m^2}\left(\frac{2m}{\pi kT}\right)^{1/2}\frac{\lambda^2}{\sqrt{3}} \cdot \Big\{\sum_{\ell_i \pm 1}\sum_{L_{i'} J_{i'}} R^2_{\ell_i,\ell_i\pm1}\widetilde{g}(x_{\ell_i,\ell_i\pm1})$$

$$+ \sum_{\ell_f \pm 1}\sum_{L_{f'} J_{f'}} R^2_{\ell_f,\ell_f\pm1}\widetilde{g}(x_{\ell_f,\ell_f\pm1}) + \Big(\sum_{i'} R^2_{ii'}\Big)_{\Delta n\neq 0} g(x_{n_i,n_i+1})$$

$$+\Big(\sum_{f'} R^2_{ff'}\Big)_{\Delta n\neq 0} g(x_{n_f,n_f+1})\Big\}, \tag{3}$$

where n and ℓ are principal and angular momentum quantum numbers, and $R^2_{\ell_k,\ell_{k'}}$, $k = i, f$ is given by

$$\Big(\sum_{k'} R^2_{kk'}\Big)_{\Delta n\neq 0} = \Big(\frac{3n_k^*}{2Z}\Big)^2\frac{1}{9}(n_k^{*2} + 3\ell_k^2 + 3\ell_k + 11) \tag{4}$$

in the Coulomb approximation.

In Eq. (3)

$$x_{\ell_k,\ell_{k'}} = E/\Delta E_{\ell_k,\ell_{k'}}, \tag{5}$$

k=i,f where $E = \frac{3}{2}kT$ is the electron kinetic energy and

$$\Delta E_{\ell_k,\ell_{k'}} = |E_{\ell_k} - E_{\ell_{k'}}| \tag{6}$$

is the energy difference between levels ℓ_k and $\ell_k \pm 1$ $(k = i, f)$. Also

$$x_{n_k,n_k+1} \approx E/\Delta E_{n_k,n_k+1}, \tag{7}$$

where for $\Delta n \neq 0$ the energy difference between energy levels with n_k and n_k+1, $\Delta E_{n_k,n_k+1}$, is estimated as

$$\Delta E_{n_k,n_k+1} \approx 2Z^2 E_H / n_k^{*3}. \tag{8}$$

In Eq. (4) the effective principal quantum number is defined by

$$n_k^* = [E_H Z^2/(E_{ion} - E_k)]^{1/2}, \tag{9}$$

Z is the residual ionic charge i.e. the charge of the rest of atom as "seen" by optical electron (for example Z=1 for neutral atoms, 2 for singly charged ions etc), and E_{ion} is the appropriate spectral series limit.

In Eq. (3) T is the electron temperature, while $g(x)$ [16] and $\tilde{g}(x)$ [8] denote the corresponding effective Gaunt factors.

In comparison with the semiclassical perturbation approach [6, 7], the modified semiempirical approach requires much less atomic input data. In fact, if there are not perturbing levels strongly violating the assumed approximation, we only need the energy levels with $\Delta n = 0$ and $\ell_{if} = \ell_{if} \pm 1$, since all perturbing levels with $\Delta n \neq 0$, needed for a full semiclassical calculation are lumped together and estimated approximately. If for the nearest perturbing levels the condition $E/\Delta E \leq 2$ is satisfied the simplified version of the Eq. (3) [9] can be used.

3. Results and discussions

We obtained Stark widths and shifts of four Te I multiplets [17] by using the semiclassical perturbation method for a perturber density of 10^{16}cm^{-3} and temperatures from 2500 up to 50000 K. As an example of the influence of Stark broadening in atmospheres of hot stars we present in Fig. 1 Stark widths for the 5125.2 Å multiplet compared with Doppler widths for a model: T$_{eff}$=10000 K, log g= 4.5 of A type star atmosphere [18]. Even if Doppler broadening dominates the line centre, Stark broadening may influence line wings. There is no experimental or other theoretical data for the comparison with the calculated Stark broadening parameters of Te I spectral lines.

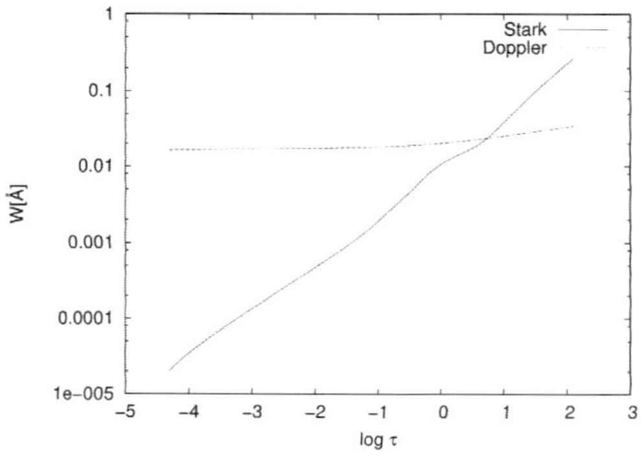

Figure 1. Thermal Doppler and Stark widths for Te I 6s ^{5}S^o - 7p ^{5}P (5125.2 Å) multiplet as functions of Rosseland optical depth for an A type star (T_{eff} = 10000 K, log g = 4.5). It is interesting that the Stark broadening mechanism is absolutely dominant in comparison with the thermal Doppler mechanism in deeper layers of the stellar atmosphere (log $\tau > 0.8$).

New Stark broadening parameters for seven strongest 4s - 4p Cr II multiplets for a perturber density of 10^{14}cm^{-3} and temperatures from 2500 up to 100000 K [19] and nine resonant 3d^5 - 3d^4 4p Cr II multiplets for a perturber density of 10^{17}cm^{-3} and temperatures from 2500 up to 100000 K have been determined within the semiclassical perturbation approach. In order to demonstrate one possibility for their usage in astrophysical plasma research, obtained results have bean applied to the analyzis of the Stark broadening influence on stellar spectral line shapes. For example, we synthesized resonant Cr II 4588.2 Å line profile using SYNTH code [20] and, DIPSO program package, for the corresponding equivalent width, for a model atmosphere with T_{eff} = 8750 K and log g =4.0 as a function of chromium abundance. One can see, that the influence of Stark broadening increases in line wings and with chromium abundance as expected. Also, one can see that in A type CP stars exist atmospheric layers where Doppler and Stark widths are comparable, so that the influence of Stark broadening is important.

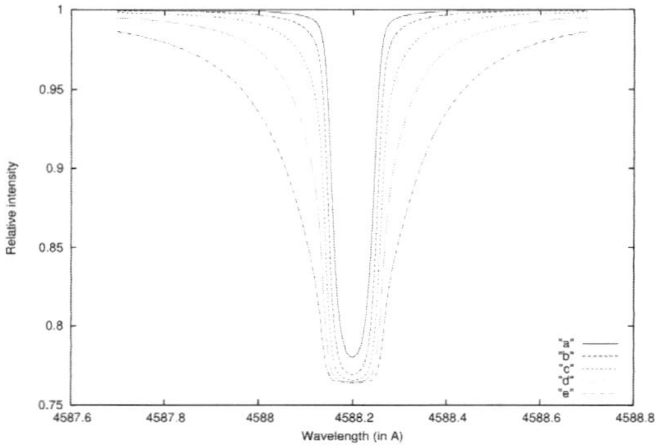

Figure 2. Comparison of the Cr II 4588.2 Å line profile ("a") without Stark broadening contribition and with this contribution for different Cr abundances log Cr/H : ("b") Solar one, ("c") -3.75, ("d") -3.25 , ("e") -2.75. The atmosphere model: T_{eff} = 8750 K and log g = 4.0.

Large differences between previous calculations of the Mn II spectral line Stark widths [21, 22] and a recent experiment in [23] are considered in [24]. In order to do so, new semiclassical perturbation calculations of the Stark broadening parameters of three spectral lines within Mn II 3d^{5}4s a^{7}S -3d^{5}4p z^{7}P^o and three within Mn II 3d^{5}4s a^{5}S - 3d^{5}4p z^{5}P^o multiplets, are performed. The influence of hfs on the Stark broadening parameters determination as a possible reason of disagreement is discussed also, on the example of Mn II 260.6455 nm and 259.4497 nm lines. Namely, measured and calculated values taking into account the hfs effect are compared. We found that this efect can not explain the large discrepancy between experiment and theory. Other possible reasons for disagreement like the configuration interaction are discussed as well. The new semiclassical perturbation calculations, more sophisticated than previous ones, performed by using the modified semiempirical theory, are in better agreement with experiment, but a large difference up to a factor 2.39 for the width, still exist. Also, the obtained results were used for the investigation of the influence of Stark broadening on Mn II spectral line profiles in DB and DA white dwarfs, see Fig. 3. It was demonstrated the importance to take into account Stark broadening mechanism for the analysis of DB white dwarf spectra. The obtained data and conclusions are of interest for a number of problems in stellar and Solar physics, like spectrum analysis and synthesis, radiative transfer and modelling of sub photospheric layers.

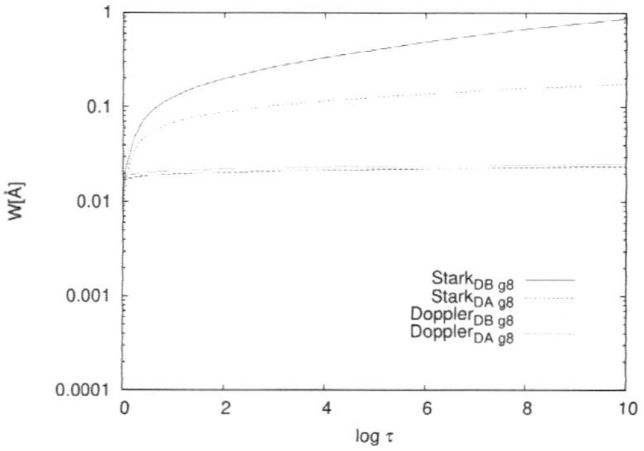

Figure 3. Thermal Doppler and Stark widths for Mn II a ^{5}S - z ^{5}P^o spectral line for a DB and DA white dwarf atmosphere model with T_{eff} = 15000 K and log g = 8, as a function of optical depth τ_{5150}.

We reviewed and compared our experimental [25] and theoretical [26] Au II Stark widths. The new measurements and calculations are in course in order to obtain a better agreement between experimental and theoretical values.

Table 1. Experimental [25] and theoretical [26] Stark widths are compared in Table 1: W_m/W_{th} ratio of experimental [25] and theoretical [26], $W_{m'}/W_{th}$ ratio of new experimental and theoretical results.

λ(Å)	W_m(Å)	W_m/W_{th}	$W_{m'}$(Å)	$W_{m'}/W_{th}$
1921.00	0.0338	1.15	0.0173	0.96
2044.54	0.0372	1.12	0.0179	0.87
2215.64	0.0424	0.38	0.0759	1.10
2263.63			0.1006	1.16
2291.41	0.0408	0.35	0.0759	1.03
2315.75	0.0398	0.33	0.0816	1.05
2533.53			0.1518	1.19
2552.67	0.0328	0.19		
2802.04	0.0438	0.24	0.1105	0.98
2819.79	0.0452	0.25	0.1108	0.97
2822.55			0.1422	1.03
2837.85	0.0454	0.25	0.1077	0.93
2846.92			0.1335	0.96
2893.26			0.1291	0.90
2954.22	0.0434	0.21	0.1055	0.85
2994.80	0.0442	0.21	0.1070	0.83

For example, the influence of Stark broadening on Cu III, Zn III and Se III spectral lines in DB white dwarf atmospheres was also investigated by [27] for 4s ^{2}F - 4p ^{2}G^o (λ=1774.4 Å), 4s ^{3}D - 4p ^{3}P^o (λ=1667.9 Å) and 4p5s ^{3}P^o - 5p ^{3}D (λ=3815.5 Å) by using the corresponding model with T_{eff} = 15000 K and $\log g = 7$ [28]. For the model atmosphere of the DB white dwarfs the prechosen optical depth points at the standard wavelength λ_s=5150 Å(τ_{5150}) are used in [28] and in [27]. As one can see in Fig. 4, for the plasma conditions in the DB white dwarf atmospheres, thermal Doppler broadening is much less important compared to Stark broadening. For example the Stark width of the Se III 3815.5 Å line is larger than the Doppler one by up to two orders of magnitude within the range of optical depths considered. Much larger Stark widths in DB white dwarf atmospheres in comparison with A type stars are the consequence of larger electron densities due to much larger $\log g$ and larger T_{eff}, so that electron-impact (Stark) broadening is more effective.

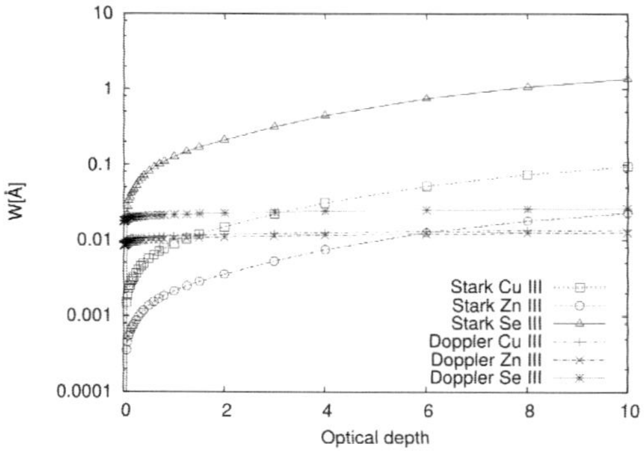

Figure 4. Thermal Doppler and Stark widths for Cu III 4s ^{2}F - 4p ^{2}G^o (λ=1774.4 Å), Zn III 4s ^{3}D - 4p ^{3}P^o (λ=1667.9 Å) and Se III 4p5s ^{3}P^o - 5p ^{3}D (λ=3815.5 Å) spectral lines for a DB white dwarf atmosphere model with T_{eff} = 15,000 K and $\log$ g = 7, as a function of optical depth τ_{5150}.

In order to see the influence of Stark broadening mechanism for In III and Sn III spectral lines in stellar plasma conditions, we have calculated Stark widths for a [18] A type star (T_{eff} = 10000 K; log g = 4.5) atmosphere model and compared them with Doppler ones. We found

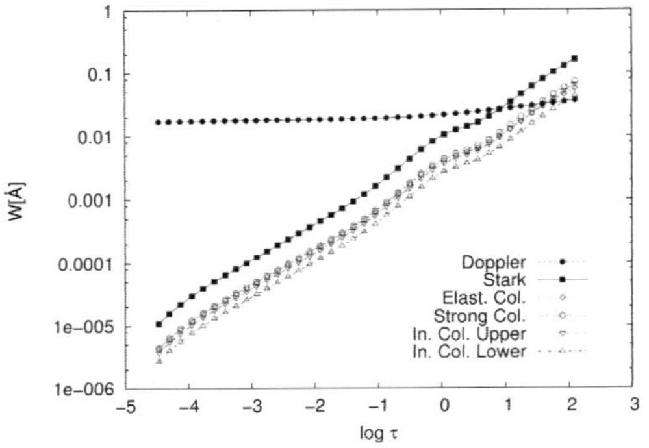

Figure 5. Thermal Doppler, total Stark width and contributions of different collisional processes to the total Stark width of Sn III 5226.2 Å for a model: T_{eff} = 10000 K, log g = 4.5 [18] of an A type star, as a function of the Rosseland optical depth. In this case, elastic and strong collisions and inelastic collision from upper levels have a similar contribution to the full Stark width as well as the similar behaviour with temperature.

that exist photospheric layers where Doppler and Stark widths are comparable and even where the Stark width is dominant and must be taken into account. Also, for the same atmosphere model, we present here in Fig. 5. Stark widths and contributions of different collision processes to the total Stark width in comparison with Doppler one.

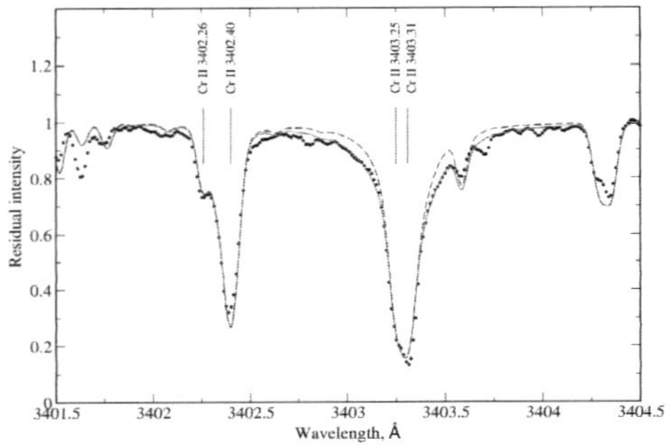

Figure 6. Comparison between the observed Cr II 3403.30 line profile (dots) and synthetic calculations with the Stark parameters from paper by [19](full line) and those from [30] (dashed line).

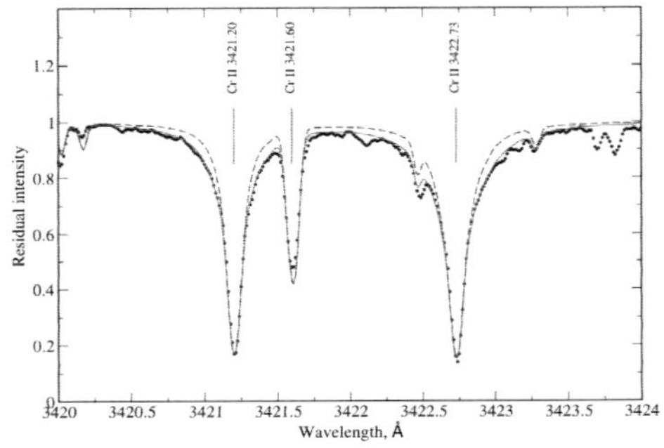

Figure 7. Comparison between the observed Cr II 3421.20, 3422.73 lines profile (dots) and synthetic calculations with the Stark parameters from paper by [19](full line) and those from [30] (dashed line).

Cr II lines in the spectrum of the Ap star HD 133792, for which careful abundance and stratification analysis has recently been performed in [29] were analyzed in [19]. HD 133792 has an effective temperature of T_{eff}=9400 K, log g=3.7, and a mean Cr overabundance +2.6 dex relative to the solar Cr abundance [29]. All calculations were carried out with the improved version SYNTH3 of the code SYNTH for synthetic spectrum calculations. Stark damping parameters were introduced in the spectrum synthesis code. The stratified Cr distribution in the atmosphere of HD 133972 derived by [29] was used. Figure 6 shows a comparison between the observed line profiles of Cr II lines 3403.30 Å and synthetic calculations with the Stark damping constants from [30] line lists and with the data of [19]. Good agreement between observations and calculations for a set of weak Cr II lines proves the use of the stratified Cr distribution, while all four strong Cr II lines demonstrate a good accuracy for the theoretical Stark broadening parameters obtained [19].

This opens up a new possibility of checking the theoretical and experimental Stark broadening results with the help of stellar spectra, which will be particularly interesting with the development of space born spectroscopy, building of giant telescopes of the new generation and increase of accuracy of computer codes for modellisation of stellar atmospheres. The Cr II lines analyzed in [19] are particularly suitable for such purpose since they have good clean wings where the influence of Stark broadening is the most important.

3.1. Acknowledgments

This work is a part of the project 146 001 "Influence of collisional processes on astrophysical plasma lineshapes", supported by the Ministry of Science and Technological Development of Serbia.

References

[1] Yuschenko A V and Gopka V F 1996 *Astron. Astrophys. Transactions* **10** 307
[2] Adelman J and Philip A G 1996 *Mon. Not. R. Astron. Soc.* **282** 1181
[3] Caliskan H and Adelman J 1997 *Mon. Not. R. Astron. Soc.* **288** 501
[4] Popović L Č and Dimitrijević M S 1996 *A&AS* **120** 373
[5] Popović L Č, Simić S, Dimitrijević M S, Milovanović N 2001, *Astropys. J. Suppl. Series*, **135** 109
[6] Sahal-Bréchot S 1969a *A&A*, **1** 91
[7] Sahal-Bréchot S 1969b *A&A*, **2** 322
[8] Dimitrijević M S and Konjević N 1980 *J. Quant. Spectrosc. Radiat. Transfer* **24** 451
[9] Dimitrijević M S and Kršljanin V 1986 *A&A* **165** 269
[10] Dimitrijević M S 1996 *Zh. Priklad. Spektrosk.* **63** 810
[11] Sahal-Bréchot S 1974 *A&A* **35** 321
[12] Dimitrijević M S and Sahal-Bréchot S 1984 *A&A* **136** 289
[13] Dimitrijević M S, Sahal-Bréchot S and Bommier V 1991 *A&A S* **89** 581
[14] Fleurier C, Sahal-Bréchot S and Chapelle J 1977 *JQSRT* **17** 595
[15] Dimitrijević M S and Popović L Č 2001 *J. Appl. Spectrosc.* **68** 893
[16] Griem H R 1968 *Phys. Rev.* **165** 258
[17] Simić Z, Dimitrijević M S, Kovačević A 2009 *New AR* **53** 246
[18] Kurucz R L 1979 *Astrophys. J. Suppl. Series* **40** 1
[19] Dimitrijević M S, Ryabchikova T, Simić Z, Popović, L Č, and Dačić M 2007 *A&A* **469** 681
[20] Piskunov N 1992 In: *Stellar Magnetism* Eds. Glagolevskij Yu, Romanyuk I 92
[21] Popović L Č and Dimitrijević M S 1997a *A&AS* **128** 203
[22] Popović L Č and Dimitrijević M S 1997b *Serb. Astron. J.* **156** 173
[23] Djeniže S, Bukvić S, Srećković A and Nikolić, Z, 2006, *New A.* **11** 256
[24] Popović, L Č, Dimitrijević, M S, Simić, Z, Dačić, M, Kovačević, A and Sahal-Bréchot S, 2008 *New A.* **13** 85
[25] Ortiz M and Mayo R 2005 *J. Phys. B: At. Mol. Opt. Phys.* **38** 3953
[26] Popović L Č, Dimitrijević M S and Tankosić T 1999 *A&ASS* **139** 617
[27] Simić Z, Dimitrijević M S, Popović L Č and Dačić M 2006 *New A.* **12** 187
[28] Wickramasinghe D T 1972 *Mem. R. Astron. Soc.* **76** 129
[29] Kochukhov O, Tsymbal V, Ryabchikova T, Makaganyk V and Bagnulo S 2006 *A&A* **460** 831
[30] Kurucz R L 1993 *CDROMs 13, 22, 23* SAO Cambridge

AUTHOR INDEX

AUTHOR INDEX